A Modern Dictionary of Geography

A Modern Dictionary of Geography

FOURTH EDITION

Michael Witherick
Visiting Fellow in Geography
University of Southampton

Simon Ross
Head of Geography
Queen's College Taunton

John Small
Emeritus Professor of
Geography
University of Southampton

A member of the Hodder Headline Group
LONDON
Co-published in the United States of America by
Oxford University Press Inc., New York

First published in Great Britain in 2001 by
Arnold, a member of the Hodder Headline Group,
338 Euston Road, London NW1 3BH

http://www.arnoldpublishers.com

Co-published in the United States of America by
Oxford University Press, Inc.,
198 Madison Avenue, New York, NY10016

British Library Cataloguing in Publication Data
A catalogue record for this book is available from the British Library

Library of Congress Cataloging-in-Publication Data
A catalog record for this book is available from the Library of Congress

ISBN 0 340 80713 X (hb)
ISBN 0 340 76210 1 (pb)

1 2 3 4 5 6 7 8 9 10

Production Editor: Wendy Rooke
Production Controller: Brian Eccleshall
Cover design: Terry Griffiths

Typeset in 9/11 Minion by J&L Composition Ltd, Filey, North Yorkshire
Printed and bound in Malta by Gutenberg Press

Preface to the 4th edition

This new edition of *A Modern Dictionary of Geography* is the outcome of a number of significant changes. One of its aims is to meet more directly the needs of sixth-formers preparing for the new AS and A2 examinations, at the same time maintaining the book's proven indispensability to geography undergraduates.

The dimensions of the Dictionary now stand at nearly 2000 full entries, plus over 400 terms dealt with briefly and embedded in the definitions of related terms. The number of illustrations has also been increased substantially.

My former colleague for over 20 years at the University of Southampton, John Small, in previous editions responsible for the entries relating to physical geography, has decided that his mantle should now fall on younger shoulders. As a consequence, I have been happy to work alongside Simon Ross on this new edition. Its preparation has involved several revision processes in order that the Dictionary should reflect the current state of geography – as ever, a highly dynamic subject. These processes have included:

- eliminating terms that have dropped from the vocabulary of modern geographers
- updating existing entries in the light of research and the general march of time
- including new terms that have entered into current usage during the last five years or so
- the inclusion of more line diagrams to assist understanding of some of the more complex terms
- coverage of terms encountered in the new AS and A2 specifications.

Since the publication of the third edition, there have been some significant changes affecting both human and physical geography. The collapse of much of the communist world, the growing concern about environmental issues and the relentless march of globalization are three examples of significant developments since 1995 that have necessitated either the redrafting of existing entries or the production of entirely new ones. What to include and what to exclude remains a perennial problem, if only because selection is ultimately driven by a degree of personal judgement. However, we hope that we are not too wide of the mark in fulfilling the needs and expectations of you, the reader.

Michael Witherick
December 2000

Introduction

In compiling this Dictionary we have been guided by a number of principles. Of these, the overriding one was to produce a book that would meet the needs primarily of pupils in advanced courses at secondary schools and colleges, or their equivalents in overseas countries, together with those of first-year undergraduates at universities and other institutions of higher education.

The most difficult decision concerned the choice of terms for definition. Given the nature of geography as a discipline, and the fact that it interfaces with a range of other subjects, it is manifestly impossible to select a vocabulary that is in any way *exclusive* to geography. Inevitably, therefore, terms that are more properly geological, economic, sociological, statistical and so on, have been included, though a conscious effort has been made to avoid 'opening the flood-gates' and to employ terms that are widely used by geographers at the level specified. More controversial still was the identification of the terms deemed to be relevant to A-level and undergraduate geographers. Quite clearly, it is impossible to compile a definitive list that would be acceptable to everyone. What we have done, as A-level examiners and teachers, is to choose terms that – in our experience – are currently in use by A-level candidates (both from home and overseas centres) and that we would expect to be understood by first-year undergraduates. We have also consulted current AS and A2 syllabuses and question papers from all the British GCE boards, referred to the indices of textbooks that are primarily intended for A-level students and first-year undergraduates, as well as sounding out the views of practising teachers of geography.

Although we have tried to be objective in our selection of terms, it is perhaps inevitable that our own particular interests and enthusiasms have had some influence on the final list. Some of these will doubtless be regarded as 'superfluous', 'too advanced', 'too elementary'; important omissions will also be identified. All that we can say is that this is *our choice*, made in good faith *at this time*. As the discipline of geography changes and develops, so undoubtedly we shall need to modify the selection for future editions. Indeed, we would like to extend an invitation to our readers to join with us in this challenge of extending, updating and refining the Dictionary. If you have any comments or suggestions for future editions, please write to us via the publishers, Arnold.

Our main hope, however, is that the Dictionary as it presently stands will provide a comprehensive guide to, and in many instances an explanation of, the principles, concepts and terminology of modern geography. We have deliberately aimed to achieve a balance between 'physical' and 'human' definitions. In some previous dictionaries of geography, there has arguably been a bias towards the former, reflecting the widespread use of 'technical' or scientific terminology in branches of the subject such as geomorphology, meteorology and hydrology. However, it is in our opinion necessary to bring out the increasing use, particularly during the past two decades, of specialized terminology on the human side of geography.

We have not attempted to define common commodities (which are adequately covered in 'standard' dictionaries); we have elected not to include esoteric, unusual or even bizarre terms (this Dictionary is not to be regarded as a jargoneer's charter!); and we have aimed to include 'local' terms only where they are also used, and known about, outside the country of origin. In some instances, we have included examples, where these were felt to illuminate further the definition and explanation of particular terms. In other instances, our view is that readers should be capable of deriving appropriate examples, both from their own first-hand experience and the reading of currently available textbooks. Ultimately, it is our hope that the Dictionary will go beyond the provision of rather 'bare', academically correct definitions, and will provide material that is interesting to read, that can be incorporated by students in essay work, and that can be used to assist revision work in preparation for examinations.

Finally, a few additional points, which will assist readers in their use of the Dictionary, need to be stressed. These are outlined in the following checklist.

A checklist for use

- A cross-referencing system is employed and is signalled when, either within or at the end of a particular entry, another term is given in SMALL CAPITALS. For such terms, a full definition is included elsewhere in the Dictionary. Consultation of these entries will then amplify, and aid the understanding of, the original entry.
- Where a term is given in italics, it means that there is no separate entry. This device is mainly used in three different circumstances: (i) where the meaning of the term is apparent from the content of the entry in which it is contained; (ii) where the meaning is explained as part of a more comprehensive entry; and (iii) where the meaning is essentially synonymous. None the less, many of these italicized terms are recorded in the alphabetical listing of the Dictionary along with the identity of their 'host' entry.
- The Dictionary contains over 200 illustrations. Where a definition has an accompanying map or diagram, the abbreviation [*f*] is given at the end of the entry. Where we think an entry might usefully be illustrated by reference to a figure associated with another definition, the location of that map or diagram is indicated by [*f* TERM].
- SI units are used throughout the Dictionary, though a full definition of these is not included on grounds of the length and complexity of the necessary tables.

List of abbreviations

BP	before the present (era)
cf	see, for purposes of comparison
cm	centimetre
ct	see, for purposes of contrast
d	depth
E	east
e.g.	for example
[*f*]	see figure attached
[*f* TERM]	see figure attached to term cited
g	gram
ha	hectare
i.e.	that is
kg	kilogram
km	kilometre
km^2	square kilometre
m	metre
m^3	cubic metre
mb	millibars
mg	milligrams
mm	millimetre
Mt	Mount/mountain
N	north
NE	northeast
NW	northwest
R	river
S	south
s	second
SE	southeast
SW	southwest
SI	Système Internationale d'Unités
w	width
W	west
UK	United Kingdom
USA	United States of America
yr	year

A

abiotic A descriptive term meaning 'without life', commonly used to describe some components of an ECOSYSTEM. Examples of abiotic factors include climate, geology and mineral matter in soils.

ablation The process by which ice and snow are lost from a glacier. It includes: (i) surface, internal and basal melting (of which the first is by far the most important); (ii) *sublimation*, which is the direct transfer of water from the solid to the gaseous state; (iii) CALVING of icebergs or smaller ice blocks where the ICE SHEET or glacier enters the sea or a lake.

ablation zone That part of a glacier or ICE SHEET lying below the EQUILIBRIUM LINE, where the ice surface is lowered by melting during the summer. The amount of ABLATION increases downglacier from the FIRN line (where net ablation is nil) to as much as 5–10 m near the snout; this is known as the *ablation gradient*. [*f* MASS BALANCE]

abrasion The processes by which solid rock is eroded by rock fragments transported by running water, glacier ice, wind and breaking waves. Characteristic products of abrasion are: (i) POT-HOLES in river beds, formed by eddying water and concentrations of pebbles; (ii) smoothed, striated and polished surfaces formed by debris frozen into the base of a glacier, or trapped between the ice and BEDROCK; (iii) basally eroded rock formations due to abrasion by SAND particles transported just above ground level by the wind; (iv) WAVE-CUT PLATFORMS – the product mainly of the impact of rock particles contained within turbulent seawater and the SWASH of breaking waves. Abrasion is most effective when the impact of the particles on bedrock is vigorous, and the particles themselves are coarse, hard and angular.

absolute humidity The amount of water vapour contained within a unit volume of air, commonly expressed in grams per cubic metre (gm^{-3}). Cold air can contain less vapour than warm air. Absolute humidity is highest near the Equator, and least over Antarctica and the central Asian land-mass in winter. Ct RELATIVE HUMIDITY, SPECIFIC HUMIDITY. [*f*]

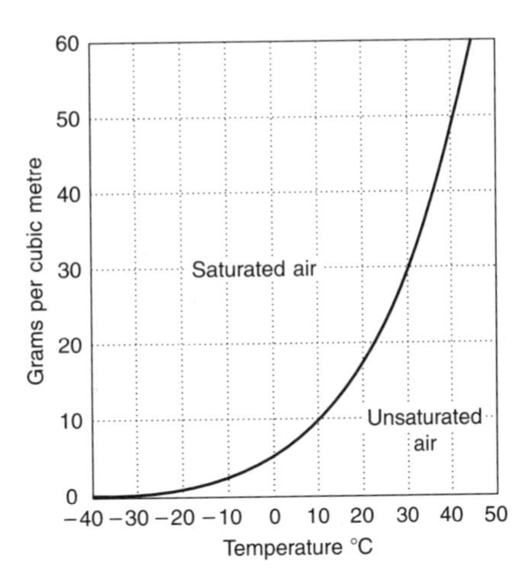

Absolute humidity

absolute instability The condition of the ATMOSPHERE in which the ENVIRONMENTAL LAPSE-RATE exceeds both the SATURATED ADIABATIC LAPSE-RATE and the DRY ADIABATIC LAPSE-RATE. If air pockets begin to rise, as a result of initial heating and convection, they will lose heat adiabatically owing to expansion, but will remain warmer than the surrounding air. They will therefore continue to rise to great heights, forming tall clouds and leading to heavy precipitation. Cf CONDITIONAL INSTABILITY.

absolute stability The condition of the ATMOSPHERE in which the ENVIRONMENTAL LAPSE-RATE is less than both the SATURATED ADIABATIC LAPSE-RATE and the DRY ADIABATIC LAPSE-RATE. If air pockets (even if very moist and subject to CONDENSATION upon cooling) are forced to rise, they will lose heat adiabatically at a rate such that they will be cooler than the surrounding air. Thus, air pockets will have a tendency to sink and ascent will only occur if the air pocket is forced to rise, for example at a FRONT or over a mountain range.

abstraction The process by which water is drawn from an underground AQUIFER for domestic, agricultural and industrial uses. Sustainable abstraction involves a balance being achieved between the pumping of water from the underground aquifer and its recharge by rainfall. Excessive abstraction may result in a falling WATER TABLE.

abyssal A term applied to the deepest parts of the ocean floor (mainly between 2200 and 5500 m), on which fine-textured deposits (*ooze*) have accumulated to considerable thicknesses over long periods of geological time.

accelerated erosion An increase in the rate of 'natural' erosive processes, such as RAINWASH, owing to the activities of people – for example, in the clearing of vegetation, construction, ploughing of fields and OVERGRAZING by domestic animals. Evidence of accelerated erosion includes gullies on hillslopes in areas subject to SOIL EROSION and the rapid erosion of previously stable stream banks. It is also associated with significant increases in the LOAD of streams (see SEDIMENT YIELD).

accelerator In economic geography, this refers to something that increases the momentum of a *boom* or *slump* (see BUSINESS CYCLE). For example, a marked expansion of overseas markets would accelerate a boom, while a slump would be hastened by a raising of taxation.

acceptable dose limit This applies to those circumstances in which an alien substance is introduced into the environment, such as the discharge of raw sewage into the sea. The limit is the greatest amount that can be released safely without having a serious detrimental effect. See ENVIRONMENTAL POLLUTION.

accessibility The ease with which one location may be reached from another. In transport studies, it is that quality possessed by a place as a result of its particular location within a TRANSPORT NETWORK. The greater the number of routes that converge on a settlement, the greater its NODALITY and, therefore, its accessibility (see COST-SPACE CONVERGENCE, NETWORK). Indicators such as the *König Number* and SHIMBEL INDEX are used to measure the accessibility of a place within a network. In economic studies, it is more about time and cost. An accessible location is one that minimizes the costs of distance and contact with linked activities. In social studies, it refers to the degree to which different social groups are able to obtain goods and services – for example, the poor have much less accessibility to good housing and luxury goods than the rich (see DEPRIVATION).

accretion A process of growth by accumulation; e.g. the enlargement of a raindrop by collision with many other tiny water droplets within a cloud, or the accumulation of mineral matter in a particular location, e.g. sand on a beach.

acculturation This occurs when a person comes into contact with a different culture, as, for example, when they move from one type of society to another. Almost inevitably, some of the habits, values, attitudes and behavioural characteristics of the society into which the move is made will be acquired. Elements of the person's original culture are thereby replaced. At an aggregate level, acculturation refers to the contact between two adjacent cultures or civilizations, whereby each influences the other by a sort of exchange process. Cf ASSIMILATION; ct INTEGRATION.

accumulation zone That part of a glacier or ICE SHEET lying above the EQUILIBRIUM LINE, on which the dominant process is the addition of snow and ice. Winter snowfall and avalanches are the main forms of accumulation and are regarded as INPUTS into glacier SYSTEMS. In the accumulation zone, net winter accumulation exceeds net summer ABLATION. The accumulation zone thus comprises a layered structure, with a series of winter accumulation layers separated by summer ablation surfaces. [*f* MASS BALANCE]

acid lava Volcanic LAVA that is rich in silica and flows slowly owing to its high viscosity. Acid lava forms steep-sided, dome-like volcanoes.

acid rain Rain contaminated by chemicals (notably sulphur dioxide, producing dilute sulphuric acid) that have been released from industrial chimneys, and in particular from coal-burning power stations. Acid rain has

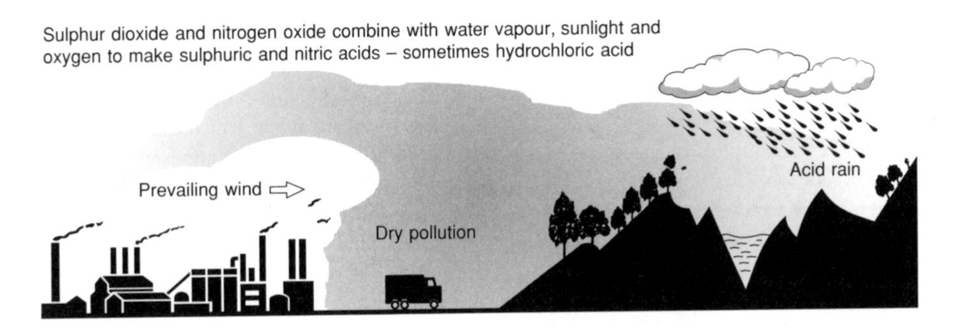

The formation of acid rain

been held at least partly responsible for: (i) the ACIDIFICATION of rivers and lakes in upland areas; (ii) the widespread destruction of fish and other wildlife; (iii) the serious degeneration of coniferous forests in many parts of Europe (such as S Scandinavia, which may have been seriously affected by rains contaminated over Britain). Action has been taken to reduce the emission of sulphur dioxide from power stations by the use of filter mechanisms and by political agreements (e.g. within the European Union). [*f*]

acid soil See pH VALUE.

acidification See pH VALUE.

action space See BEHAVIOURAL ENVIRONMENT.

active layer In PERIGLACIAL conditions, where PERMAFROST exists, only the upper layer of ground thaws in summer. This upper layer, which is affected by summer thawing and winter freezing, is the active layer. Its lower limit is the permafrost table, which forms an impermeable surface and causes the active layer to be poorly drained. At its maximum, the active layer may reach a depth of 3–6 m, depending on summer temperatures, the duration of the thaw season, soil composition (GRAVEL favours deeper thawing than peaty soils), SOIL MOISTURE and content, and the density of the plant cover. Within the active layer, processes such as SOLIFLUCTION and FROST HEAVE can be highly effective. Engineering structures can become unstable on active layers and, where possible, piles are driven into the permafrost or structures are raised above the ground surface.

activity rate The proportion of the population in the working age group (usually 15 to 64 years for men and 15 to 59 years for women) who are registered as employed or who are unemployed but seeking work.

adiabatic The change of temperature in a gas that experiences compression (leading to heating) and expansion (leading to cooling), without any exchange of heat from outside. In the Earth's ATMOSPHERE, rising and descending air pockets will be affected by adiabatic changes. See DRY ADIABATIC LAPSE-RATE and SATURATED ADIABATIC LAPSE-RATE.

administrative principle One of three *principles* underlying Christaller's CENTRAL PLACE THEORY and governing the spatial arrangement of a central place relative to its MARKET AREA. It applies where advanced systems of centralized administration have developed and where six centres of a given order fall entirely within the hexagonal market area of a higher-order central place. This arrangement, with a K-VALUE of 7, ensures that there is no shared allegiance. It thus avoids the unsatisfactory situation of one settlement being located within the administrative area of more than one higher-order central place. Ct MARKET PRINCIPLE, TRAFFIC PRINCIPLE. [*f*]

adret A hill-slope that, in the Northern Hemisphere, faces southwards or southwestwards, and thus receives the maximum amount of sunshine and warmth. It is the opposite side to the UBAC (the shady side). On adret slopes the tree line may be significantly higher, and certain geomorphological

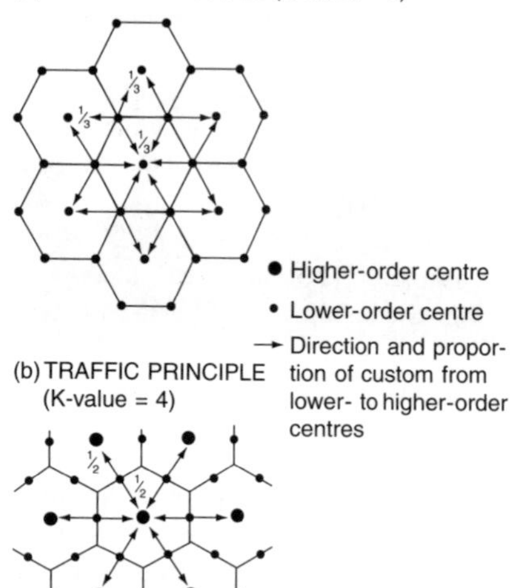

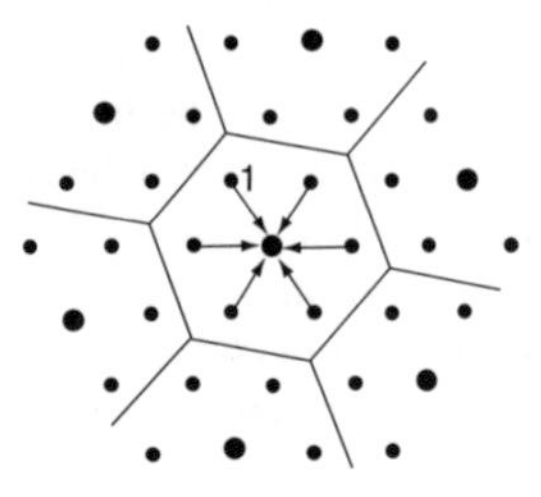

Three principles of central place theory

processes more active (e.g. FREEZE-THAW WEATHERING in a PERIGLACIAL climate). The cultivation of crops such as vines may be more successful on adret slopes because conditions promote ripening.

advanced countries See MEDC and FIRST WORLD.

advection fog FOG developed in air that is moving in a horizontal direction. The air, which is initially warm and moist, is cooled to DEW-POINT as it passes over a cold land or sea surface. Advection fog forms in mid-latitudes in winter (for example, when tropical maritime air crosses a land-mass previously cooled under anticyclonic conditions), and in spring and early summer (when very warm tropical air is cooled by contact with a relatively cold sea surface, giving *sea fog*). It is also particularly common at the convergence of warm and cold ocean currents (e.g. off Newfoundland, where warm air from above the Gulf Stream drifts over the Labrador Current, to give up to 100 days of fog each year).

aeolian A term applied to the action of wind. Aeolian transport is an important coastal process involving the transportation of sand and the construction of sand dunes. It is also important in arid environments where rocks may become blasted by the sand.

aerial photograph The term normally refers to a photograph, vertical or oblique, taken from an aircraft, but might also include the images recorded from an orbiting satellite (see REMOTE SENSING). Among other things, aerial photographs may be used for mapping (see PHOTOGRAMMETRY) and for the general study of landforms and landscape change.

aerobic In the biological sense, a term referring to organisms living in the presence of free oxygen, specifically with reference to those in the soil. Ct ANAEROBIC.

affluent society A term used to describe those MEDCs that have benefited from long periods of continuous economic growth, and in which the general level of prosperity allows the population at large to enjoy a good QUALITY OF LIFE and a high level of WELL-BEING (see also WELFARE). Most people are able to purchase a wide range of goods and services over and above their basic subsistence needs.

afforestation The deliberate planting of trees, usually where none grew previously or recently, as by the FORESTRY COMMISSION on the heathlands and moorlands of Britain for much of the 20th century. Where the planting takes place on areas of cleared woodland, then it would be more appropriate to refer to it as *reforestation*. Afforestation may be used as part of a programme to reduce flood hazards in a drainage basin because increased INTERCEPTION slows down the transfer of precipitation to the ground.

aftershock Vibration of the Earth's crust that occurs following an EARTHQUAKE, and results from minor adjustments of rocks along a fault-line after the main rupture. Aftershocks may continue for hours, days or even longer periods of time, and may cause considerable destruction and loss of life where buildings have been weakened by the initial earthquake. There can be hundreds of aftershocks after a major earthquake.

age dependency See DEPENDENCY.

age–sex pyramid A frequency distribution or HISTOGRAM of the population of a specific area, constructed in 1-, 5- or 10-year age groups, with males on one side, females on the other. This usually takes the form of a pyramid, with the base representing the youngest group and the apex the oldest. The horizontal bars are drawn proportional in length to either the percentage of the population or the actual number in each age group. [*f*]

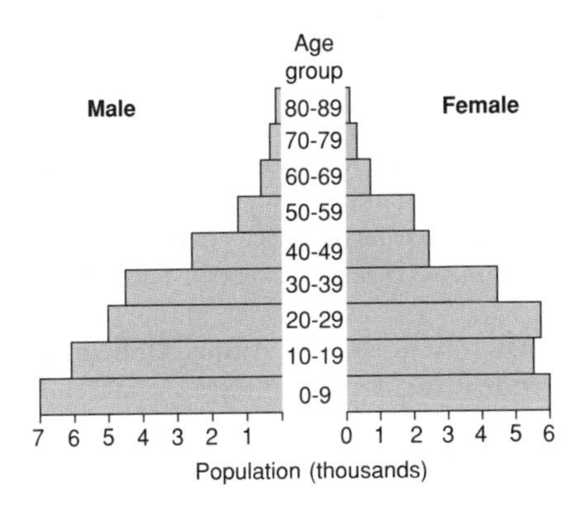

Age–sex pyramid

agglomeration In ECONOMIC GEOGRAPHY, agglomeration refers to the clustering of activities and people at nodal points, as for example in towns and cities. This clustering is prompted by CENTRIPETAL FORCES in spatial organization and by the need to achieve AGGLOMERATION ECONOMIES. See also CENTRALIZATION, POLARIZATION.

agglomeration economies The potential savings to be made by a firm as a result of locating within an AGGLOMERATION. These savings occur because: (i) the firm is able to share with others in the agglomeration, rather than bear on its own the full costs of such items as PUBLIC UTILITIES and specialist services (e.g. legal, financial and advertising); (ii) agglomeration means that distances, and therefore transport costs, are minimized by those firms between which there is some form of LINKAGE; (iii) there are COMMUNICATIONS ECONOMIES. The scale of potential savings may be thought of as being directly proportional to the scale of agglomeration. These economies may also be referred to as EXTERNAL ECONOMIES. See also LOCALIZATION ECONOMIES.

aggradation A term used loosely to describe the building up of SEDIMENT by rivers and wave action, hence an *aggraded river valley*, and *aggraded beach profile*. More strictly, the term refers to DEPOSITION carried out to restore or maintain the condition of GRADE. For example, an influx of sediment may cause a stream to become overloaded; it will therefore deposit sediment and steepen its course (*aggrade*), thus increasing its energy and capacity to transport the increased LOAD.

agribusiness A large-scale farming system run on business lines by a relatively small number of companies, most of them TNCS. Besides farming, business interests extend 'upstream' to include producing farming inputs such as seeds, chemicals, feedstuffs and machinery, and 'downstream' to food-processing industries and marketing. Examples of such companies are Monsanto, Unilever and Nestlé. Cf FACTORY FARMING.

agricultural geography The description and explanation of spatial variations in the character and productivity of AGRICULTURE. In terms of description, land use is the most conspicuous aspect of that character. Explanation requires reference to such factors as the behavioural environment of farmers' decision-making, economic rent, government intervention, land tenure, etc. Today, there is increasing interest in the production side, particularly in the adequacy or otherwise of food supply (see FOOD PRODUCTION CHAIN).

agricultural chain See FOOD PRODUCTION CHAIN.

agriculture Used in a wide sense to include the growing of crops and the rearing of livestock – the whole science and practice of farming. However, some restrict the term to the growing of crops alone. See ARABLE FARMING.

agrochemicals Chemical inputs in farming that are used to increase productivity by fertilizing the soil, and controlling pests and diseases (*pesticides*, *herbicides* and *fungicides*). There is increasing concern about their use, with mounting evidence of damage to the environment, wildlife and human health. See BIOLOGICAL CONTROL.

agroforestry The integrated use of land for agriculture and forestry. For example, the phased planting and felling of WINDBREAKS in agricultural areas prone to wind erosion, or the planting of trees on farmland to raise soil fertility by nitrogen fixing and the supply of leaf litter. It is being encouraged in those LEDCs where the search for fuelwood is causing the serious depletion of timber resources.

A-horizon The uppermost layer of topsoil in a well-developed or mature SOIL PROFILE. It is characterized by well-weathered PARENT MATERIAL and a relatively large amount of HUMUS. The A-horizon is generally considered to be fertile and well suited to agriculture. It may be subject to the processes of LEACHING or ELUVIATION in regions with relatively high rainfall. This causes the removal of some of the nutrients down through the soil profile to leave behind a nutrient-poor eluviated horizon within the main A-horizon. Immediately above the A-horizon is the O-HORIZON, which comprises the rotting humus that will slowly be incorporated into the A-horizon by the action of worms and other organisms. [*f* SOIL PROFILE]

aid Assistance extended by the more wealthy nations (see MEDCs) to the LEDCs of the THIRD WORLD mainly to encourage DEVELOPMENT and to help overcome the obstacles to it. Aid can take a variety of forms, from the transfer of capital, technology and expertise to the granting of loans and educational scholarships, from assistance with military defence to the setting up of training programmes. Three main types of aid are recognized: (i) *bilateral aid* is arranged directly between two countries (e.g. from the UK to some of its former colonies); (ii) *multilateral aid* involves donor countries giving to international organizations, such as the WORLD BANK, IMF and UNO, which then distribute the aid to needy countries; (iii) *voluntary aid* is provided by organizations such as Oxfam, CAFOD and Comic Relief, which collect money from the public and then spend it on specific projects. Bilateral aid is usually granted with 'strings attached' (e.g. interest repayments, the supply of primary goods at preferential rates) and in this way the donor country is frequently able to extend its economic and strategic influence, as well as increase the general level of dependence of the receiving country (see DEBT RELIEF, NEO-COLONIALISM). Aid programmes have often fallen short of achieving what was originally intended, often because of inadequate co-ordination. For example, while medical aid may have been very effective in terms of reducing levels of mortality, food production programmes often have been unable to keep pace with the resulting increase in population. Programmes have also failed because they have not involved the transfer of APPROPRIATE TECHNOLOGY. See also BRANDT COMMISSION.

AIDS (Acquired Immune Deficiency Syndrome) A fatal disease currently spreading at an EXPONENTIAL GROWTH RATE and which already seriously threatens populations in certain parts of the world, most notably in sub-Saharan Africa. Elsewhere, India and Russia are experiencing galloping rates of infection. In essence, AIDS starts with victims contracting what is known as the human immune-deficiency virus (HIV). This progressively destroys the human body's ability to combat disease. HIV is spread in a variety of ways, such as contact with contaminated blood, sexual intercourse and the sharing of needles by drug addicts. At present, much research is being undertaken to discover a cure; as yet, one has not been found; in the meantime, over 35 million people have become infected by AIDS and it is generally agreed that 'things will get much worse before they get better'.

air mass A large and essentially homogeneous mass of air, often many thousands of km^2 in area, characterized by more or less

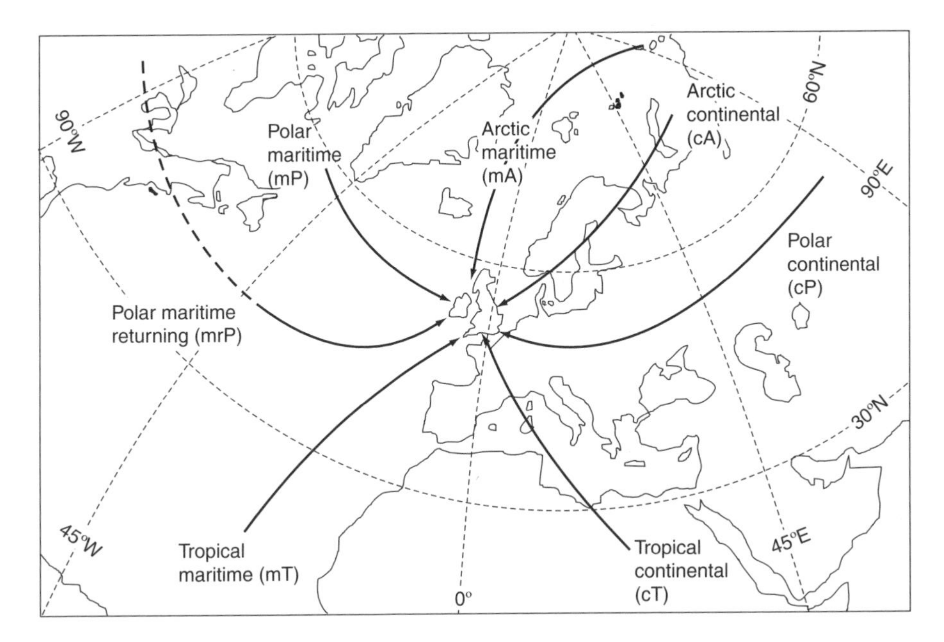

Air masses affecting the UK

uniform temperature and humidity. Air masses originate in *source-regions* (normally large ANTICYCLONES), where they are able to derive their principal characteristics and level of STABILITY from the underlying land or sea surface. Subsequently, the air masses migrate over large distances, as part of the Earth's atmospheric circulation. As they move, their humidity and temperature characteristics, and their stability, become modified and their initial characteristics are considerably altered. Air masses may be broadly classified into four types: (i) cold and dry (*polar continental*); (ii) cold and moist (*polar maritime*); (iii) warm and dry (*tropical continental*); (iv) warm and moist (*tropical maritime*). The UK is affected by several different air masses, which helps to account for the varied nature of the weather experienced in the UK. [*f*]

albedo The reflectivity of a surface, often expressed as a percentage of the solar radiation that falls on the Earth's surface. The average global albedo is about 34% (i.e. that percentage of the total solar radiation reaching the Earth's surface is reflected back into the atmosphere). However, albedo values vary greatly from place to place depending on the precise nature of the surface (ice, snow, different types of vegetation, etc.). The albedo for fresh clean snow exceeds 80%, but is much reduced for coarsely crystalline glacier ice (50%). For grassland, the albedo ranges between 30% and 20%, and for a dark peaty soil it is less than 10%. Water generally has a low albedo (between 5% and 10%) unless the Sun is at a low angle, when the albedo can be in excess of 50%.

alluvial fan A fan-shaped mass of ALLUVIUM (SAND, GRAVEL, COBBLES and sometimes BOULDERS) formed where a rapidly flowing stream leaves a steep and narrow valley and enters a lowland or broad valley. At such points a reduction in gradient, and thus stream velocity, occurs, causing overloading and DEPOSITION. Alluvial fans form in many different locations – for example, where a small tributary valley joins a major glacial trough, or where a desert stream passes through the MOUNTAIN FRONT to the PIEDMONT ZONE.

alluvium This describes the SEDIMENTS laid down by streams. Alluvium is unconsolidated material forming features such as ALLUVIAL FANS, FLOOD PLAINS, RIVER TERRACES and DELTAS. The most common constituents are CLAY and SILT (from the SUSPENDED SEDIMENT LOAD of the stream) and SAND and GRAVEL (from the BED

LOAD of the stream). Alluvium is generally regarded as being very fertile and it has the potential to form rich farmland.

Alonso model A model developed by Alonso in the 1960s to explain the paradox observed in many cities that poor people tend to live close to the city centre on high-value land, while the rich occupy cheaper land close to the city margins (see BID-RENT CURVE, URBAN DENSITY GRADIENT). The explanation is based on the assumption that the income of a household is consumed by three basic costs: (i) subsistence (food, etc.); (ii) housing and (iii) COMMUTING (assuming that, for most people, their place of work is located in or near the city centre). Poor households can make a saving on these costs by: opting for an inner-city residential location (so as to be close to their work); limiting the amount of high-cost space occupied (i.e. by living at high densities); occupying older and often substandard housing, and by accepting MULTI-FAMILY OCCUPATION of dwellings. Conversely, rich families are assumed to have large space requirements. Because they can afford higher commuting costs, they are also able to purchase large amounts of the lower-value land to be found at the edge of the city.

alp A high-level bench or gently sloping area standing above a deep U-SHAPED VALLEY. Alps are sometimes interpreted as the remaining parts of preglacial valleys, left upstanding as a result of intense glacial overdeepening. However, many are mantled by glacial deposits indicating their former occupation by glaciers. Alps often provide sites for villages and temporary settlements (*mayens*), which are occupied during the early summer months when the lower slopes of alps are used for haymaking. The higher parts of alps (*alpages*) are used for the summer pasturing of cows and sheep, following the clearance of the winter snow cover.

alpine glacier A long, tongue-like valley glacier occupying a clearly defined mountain valley, typical of the European Alps. The glacier is nourished by an ACCUMULATION ZONE (or FIRN basin) which may comprise a number of coalescent CIRQUES or a high-level ice field. The glacier may descend steeply from its source by way of an ICE FALL, or possess a relatively smooth unbroken long-profile. Alpine glaciers are usually *active glaciers*, owing to the large winter snowfalls associated with high mountains and the resultant considerable inputs of ice, which passes quite rapidly through the glacier to the melt zone (ABLATION ZONE) at lower altitudes.

alternative energy Renewable sources of energy that offer an alternative to FOSSIL FUELS and NUCLEAR POWER. These include GEOTHERMAL HEAT, and solar, tidal and WIND POWER. Given that fossil fuels are non-renewable, and given the problems associated with nuclear power, many countries are now researching the possibilities of making greater use of these alternative energy sources. As yet, however, these sources meet only a small proportion of total energy demand.

alternative technology Technology, intended for LEDCs, which is labour-intensive, low-cost and makes use of local renewable resources and skills. The aim is to maintain harmony with the ENVIRONMENT. See SUSTAINABLE DEVELOPMENT.

altiplanation A PERIGLACIAL process, involving FREEZE-THAW WEATHERING and SOLIFLUCTION, that produces step-like features (*altiplanation terraces*) and flattened hill-tops. Where a slope is underlain by rock of variable resistance to frost action, selective weathering will attack the weaker strata, forming ledges on which snow banks can accumulate.

amenity A feature of the ENVIRONMENT that is perceived as being pleasant and attractive. In current geographical usage, the term tends to be applied to something that has aesthetic, physiological or psychological benefit rather than direct monetary value – for example fine scenery, an equable climate, open space, privacy.

anabatic wind A local breeze that blows upslope during the day in regions of high RELIEF. When intense solar radiation warms the ground surface (often on the ADRET side of a valley), the air above is heated by conduction and starts to rise up the valley as convection currents. This convectional activity leads to a light and irregular drift of air up

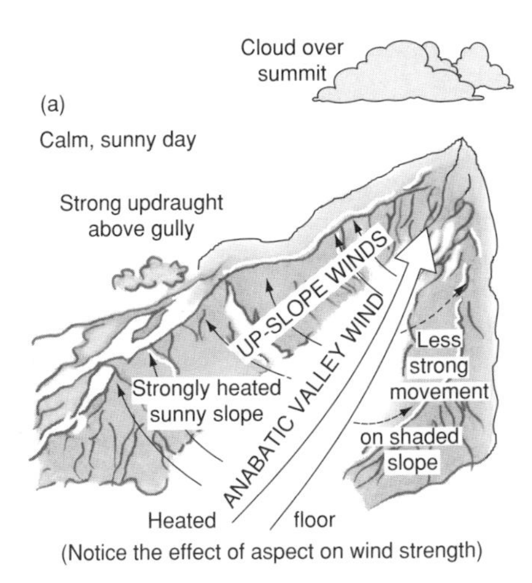

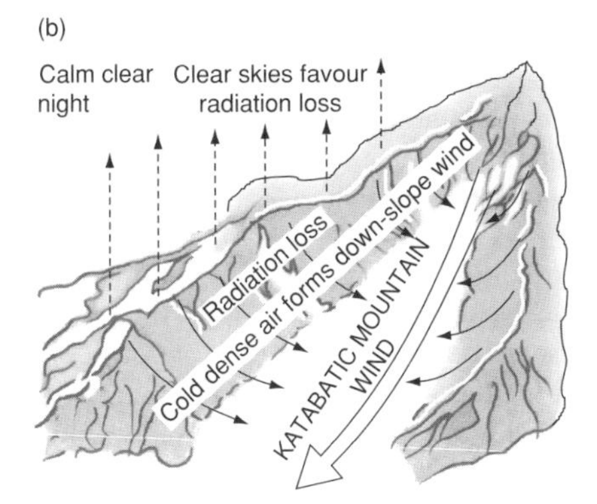

(a) Anabatic wind and (b) katabatic wind

the mountain slope, and may promote the formation of CUMULUS CLOUD in mountainous areas. The anabatic wind is rarely as strong as the down-valley KATABATIC WIND. [*f*]

anaerobic In the biological sense, organisms living in the absence of free oxygen. An anaerobic soil is thus in an airless state, notably when waterlogged. See GLEY SOIL, PEAT.

analysis of variance A statistical technique used to test whether a series of samples differs significantly with respect to some defined property. The technique compares within-sample differences with between-sample differences, with SIGNIFICANCE TESTS being used to measure the degree of dissimilarity. If between-sample differences are significantly greater than within-sample differences, it can then be assumed that, in terms of the defined property, the sample represents a distinctive group or CLASS.

anastomosing A term sometimes used for a stream in which numerous individual channels are continually separating and rejoining. See BRAIDED STREAM.

anchor tenant A term used in RETAILING to describe a shop that, because of its reputation, may be expected to draw large numbers of customers to a shopping centre and so benefit the smaller and less well-known shops nearby. For example, in launching many new shopping centres (see RETAIL PARK) in the UK, local authorities have tried to persuade firms such as Marks & Spencer, Sainsbury and Tesco to become anchor tenants, often by offering them incentives like cheap land or low rents.

anemometer An instrument used to measure wind speed, commonly comprising three cups attached to a central pivot that rotates to record the wind speed. Wind speed is usually given in knots or km per hour.

angle of repose The natural angle of rest of fragments of rock occupying a slope. The fragments may be derived either from WEATHERING of the underlying rock, or may have fallen on to the slope from a FREE FACE above to build up as SCREE below. The precise angle of repose is determined by the size and shape of the fragments. Where these are large and angular, and 'wedge' into each other, the slope will be steep (in excess of 35°); but where they are small and rounded, and intergranular friction reduced, the slope will be more gentle. The presence or absence of water is also important. Where this occurs in large quantities, friction is much reduced, and flowage will occur, thereby reducing the angle of repose.

annular drainage A DRAINAGE PATTERN in which the tributary streams follow arcuate courses determined by lines of weakness in the underlying rock structure. Annular drainage is thus characteristic of dissected

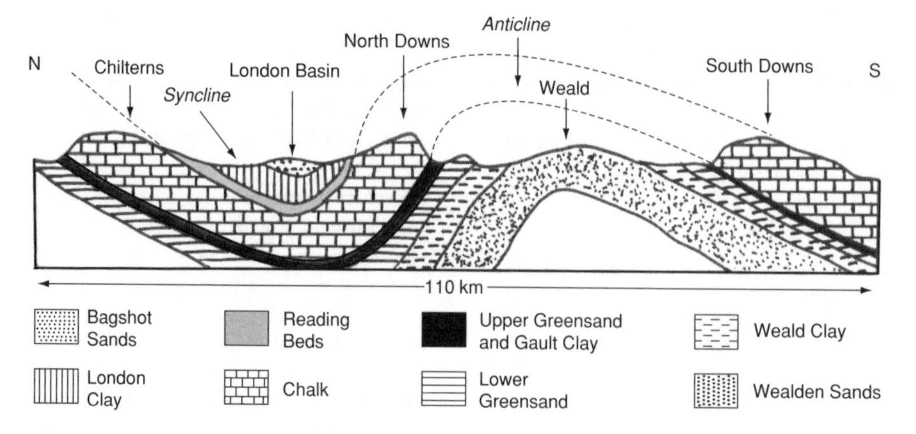

Anticline and syncline

domes comprising alternating hard and soft rocks.

anomaly The departure of any element or feature from uniformity or from a normal state, used particularly in meteorology in connection with temperature, and in oceanography in connection with salinity. For example, a *temperature anomaly* is the difference in °C between the mean temperature (reduced to sea-level) for a meteorological station and the mean temperature for all stations in that latitude. The result may be either *positive* (higher than average) or *negative* (lower than average).

antecedent drainage A process of drainage development in which an ancient river is able to maintain its course across more recently folded or faulted structures without being 'diverted' by the new structures. A prerequisite for this is that the river's capacity for downcutting must equal or exceed the rate at which the new structure grows upwards across its path.

anthropogeomorphology The study of the influence of human activities on the physical landscape and the processes that shape it; for example, the effects of agriculture, dam construction and deforestation on processes of fluvial erosion, transport and sedimentation. At the coast, there may be adverse effects of groyne and sea-wall construction on beach sediment transportation and on coastal (e.g. the harbour constructions at Newhaven, East Sussex, reduced the supply of beach sediment to Seaford and increased the risk of flooding). Deforestation and construction on slopes can increase slope instability and contribute to the LANDSLIDE hazard. A further example is the creation of THERMOKARST as a result of road- and house-building, and the stripping of surface gravel from PERMAFROST.

anticline An upfold in rocks resulting from compressive stresses in the Earth's crust. The strata DIP in opposite directions from the central line, or AXIS, of the anticline. Anticlines may be symmetrical or asymmetrical. In S England, a number of anticlines and SYNCLINES (downfolds) were formed during the Alpine OROGENY some 20–50 million years ago. [*f*]

anticyclone A large area of high atmospheric pressure, usually stationary or slow-moving. It is normally associated with widely spaced isobars, resulting in light and variable breezes or calm conditions. Anticyclones often result from the high-level convergence of air, leading to large-scale subsidence, and causing relatively high pressure on the ground. Very cold ground temperatures, such as those recorded over continental Europe and Asia in the winter, can also result in the formation of an anticyclone. As the air above the ground is cooled by conduction, it becomes dense and sinks to the surface. In Great Britain, anticyclones give stable weather conditions with high amounts of sunshine and little precipitation; in summer the weather will often be very warm, but in the winter there may be frosts as heat escapes from the ground surface. Occasionally the low temperatures in the winter may promote condensation, leading to the formation of FOG and STRATUS cloud giving the dull, dreary weather known as *anticyclonic gloom.*

anvil cloud A very high *cumulonimbus cloud*, in which the topmost parts spread out (like an anvil shape) at the base of the STRATOSPHERE in the direction of the high-level winds. Cumulonimbus clouds result from highly unstable atmospheric conditions and rapid convectional uplift. They give rise to heavy showers of rain, hail, sleet or snow, and are often associated with THUNDERSTORMS.

AONB (Area of Outstanding Natural Beauty) Areas in England and Wales that, for reasons of scenery, interest and AMENITY, enjoy special protection under the terms of the National Parks and Access to the Countryside Act 1949. AONBs are generally smaller than NATIONAL PARKS and are the responsibility of local planning authorities, which have powers for 'the preservation and enhancement of natural beauty'. Today there are 41 AONBs in England and Wales (including the Cotswolds, the South Downs and the Cornish coast), and they cover around 15,500 km^2 (about 10% of the total land area).

apartheid A policy of separate development involving planned racial segregation and spatial reorganization pursued, for example, in S Africa for over 40 years in order to ensure White domination over the non-White populations. The policy, launched in 1948, forbade the mixing of races through marriage, promoted the residential segregation of races and yet at the same time sought to ensure the supply of non-White labour to support the White-controlled ECONOMY. In pursuit of the objective of residential segregation, African Homelands (*Bantustans*) were established in rural areas in which Black Africans were able to exercise some of the political rights denied them elsewhere in the country. The policy of apartheid has now been abandoned and its associated structures dismantled.

appropriate technology A term used increasingly in the context of AID and SUSTAINABLE DEVELOPMENT when the know-how and equipment provided by donor nations are properly suited to the conditions prevailing in the receiving country. For example, in many LEDCs facing serious food shortages, it would be more appropriate to help improve existing farming practices and the implements used rather than introduce a totally alien Western mode of farming. See also ALTERNATIVE TECHNOLOGY.

aquaculture The management of water ENVIRONMENTS for the purpose of increasing the production and harvesting of organic matter (both plant and animal), as for example with *fish farming* in rivers, lakes and on the continental shelf. Although large-scale aquaculture is still in its infancy, it is expected to make an increasingly significant contribution to world food production. In India, shrimp farming has resulted in the removal of large areas of coastal mangrove forests. This has increased the risk of flooding from cyclones as mangroves hold sediment together and build up protective beaches.

aquifer A PERMEABLE rock, such as LIMESTONE or SANDSTONE, that is capable of holding and transmitting underground water. GROUNDWATER in an aquifer is an important source of water for human use and is usually tapped by wells; it may also escape to the surface naturally, by way of a SPRING where the WATER TABLE intersects the Earth's surface. An aquifer is recharged by rainwater, which percolates through pores and joints in the rocks. If demand exceeds recharge, the water table will fall and in some coastal areas this has led to the seepage of seawater into the aquifer. This is a form of pollution requiring expensive treatment before the water can be used.

arable farming A type of AGRICULTURE in which the emphasis is on the cultivation of plant crops (cereals, vegetables, grass and other animal feedstuffs). Ct PASTORAL FARMING.

arch A natural 'door' through a projecting mass of rock. The most common and spectacular arches are formed where the sea, taking advantage of weaknesses afforded by JOINTS, BEDDING PLANES or FAULTS, erodes caves on either side of a narrow promontory. These are gradually enlarged and coalesce to form a small passage, which is in turn transformed into a fully developed arch.

arctic smoke A phenomenon sometimes observed in very high latitudes, where the sea's surface appears to be giving off smoke (actually steam). It develops where the sea temperature is above freezing but the overlying air is much colder. As water vapour is evaporated from the sea surface into the atmosphere, it is immediately condensed into tiny water droplets, forming a low-level mist. Arctic smoke is thus a local form of FOG.

arcuate delta A river DELTA that extends into the sea of a lake in a fan shape, with the outer edge of the delta having a rounded outline (for example, the Nile delta in Egypt). Small arcuate deltas may also be developed where heavily loaded streams enter lakes.

Area of Outstanding Natural Beauty See AONB.

arête A narrow sharp-crested ridge resulting from the headward extension of neighbouring glacial CIRQUES. As the headwalls of the cirques are attacked by glacial plucking, FREEZE-THAW WEATHERING and rock collapse, the intervening upland becomes increasingly narrow and in time takes the form of a knife-edged ridge. Continual headwall recession may lead to the formation of a COL, giving a *breached arête*, across which the cirque glaciers can become joined.

arithmetic mean Sometimes referred to as the *average*. It is found by summing all the values in a set of data and dividing that total by the number of values. Cf GEOMETRIC MEAN, HARMONIC MEAN, MOVING AVERAGE. [*f* MEDIAN]

arithmetic scale See LOGARITHMIC SCALE.

artesian basin A large synclinal structure, comprising an AQUIFER (or series of aquifers) sandwiched between overlying and underlying IMPERMEABLE strata. Rainwater percolates into the ground at the margins of the basin, where the permeable rocks are exposed, and migrates down-DIP towards the AXIS of the artesian basin. The GROUNDWATER lying at depth here is under great hydrostatic pressure, and when wells are bored through the overlying impermeable stratum, the groundwater will rise under its own pressure to the surface. The London Basin is a good example, but because much of its artesian water has now been withdrawn, resulting in a reduction in pressure, artificial pumping is now necessary in order to abstract water. [*f*]

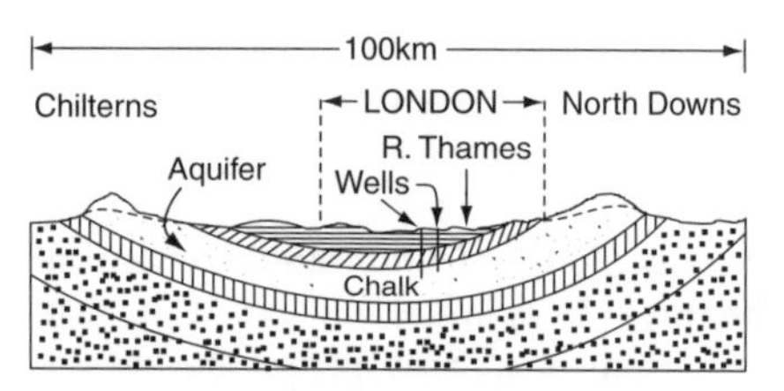

Artesian basin of the Chalk syncline beneath London

artesian well See ARTESIAN BASIN.

artificial recharge The technique of increasing GROUNDWATER supplies by, for example, the pumping of river water into wells, or the impounding of surface water to enable it to soak into underlying rocks. In alluvium-floored WADIS in desert areas such as Saudi Arabia, transverse 'low earth dams' are constructed, to impound water during infrequent FLASH FLOODS. This water sinks rapidly into the permeable alluvium, where it is stored and withdrawn from shallow wells as and when required.

ASEAN An abbreviation for the Association of South-East Asian Nations, which was set up in 1967 by the governments of Indonesia, Malaysia, the Philippines,

Singapore and Thailand to improve regional security in a somewhat troubled quarter of the world (bearing in mind the former conflict between Indonesia and Malaysia, the Vietnam War, the Vietnamese invasion of Cambodia, as well as the potential external threat posed by China). The membership of ASEAN was increased in 1984 when it was joined by Brunei; while Vietnam joined in 1995, Laos and Myanmar in 1997, and Cambodia in 1999. ASEAN is not a military alliance, but it does exert considerable diplomatic pressure in the affairs of SE Asia. It receives much support from Japan as its major trading and investment partner. Member countries have agreed to become the ASEAN Free Trade Area (AFTA) in 2003.

ash Fine powdery material emitted during a volcanic ERUPTION. The ash is often so fine (the particles may be less than 0.25 mm in diameter) that it can be carried by the wind over vast distances. During the catastrophic Mt St Helens eruption of 18 May 1980, a violent explosion blew a large cloud of ash to a height of some 18 km; the eruption continued for 9 hours, producing further quantities of ash that fell to blanket parts of Washington, N Idaho, and W and central Montana. It is estimated that several km^3 of ash were expelled, and the ash cloud eventually crossed to the east coast of the USA. *Ash cones*, or *ash volcanoes*, are formed where the ash emitted accumulates around the vent; such cones are usually concave in profile and relatively gentle-sided. Ash represents a major volcanic hazard, causing buildings to collapse as the ash piles up on their roofs. It can combine with water to form damaging mudflows called LAHARS. The eruption of Mt Vesuvius in AD79 blanketed the town of Pompeii with 3 m of ash and over 2000 people were killed, mostly by ash asphyxiation.

Asian Tigers The name given to the four Pacific Rim countries of Hong Kong, Singapore, South Korea and Taiwan, which during the last quarter of the 20th century experienced remarkable ECONOMIC GROWTH. This was based initially on domestic manufacturing; later on, SERVICES and FOREIGN INVESTMENT played their part. In the late 1990s, the economies were shaken badly by financial scandals and political corruption. The resulting slump (see BUSINESS CYCLE) was described as a bout of *Asian flu.*

aspect The direction in which a slope faces. This has effects on the climate of the slope in terms of total INSOLATION received, exposure to rain-bearing winds, amount and duration of frost and snow cover, etc. These, in turn, may significantly affect the operation of geomorphological processes (see ASYMMETRICAL VALLEY), and the degree of development of certain landform types. Settlement and land use are also greatly affected by aspect, particularly in steep mountain valleys (for example, in the European Alps, where south-facing slopes are more favoured due to higher temperatures and longer hours of sunshine).

assembly costs The TRANSPORT COSTS incurred by a manufacturing firm in bringing together its raw material requirements; sometimes referred to as *collection costs.* See also PROCUREMENT COSTS.

assembly-line production A serial arrangement of workers and machinery for passing work on from stage to stage in the assembly of a manufactured product, as widely adopted in the motor vehicle industry, for instance. Cf AUTOMATION, DESKILLING, MASS PRODUCTION.

assimilation The process by which different groups within a community (distinguished on the basis of criteria such as affluence, economic status, race or religion) intermingle and become more alike. The process particularly applies to the integration of immigrant MINORITY groups, as for example to New Commonwealth immigrants in Britain. Assimilation may take a number of different forms, such as intermarriage, adopting the values and attitudes of the community at large (see ACCULTURATION), contributing to the cultural life of that community or becoming proportionately represented in all strata of the social and occupational hierarchies.

assisted area A term used in Britain to describe those parts of the country that benefit from various forms of government help as part of regional policy (see DEVELOPMENT AREA, INTERMEDIATE AREA).

Association of South-East Asian Nations See ASEAN.

asthenosphere The uppermost zone of the Earth's mantle, lying at a depth usually within the range 60–200 km. Within it, the rocks are probably close to melting point, as a result of the concentration of heat from radioactive decay. The asthenosphere therefore has a 'plastic' quality, allowing slow flowage to occur under high pressure. It is envisaged, in the theory of PLATE TECTONICS, that movement of the overlying lithospheric plates (see LITHOSPHERE) is facilitated by the 'plastic' asthenosphere.

asylum seekers See POLITICAL ASYLUM.

asymmetrical valley A valley whose slopes on one side are steeper than those on the other. Many asymmetrical valleys in Britain and Europe are attributed to past PERIGLACIAL conditions, when differential exposure of the valley slopes to solar radiation and/or snow-bearing winds resulted in differential DENUDATION by frost action and SOLIFLUCTION. The *active slopes* (in many instances those facing to the southwest) were modified more rapidly than the *inactive slopes* (often facing northeastwards, and remaining frozen and snow covered), resulting in asymmetry. [*f*]

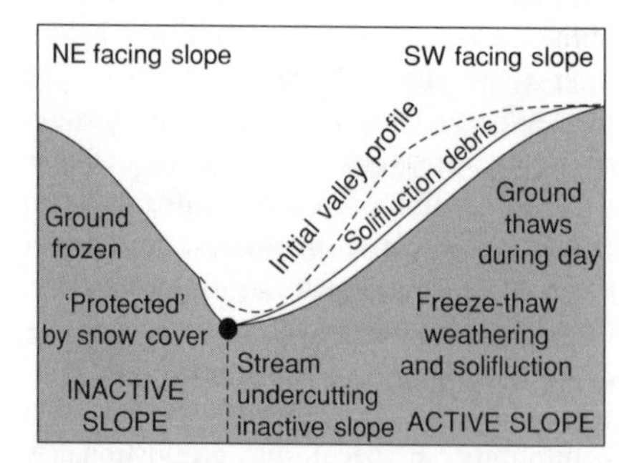

Asymmetrical valley formed under periglacial conditions

Atlantic-type coast A strongly embayed type of coastline, which is formed where the trend of mountain ridges and major lowlands is at right angles to the coastline as a whole (for example in SW Ireland, S Wales, and the Brittany peninsula). Also referred to as a *discordant coastline*. Ct PACIFIC-TYPE COAST.

atmosphere The layer of gases (mainly nitrogen 78% and oxygen 21%), liquids and solids (e.g. dust, ash) that surrounds the earth, and which is held in place by gravity. See also STRATOSPHERE and TROPOSPHERE, which are major subdivisions of the atmosphere.

atoll A coral REEF, surrounding a central lagoon, commonly found among the islands of the S Pacific. It is believed that atolls have been formed above former islands (sometimes volcanoes); as these have become submerged, either as a result of subsidence of the sea-floor, or rises of sea-level (including that at the close of the PLEISTOCENE period), they provided a base or 'platform' for upward coral growth. The corals have been most active on the outer walls of these reefs, which has thus grown not only upwards but outwards. The inner, inactive sides of the reef mark the boundaries of the enclosed, or partially enclosed, lagoon.

attribute Used in STATISTICS to denote a feature that is confined to the *nominal* scale (see NOMINAL DATA); i.e. the feature is either present or absent. For example, a city may or may not display the following attributes: an airport; a riverside location; a population greater than 1 million.

attrition The process whereby the LOAD particles of rivers, winds and waves are reduced in size and become more rounded, as a result of continual impacts between individual particles. The effects of attrition on particle size in a river can be illustrated by data from the Mississippi, where 38 km from the source of the river the mean diameter of particles on the river bed is 210 mm; at 120 km it is 80 mm; at 2080 km it is 0.29 mm; and at 5600 km it is 0.16 mm. Attrition is most effective in AEOLIAN (wind) environments, often resulting in almost spherical particles that have been extremely well sorted. Water, for example in a river, is a less effective medium because, in coating particles, it tends to cushion the effects of attrition, thereby reducing the force of the impacts. Sediments carried by ice rarely collide with each other, so tend to remain angular and poorly sorted (see TILL).

automation A system of automatic machine control extending over an entire series of manufacturing operations. Such systems are now widely used in manufacturing – for example, in the motor vehicle industry. Cf ASSEMBLY-LINE PRODUCTION, DESKILLING, MASS PRODUCTION.

autonomy The power or right of self-government; the attribute of an independent state or organization. See NGO, QUANGOS.

autotroph A self-feeding organism (e.g. a plant) capable of combining solar energy with salts derived from the soil to form sugars via *photosynthesis*. Autotrophs form the first TROPHIC LEVEL in the FOOD CHAIN.

avalanche The rapid descent of a large mass of rock, ice and snow (sometimes all three) down a steep mountain slope. Avalanches occur most commonly in winter and spring. Snow avalanches in particular form either from large masses of recent uncompacted snow occupying the mountain side, or from partially thawed layers of older snow during warm spells of weather (e.g. with the onset of FÖHN WINDS in the Alps). Avalanches are also frequently triggered by human activity, such as skiing or snowboarding. Many avalanches develop along well-used *avalanche tracks*; these can be mapped and precautions taken (such as restrictions on new building, avalanche sheds over roads and railways, and tunnel construction) to minimize damage. However, avalanches sometimes follow previously unused paths, especially in areas of deforestation, and thus can constitute a serious hazard to life and property.

average See ARITHMETIC MEAN.

axial belt A CORRIDOR along which is concentrated much of a country's population, economic wealth and URBAN development; the axial belt of Britain, for example, extends from the Greater London area northwestwards to the Merseyside and Manchester conurbations, and includes Milton Keynes as well as the West Midlands conurbation. The Japanese equivalent runs along the southern coastlands of Honshu, from Tokyo westwards, and includes the major cities of Yokohama, Nagoya, Osaka and Kobe, possibly extending as far as N Kyushu. See also CORE, MEGALOPOLIS.

axis The central line of a geographical fold structure, such as an ANTICLINE or SYNCLINE. In an anticline, the axis marks the 'crest' of the fold, from which the strata DIP away either side; in a syncline the axis marks the 'trough', from which the strata rise on either side.

azonal soil A soil that has undergone limited development and is characterized by an absence of well-developed SOIL HORIZONS. Azonal soils are not associated with particular climatic vegetation zones (ct ZONAL SOIL). Azonal soils include *lithosols* (formed on SCREE and glacial MORAINES, usually on steep slopes), *regosols* (on dry SANDS and GRAVELS in deserts) and *alluvial* soils (on lowlands prone to flooding).

B

backshore That part of a beach lying above the high-water mark and normally beyond the reach of wave action. Ct. FORESHORE

backward linkage See LINKAGE.

backwash The return flow of water down a BEACH, after a breaking wave has sent SWASH up the beach. The backwash is most powerful with plunging breakers (DESTRUCTIVE WAVES), when the steep waves crash down on to the beach with a strong overturning motion. With CONSTRUCTIVE WAVES, the backwash is much less powerful as most of it soaks away as it returns down the beach.

backwash effect A term used by Myrdal in his theory of CUMULATIVE CAUSATION to describe the spatial concentration of resources and wealth in the CORE or centre at the expense of the PERIPHERY. Cf POLARIZATION; ct SPREAD EFFECT. See also CORE-PERIPHERY MODEL.

backwoods A term first used in the USA to denote sparsely settled, partially cleared land; generally an area of PIONEER SETTLEMENT. It is commonly applied to any sparsely settled area remote from an URBAN centre It is also used colloquially in the derogatory sense of areas that are regarded as being out of touch and therefore, by implication, backward.

badland A landscape made up of a maze of steep-sided gullies, which are difficult to cross and too steep for cultivation. Such areas have been intensely dissected by surface streams and rivulets. Badlands develop owing to a combination of rock impermeability, sparse vegetation cover, rapid surface RUN-OFF from brief but heavy rainstorms, and poor farming practices. They are particularly characteristic of semi-arid regions (for example, South Dakota, USA).

bajada An alluvial formation in a semi-arid or arid region. In SW USA, bajadas comprise fans of BOULDERS, GRAVELS and SANDS formed at the base of the MOUNTAIN FRONT. The alluvial fans are deposited by streams that occasionally flow out of narrow canyons after storms have occurred. As the streams enter this flatter zone, called a PEDIMENT, they lose velocity and transporting power, and deposit BED LOAD in large quantities.

balance of payments The relation between the *payments* of all kinds made from one country to the rest of the world and that country's *receipts* from all other countries. It takes into account trade in both commodities (*visible trade*) and services (*invisible trade*). For much of the postwar period, Britain has faced a balance of payments deficit in that payments have exceeded receipts. Cf BALANCE OF TRADE.

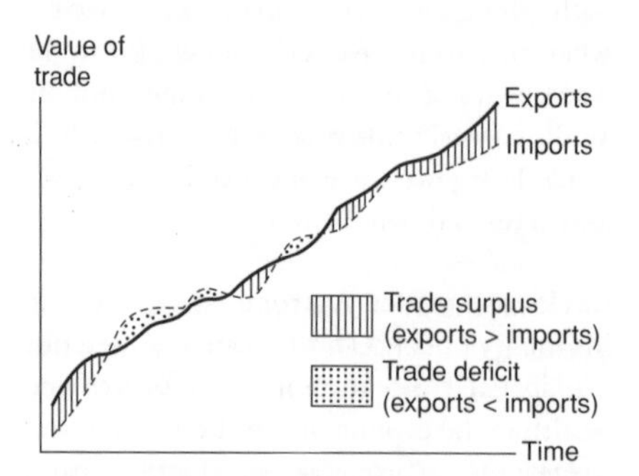

Balance of trade

balance of trade The relation between the value of the EXPORTS and IMPORTS of any country. This is referred to as being favourable when exports exceed imports and adverse when the balance is reversed. In the strict sense, the term should be limited to *visible trade* (i.e. trade in goods and commodities) and should exclude *invisible earnings* (derived from services such as banking, insurance and TOURISM). If these invisibles are included, then the term BALANCE OF PAYMENTS should be used instead. [*f*]

bankfull The state of a river's flow, or DISCHARGE, at which the CHANNEL is completely filled from the top of one bank to the other. Beyond this point the channel cannot cope, and overbank flow occurs (see FLOOD).

bar (i) A linear deposit of shingle, sand or mud usually aligned roughly parallel to a coastline and submerged either wholly or partly by the sea. (ii) A deposit of sand or mud in a river channel (see RIFFLE). (iii) A unit of atmospheric pressure.

bar diagram A diagram consisting of a series of bars or columns proportional in length to the quantities they represent. They may be either *simple* (where each bar shows a total value) or *compound* (where each bar is subdivided to show the composition of the total value, e.g. the commodity structure of a trade total). The bars may be placed vertically, horizontally or in pyramidal form. The first of these options is to be preferred when representing a set of values taken over a period of time, e.g. annual production figures covering, say, a 35-year period. [*f* AGE–SEX PYRAMID, *f* BINOMIAL DISTRIBUTION]

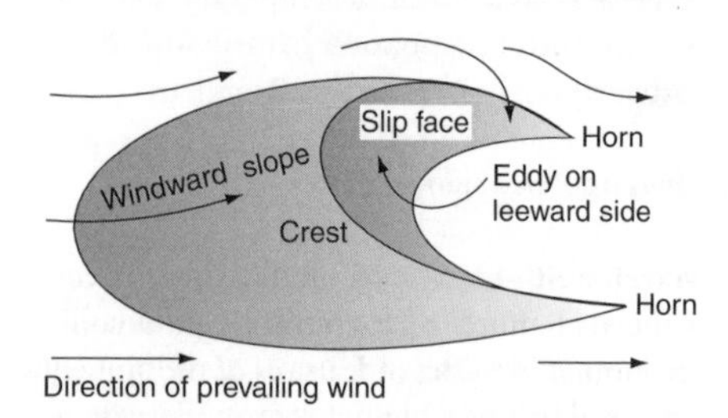

Barchan

barchan One of the most common types of SAND DUNE found in deserts. The barchan is a crescentic dune formed at right angles to the wind, particularly where this blows consistently from one direction. The dune begins as a small mound of SAND, formed on the lee side of an obstruction to air flow, such as a rock or bush. Once in existence, the dune will continue to trap sand blown in by the wind. This fresh sand will be transported up the windward slope, over the dune crest, and on to the lee face of the dune; this will cause downwind migration of the dune. The rate of migration will be slowest at the centre of the dune (where it is highest) and more rapid at its extremities (where it is lowest); as a result, the dune will develop 'horns' pointing downwind. In profile, barchans are asymmetrical. The windward slope is gentle, but the lee slope is continually steepened by inputs of sand. However, the lee slope angle does not usually exceed 34° (the angle of repose of dry sand); any tendency for the dune face to steepen beyond this is countered by slippage of sand (hence *slip face*). [*f*]

barometric pressure See PRESSURE GRADIENT.

barrage A large structure, usually of concrete, sometimes of earth, built across a river usually to hold back a large body of water for IRRIGATION or for supply to domestic and industrial users. A barrage may also be used in flood control schemes (e.g. the Thames Barrier in London). Some draw the distinction between a barrage and a *dam*, on the basis that the former is not associated with the generation of HEP (hydro-electric power) and that it is concerned with annual rather than perennial water storage. A *tidal barrage* is a controlled barrier across an estuary or an arm of the sea. The water of an incoming tide is held at high tide as a temporary reservoir behind the closed barrier. After the tide has fallen, the barrier is opened and the water gradually released. The energy derived is then used to generate electricity.

barrier effect May be used in two different contexts in human geography: (i) to describe the impeding effect of features of the physical ENVIRONMENT (mountains, deserts, gorges) on transport and communication; (ii) to describe the resistance of people to innovation for a variety of possible reasons (lack of CAPITAL, dislike of risk, distrust). See SPATIAL DIFFUSION.

barrier island A low sandy island, usually forming one of a series of islands running parallel to the mainland and separated from it by a tidal lagoon. These were formerly attributed to the emergence of submarine bars, owing to a fall in sea level. But it is now known that barriers are largely POSTGLACIAL features, formed during the past few thousands of years under conditions of rising sea-level. As the sea transgressed a low RELIEF surface, covered by abundant SANDS, ridges were formed by CONSTRUCTIVE WAVE action. These eventually became so large that they could no longer be pushed shorewards, and the areas inland of the ridges were inundated to give lagoons. Subsequently, the barrier islands have either been further built up by wind DEPOSITION (hence their occupation by large DUNE systems) or are being 'washed over' and eroded by large waves.

barrier reef A large coral reef running parallel to the coastline, from which it is separated by an extensive lagoon. The most famous example is the Great Barrier Reef of E Australia, which extends for over 2000 km. This is a massive and complex structure, comprising an outer line of coral reefs, innumerable *cays* (accumulations of coral sand just breaking the sea surface) and larger islands lying some 50–200 km offshore, together with many inner reefs.

basal sapping EROSION concentrated along the base of a slope, causing undermining and recession of that slope. Basal sapping may be particularly active in some tropical environments: (i) where laterite-capped slopes are undermined by SPRINGS and seepages in underlying weathered SANDS and CLAYS; (ii) at the foot of SCARPS and INSELBERGS, where concentrated moisture (from RUN-OFF on the slopes, or GROUNDWATER held by TALUS) results in rapid chemical decomposition; (iii) the recession of the backwalls of CIRQUES may also involve basal sapping, whereby disintegrated rocks are incorporated within the ice and removed (see BERGSCHRUND hypothesis and JOINT-BLOCK REMOVAL).

basal slip, sliding See GLACIER FLOW.

basal surface of weathering The lower limit of a deeply weathered REGOLITH, developed mainly in humid tropical environments but also found elsewhere. There is usually a very rapid change at the basal surface from rotted rock above to solid rock below. In the tropics, the basal surface may lie at depths of 30–60 m beneath the ground surface. Here the prevailing high temperatures and abundant GROUNDWATER have promoted intense CHEMICAL WEATHERING, and removal of the resultant weathered material has often been hindered by the dense vegetation cover. In other areas the weathered material has been removed by stream action to reveal the basal surface as a bare rock platform or series of low, rounded hills. See PEDIMENT, BORNHARDT. [*f* RUWARE]

basalt A fine-grained, dark-coloured IGNEOUS ROCK. Basaltic LAVAS are basic, have a low melting point and low viscosity, and flow freely to cover wide areas. When extruded from extensive and numerous fissures, basalt may inundate the pre-existing landscape (hence *flood basalts*).

base flow That part of the DISCHARGE of a stream that maintains a near constant steady flow provided by the gradual seepage of water into a river channel, rather than that which is directly related to a storm event. On a HYDROGRAPH, it accounts for the steady low flow either side of the hydrograph peak. Base flow is a very slow form of water transfer within a drainage basin and involves the gradual seepage of GROUNDWATER through the rocks. The rate of flow will be greatest after winter recharge of groundwater supplies. It will be reduced during a period of drought or if an excessive amount of water is pumped for industrial, agricultural or domestic purposes (see ABSTRACTION). The concept of base flow can also be applied to a glacier. During a period of intense surface ABLATION, much meltwater will enter a glacier, recharging the water held by cavities within the ice and speeding up outflow at the glacier snout. When ablation ceases (in a cold period, or during winter), drainage of water stored within the glacier will continue to provide base flow to the PROGLACIAL stream.

base-level of erosion The theoretical limit, usually regarded as sea-level, below which rivers cannot erode their courses. In other words it represents the lowest level to which a fluvially dissected land surface can be lowered. Base-level is sometimes seen as a strictly horizontal limit to EROSION, but it is more logical to view it as a very gently sloping surface, since rivers do require some gradient over which to flow. Thus a PENEPLAIN, the product of many millions of years of erosion, will stand slightly above the theoretical base-level, and will possess gentle gradients along river courses and across interfluves. The term base-level has also been used in other contexts. For example, *marine base-level* is the lower limit of wave erosion. It might be assumed that this coincides with the intertidal zone, where wave break is concentrated and extensive WAVE-CUT PLATFORMS are found.

base year The datum used in investigating a time series of data. The chosen base year is usually assigned a value of 100. All the other values in the series are converted into index numbers related to that value of 100.

basic industry See INDUSTRY.

basic lava See BASALT.

basin and range A type of geological structure comprising large and often tilted FAULT BLOCKS (*ranges*) separated by downfaulted blocks and/or the downtilted margins of fault blocks (*basins*).

batholith A very large, dome-like mass of intruded IGNEOUS ROCK. Initially, the intrusion takes place at a great depth within the Earth's crust, in association with earth movements and mountain building. However, after a long period of DENUDATION, the upper surfaces of the batholiths are eventually exposed at the surface, as on Dartmoor, Bodmin Moor and the smaller granite moorlands of SW England.

bay bar A bank of SAND OR SHINGLE, extending across a bay from one enclosing headland to the other. Inland from the bar there is commonly a lagoon that, if the bar is breached, will become tidal. It is believed that some bay bars result from the convergence of SPITS, growing in opposite directions from each end of the bay. An early stage of this may be seen in Poole Bay, Dorset, S England, where the sand BEACH at South Haven peninsula has extended northeastwards and that at Sandbanks southwestwards (though the strong tidal currents from Poole Harbour have prevented linkage). An alternative mode of formation involves the growth of a single spit across a bay where the LONGSHORE DRIFT is unidirectional. However, it is likely that most bay bars have resulted from the onshore migration of OFFSHORE BARS, with the aid of a rising sea-level. For example, Chesil Beach may have been formed initially as a spread of SHINGLE on the exposed floor of Lyme Bay, S England, during the last glacial period. As the sea-level rose during the early POST-GLACIAL period, the shingle was fashioned by wave action into a bar, which was slowly driven landwards by the process of 'overtopping'.

battery farming A particularly intensive type of FACTORY FARMING in which poultry, in particular, are reared in cages under cover, and are fed and watered automatically.

bay-head beach A small SAND or SHINGLE beach occupying part of a bay protected by projecting headlands on either side. The BEACH may be offset towards one end of the bay, particularly when waves approach the bay obliquely. There may also be some sorting of beach material, with the larger COBBLES concentrated at the 'downdrift' end of the beach, the finer material having been removed by wave action.

bazaar economy Prevalent in LEDCs, where many commercial transactions are conducted on a person-to-person basis, mainly in a public market or bazaar. Such transactions typically involve bargaining and bartering, with goods and services being acquired by exchange rather than for cash.

beach An accumulation of SAND and/or SHINGLE found between the highest point attained by storm waves and the lowest tide-level. The beach material is deposited by breaking waves, possibly with the aid of tidal currents. Constructive action is mainly effected by the SWASH, and destructive action by the BACKWASH. The detailed form of the beach represents an ever-changing balance between these processes. Most beaches are concave in profile, and comprise an upper section of coarse material (GRIT and pebbles), with a steep gradient towards the sea, and a lower section of sand, or even mud, with a much gentler gradient. The upper beach is also diversified by ridges (see BERM), and the lower by longitudinal sand ridges separated by shallow depressions (*ridge-and-runnel*). *Beach cusps* are small, regularly spaced embayments, usually developed on the face of the shingle beach or at the junction between the shingle and sand.

beaded esker See ESKER.

Beaufort wind-scale A scale of wind force, initially devised by Captain Beaufort

Scale	Wind	Average wind speed (km hr)	Effects in inland situations
0	Calm	1	Smoke rises vertically
1	Light air	3	Wind direction shown by smoke
2	Light breeze	9	Wind felt on face; leaves rustle
3	Gentle breeze	16	Leaves and twigs in constant motion
4	Moderate breeze	24	Raises dust and loose paper
5	Fresh breeze	34	Small trees in leaf sway
6	Strong breeze	44	Large branches in motion
7	Moderate gale	56	Whole trees in motion
8	Fresh gale	68	Twigs break off trees
9	Strong gale	81	Slight structural damage
10	Full gale	95	Trees uprooted
11	Storm	110	Widespread damage
12	Hurricane	>121	Devastation

Beaufort wind-scale

for use by seamen to standardize subjective terms such as 'light breeze', 'fresh breeze', 'gale' and 'hurricane'. The basis is a numerical scale, from 0–12, in which each number coincides with a descriptive title. The same scale has been adapted for use on land, according to the effects of different wind speeds on smoke, trees and buildings. [*f*]

bed load The solid rock particles that are transported along the floor of a river CHANNEL by rolling, sliding and SALTATION. An alternative term is *traction load.* The rate of movement of the particles is less than that of the water (which provides the hydraulic force), but there is considerable variation in the speed of individual components of the bed load. At a particular flow rate (the *erosion velocity*) particles of a given size are set in motion; as the EROSION velocity increases, larger and larger particles will be moved (see COMPETENCE). However, it has been noted that, once in motion, large grains may actually move more rapidly than small ones, and that particle shape is an important factor (rounded particles move more readily than flat or angular particles). The amount of bed load transported by a stream will vary greatly with time, as volume and velocity fluctuate. In times of severe flooding, even large BOULDERS will be moved, although under conditions of normal DISCHARGE the same stream may be capable of moving only SAND or fine GRAVEL.

bedding plane The surface separating individual layers of a SEDIMENTARY ROCK such as LIMESTONE, CHALK or SANDSTONE. Bedding planes often take the form of cracks along which underground water can move in a down-DIP direction. They also constitute lines of weakness that can be exploited by WEATHERING processes (for example, FREEZE-THAW WEATHERING), so that well-bedded rocks tend to be relatively unresistant (see MASSIVE).

bedrock Solid unweathered rock, underlying the SOIL or REGOLITH.

behavioural environment That part of the *perceived environment* (see ENVIRONMENTAL PERCEPTION) that influences individual behaviour and decision-making, and to which behaviour is directed. It is regarded by some as being synonymous with *action space, awareness space* and *task environment.*

behavioural geography An aspect of, or approach to, HUMAN GEOGRAPHY that is particularly concerned with the ways in which people perceive, respond to and affect their surroundings. See MENTAL MAPS, SPATIAL PREFERENCE and SPATIAL DIFFUSION.

behavioural matrix A framework devised by Pred for the analysis of locational DECISION-MAKING, in which decision-making is seen as a function of two things: (i) the quantity and quality of perceived information that is available to a person; (ii) the ability of that person to make use of such information. These functions provide the two axes of the MATRIX. Given these two dimensions, it is reasonable to suppose that a business-owner with limited information but great ability would choose a location for his FIRM that is different from that selected by another with extensive information but limited ability. Furthermore, it is assumed that, over time, decision-makers accumulate more and better information and become more skilled in its use.

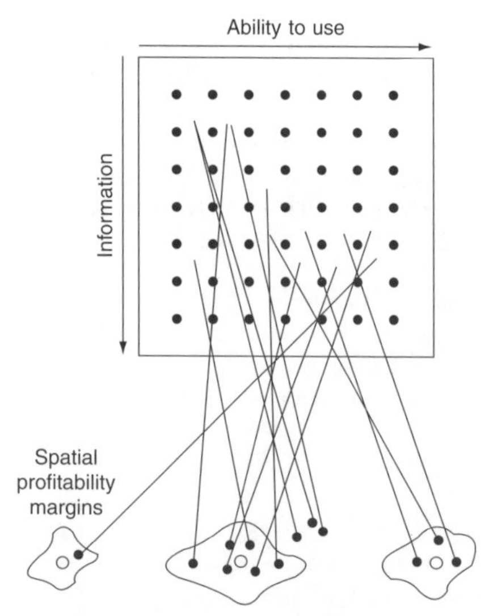

Behavioural matrix and locational choice in an industrial situation

As a result, they should move downwards and to the right in the matrix.

In the lower part of the accompanying figure, the location of 13 firms is shown with reference to three areas bounded by the SPATIAL MARGINS to profitable operations. Each firm is connected by a line to the place in the behavioural matrix above that best summarizes its situation as regards information and the ability to use that information. Those firms located towards the bottom right of the matrix have, in general, chosen locations close to the optimum in each of the three areas, while of those firms with limited information and ability, three have taken up unprofitable locations outside the spatial margins. [*f*]

beneficiation The process by which low-grade mineral ores (e.g. bauxite, copper, iron) are concentrated at the site of extraction in order to save the cost of transporting bulky waste material.

Benioff zone A linear earthquake zone, named after H. Benioff, developed at a destructive, or convergent, plate margin (see PLATE TECTONICS). The earthquakes are actually generated along a sloping plane of friction, where one lithospheric plate is subducted beneath another overriding lithospheric plate, resulting in the formation of an OCEAN-FLOOR TRENCH/ISLAND ARC system. The foci of the resultant earthquakes are relatively shallow close to the trench, but become increasingly deep-seated as the Benioff zone slopes away from the trench floor, beneath the island arc. The intense pressures in this zone may lead to the melting of the lithospheric plate and the formation of a body of MAGMA. This may escape to the surface to form volcanoes (e.g. island arc volcanoes).

Bergeron-Findeison hypothesis A theory, first developed by Bergeron in 1933, to explain the formation of rainfall from a cloud resulting from the ascent of air above the FREEZING LEVEL in the atmosphere. Such a cloud will comprise a mixture of supercooled water droplets (from condensation below the freezing level) and ice crystals (from condensation above the freezing level). The ice crystals grow at the expense of the water droplets (as VAPOUR PRESSURE is lower over ice than over water droplets) and a transfer of water vapour can take place to the neighbouring ice crystals to form snowflakes which, as they fall below the freezing level, will melt to form raindrops.

bergschrund Literally meaning 'mountain crack', a large CREVASSE in ice running around the upper part of a CIRQUE glacier and developed as the glacier pulls away from the cirque headwall and subsides, with the result that the ice surface is higher above the bergschrund than below. The bergschrund is sometimes open, and sometimes (particularly in the early part of the ABLATION season) bridged by snow. It may be developed wholly within the glacier, or may penetrate to the SUBGLACIAL surface. In the latter case, the presence of angular fragments broken from the bedrock was evidence used to support the *bergschrund hypothesis*, which postulated that temperature changes within the crevasse could cause frost shattering and thus aid the process of JOINT-BLOCK REMOVAL. However, measurements of temperature within bergschrunds have shown little deviation from 0°C, so discrediting the hypothesis. [*f* CIRQUE]

berm A nearly horizontal, or gently landward-sloping ridge at the crest of a BEACH. The seaward edge of a berm is marked by a sudden change of slope to the *beach-face*, which descends quite steeply towards the sea, particularly on SHINGLE beaches. The berm consists of materials that have been thrown up by breaking waves, mainly under storm conditions.

best-fit line As used in REGRESSION ANALYSIS, this is the line that best fits the trend of a scatter of points plotted on a GRAPH – i.e. it is the *regression line*. It is usually determined by the LEAST SQUARES method. [*f* LEAST SQUARES]

bevelled cliff See SLOPE-OVER-WALL CLIFF.

B-horizon The subsoil layer beneath the A-HORIZON in a fully developed or mature SOIL PROFILE. It is characterized by a less advanced degree of WEATHERING of constituent minerals, a reduced humus content and some degree of enrichment by compounds washed down from above. It is sometimes referred to as the *illuvial horizon* (see ILLUVIATION), and is often more yellow, brown or red than the A-horizon. In soils where LEACHING and ELUVIATION of the A-horizon are intense, illuviation in the lower part of the B-horizon may be considerable, leading to the formation of HARDPAN. [*f* SOIL PROFILE]

bias Error or distortion in a data set caused by such things as faulty SAMPLING procedures, poor questionnaire design and interviewer prejudice.

bid-rent theory An economic theory, providing the basis of a number of geographical models (see VON THUNEN'S MODEL, CONCENTRIC ZONE MODEL), which states that rent or land values decrease with increasing distance from a centre or nodal point (i.e. they show DISTANCE DECAY). This may be seen as applying as much to agricultural land (where distance from market is deemed to be crucial) as it does to URBAN land (where distance from the TOWN or CITY centre is considered to be significant).

A *bid-rent curve* shows the theoretical effect of this increasing distance from a centre on the value or rent of land. In the case of a city, land is most expensive at the centre because competition for space is keenest in this the most accessible part of the city, and because land here is most scarce. As the demand for land decreases away from the centre, and as land becomes more plentiful, so bid-rents fall. In other words, the bid-rent curve shows a downward slope away from the centre, as the rents or land values that people and businesses are prepared to pay decrease with distance.

Different LAND USES show different bid-rent curves, because they differ in terms of their bidding power on the land market and in terms of their tolerance of increasing distance from the centre. The accompanying figure shows the bid-rent curves for three urban land uses. Because retailing is, in general, a strong bidder (it is a capital-intensive user of space) and because it relies greatly upon a central, accessible location, so the bid-rent curve pitches high at the city centre and dips steeply with increasing distance from the centre. By superimposing the bid-rent curves of different land uses, it becomes possible to delimit concentric zones, in each of which a particular activity may be expected to become the dominant land use. [*f*]

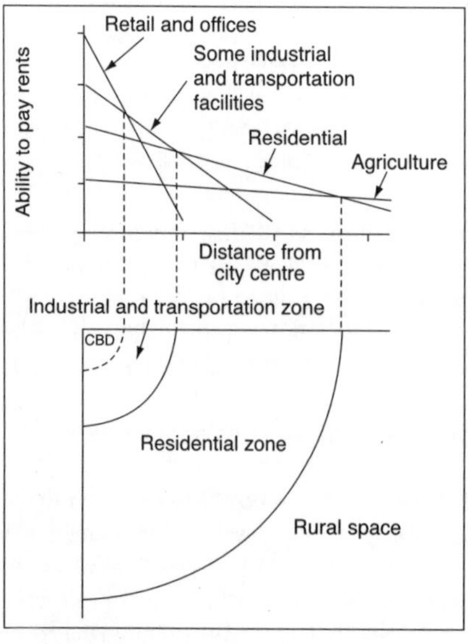

Bid-rent curves

bifurcation ratio A STATISTIC used in drainage basin morphometry, in conjunction with STREAM ORDER, to define drainage networks and assist in the formulation of laws of drainage basin form. The bifurcation ratio states the relationship between the number of streams of one order and the number of the next higher order. For example, if in one drainage basin there are 231 1st-order streams and 77 2nd-order streams, and in another there are 96 1st-order streams and 24 2nd-order streams, the bifurcation ratios will be 3.0 and 4.0 respectively. The higher number of stream junctions in the second basin is a measure of its greater complexity of form.

bilateral See AID, MULTILATERAL.

binary pattern Used in the analysis of settlement size frequencies to describe the situation where the upper end of the settlement HIERARCHY is dominated by a number of SETTLEMENTS of a similar size. It is a pattern to be expected where a federal system of government prevails (e.g. in the USA and Switzerland), with each member state having its own capital city. See CITY-SIZE DISTRIBUTION; ct LOGNORMAL DISTRIBUTION, PRIMATE CITY.

[*f* RANK-SIZE RULE]

binomial distribution This is one of the most common PROBABILITY distributions. It is associated with the repetition of events, in an independent trial situation, where there are only two possible outcomes (as, for example, with the result of tossing a coin or the sex of a newly born child). Take the latter instance. If the probabilities associated with different numbers of girls occurring in a six-child family are plotted on a HISTOGRAM, then the following characteristics of a binomial distribution may be noted: (i) the distribution is symmetrical; (ii) the greatest possibilities cluster around the MEAN (in this case three girls); (iii) the greater the deviation from the mean, the smaller the probability of that number occurring. [*f*]

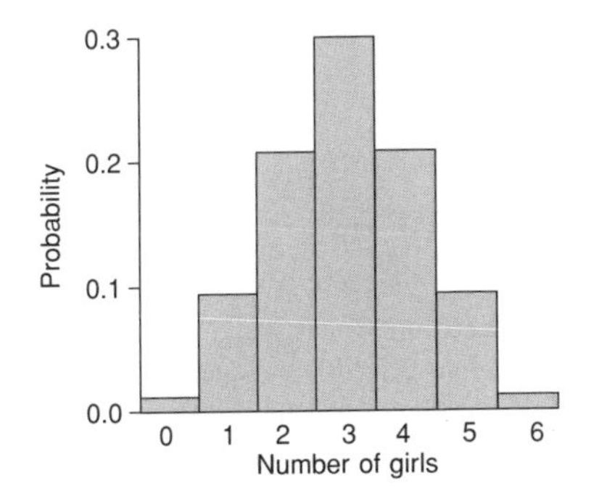

Possibilities associated with different numbers of girls in a six-child family

biodiversity A term describing the variety of species, both flora and/or fauna, contained within an ECOSYSTEM. Some ecosystems (such as a tropical RAINFOREST) are characterized by very high biodiversity, while others (such as BOREAL FOREST) have much lower biodiversity. A serious human impact on many ecosystems has been to reduce biodiversity through the extermination of species; the growing number of ENDANGERED SPECIES threatens to reduce biodiversity still further.

biofuel ENERGY derived from BIOMASS. It can range from the simple burning of fuelwood, dung and crop residues to the extraction of landfill gas.

biogas Methane gas obtained from animal dung, human excreta and crop residues. It can be used as a cooking or engine fuel, and can even be used to drive high-efficiency gas turbines to generate electricity.

biogeography The study of the spatial distribution of plants (*phytogeography*) and animals (*zoogeography*), and changes to those distributions over time (see VEGETATION SUCCESSION).

biological control The control of pests and weeds, not by the application of chemicals, but through the use of natural predators, parasites, and disease-carrying bacteria and viruses. Advantages of this method are the non-accumulation of harmful chemicals in the FOOD CHAIN (see INDICATOR SPECIES), and, in theory at least, the absence of side-effects on other species. However, when the pest being controlled constitutes prey for other species, this may be unavoidable; thus, when the disease myxomatosis was introduced into Britain in the mid-1950s to reduce rabbit numbers, buzzard populations in some areas fell by approximately 50%. See AGROCHEMICALS.

biological (biochemical) oxygen demand See BOD.

biological weathering Also known as organic or biotic weathering, this is the breakdown of rocks by the activities of plants and animals. The action can be purely physical (as in the case of tree roots that penetrate rock JOINTS and prise the rock apart, acting in much the same way as FREEZE-THAW WEATHERING), but is likely to be more effective and widespread when involving chemical changes. When plant materials rot, humic acids are released and these assist chemical processes. This is of considerable importance in tropical DEEP WEATHERING. The decay of plant and animal remains within the soil, plus respiration from roots, may significantly increase the carbon dioxide content, thus accelerating weathering by CARBONATION. In some limited areas (for example, offshore islands occupied by large seabird colonies) even weathering by animal excreta may be active.

biomass The total organic matter contained by plants and/or animals, usually expressed in terms of oven-dry weight per unit area. Plants, both living and decaying, constitute by far the greatest proportion of the Earth's biomass. Animal biomass (*zoo-biomass*) is very small by comparison and much of it consists of micro-organisms in the soil. *Plant biomass* varies considerably with climate and vegetation type. In the PRAIRIE grasslands of N America, maximum biomass (at the height of the growing season) is approximately 10,000 kg ha^{-1}; in the tropical SAVANNA it is 60,000 kg ha^{-1}; and in deciduous oak FOREST it is 250,000 kg ha^{-1}.

biome A world-scale ECOSYSTEM usually defined by its dominant form of vegetation, and usually associated with a particular climate. Tropical grassland and SAVANNA is one example of a biome. The woodlands and grassy plains of E Africa (e.g. the Serengeti Plains), with their vast herds of browsing and grazing animals (elephant, giraffe, wildebeest, zebra and gazelle), and their associated predators (lion, leopard, cheetah, hyena and jackal), correspond to a semi-arid tropical climate. Other examples of major world biomes include tropical RAINFOREST, TUNDRA, and Mediterranean woodland and scrub.

biotechnology A term applied to a wide range of activities united by the fact that each harnesses one or more of the special abilities of living cells. It is at present an area of considerable R&D directed towards such specific fields as GENETIC ENGINEERING, microbial mining, and the discovery of new antibiotics and vaccines. It is currently classified as one of the leading HIGH-TECHNOLOGY INDUSTRIES. See GM FOOD.

biotic A descriptive term for the living organisms, plants and animals (*biota*) within an ECOSYSTEM. Ct ABIOTIC.

bird's-foot delta A DELTA in which narrow banks of SEDIMENT line a branching network of river channels (see DISTRIBUTARY) to form a feature resembling a bird's foot (as in the Mississippi delta).

bi-polar test A method used to assess people's attitudes towards a particular phenomenon (e.g. the building of a new bypass). It involves identifying the two extremes of attitude on a particular aspect (e.g. the bypass will be bad or good for business in the town centre); an individual's opinion on that particular point can then be scored somewhere between these two extremes and so used to reflect the strength of feeling. For the test to be of value, a number of different aspects need to be taken into account.

birth control The deliberate control of population growth by various means (such as contraception, sterilization and abortion) that seek to reduce the BIRTH RATE; also referred to as *family planning*. While the need to control population numbers is widely recognized, not just in those countries suffering from OVERPOPULATION, and while some governments (e.g. those of India and China) have introduced birth control programmes, it is not a practice that receives universal approval and adoption. In some instances, there is strong opposition that derives from deeply held moral and religious beliefs; in other cases, birth control is inhibited by the persistence of traditional attitudes about large families and by inadequate knowledge of contraception.

birth rate The most widely used measure of the fertility of a POPULATION is the *crude birth rate*. This ratio between the number of births in a single year and the total population is expressed as a number per 1000. A more refined figure for studying fertility is the *standardized birth rate*, in which age and sex anomalies of a particular population are smoothed out by comparison with a hypothetical standard population. Generally speaking, the crude birth rate will be higher than the standardized rate. Birth rates in MEDCS are for the most part low, usually below 20‰, but are high in many LEDCs, often in excess of 50‰. It is tempting to assume that birth rates and economic development are in some way linked. However, all that is certain is that fertility tends to decline in countries where living styles become more 'westernized'. See DEMOGRAPHIC TRANSITION, FERTILITY RATIO.

black-earth See CHERNOZEM.

black economy See INFORMAL SECTOR.

blanket bog An area of bog (waterlogged spongy ground, occupied by sphagnum, cotton grass, etc., which decays to form highly acid PEAT), developed under conditions of high rainfall and forming extensive areas of low-relief landscape. Many blanket peat deposits in Britain are relict features, dating from the Atlantic stage (approximately 7500–4500BP) of the POSTGLACIAL period, when the climate was milder, cloudier and much wetter than it is today. Peat boglands are important wetland areas, and form natural habitats for many species of flora and fauna. Some areas are under threat from drainage (for agriculture) and peat cutting (for fuel and garden centres), and organizations such as the RSPB (the Royal Society for the Protection of Birds) are campaigning for greater protection of these unique wetland environments.

blight See URBAN BLIGHT.

bloc See TRADE BLOC.

block diagram A drawing that, by the use of perspective, gives a three-dimensional view. Most widely used in the depiction of landforms. [*f* BREACHED ANTICLINE]

block disintegration The breakdown of rock into large blocks by both MECHANICAL WEATHERING and CHEMICAL WEATHERING. The process depends on the existence of lines of weakness (fissures, JOINTS and BEDDING PLANES) that can be penetrated by WEATHERING agents, particularly acidic rainwater. One major form of block disintegration results from the freezing of water that has entered cracks in the rock. The consequent expansion by 9% in volume as the water freezes to form ice causes the wedging apart of cracks and, eventually, the physical disintegration of the rock. However, in warm, humid climates acid rainwater can, by processes such as HYDROLYSIS, open up joints in rocks such as GRANITE, and again lead to block disintegration. In this instance the resulting blocks usually show evidence of rounding, whereas a purely physical process such as ice wedging produces sharply angular, joint-bounded blocks.

block faulting The division of an area by faulting into elevated and depressed blocks. The upraised blocks form PLATEAUS, ESCARPMENTS and ridges (see BLOCK MOUNTAIN). The lowered blocks form fault-troughs, bounded by FAULT SCARPS (see GRABEN, RIFT VALLEY). [*f* FAULT]

block mountain An upland massif associated with a raised block and demarcated by a FAULT or faults (see BLOCK FAULTING, HORST)

blocking high An ANTICYCLONE that remains stationary over a period of several days or even weeks, thus holding back or diverting approaching FRONTAL DEPRESSIONS, and maintaining a period of fine, dry weather. During the 16-month period May 1975 to August 1976 anticyclones were frequent over and in the vicinity of the British Isles, constituting in effect a long-term blocking high. The mid-latitude JET STREAM bifurcated, one arm passing between N Scotland and Iceland, and the other towards Spain. Surface depressions were in turn 'steered' around the block, following the arms of the displaced jet. The result was that this was the driest 16-month period in the British Isles since records began in 1727.

blowout (i) A localized area of EROSION (more strictly DEFLATION) resulting from wind action, particularly in coastal SAND DUNES (though blowouts may also occur in desert DUNES, in SANDSTONE areas in deserts, and in unprotected peaty SOILS, as in the Fenlands of E England). In coastal dunes the blowouts are most commonly associated with a reduction in vegetation cover. This may

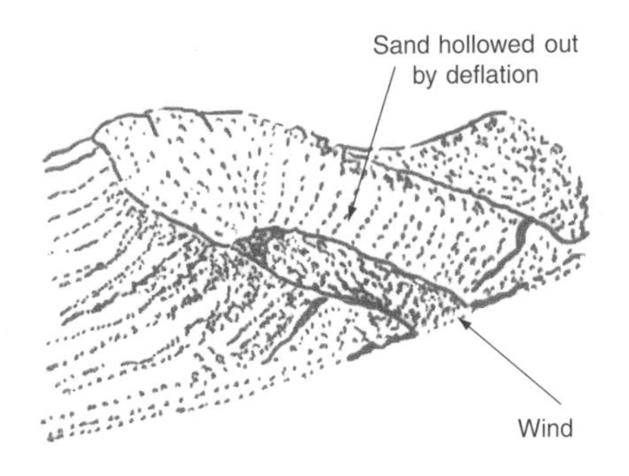

Developing blowout

come about naturally if the supply of fresh sand is reduced (dune grasses thrive only with the continual addition of fresh supplies of SAND) or it may be the result of human trampling, or other forms of coastal development. Conservation work is often implemented to re-establish vegetation and to stabilize dunes. [*f*] (ii) The term is also used to describe the explosive effect of rising oil or gas at a well that is insufficiently capped or controlled.

blue-collar worker A person engaged in manual work, as distinct from a *white-collar worker*, who is employed in non-manual work, most frequently in an office.

bluff A relatively steep slope, frequently resulting from lateral undercutting by a river, and most often applied to the margin of an extensive FLOOD PLAIN.

BOD The acronym for biological, or biochemical, oxygen demand, which describes the amount of dissolved oxygen required for the decomposition of organic material in polluted water, measured in terms of milligrams of oxygen consumed per litre of water at 25°C, measured over a 5-day period. A typical BOD value for water contaminated with domestic sewage is 300–400. It follows that polluted streams are characterized by low values of dissolved oxygen content (since this is used up as BOD increases). A dissolved oxygen content of less than 5 mg l^{-1} of water is sometimes taken as a critical threshold indicative of serious pollution.

bolson A basin of inland drainage, often the product of downfaulting, in SW USA. The basin is partially filled by ALLUVIUM, sometimes to depths of several hundreds of metres, which has been washed in by ephemeral streams draining surrounding uplands. The central part of the bolson may be occupied by a temporary lake (PLAYA), or extensive salt encrustations resulting from the evaporation of previous lakes.

boom See BUSINESS CYCLE.

border An area or zone lying along each side of the BOUNDARY between one STATE and another; usually synonymous with FRONTIER.

bora A locally strong and gusty cold wind that blows from the Balkans towards the eastern Adriatic, particularly in winter. It is more apparent when a deep FRONTAL DEPRESSION over the Mediterranean causes a steep PRESSURE GRADIENT with the winter ANTICYCLONE over Russia, causing air to be drawn towards the Adriatic.

bore A 'wall', or wave, of broken water moving upstream in a progressively narrowing ESTUARY subjected to a wide tidal range. At high SPRING TIDES, the advancing water is constricted by the shape of the estuary, retarded by friction at the base as the estuary becomes shallower inland, and impeded by out-flowing river water.

boreal forest The largely coniferous FORESTS occupying vast areas of N America and Eurasia mainly between the latitudes 45°N and 75°N, offering a good example of a BIOME. Climatic conditions here are harsh, with cold winters and brief summers. There is a short growing season, always of under 6 months and sometimes of only 3 months, and rainfall is low (up to 500 mm yr^{-1}) but adequate for tree growth. Over much of the boreal forest zone the trees are evergreen; species such as fir, pine and spruce are dominant. These are adapted to the environmental conditions (e.g. the short and flexible branches shed heavy snow, and the small needle-like leaves reduce transpiration during winter, when freezing of the soil imposes a physiological drought).

bornhardt A dome-like INSELBERG, frequently composed of GRANITE, found particularly, but not exclusively, in tropical regions. The hill is mainly shaped by large-scale EXFOLIATION, involving the detachment of sheets of rock often several metres in thickness. This is the result of PRESSURE RELEASE (or DILATATION). [*f* RUWARE]

Boserup's theory A theory concerning POPULATION and economic DEVELOPMENT. Whereas MALTHUS' THEORY OF POPULATION GROWTH

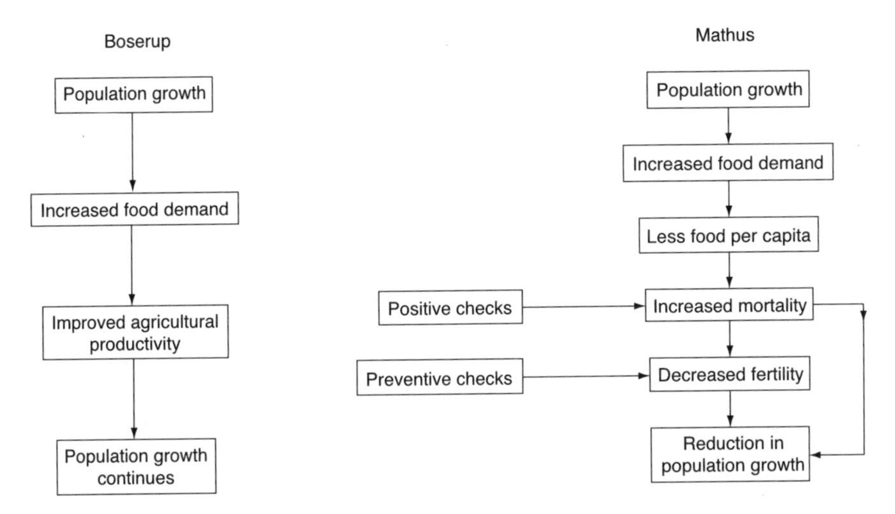

Boserup's and Malthus's views of population growth and food supply

held that food supply limited population, Boserup (1965) suggested that in a pre-industrial society an increase in population stimulates a change in agricultural techniques so that more food can be produced. The essence of her theory lies in the old adage that 'necessity is the mother of invention'. [*f*]

bottom-up development A theory of DEVELOPMENT that calls for funds and projects to be aimed at the rural poor in an effort to reduce poverty and meet basic human needs in LEDCS. It is the opposite to *trickle-down* or *top-down development*, which was the main theory behind foreign AID until the late 1960s.

boulder A large fragment of rock with a diameter in excess of 200 mm. Boulders are initially detached from the BEDROCK by a process such as BLOCK DISINTEGRATION and are then modified by TRANSPORTATION processes, becoming reduced in size and less angular.

boulder clay (till) An unstratified mass of poorly sorted glacial deposits comprising large, often angular stones and even boulders, held together by a 'clayey' matrix. Boulder clay forms a superficial deposit that covers many areas of lowland glacial DEPOSITION, and may attain a thickness of hundreds of metres. The deposit comprises mainly GROUND MORAINE formed by the basal melting of debris-rich ice. Running water, however, plays no part in the deposition of boulder clay, hence its unstratified nature and the dominance of angular rock particles. Boulder clay is a weak and unconsolidated deposit that is readily eroded, particularly when exposed at the coast.

boundary The dividing line between one political STATE and another. More widely used in geography to denote the division between discrete entities (e.g. geological outcrops, climatic types, economic regions, social areas). Cf BORDER, FRONTIER.

bounded rationality A concept of BEHAVIOURAL GEOGRAPHY put forward as a reaction to the established view of the decision-maker as a rational economic person. Instead, it is argued that: (i) the information on which decision-making is based is not freely available, but is constrained by time, financial resources and PERCEPTION; (ii) decision-makers have a limited capacity to process any such information that they acquire. Although decision-makers may strive to act rationally, they will inevitably be constrained (bounded) by their own BEHAVIOURAL ENVIRONMENT and that, correspondingly, they will be content to adopt suboptimal solutions. See BEHAVIOURAL MATRIX, SATISFICER CONCEPT, SUBOPTIMAL LOCATION.

bourne A temporary stream that occasionally flows in a CHALK dry valley. During winter, the WATER TABLE rises, owing to the

PERCOLATION of rainwater. If it reaches the surface on the valley floor, it then flows in the manner of a normal stream until the water table retreats below ground once more.

BP A widely used abbreviation for 'before the present day'.

braided stream A stream in which there is not one single CHANNEL, but a series of small interconnecting channels (some used continually, some used only under conditions of high DISCHARGE) separated by small bars or larger, stable and vegetated islands. Braided streams occur: (i) in areas where the channel banks are easily eroded (for example, where they are composed of loose SANDS and GRAVELS); (ii) where the discharge is highly irregular (as in glacial meltwater streams, which experience both seasonal and diurnal flow variations). Braided stream channels are hydraulically inefficient, and are characterized by steep longitudinal gradients. These are necessary to promote the velocity needed to move the water and bed load through the numerous channels.

branch plant A subordinate or subsidiary division of a business, usually established to meet an increasing demand, and often located away from the parent company but near to the new market or some cheap RESOURCE. For example, Japanese firms have established branch plants in Britain in order to increase their sales of products such as cars and household goods. Not only are there cost advantages to be gained by the firm manufacturing in Britain rather than shipping the finished goods there from Japan (advantages that are increased still further by the availability of financial help should the branch plant be located in a DEVELOPMENT AREA), but sales of Japanese goods made in Britain are also a substitute for direct imports from Japan. As such, therefore, they help to reduce the embarrassingly large trading surplus that Japan has with the UK – i.e. they have helped reduce TRADE FRICTION. See EXPORT SUBSTITUTION.

Brandt Commission The Brandt Commission (properly known as the *Independent Commission on International Development Issues*) was set up in 1977 by Willy Brandt at the suggestion of the President of the World Bank. The Commission consisted of 18 distinguished politicians and economists from all major regions of the world except the communist bloc. Its aim was to examine the consequences for LEDCS of changes in international relations and in the world ECONOMY, particularly as regards such vital issues as food supply, energy, finance and trade. The first report of the Commission was published in 1980, entitled *North–South: a Programme for Survival.* The word 'North' in the title refers to the advanced, industrial nations of the temperate world (see DEVELOPED WORLD, MEDC), the word 'South' to the less-developed countries of the tropics and subtropics (see THIRD WORLD, LEDC). [*f*]

The programme of priorities for the 1980s and 1990s set out by this report included: (i) a massive increase in the transfer of resources from North to South; (ii) agreeing a global energy strategy between oil producers and oil customers; (iii) increasing food production in the South by massive investments in agricultural projects; (iv) reducing poverty through ensuring a more equitable distribution of income and employment opportunities; (v) setting up an effective international monetary system and generally improving the conditions of trade and manufacturing for the South; (vi) building up the production systems of the poorest countries of the South through large-scale investment in the development of NATURAL RESOURCES and infrastructure, thus making those countries more self-sufficient; and (vii) making the less-developed countries more aware of the prob-

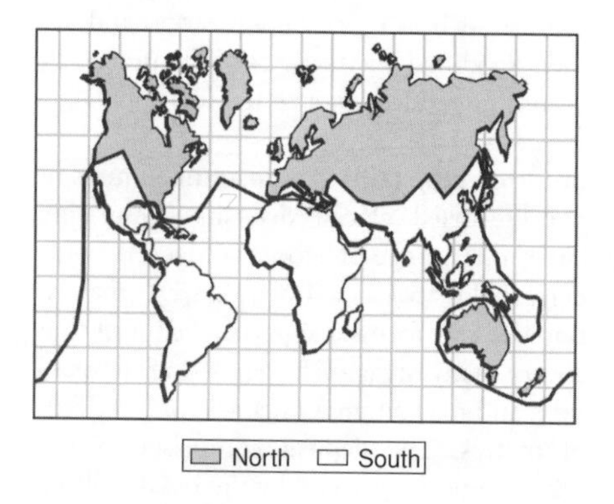

Brandt's North–South divide

lems of population growth and environmental issues.

In 1983 the Commission published its second report, under the title of *Common Crisis North–South: Cooperation for World Recovery*. It acknowledged the deteriorating situation with regard both to relations between industrialized and developing countries and to the outlook for the global economy as a whole. The failure of the international community to tackle its most serious problems was also highlighted.

breached (denuded) anticline An ANTICLINE where EROSION has been concentrated along the fold AXIS, to result in an elongated valley (*anticlinal vale*) bounded by inward-facing ESCARPMENTS. The formation of breached anticlines represents an early stage in the INVERSION OF RELIEF. The initial concentration of erosion along the crest of an anticline reflects weakening of the rock by tensional stresses; this leads to the formation of JOINT systems that can be exploited by streams. The further development of the feature, with its associated escarpments, is due to DIFFERENTIAL EROSION of hard and soft rocks within the core of the anticline. [*f*]

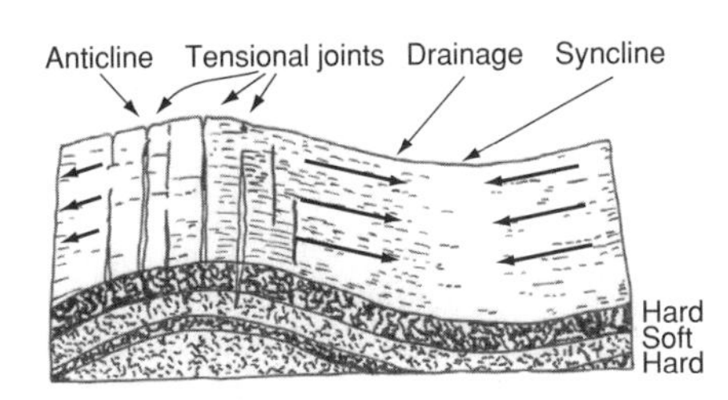

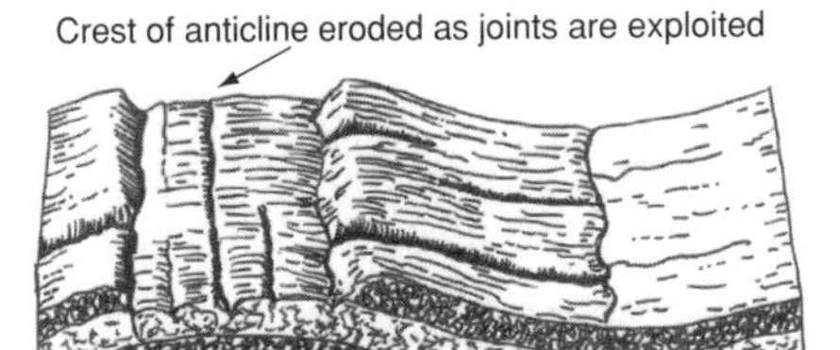

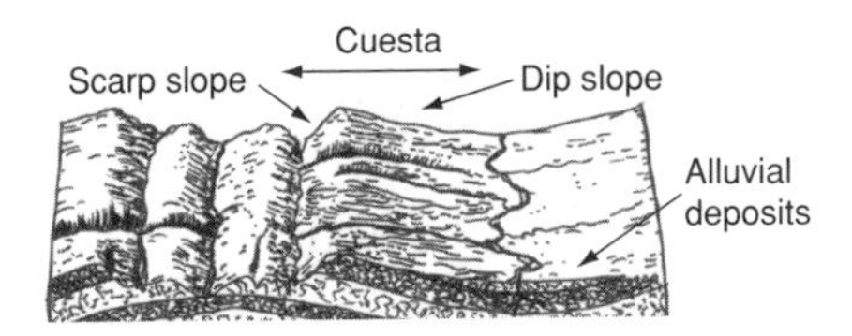

Breached anticline

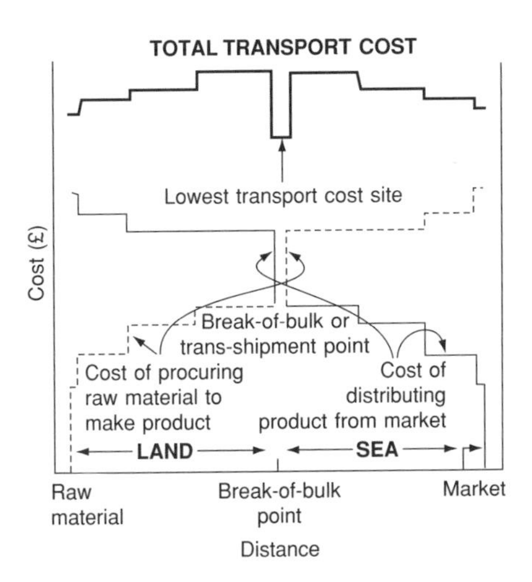

Hoover's analysis of a break-of-bulk point as a site of lowest transport costs

break-of-bulk point This is where cargo is transferred from one mode of transport to another, as at a PORT or railway station. Such points are significant in terms of economic location in that they offer potential savings in TRANSPORT COSTS. From the accompanying figure it can be seen that by processing RAW MATERIALS at the break-of-bulk point, savings are made for the following reasons: there is no transfer of raw materials from ship to rail, so no TRANSHIPMENT costs are incurred; there is no loss of benefit due to tapering FREIGHT RATES. The significance of break-of-bulk points as lowest transport cost locations was stressed in Hoover's *Theory of the Location of Economic Activity*. [*f*]

break of slope A clearly visible, sharp change of steepness in a slope profile or river long-profile (see KNICKPOINT). Breaks of slope often result from geological influences (a change from hard to soft rocks) and the mechanisms of slope recession (for example, the rapid retreat of a steep cliff segment 'consuming' a gentler slope element above). See SLOPE-OVER-WALL CLIFF.

break-point bar A SAND bar formed, on coasts of very shallow gradient, at the line of wave break. They are thought to be formed by the breaking of steep waves that cause a seaward drift of sand inside the break-point, where the resultant accumulation forms the bar.

breaking-point theory This is part of Reilly's LAW OF RETAIL GRAVITATION, and is concerned with the delimitation of the MARKET AREAS of neighbouring CENTRAL PLACES. In essence, it is a form of GRAVITY MODEL. If two central places are of the same size or status, theory states that the *breaking-point* will be exactly halfway between them. If, however, they are not of equal size or status, the larger or more important will probably exert a greater attraction to customers in the intervening area than the other, and the breaking-point will therefore be nearer the latter. To find the exact position of the boundary, the following formula is applied:

$$\text{distance of breaking-point from A} = \frac{\text{distance between A and B}}{1 + \sqrt{\frac{PB}{PA}}}$$

where *PA* and *PB* are the populations or CENTRALITY values of the two settlements in question.

breccia A SEDIMENTARY rock composed of broken, angular fragments, which is often held together by finer SEDIMENTS such as sand or clay.

bridging-point A point at which a river is, or could be, bridged; an important factor in the location of early SETTLEMENT, as indicated by numerous place names in every language involving '-bridge' and '-ford' elements. Of particular importance is the lowest bridging-point and its relationship to the upstream limit to navigation. In the past, these often coincided and gave added significance in terms of the growth of settlement (e.g. the RIA-head towns of SW England). However, it is necessary to appreciate that both the lowest bridging-point and the limit to navigation are not fixed. Their positions change over time with increasing technology and as a result of the silting of rivers.

Bronze Age A major phase in the development of human culture, succeeding the PALAEOLITHIC, MESOLITHIC and NEOLITHIC periods. In the early stages of the phase, copper was used in its pure form for adornment, starting in Mesopotamia *c.* 4000BC. Its use spread during the next millennium. About 3000BC the alloy of copper and tin (i.e. bronze) was discovered, and the Bronze Age reached its height in the second half of the last millennium BC; the gradual superseding of bronze by iron heralded the advent of the IRON AGE. In Britain, the Bronze Age lasted from *c.* 2000BC to the 6th century BC.

brown (forest) soil A ZONAL soil type characteristic of the deciduous woodland areas of the middle latitudes. Here, trees such as oak and beech provide abundant supplies of leaf litter that is relatively high in base content (giving *mull humus*). As a result, acidification is less pronounced than in coniferous forest areas, where *mor humus* is dominant. Owing to the moderate rainfall, ELUVIATION and LEACHING do take place, though not excessively so. The SOIL PROFILE of a brown soil is less clearly divided into horizons than that of, say, a PODSOL. Beneath a surface A^o layer of leaves and mull humus, the A-HORIZON is brown and weakly eluviated. It passes downwards, through an ill-defined transition zone, into a B-HORIZON that is pale brown and moderately enriched (e.g. by sesquioxides of iron washed in from above). Brown soils are generally regarded as being fertile and well suited to arable farming.

brownfield site Land that has been used, abandoned (DERELICT LAND) and now awaits some new use. Commonly found in URBAN areas, particularly in the inner city where land has been made vacant by factory closures, and the DECENTRALIZATION of people and activities to the SUBURBS and beyond. Brownfield sites are commonly occupied by derelict buildings and favoured for dumping rubbish. As such, they are bad for the image of the inner city. It is argued that redeveloping these sites should have priority over the development of GREENFIELD SITES in the countryside. The point is made that such sites are cheaper and spare the countryside, and that their use will help revive inner-city areas.

BSE The acronym for bovine spongiform encephalopathy, commonly referred to as 'mad cow disease'. It came to prominence in the late 1980s following a major outbreak in the UK. The repercussions were a wholesale slaughtering of cattle, a loss of valuable overseas meat markets and financial ruin for many farmers. Even more worrying is the fact that eating contaminated meat can cause BSE to be passed on to humans in the form of what is known as new-variant CJD (Creutzfeld Jakob Disease).

builder and user costs A concept that arises in the analysis and planning of TRANSPORT NETWORKS, and that concerns the relative costs of building the LINKS of a NETWORK compared with the costs of using the network. Taking the hypothetical situation in the accompanying figure, in (A) there is a need to construct a network linking the five places. In (B) the network is designed to minimize user costs, so there is a direct link between each and every place; clearly high builder costs are implied. In (C) the network seeks to minimize builder costs (i.e. by reducing the overall length of network to the bare minimum); but it is evident that, for example, a journey between points A and B will involve much higher user costs than in (B). [*f*]

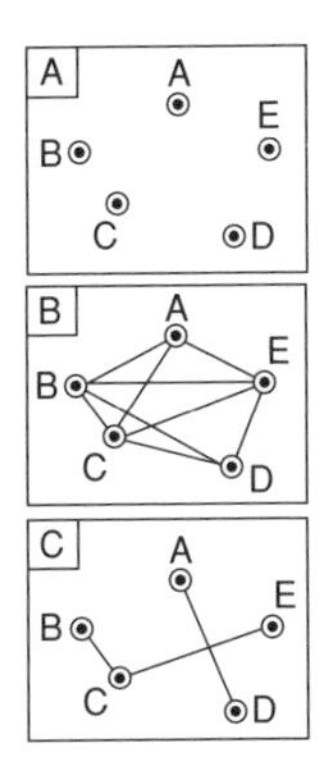

Alternative route networks to connect five places

built-up area The artificial environment of a SETTLEMENT, particularly its buildings, land uses and transport routes. Cf TOWNSCAPE.

Burgess model See CONCENTRIC ZONE MODEL.

bush fallowing See SHIFTING CULTIVATION.

business climate See ECONOMIC ENVIRONMENT.

business cycle Recurrent ups and downs in the level of business in the ECONOMY of a region, country or the world (hence *global business* cycle) – a regular alternation of *boom* (growth and prosperity) and *slump* (recession and hardship). Also referred to as a *trade cycle*. See ACCELERATOR, KONDRATIEFF CYCLES.

business park An area of land, usually located near the RURAL-URBAN FRINGE and with good ACCESSIBILITY, set aside for office development and high-tech companies involved in R&D or producing such things as computer software. In most parks, the creation of a pleasant working environment has been an important objective. Cf RETAIL PARK, SCIENCE PARK, TRADING ESTATE.

butte A steep-sided, flat-topped, isolated hill (less extensive than a MESA) which has become separated from an adjacent PLATEAU by stream EROSION, and has subsequently been reduced by slope retreat. Buttes are typical of horizontal or near-horizontal sedimentary structures, with a hard layer of rock acting as a CAP-ROCK. The slope profiles of buttes are often irregular, but commonly have an upper FREE FACE (associated with the cap-rock) and lower rectilinear segments often comprising TALUS accumulations or less resistant rocks. They are characteristic landforms in semi-arid regions of the USA.

Buys Ballot's law A law, proposed in 1857 by the Dutchman Buys Ballot, stating that if an observer stands with his back to the wind in the Northern Hemisphere, atmospheric pressure is low to the left and high to the right. In the Southern Hemisphere, the reverse holds true. The phenomenon is due to the operation of the CORIOLIS FORCE.

C

calcareous Descriptive of a rock or SOIL containing calcium carbonate ($CaCO_3$). Thus LIMESTONE and CHALK are referred to as *calcareous rocks*. An example of a *calcareous soil* is a RENDZINA, which develops on chalk.

calcification A soil-forming process that results in the formation of PEDOCALS in regions where evaporation is greater than rainfall over much of the year, so that bases are not leached from the soil in significant quantities. During the summer, calcium carbonate is raised in SOLUTION from the parent material or rock and, after evaporation of the moisture, collects in the B-HORIZON. Soils associated with calcification are usually well structured, easy to plough and highly prized for AGRICULTURE, especially cereal cultivation.

caldera A large CRATER or complex depression resulting from a powerful explosion that removes the upper part of a volcanic cone, or the collapse of the central part of the volcano as LAVA is expelled from an interior reservoir. The former type is exemplified by Krakatoa in the Sunda Straits between Java and Sumatra, which was largely destroyed by a massive explosion in 1883. An example of the latter type is Crater Lake, Oregon, formed by the collapse of nearly 120 km^3 of solid rock into the underlying magma. [*f*]

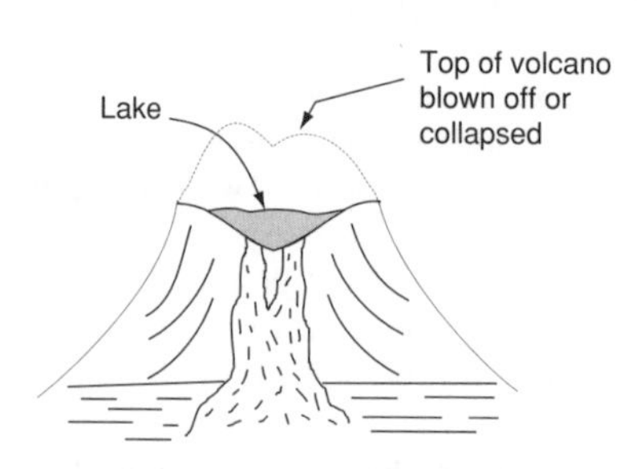

Caldera

call centre An office that deals by telephone with customer transactions such as insurance claims, travel ticket reservations and mail order. Since these organizations are relatively 'footloose', they are increasingly attracted to rural areas by lower-cost labour and office space. At present, call centres in the UK employ over a quarter of a million people (about 3% of the labour force) and the number of call centres continues to increase quite rapidly.

calving The formation of ICEBERGS and small masses of floating ice from the break-up of either a thick layer of sea ice (for example, at the margin of the Ross Ice-Barrier in Antarctica) or glacial ice that terminates in water (for example, 'outlet' glaciers from the Greenland ICE CAP that extends to the sea, or VALLEY GLACIERS in mountainous regions, which enter lakes). Calving represents a loss from the glacier system and is a form of ABLATION.

canopy The 'layer' of foliage in a woodland, formed by the crowns of fully grown trees, and often effectively shading the ground beneath. Indeed, it is calculated that the light intensity at the floor of a tropical RAINFOREST may be only 1% of that outside the forest. A well-developed canopy will therefore restrict the growth of small trees and shrubs, so that forest regeneration may be possible only at points where the canopy has been breached, owing to the death and fall of individual trees. The canopy reduces the amount of precipitation reaching the ground surface (see INTERCEPTION) and it has a significant role to play in the DRAINAGE BASIN hydrological cycle.

CAP The acronym that stands for the Common Agricultural Policy, the agricultural policy of the EU that originally had the following aims: (i) to increase agricultural productivity; (ii) to ensure a fair standard of living for the agricultural population of the EU; (iii) to stabilize markets; (iv) to guarantee regular supplies of agricultural products; (v) to ensure reasonable prices of supplies to consumers. Progress has been made towards the first four objectives of CAP, but at the expense of the fifth. Consumers have had to pay for the success of the farming sector, not only through higher food prices (almost invariably higher than world prices), but also through taxes that finance an EU budget predominantly devoted to supporting farmers. A controversial part of CAP is the system of *intervention prices* set for cereals, milk, sugar, meat, fruit and vegetables, table wine and fish products. Whenever the market price falls to this level, the entire production is bought in at the intervention price. The intervention price for each commodity is fixed each year by the EU agriculture ministers, all of whom are under pressure from their farming communities to set it as high as possible. The result has been a constant stimulus to overproduction and, from time to time, enormous stocks of a particular product have accumulated, giving rise to journalistic labels like *butter mountain* and *wine lake*. This has led to the introduction of SETASIDE and other surplus-reducing schemes (see EXTENSIFICATION).

cap-rock A resistant layer of rock that protects underlying weaker rocks, giving rise to PLATEAUS and associated landforms (MESAS, BUTTES), and CUESTAS. [*f* CUESTA]

capacity The ability of a stream to move its SEDIMENT load by processes such as BED LOAD flow (traction) and SUSPENSION. Capacity is said to vary according to the *third power* of stream velocity; in other words, if the velocity is doubled, capacity will increase 8 times, and if it is trebled, 27 times. However, this should be treated only as an approximate rule, for much will depend on the calibre of the available LOAD. If the river bed is strewn with large angular BOULDERS, even a large increase in velocity may not result in movement, simply because the critical erosion velocity (see HJULSTROM CURVE) will not have been attained. Conversely, where the bed is occupied by abundant loose SAND particles, the stream can become 'fully loaded' without difficulty since the sand particles have very little coherence.

capital (i) The chief city of a country or province, normally the seat of government. In some instances, however, especially where a federal system of government obtains, the capital may not be the largest city (for example, Canberra in Australia, Washington DC in the USA and Ottawa in Canada).

(ii) In ECONOMIC GEOGRAPHY, capital is one of the three FACTORS OF PRODUCTION, the others being LABOUR and land. In its broadest sense, the term refers to all those things made by people for use in the production process, but a distinction is usually drawn between *financial capital* and *capital equipment* (also known as *capital goods*). *Financial capital* means the stock or source of money used to finance an enterprise. *Capital equipment* refers to those human aids to further production such as machinery and buildings, plant and equipment. One important difference between these two types of capital is that financial capital is an essentially mobile commodity, while capital equipment may often be fixed in location. Indeed, capital equipment is often referred to as fixed capital and, as such, is an important contributory factor to INDUSTRIAL INERTIA. See also RISK CAPITAL.

capital equipment See CAPITAL (ii).

capital goods See CAPITAL (ii).

capitalism A politico-economic system characterized by private or corporate ownership of CAPITAL and by private profit. A mode of production in which investments are determined by private decisions rather than by state control, and prices, production and the distribution of goods are determined mainly by free market forces. The societies in which capitalism prevails typically show a clear-cut CLASS system. While the capitalist system flourishes in most parts of the so-called FIRST WORLD (e.g. Brazil, USA, Britain, Japan and Australia), in all such countries the free market is disrupted by varying degrees and forms of GOVERNMENT INTERVENTION. Ct COMMUNISM, SECOND WORLD.

capture See RIVER CAPTURE.

carbon-14 dating See RADIO-CARBON DATING.

carbon cycle This circulation might be seen as revolving around the store of carbon dioxide in the atmosphere. That amount is added to by the burning of FOSSIL FUELS, the respiration of plants and animals; it is reduced by PRECIPITATION and PHOTOSYNTHESIS. Carbon dioxide is also stored in the remains of dead organisms that gradually accumulate on land and in water bodies to form soil and peat, which may eventually be converted into CALCAREOUS rocks and fossil fuels. There is widespread concern that the scale of burning of fossil fuels is now such that it is seriously upsetting the carbon cycle and significantly increasing the amount of carbon dioxide in the atmosphere. This is thought to be the major cause of GLOBAL WARMING. [*f*]

carbonation A process of CHEMICAL WEATHERING whereby rocks containing calcium carbonate (such as LIMESTONE and CHALK) are attacked by rainwater containing dissolved carbon dioxide and therefore acting as a weak carbonic acid. The product of the

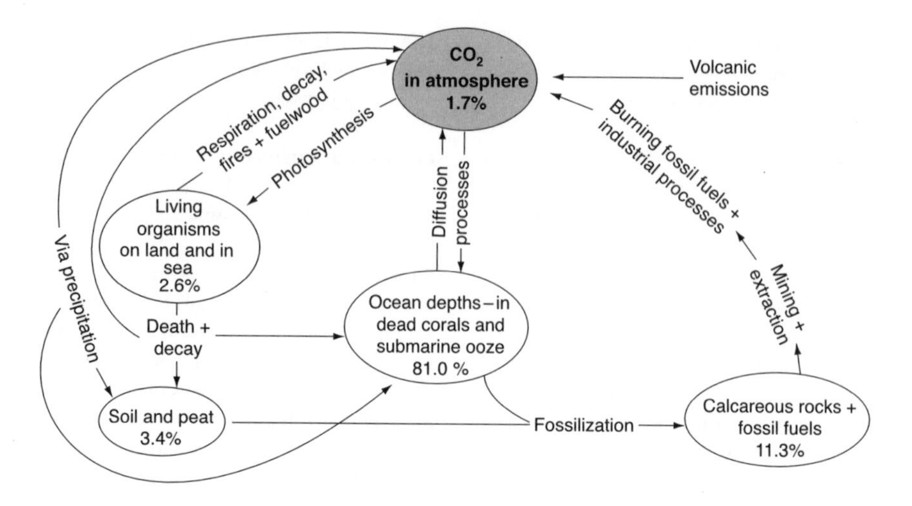

Carbon cycle

reaction is calcium bicarbonate, which is removed in SOLUTION, though it may be re-precipitated elsewhere as TUFA around SPRINGS or as STALACTITES and STALAGMITES in limestone caverns. It is a major process in limestone areas, assisting the formation of features such as LIMESTONE PAVEMENTS, DOLINES and underground passages. It is also responsible for some spectacular limestone landforms in humid tropical regions (see COCKPIT KARST, TOWER KARST). Here the lower solubility of carbon dioxide in the warm waters is more than offset by the sheer volume of available water from the large annual PRECIPITATION and the release of humic acids from decaying vegetation.

cardboard city An area of an MEDC town or city (usually in or near the centre) where homeless people gather at night under makeshift shelters made from discarded cardboard boxes and other packing materials. Ct SHANTY TOWN.

CARICOM The Caribbean Community and Common Market was set up in 1973 to promote regional cooperation and greater economic integration in the Caribbean, thereby serving to rectify the prevailing situation of small-scale, slow DEVELOPMENT and of the dominating influence of the USA.

carnivore An animal that derives most of its food from eating other animals, e.g. the lion, which lives on prey such as zebra and wildebeest on the African SAVANNA. Carnivores represent the third TROPHIC LEVEL in a FOOD CHAIN.

carrying capacity The maximum number of 'users' that can be supported by a given RESOURCE or set of resources. For example, the greatest number of livestock that can be adequately fed on the output of a given area of pasture, the amount of plant life sustained in an ECOSYSTEM, the visitor capacity of a recreational area, or the number of vehicles that can move along a road without undue impedance. [*f* LOGISTIC CURVE]

cartel A group of FIRMS entering into an agreement that involves setting mutually acceptable prices for their products, as well as OUTPUT and investment QUOTAS. Since the general effect of cartels is to restrict output, raise prices and create a MONOPOLY situation, they have been declared illegal, for example, in the UK and the USA, and by the World Trade Organization (WTO).

cartogram See TOPOLOGICAL MAP.

cartography In its widest sense, the representation and communication of spatial information in the form of MAPS. More recently, some have argued that it is not just the construction of maps, but also a discipline concerned with the scientific development and improvement of techniques to be used in this communication of spatially related data. Cf GRAPHICACY.

cascading system A natural system (see GENERAL SYSTEMS THEORY) made up of a chain of subsystems that are linked by a 'cascade' of mass or energy, such that the output from one subsystem becomes the input for another. Thus, within the broad fluvial geomorphic system, valley-side slopes can be viewed as a subsystem with an output of water and sediment. The latter becomes an input to a stream flowing at the slope base and, in turn, the output of the stream subsystem can become an input into the coastal subsystem at the point where the stream flows into the sea.

cash crop A crop grown for sale rather than for consumption by the grower (ct SUBSISTENCE AGRICULTURE), e.g. cocoa in Ghana, rubber in Malaysia, sisal in Tanzania, cereals in Canada.

caste A Hindu hereditary social group in which all members are socially equal, united by religion and, in some instances, follow the same trade. A person remains a member of the caste into which they are born and is usually debarred from social intercourse with persons of other castes. The caste system thus inhibits SOCIAL MOBILITY and, although less rigidly enforced nowadays, it is still a factor that must be considered in the economic development of India. Ct CLASS.

catastrophism The geological concept, widely held up to the beginning of the 19th century, that the Earth's features are the product of sudden catastrophic events, rather than slow processes of crustal movement, weathering, erosion, transport and deposition acting over long periods of geological time (see UNIFORMITARIANISM). The Biblical Flood would have been regarded as the prime example of such a catastrophe.

catch crop A fast-growing crop grown between two main crops in a rotation, or between the rows of a main crop, or in place of a failed crop. The principle is thus to 'snatch' a quick crop (e.g. buckwheat) on land that would otherwise be temporarily unproductive.

catchment The area drained by a river, defined by a surrounding WATERSHED (note that in the USA, the catchment is itself referred to as the watershed of the river). Within the catchment all surface RUN-OFF, THROUGHFLOW and GROUNDWATER will eventually find its way into the river. Complications occur in PERMEABLE rocks where the surface catchment, as defined by the visible watershed, may not coincide exactly with the *underground* catchment. The latter may, for reasons of geological DIP, extend beyond the surface divide, so that groundwater may be abstracted from beneath an adjoining river basin. In this way one river may 'underdrain' its neighbours.

catena A sequence of SOIL changes corresponding to positions on a slope (typically the side of a valley) owing to variations in WEATHERING, transportational processes and SOIL MOISTURE conditions on the slope. On the flat land above a valley-side slope, thick soils will tend to develop. However, on the slope itself, transport mechanisms, such as wash and creep, will move material downslope to accumulate in the valley bottom as colluvial soils. Here, the soils may be temporarily or permanently waterlogged, leading to the process of 'gleying' (see GLEY SOIL). In addition to the physical movement of soil, LEACHING will transfer nutrients downslope leaving behind more acidic conditions near the top of the slope. [*f*]

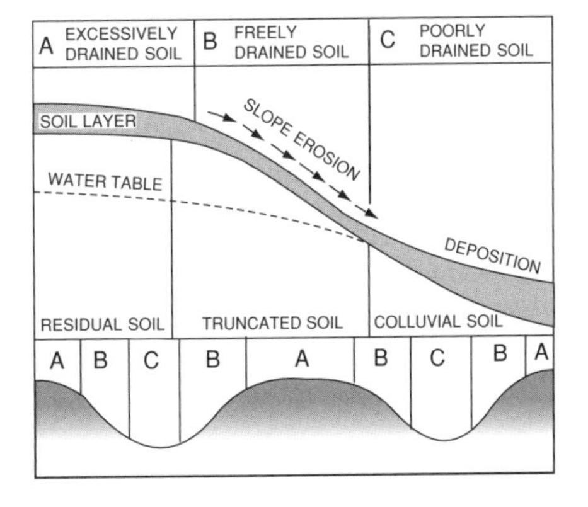

Catena, showing soil development across a series of ridges and valleys

cation exchange A process whereby *basic cations* (plant nutrients) – such as calcium, sodium and potassium – can be 'exchanged' between the surface of a clay or humus particle and the soil solution (primarily water). Basic cations can also be exchanged (removed) from the soil particle or the soil solution by plant roots seeking nutrients for growth. Over time, the latter exchange will progressively deplete a soil of its bases, thus making it more acidic and less fertile.

cavitation A process of fluvial erosion commonly occurring at a WATERFALL or RAPIDS, where bubbles of air within the water implode, sending out shock waves that can exert considerable stresses, particularly within confined spaces such as cracks and joints, in the channel sides. Together with the process of CORRASION, it contributes to the formation of such fluvial features as POTHOLES. Cavitation is thought to be very effective in streams flowing under considerable hydrostatic pressure beneath glaciers.

CBD The acronym for the central business district, the commercial centre of a town or city in which CENTRAL BUSINESSES are concentrated. Because these businesses are united in their need for ready access to clients and employees, the CBD is characteristically the most accessible part of the town or city and its HINTERLAND. This is reflected in the high pedestrian and vehicular traffic flows. The highest urban land values and rents also prevail here, because of the keen competition for sites and premises in this area of enhanced ACCESSIBILITY, while the typical resort to vertical development is one way of satisfying this demand and of increasing the amount of space available. Often the CBD will coincide with the historic nucleus of a town or city, but there are examples (e.g. Southampton and Aberystwyth) where the CBD has literally moved away from the nucleus in search of a more accessible and spacious location within the evolving BUILT-UP AREA. Such 'migrating' CBDs characteristically show two distinct marginal zones, a ZONE OF ASSIMILATION and a ZONE OF DISCARD.

Even where the location remains rooted, the CBD is a highly dynamic part of the URBAN structure. There is constant rebuilding; and, as the town or city grows, so does the extent and capacity of its CBD. At the same time, there is a progressively finer spatial sorting of the different types of central business within it. At first, it is a matter of the broad categories of central business (RETAILING, WHOLESALING, professional and personal services) becoming segregated into *quarters*, as was evident in the towns of medieval Europe. Normally, there is an *inner core* that contains department stores, specialist shops, commercial offices and high-rise buildings. Beyond this lies the *outer core*, which

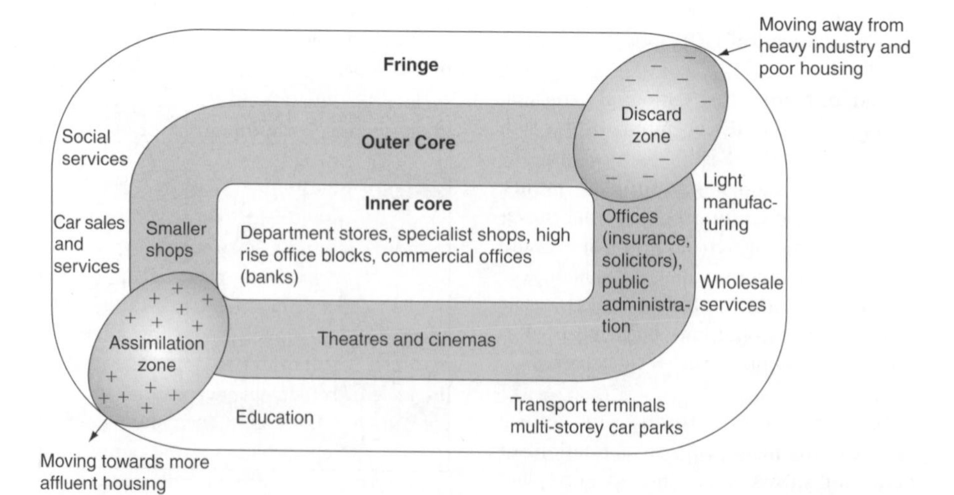

CBD

embraces a mix of less competitive land uses, and this in turn gives way to the *fringe*, where residential areas have been invaded by the least competitive of the central businesses. In major cities. the spatial sorting becomes so fine that, literally, individual streets specialize in particular types of retailing, wholesaling, professional and personal services. This acute concentration of competing firms in the same line of business is undoubtedly beneficial. The area develops a widespread reputation which, together with the fact that *comparative shopping* is readily facilitated, means the attraction of large numbers of potential customers and high levels of business.

In many cities of the world today, the status of the CBD is beginning to decline. The major cause of this is the DECENTRALIZATION of retailing and offices to suburban and edge-of-city locations. See BUSINESS PARK, RETAIL PARK. [*f*]

census The collection of demographic, economic and social information about all or a sample of the people in a defined area at a particular time. National censuses have been undertaken since the 18th century. To be of real value, censuses need to be taken at regular intervals (once every 10 years is a widely observed norm) and to collect the same types of data for the same system of spatial units. By observing these three criteria, the results of successive censuses can be compared and trends of change more readily identified. As regards the second criterion, the UN has recommended that all national censuses should enumerate the following: total population; age, sex and marital status; place of birth and nationality; literacy and educational attainment; family and household structure; fertility; rural or urban residence. There are two different census methods. The *de facto* approach records population where it is found at the precise time of the census (e.g. as in Britain), whilst the *de jure* approach involves recording people according to their usual place of residence.

central businesses CENTRAL PLACE FUNCTIONS commonly found in the central area (CBD) of a town or city, particularly the RETAILING of higher-order goods, professional services (financial, legal, medical), personal services (hairdressers, dry-cleaners, travel agents) catering and entertainment, WHOLESALING and certain types of specialist industry (e.g. printing and publishing, fashion clothing). Some geographers would restrict the term to include only profit-making activities (as above), while others take a wider view and would include activities falling under the broad heading of PUBLIC UTILITIES, such as local government offices, libraries and museums, schools and colleges, sports centres, emergency services, etc. Central businesses typically require a central location in order to maximize access to customers drawn from all parts of the built-up area and from the surrounding MARKET AREA, and the assembly of labour for what are characteristically labour-intensive activities. Some businesses are drawn to the central city because they perceive it as offering a prestige location. EXTERNAL ECONOMIES also play a part in the CENTRALIZATION of most types of central business. Generally speaking, high levels of turnover and large profit margins enable central businesses to afford the high land values and rents that prevail in such central and accessible locations.

central eruption An eruption of volcanic LAVA, cinder or ASH from a single vent or group of closely spaced vents (ct FISSURE ERUPTION), resulting usually in the formation of a volcanic cone. Within an area, a number of central eruptions may occur along a FAULT-line (or series of faults), giving rise to a line of volcanic cones. Most classic volcanoes, such as Mt Pinatubo in the Philippines, are examples of central vent volcanoes.

central place Any SETTLEMENT providing goods and services for the benefit of a surrounding tributary area (MARKET AREA) that might comprise both RURAL districts and smaller, dependent settlements. While most villages, towns and cities function as central places, it is important to appreciate that as the scale of settlement increases, so individual settlements develop their own internal systems of central places. For example, the street-corner store and the suburban shopping PRECINCT might be seen as making up different orders of central place within the

BUILT-UP AREA of a town or city. See CENTRAL PLACE HIERARCHY, CENTRAL PLACE THEORY.

central place functions Activities, mainly within the TERTIARY SECTOR, that market goods and services from CENTRAL PLACES for the benefit of local customers and clients drawn from a wider MARKET AREA. Typical functions include RETAILING, WHOLESALING, professional and personal services, entertainment, as well as a range of activities included under the heading of PUBLIC UTILITIES (see also CENTRAL BUSINESSES). An essential part of CENTRAL PLACE THEORY is the recognition that all central place functions and their individual outlets (shops, offices, etc.) may be classified into distinct ORDERS depending on their THRESHOLD and RANGE values.

central place hierarchy In his statement of CENTRAL PLACE THEORY, Christaller envisaged the central place system of a region or country as being a vertical class system or hierarchy, in which central places might be classified according to the THRESHOLD and RANGE values of their CENTRAL PLACE FUNCTIONS (i.e. that each class or order in the hierarchy would be characterized by a certain order and type of central place function). Thus high-order central places distinctively perform functions with high threshold and range values and yet, at the same time, possess all those functions that characterize lower orders of central place. Another feature of the hierarchy shown by the accompanying figure is that the number of central places in each order or class decreases with increasing status, but the number of DEPENDENT PLACES increases. As a result, the central place hierarchy is best imagined as a step-sided pyramid, with the number of central places belonging to each class being in some fixed arithmetic ratio determined by the K-VALUE of the system. (See also *f* CONTINUUM; *f* URBAN HIERARCHY). [*f*]

central place theory A major theory within SETTLEMENT geography, first postulated by Christaller as a result of observations made in the late 1920s of the settlement system of S Germany. In this theory, he tried to explain the size and distribution of settlements in terms of the marketing of goods and services. Four basic ideas underlie central place theory. (i) Most settlements, to varying degrees, function as CENTRAL PLACES, providing goods and services (see CENTRAL PLACE FUNCTIONS) to a surrounding MARKET AREA. (ii) Central places vary in the range and quality of the goods and services they provide, and these variations provide a basis for classifying the central places of an area into distinct orders or status classes (see CENTRAL PLACE HIERARCHY). (iii) The most efficient spatial organization is achieved if the central places are located on a lattice of equilateral triangles (so that each central place is equidistant from six neighbours) and with each central place serving a hexagonal market area. This eliminates the problems of underlap and overlap between adjacent market areas (as shown in the accompanying figure). The mesh of hexagonal market areas is seen as assuming three basically different relationships to the triangular settlement

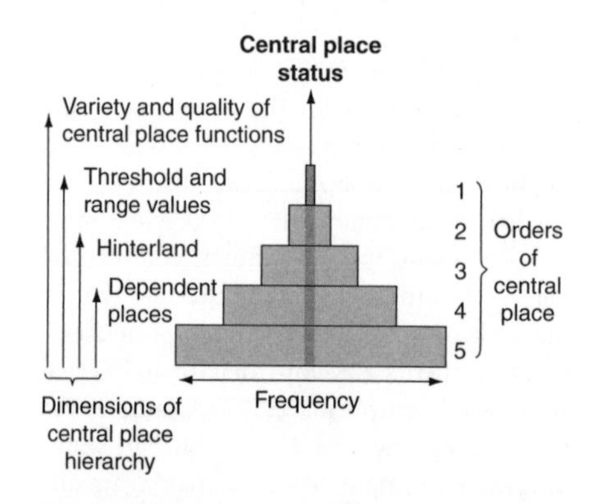

Characteristics of the central place hierarchy

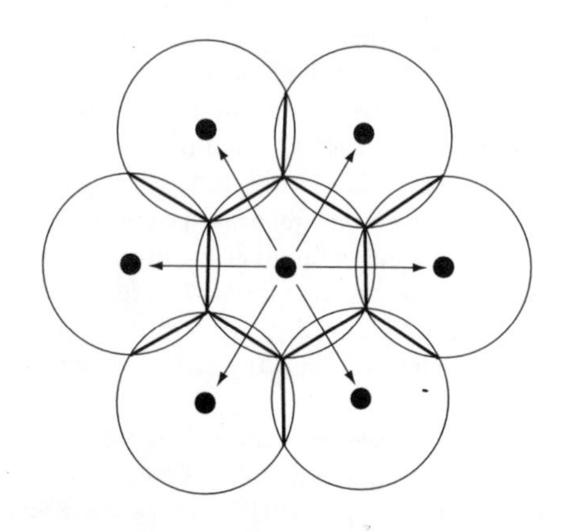

The lattice of central places and their hinterlands

lattice, each variant having a distinctive K-VALUE and each guided by a different *principle* (see ADMINISTRATIVE PRINCIPLE, MARKET PRINCIPLE). (iv) It is assumed that consumers will use the nearest centre offering goods and services (NEAREST-CENTRE HYPOTHESIS) and that the entry of suppliers into the central place system is organized so that the number of outlets and centres is minimized (*profit maximization hypothesis*). [*f*]

central planning See PLANNING, SECOND WORLD.

central tendency The tendency of values of individual items within a set of data to cluster about a particular value or values. A variety of statistical measures may be calculated to describe or summarize in one value the central tendency of a whole set of values, e.g. ARITHMETIC MEAN, GEOMETRIC MEAN, MEDIAN, MODE. These measures of central tendency constitute a major category of DESCRIPTIVE STATISTICS. See also MEAN CENTRE, MOVING AVERAGE.

centralization The tendency for people and economic activities to become concentrated at specific points, as for example by URBANIZATION and by the BACKWASH EFFECT. In the case of people, the process is encouraged by factors such as the perceived opportunities of urban settlements and core regions. For economic activities, the existence of a localized market, ease of contact with linked activities and other EXTERNAL ECONOMIES are likely to play a part. [*f* URBANIZATION]

centrality The state of being in a central situation; the degree to which a place acts as the focal point for an area. Towns and cities can generate centrality in a variety of ways, for example by becoming the nodal point in a converging TRANSPORT NETWORK (see also ACCESSIBILITY), by acting as a centre of employment, by being a seat of local or regional government or by functioning as a CENTRAL PLACE. As a general rule, there is a direct correlation between centrality and the size and status of a settlement.

centre firm See DUAL ECONOMY, MULTINATIONAL CORPORATION.

centre-periphery model See CORE-PERIPHERY MODEL.

centrifugal forces Used in the context of URBAN growth to denote those forces that encourage the outward movement of activities and people. Such forces are the result of interacting and reciprocal 'push' and 'pull' factors. High rents, traffic congestion, noise and pollution in the town or city centre, together with poor housing and inadequate services in adjacent areas, are among those considerations having a propelling effect, while lower land values, easier vehicular movement, proximity to the open countryside, modern housing and better services figure among the complementary factors attracting activities and people towards the margins of the BUILT-UP AREA – in short, the forces that give rise to SUBURBANIZATION. See also DECENTRALIZATION; ct CENTRIPETAL FORCES.

centripetal drainage A pattern of streams converging on a central lowland from surrounding highlands. Centripetal drainage is found in many desert areas – for example in the western USA – where downfaulted blocks give rise to *basins of internal drainage*. It is also found in parts of the African RIFT VALLEY, where the rift floor has been divided into 'compartments' by a combination of faulting and volcanic activity.

centripetal forces Used in the context of URBAN growth to identify those forces that encourage the CENTRALIZATION of businesses and services, particularly in the CBD. The magnetism of the town or city centre to such activities comes from 'pull' factors such as good ACCESSIBILITY to clients and labour, locational prestige, LINKAGES, AGGLOMERATION and COMMUNICATIONS ECONOMIES. Ct CENTRIFUGAL FORCES.

CFCs Short for chlorofluorocarbons – gases used as a repellent in aerosols and in the cooling systems of fridges. When these escape into the atmosphere, they readily absorb long-wave radiation, thereby increasing the GREENHOUSE EFFECT and contributing towards GLOBAL WARMING. The Montreal Protocol (1987) stipulated that there should be a 50% reduction in the global production of CFCs by 1999. The EU has now agreed on a 100% ban, but the USA appears much less committed.

chain migration See MIGRATION CHAIN.

chalk A relatively weak type of LIMESTONE, characteristically white in colour, that contains a very high proportion (often greater than 95%) of calcium carbonate. Chalk can, however, contain some important impurities, most notably nodules of FLINT (a silica precipitate). Chalk is a highly porous rock, but owes its permeability less to the numerous pores than to the presence of JOINTS and BEDDING PLANES, which allow the passage of GROUNDWATER. At present, it gives rise to largely waterless uplands and, where the DIP of the rock is gentle, CUESTA landscapes are created, with their characteristic SCARP and dip slopes. Chalk landscapes are also characterized by networks of DRY VALLEYS, COMBES and BOURNES. Chalk typically forms a thin, poorly developed soil (due to its vulnerability to CARBONATION weathering) called a RENDZINA. However, chalklands are used extensively for agriculture and they are much valued habitats for rare wild flowers, such as orchids.

channel The entrenched part of a valley floor occupied either temporarily or permanently, and either in part or in full, by the flowing water of a river or stream. However, under conditions of high DISCHARGE the flow may not be contained within the channel, but may spill over (see FLOODING) on to adjacent land. Channels vary greatly in size and form (both in plan and cross-profile). Broad, shallow channels are developed in streams with widely fluctuating discharge and where the channel banks are composed of unconsolidated SANDS and GRAVELS. Deep, narrow channels are associated with a more regular discharge and more solid bank materials such as CLAY. The geometry of channels can be expressed in terms of: width, depth, cross-sectional area, longitudinal slope, WETTED PERIMETER, HYDRAULIC RATIO and ROUGHNESS. *Channel flow* is the RUN-OFF of surface water within a well-defined channel, rather than spread over a large area (see SHEET FLOW (FLOOD)).

chelation A process of CHEMICAL WEATHERING that relates specifically to acidic organic liquids passing through the soil. The acids are derived from surface vegetation (pine needles are particularly acidic) and from acids within the soil. The resultant soil liquid can readily decompose the underlying bedrock. As acids pass through the soil, the process of *podsolization* will be promoted.

chemical weathering The decomposition of rock minerals *in situ* (in their original position) by agents such as water, oxygen, carbon dioxide and organic acids, where a chemical change occurs. Among the most important individual processes of chemical weathering are CARBONATION, CHELATION, HYDROLYSIS, OXIDATION and SOLUTION. Chemical weathering is at its most effective in humid tropical climates where chemical reactions are favoured by high temperatures (see VAN'T HOFF'S RULE), abundant SOIL MOISTURE, and the generation of humic acids by decaying vegetation. However, it is now known to be active even in cold Arctic climates, where carbon dioxide becomes concentrated in snow banks, and organic acids are associated with bog vegetation (carbon dioxide is more soluble in cold conditions than warm). Although a clear distinction is often made between chemical and PHYSICAL WEATHERING, the two types frequently work together in causing the break-up of rock.

chernozem A type of ZONAL SOIL developed in mid-latitude continental grasslands, such as the Russian STEPPES or the N American PRAIRIES (see also PEDOCAL). The dominant soil-forming process is CALCIFICATION, leading to the PRECIPITATION of calcium carbonate in the B-HORIZON. Soil fertility is high, owing to the fact that although the grasses utilize the bases in the soil, they die down each year and, in the process of decay, restore these bases to the soil. Soil bacterial activity is not too rapid, so that the soil HUMUS content remains high, with the humus being well distributed through both the A-HORIZON and the B-horizon. As a result chernozems are often dark in colour, forming the so-called *black earths* (as in the Ukraine). In texture, chernozems are loose and crumbly and this, together with their good drainage, facilitates

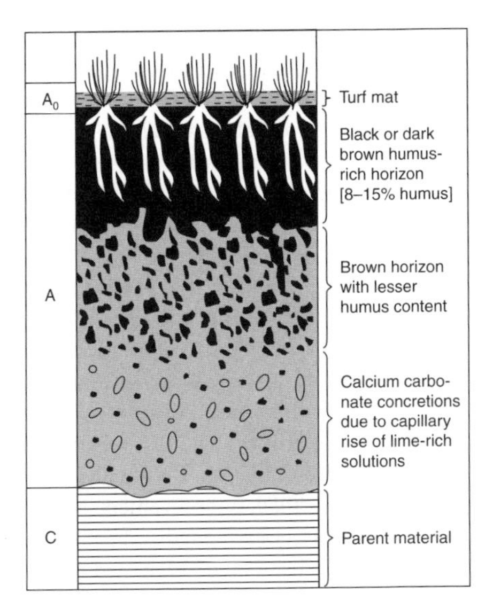

Chernozem soil profile

ploughing. They are particularly well suited to cereal cultivation. [*f*]

child labour International agreement sets 13 as the lowest age when children can work. However, official figures indicate that at least 200 million young children under the age of 15, mainly in LEDCS, are working to support themselves and their families or to pay off family debts. The actual figure may be twice as high. To an employer, child labour has the advantage of being docile, fast, agile and, above all, cheap and dispensable. To the child labourer, the disadvantages include long hours of work in bad, often unsafe conditions for little reward, and losing out on formal education.

chinook A very dry, warm wind that blows down the eastern slopes of the Rockies in W Canada. The chinook comprises air that has experienced forced ascent of the western (windward) slopes of the mountains. This air is initially humid and cools at the SATURATED ADIABATIC LAPSE-RATE once the DEW-POINT has been reached. CONDENSATION and subsequent PRECIPITATION over the mountains gradually deprive the air of much of its moisture content. As a result, when descending the lee slopes of the Rockies, it is warmed relatively rapidly, ADIABATIC LAPSE-RATE. The wind can produce very swift rises of temperature. In spring, the Chinook can lead to the rapid disappearance of winter snowfall, and it can sometimes be a contributory cause of AVALANCHES. In the European Alps, the chinook wind is called a FÖHN WIND.

[*f* FÖHN WIND]

chi-squared test A NON-PARAMETRIC TEST used mainly in the simple comparison of two FREQUENCY DISTRIBUTIONS or to determine whether the observed frequencies (O) of a given phenomenon differ significantly from the frequencies that might be expected (E) according to some assumed hypothesis (usually a NULL HYPOTHESIS), where:

$$x^2 = \sum \frac{(O - E)^2}{E}$$

For example, does the number of farms found in a given area between certain specified altitudinal limits (0–500 m, 500–1000 m, 1000–1500 m and 1500–2000 m) correspond to the amount of land occurring between those same attitudinal limits? The greater the value of x^2, the greater the difference between the two frequency distributions. In this instance, it might point to the conclusion that altitude has a significant effect on the frequency of farms. However, to be certain about this SIGNIFICANCE, it would be necessary to check the x^2 value in published x^2 tables, which take into account the appropriate DEGREES OF FREEDOM and indicate the CONFIDENCE LEVELS. [*f*]

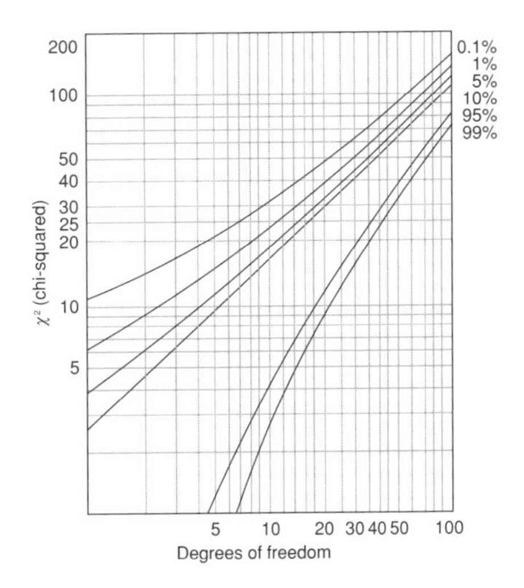

Graph for the chi-squared test

choropleth map A map that represents spatial data by means of a scheme of tonal shadings showing different degrees of density. This cartographic technique is widely used to show the distribution of population as expressed in terms of the number of persons per unit area and as related to a scheme of spatial subdivisions (e.g. parishes, local government areas, counties or regions). Particularly crucial in the construction of a choropleth map is breaking down the range of density figures into meaningful classes (how many classes should be recognized and on what basis?) (see CLASS INTERVAL) and selecting the tonal sequence that most effectively portrays the different classes of density. [*f* DESERTIFICATION]

Christaller See CENTRAL PLACE THEORY.

cinder cone A volcanic cone composed of small fragments of solidified LAVA (up to about 5 mm in diameter) formed during explosive ERUPTIONS. The 'cinders' result from the cooling of molten lava that has been projected upwards from an active central vent into the Earth's ATMOSPHERE.

circulation (i) In physical geography, circulation refers to the movement of such things as OCEAN CURRENTS and AIR MASSES. Circulation often involves cyclic cells – for example, the HADLEY CELL in the context of atmospheric circulation. (ii) In human geography, it relates to the MOBILITY of people and to the movement of CAPITAL, LABOUR, goods and services within an ECONOMY. In population geography, circulation is defined as any movement of people that does not involve a change of residence (e.g. COMMUTING, shopping, day-tripping), and is thereby distinguished from MIGRATION.

cirque (cwm, corrie) A large mountain hollow or depression, sometimes likened to the shape of an armchair, resulting from EROSION by a small glacier (*cirque glacier*). Cirques are often semi-circular in plan, and (in the Northern Hemisphere) face predominantly towards the northeast. They have steep and precipitous head- and side-walls, resulting from JOINT-BLOCK REMOVAL and FREEZE-THAW WEATHERING; they have relatively smooth and gentle floors, resulting mainly from ABRASION and, at the exit from the cirque (the lip), rocky bars that may be occupied by small TERMINAL MORAINES. Many cirques are of basin form, and contain lakes (tarns).

Cirques have been considerably modified by POSTGLACIAL weathering, and significant SCREE deposits can obscure some of the features identified above. It is believed that cirques result from the modification of preglacial valley heads and depressions. In these sheltered locations snow patches formed and, by the process of NIVATION, gave rise to enlarged hollows and thus promoted further snow accumulation. Eventually, small glaciers were formed, ice movement began, and the hollows were extended and deepened by true glacial erosion. One suggested mechanism to explain the often significant overdeepening evident in many cirques is that the ice 'pivots' about a central point, resulting in an arcuate, rotational movement as mass is added in the ACCUMULATION ZONE and then removed from the ABLATION ZONE. Such a process would also help to account for the characteristically concave long-profile of a cirque. See ASPECT. [*f*]

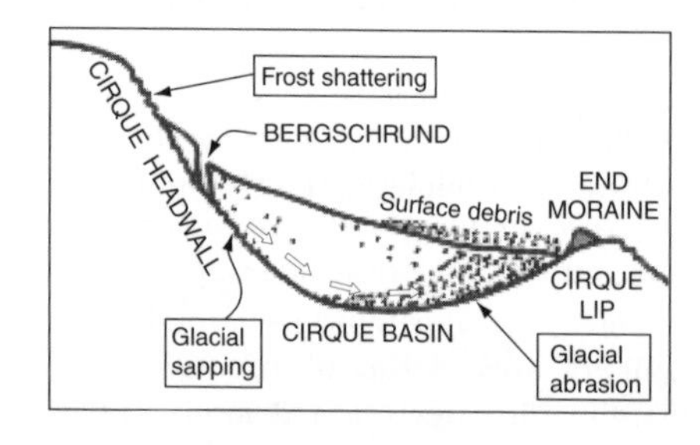

Cross profile of a glacial cirque with bergschrund

cirrus cloud A high-altitude (at up to 12,000 m), wispy cloud composed of very small ice particles and often indicative of good weather. However, if the cirrus gradually thickens into a continuous layer (*cirrostratus cloud*) it may well herald the approach of the WARM FRONT of a depression, with its associated rain. Sometimes, cirrus clouds are drawn out into so-called 'mare's tails' by strong winds in the upper ATMOSPHERE. [*f* CLOUD]

city A large SETTLEMENT having a population of diverse skills and characteristics, lacking self-sufficiency in the production of food, usually depending on manufacturing and commerce to satisfy the wants of its inhabitants, and providing goods and services for the benefit of areas lying outside it (CITY REGION, MARKET AREA). Legally, the term is given to a large town specifically incorporated by charter. In Europe, the term was formerly accorded to any settlement containing a cathedral. In the USA today, the term is very liberally used to designate quite modest URBAN settlements.

city region The tributary area of a city, the area around it that is functionally bound to it in a variety of different ways (see MARKET AREA). For example, the city region will reflect the labour recruitment area of city-based FIRMS, the tributary or market area of the city's CENTRAL PLACE FUNCTIONS or the area administered by the city as a seat of regional or local government. As such, it is a good example of a FUNCTIONAL REGION.

city-size distribution Possibly more appropriately referred to as the *settlement-size frequency distribution*, this is the frequency with which settlements within certain prescribed size ranges (e.g. 0–10,000, 10,000–25,000, 25,000–50,000, etc.) occur in a given country or region. Such FREQUENCY DISTRIBUTIONS tend to taper because there is normally an inverse relationship between settlement size and frequency, i.e. large settlements will be less frequent than smaller ones. City-size distributions may be broadly classified into three types: BINARY PATTERN, LOGNORMAL DISTRIBUTION and PRIMATE CITY (see also RANK-SIZE RULE). Berry produced a simple graphic model of the evolution of city-size distributions, changing from a state of considerable *primacy* to a *lognormal* situation. He suggested that as a country becomes more economically, socially and politically developed, so its city-size distribution becomes more lognormal (i.e. there is progressive DEVOLUTION). [*f*]

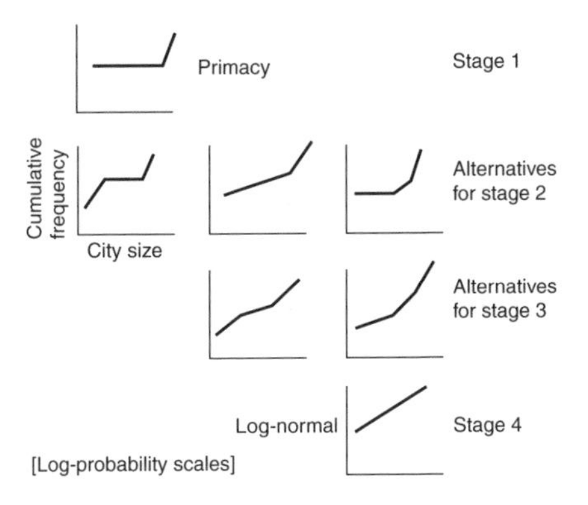

Berry's model of the evolution of city-size distribution

Clarke-Fisher sector model See DEVELOPMENT-STAGE MODEL.

class Any set of persons or things distinguished from others by some quality or qualities. In HUMAN GEOGRAPHY, the term is used to describe a group of people who are of similar social status, income, background or culture, or who are involved in broadly comparable types of employment (see SOCIO-ECONOMIC STATUS). Marx divided society into two great classes: the *bourgeoisie* (the capitalists) and the *proletariat* (the labourers). Others, however, tend to distinguish classes on the basis of economic criteria, most of which are related to the ability to purchase goods and services.

class interval The thresholds that are used in the CLASSIFICATION of data. For example, a class interval of 10,000 might be used in classifying towns and cities on the basis of population size. Equally, a class interval of 100 m might be adopted when producing a contour map. The definition of class intervals is an important stage in the preparation of a CHOROPLETH MAP.

classification The systematic grouping of objects or events into classes on the basis of properties or relationships that they have in common (see CLASS). The classification of phenomena is widely recognized as the first basic step taken by most sciences. In geography, the diversity of phenomena available for classification is immense (coasts, climates, cities, etc.). It is important to realize, however, that the aim of classification is to impose some sort of order on the phenomena under investigation. This may be achieved by two fundamentally different strategies: (i) by progressively subdividing a POPULATION into increasingly smaller and finer classes (*disaggregation*); (ii) by progressively agglomerating like individuals into increasingly larger and coarser classes (*aggregation*).

clay Very fine mineral particles with a diameter of less than 0.002 mm (see SILT and SAND). Clay is also a common type of SEDIMENTARY ROCK, formed by the compaction of mud deposits and with no clearly defined structure. When wet it is virtually IMPERMEABLE, since the minute pore spaces between individual clay particles are occupied by water held by surface tension, thus preventing the downward passage of water. As a result, in humid climates, clay outcrops support considerable surface drainage and are relatively rapidly eroded by streams into *clay vales*.

clay-humus complex A substance in the SOIL (sometimes referred to as a *colloid*) formed when HUMUS enters into a complex relationship with CLAY minerals. The clay-humus complex performs several vital functions: it absorbs water and thus improves the water-retaining capability of the soil; it expands and shrinks with wetting and drying (and thus helps to 'open up' the soil and improve ventilation); and, most important of all, soluble base nutrients in the soil link up and exchange molecules with the clay-humus complex, thus delaying the loss of nutrients by LEACHING and helping to maintain soil fertility.

clay-with-flints An unstratified mass of CLAY, with broken FLINTS, capping parts of the CHALK country in S and E England.

cleavage A property of a mineral or a rock whereby it splits into regular shapes along lines of structural weakness known as cleavage planes (see SLATE).

cliff A rocky, often near-vertical face developed in mountains, where slopes have been oversteepened by glacial EROSION or attacked by intense frost shattering, and at the coast (to form *sea-cliffs*) where marine UNDERCUTTING has been active. With sea-cliffs, a distinction can be made between *active cliffs*, which are usually steep and are experiencing current wave erosion at the base and subsequent falls of rock, and *dead or inactive cliffs*, where wave action has been halted by a fall of sea-level, extensive sedimentation at the cliff-foot, or the introduction of a coastal protection structure, such as a sea wall, which prevents the sea reaching the base of the cliff. Dead cliffs will gradually decline in angle and become degraded as a result of WEATHERING and MASS MOVEMENT. They will become covered by a layer of REGOLITH and will, eventually, be extensively vegetated.

The profile of active sea-cliffs depends very much on rock type and geological structure. A solid rock such as CHALK or LIMESTONE will tend to give vertical cliffs, while unconsolidated SANDS will form low-angled cliffs with many slumps and falls. The angle of rock DIP is also important. Where this is landwards the cliff is usually stable and precipitous, but where it is seawards the cliff is less stable and relatively gentle. The presence of structural features such as FAULTS influences the detailed forms of sea-cliffs, since these provide lines of weakness that are selectively eroded into caves, GEOS and ARCHES.

climate The average weather conditions of a place over a long period of time, usually 30 years or more. Climate should not be confused with the WEATHER, which is the condition of the atmosphere at a particular time, or over a short period of time (hours or days). It is possible to identify whole regions of the world as having similar climatic characteristics (usually relating to temperature and precipitation) and a global map of climatic zones can be produced. Such a map can be used to postulate interesting relationships between climate and other natural phenomena such as vegetation, weathering, and geomorphological regions (see CLIMATIC CHANGE). Other types of climate can be identified, such as a MARITIME or CONTINENTAL CLIMATE, which relate to the proximity of the sea.

climatic change A change in the climate of a region or a place, usually reflected in the mean annual temperature or mean annual PRECIPITATION, over a time-period that can vary from very long term to short term. For example, during the late Tertiary era there was a cooling of temperature in the British

Isles that led to the onset of the PLEISTOCENE glacial periods. Within the Pleistocene itself, there were many fluctuations of temperature resulting in alternating glacial and interglacial conditions. Fluctuations of climate in the postglacial period include the warming of the climate in W Europe since the Little Ice Age (AD1500–1850), and the relatively recent high average temperatures in Britain and many other parts of the world in the latter part of the 20th century.

Climatic changes have been attributed to many causes, such as CONTINENTAL DRIFT, variations in amounts of solar energy received from the Sun, and changes in the composition of the atmosphere, due to volcanic activity and the amount of carbon dioxide and other 'greenhouse' gases released from vegetation and other sources, such as the burning of FOSSIL FUELS. It is now widely believed that human activities are having an increasing effect on climate, owing to disruption of the hydrological cycle, large-scale destruction of forests, and atmospheric pollution from power stations, motor vehicles, etc. See GLOBAL WARMING, GREENHOUSE EFFECT and MILANKOVITCH'S THEORY.

climatic climax vegetation A widely held concept in BIOGEOGRAPHY, based on the assumption that, given sufficient time, the 'natural' vegetation of an area will come to comprise a wide range of plants fully adapted to the prevailing climatic conditions. Once established, the climatic climax vegetation will theoretically remain unaltered unless the climate – as the principal control both of the vegetation and the SOIL – itself undergoes change. All vegetation, from its initial colonization of an area, experiences a sequence of changes known as VEGETATION SUCCESSION until the ultimate vegetation cover, the climax vegetation, is established. At present, climatic climax vegetation is by no means found everywhere over the Earth's surface. Much apparent 'natural' vegetation has been removed or greatly modified by people (see PLAGIOCLIMAX), both intentionally and unintentionally.

climatology The scientific study of climate or 'average weather'. Climatology embraces: the distribution and regional patterns of climatic elements and types; regional and seasonal changes in atmospheric pressure, winds and weather patterns (*dynamic climatology*); past and present changes of climate; the effects of climate on man (*applied climatology*).

clinometer An instrument used to measure slope angles or the DIP of rocks.

clint An upstanding block of LIMESTONE, bounded by fissures (grikes) formed by the deepening of joints by solution processes (see carbonation). Clints are characteristic features of LIMESTONE PAVEMENTS.

closed system See GENERAL SYSTEMS THEORY.

cloud A visible mass of condensed water droplets, suspended in the atmosphere and resulting from the cooling of a body of air, usually as a result of free ascent under conditions of INSTABILITY, or forced ascent above mountains or frontal surfaces. Clouds vary greatly in height of formation, form and scale. See ANVIL CLOUD, CIRRUS CLOUD, CUMULUS CLOUD, NIMBUS, STRATUS. [*f*]

cloud seeding The dropping from aircraft of particles of dry ice, silver iodide or other substances into clouds, in an attempt to stimulate PRECIPITATION. Cloud seeding has been used in the past to attempt to reduce the impact of HURRICANES on coastal regions, but it is a procedure that has now been largely discredited.

Club of Rome A group (formed in 1968) of economists, managers, philosophers and scientists drawn from non-communist countries, for the purpose of understanding the 'workings of the world as a finite system and to suggest alternative options for meeting critical needs'. The major problems that have so far received attention include: the gap between rich and poor nations, and rich and poor regions; the pollution of the physical environment; urban planning; unemployment; inflation; and law and order. The results of the first major investigation conducted by the Club were published in a book entitled *The Limits to Growth* (1972). In this, they sought to explore the limitations and

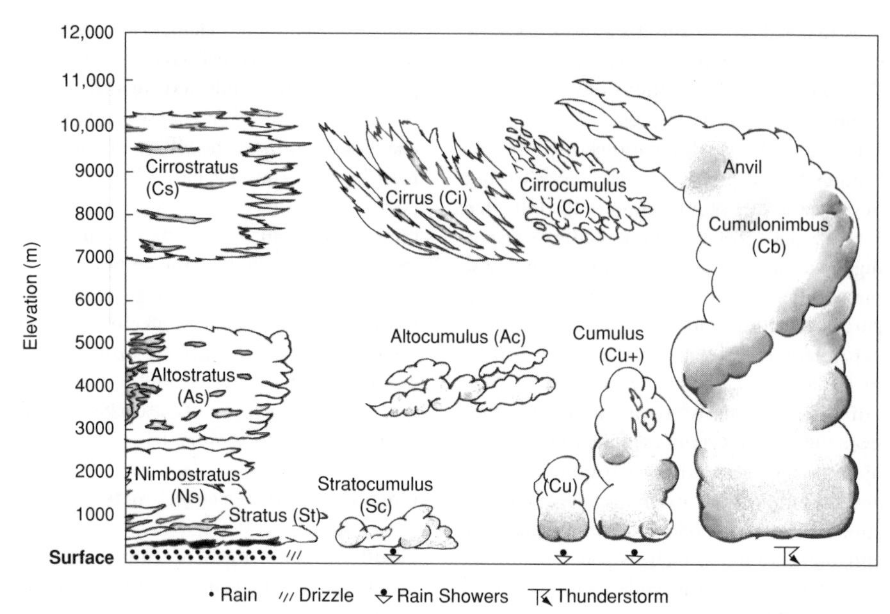

Cloud classification by height, vertical development and form

difficulties that demographic and economic growth would encounter in the future if present trends continued. See also LIMITS TO GROWTH.

cluster analysis A form of statistical analysis that seeks to discover whether or not the individuals in a POPULATION fall into distinct groups or clusters. For example, it might be used to determine whether the CENTRAL PLACES of a given region fall into distinct hierarchic groupings (see CENTRAL PLACE HIERARCHY).

clustered settlement See NUCLEATED SETTLEMENT.

coastal platform See WAVE-CUT PLATFORM.

cobble A rounded or partially rounded stone, of the type frequently found on coastal BEACHES (hence *beach cobble*), in TERRACE deposits and in the beds of rivers. The diameter of a cobble lies within the range 60–200 mm; it is thus larger than GRAVEL but smaller than a BOULDER.

cockpit karst A type of tropical KARST landscape in which steep, conical hills rise above 'cockpits' (deeply incised DOLINES, formed by intense SOLUTION of LIMESTONE). The floors of the 'cockpits' are often covered by ALLUVIUM. Cockpit karst (also referred to as *kegelkarst*) is particularly well developed in Jamaica. See also TOWER KARST. [*f*]

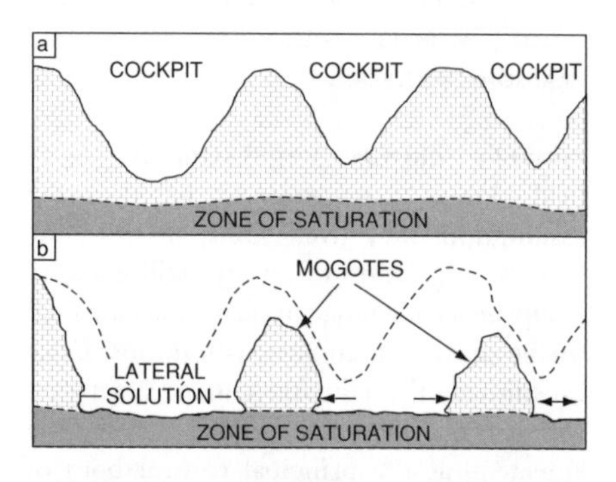

Formation of (a) cockpit karst and (b) tower karst (with mogotes)

coefficient of concentration See GINI COEFFICIENT.

coefficient of determination See REGRESSION ANALYSIS.

coefficient of localization See INDEX OF DISSIMILARITY.

cognitive map See MENTAL MAP.

cohort A group of persons united by the fact that they are at the same stage in the LIFE CYCLE. It may be that the persons were born or married in the same year or left school at the same time, for example. A unit frequently employed in demographic investigations; hence *cohort analysis.*

col (i) A well-defined depression (otherwise known as a pass, saddle or wind-gap) in a mountain range, hill ridge or CUESTA. Cols are often interpreted as marking former stream courses (as in hypotheses of RIVER CAPTURE, where the col lies along the projected course of the captured stream, beyond the 'elbow of capture'). However, cols may be formed in several ways, and should be regarded as the product of streams only in the presence of corroborative evidence, such as fluvial GRAVELS. For example, in cuesta landscapes substantial SCARP recession can lead to the beheading of dip-slope valleys, thus forming a gap in the scarp face. Erosion on opposite sides of a ridge can lead to localized lowering of the ridge summit as in the headward extension of 'back-to-back' CIRQUES.

[*f* RIVER CAPTURE]

(ii) The term is also used in a meteorological sense, to define an area of relatively low pressure (but not a DEPRESSION) marked by a general absence of ISOBARS, and therefore representing calm conditions, between two ANTICYCLONES.

cold front The clearly defined boundary between a warm and a cold air mass, where the latter is advancing and undercutting the warm air (as at the rear of the WARM SECTOR of a FRONTAL DEPRESSION [*f*]. The gradient of the cold front is usually much steeper than that at the WARM FRONT and the rapid ascent of warm moist air at the FRONT causes rapid cooling, CONDENSATION and the development of CUMULONIMBUS CLOUDS. These give heavy showers, or heavy rain (and occasionally SLEET or HAIL) over a relatively short period of time. At ground level the passage of a cold front is marked by a sudden drop in temperature, and a veering of the wind, most often from southwesterly to northwesterly (in the Northern Hemisphere). See also LINE-SQUALL.

cold (polar) glacier A glacier characterized by temperatures well below freezing point (ct WARM GLACIER, with temperatures at 0°C or PRESSURE MELTING POINT) and most commonly occurring in high latitudes. In cold glaciers, surface and near-surface temperatures may be very low indeed (−20 to −30°C), so that there is little surface melting and no internal meltwater drainage system. At its bed, a cold glacier will be frozen to the underlying rock, so that basal sliding (see GLACIER FLOW) and thus ABRASION cannot occur – though it is possible that plucking may be enhanced. In cold glaciers, there is normally an increase in temperature with depth (due to increased pressure) and, as a result, in less severe sub-polar environments (as in Spitzbergen), the basal ice may be at 0°C, thus giving rise to a *warm-based cold glacier.*

cold occlusion See OCCLUDED FRONT.

collection costs See ASSEMBLY COSTS.

collective farming A type of agricultural organization started in Russia after the Bolshevik Revolution and subsequently practised in other former communist countries (e.g. Hungary, Bulgaria and Poland). Today, it is still practised in China and N Korea. While the land is state-owned, each collective is leased to a large group of workers (often over several hundred) who run it as a single farm-holding and who share its profits. There is a need to distinguish between collective farms and *state farms*, the latter being government-run farms where the workers are state employees rather than shareholders. See also COLLECTIVISM, SECOND WORLD; cf KIBBUTZ.

collectivism Political and economic systems based on central PLANNING by the STATE and on cooperation by its citizens. Cf COMMUNISM, SECOND WORLD.

colloid See CLAY-HUMUS COMPLEX.

colluvial soil See CATENA.

colonialism The system in which one country, an imperial power, uses various economic, military, political and social policies to control areas and peoples outside its immediate boundaries. As exercised by Britain over large areas of Africa and Asia, and over the Caribbean during the 19th century. See also DEPENDENCE, IMPERIALISM, NEOCOLONIALISM.

colony (i) In BIOGEOGRAPHY the term refers to a group of closely associated, similar organisms, such as in a coral colony.

(ii) Originally a body of settlers and the territory they occupied away from their native land, usually relatively undeveloped and thinly populated. Most colonies have been founded for strategic or economic motives; their establishment played an important part in the European exploitation and settlement of Africa, the Americas and Australasia. In a political sense, a colony is subject to control by the mother country, though often enjoying some measure of self-government. Few colonies exist today in that the prevalence of anti-colonial feeling and the course of economic development have led, in the postwar period, to many such territories achieving independence. See COLONIALISM.

combe See COOMBE.

comfort zone The range of temperature and RELATIVE HUMIDITY that is physiologically most comfortable to human beings. In England this is around 15°C (60°F), with a relative humidity of 60%. As temperature rises, the relative humidity should be lower for comfort. A broad comfort zone for mid-latitudes is defined by dry-bulb temperatures of 20–25°C and relative humidity of 25–75%. See SENSIBLE TEMPERATURE.

command economy See COMMUNISM, PLANNED ECONOMY, SECOND WORLD.

commercial agriculture Agricultural practices producing commodities (crops, livestock by-products) that are sold for profit. Notable types of commercial agriculture include dairy farming, grain farming, HORTICULTURE, livestock rearing and PLANTATION agriculture. See also CASH CROP; ct SUBSISTENCE AGRICULTURE.

commercial centre See CBD.

comminution The gradual breakdown of rock material by WEATHERING or EROSION processes to form progressively smaller particles. As debris is being transported by rivers, glaciers and wave action, the particles may collide with each other, or with the rock surface, causing them to be gradually reduced in size until the SAND or SILT fraction becomes predominant (see ATTRITION).

Common Agricultural Policy See CAP.

common market An agreement by countries to establish a single market over their combined areas so that there are no restrictions on the movement of goods and labour between them. The agreement is also likely to involve a single policy as regards trade with countries outside the common market in the form of agreed TARRIFS, QUOTAS and incentives. See EU.

commonwealth A voluntary association of self-governing territories for purposes of mutual benefit (e.g. defence, trade), as in a *federation* or in the *British Commonwealth* (comprising the UK and many of its former colonies, which are now independent states but are still linked by ties of history, sentiment and national interest).

commune Essentially a group of people living and working together to protect and promote their own interests.

communications The means of communicating; the media through which information and ideas are passed (e.g. newspapers, radio and television, telephone, fax, e-mail, the Internet). Ct TRANSPORT.

communications economies One of several possible EXTERNAL ECONOMIES – in this instance the potential savings that result from efficient transfers of information between FIRMS, such as might be expected where linked firms are juxtaposed in the same agglomeration (see AGGLOMERATION ECONOMIES and IT).

communism A social and political doctrine based on Marxist socialism (see MARXISM, SECOND WORLD) that interprets history as a relentless class war eventually resulting in the victory of the PROLETARIAT. It involves the shared ownership of the means of production and distribution (see COLLECTIVISM), and the establishment of a classless society. A totalitarian system of government, which prevents the amassing of privately owned goods and in which the STATE, as owner of the major industries and acting through the medium of a single authoritarian party, controls the economic, social and cultural life of the country (as in China). Ct CAPITALISM.

community A set of interacting but often diverse groups of people found in a particular locality. Although the term implies groups bound together by common ties and in harmony, a significant aspect of many communities is of strongly differentiated groups, whose particular interests and values may conflict. See also PLANT COMMUNITY.

commuting Travelling, usually on a daily basis, to and from a place of work that is located some distance from a person's home. Commuting from the SUBURBS is a characteristic feature of many cities, since most employment is concentrated in or near the city centre. The mere fact that commuter flows converge on city centres almost inevitably makes for traffic congestion, while transport systems have to be developed in order to cope with swollen levels of traffic demand that persist only for a short while, i.e. during the *rush hours* that mark the beginning and end of the working day. As a city grows, so the volume and distance of commuting tend to increase. Ct COUNTERURBANIZATION.

company sector See PRIVATE SECTOR.

comparative advantage The principle that areas will produce those items that they are best suited to producing. It is basic to the explanation of REGIONAL SPECIALIZATION. The principle is commonly adopted in AGRICULTURAL GEOGRAPHY in order to explain why areas tend to specialize in particular types of agricultural production rather than trying to become self-sufficient. The COMPARATIVE ADVANTAGE of one agricultural area over another as regards a particular product might stem from favourable physical conditions or from the existence of a low-cost production system making efficient use of modern technology. See COMPLEMENTARITY, TRADE.

competence The ability of a stream to transport individual particles of a given size (ct CAPACITY). Theoretically there is, for a particular stream velocity, a maximum size of particle that can be moved. As velocity increases, competence also increases according to the so-called *sixth-power law*, which states that a doubling of velocity will increase competence by 26 (that is, by a factor of 64) and a quadrupling of velocity by 46 (4096). However, this must be regarded as a rough-and-ready rule when large fragments are involved because, if these are angular, they may become wedged against each other making transport difficult, whereas rounded BOULDERS, of an equivalent weight, may be more readily moved. See HJULSTROM CURVE.

complementarity One of three basic principles relating to SPATIAL INTERACTION, to the movement of people and commodities (cf INTERVENING OPPORTUNITY THEORY). The principle states that for interaction or movement to occur between two places, there must be an initial demand–supply relationship between them, i.e. one place must want what the other has to offer and the latter must be prepared to supply it. See TRADE.

components of change A framework for the investigation of change in the employment structure of an area during a defined period. Change is analysed in terms of the following components: (i) *in situ* changes in employment resulting from the expansion or contraction of FIRMS in the area; (ii) *birth* and *death* changes (i.e. changes in employment resulting from the opening of new enterprises and the closure of others); and *migration* changes resulting from the movement of firms out of, and into, the area under investigation.

composite cone A volcanic cone showing a crude stratification owing to the alternate

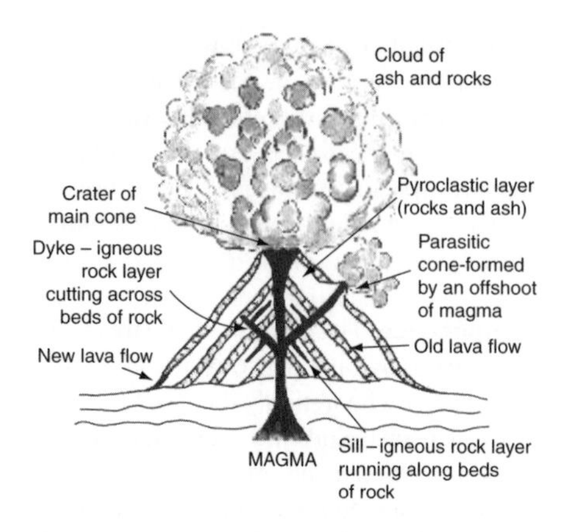

Main features of a composite volcano

DEPOSITION of layers of ASH, cinder and LAVA (also referred to as a *strato volcano*). Some of the world's largest volcanoes (such as Mt Etna in Sicily, Mt Hood in Oregon, USA, and Mt Fujiyama in Japan) are composite cones that have been constructed through a series of ERUPTIONS over a long period of time. [*f*]

comprehensive redevelopment A term used in URBAN PLANNING when a sizeable tract of the BUILT-UP AREA (often in the INNER CITY) is completely cleared and then rebuilt to accommodate either the original uses in more efficient and modern structures or the conversion of the area to completely different uses. Comprehensive redevelopment has sometimes been scathingly referred to as the 'bulldozer approach' to the refurbishing of tired urban fabric, desirable in terms of the ECONOMIES OF SCALE achieved, but highly undesirable in terms of the disruption caused, particularly to areas of old housing. In Britain, the rehabilitation of old residential areas is being increasingly undertaken by *improvement* rather than redevelopment (see URBAN RENEWAL), while the size of individual schemes nowadays is much smaller as compared with the large comprehensive redevelopment schemes that characterized urban renewal in the late 1950s and 1960s.

compressing (compressive) flow A type of GLACIER FLOW in which there is a decrease of velocity in a downglacier direction. Compressing flow is normally found towards the glacier snout (where it may be accentuated by the presence of DEAD ICE), and may be indicated by prominent shear planes within the ice. However, it may occur at any point where ice flow is impeded (for example, on the upglacier side of large BEDROCK obstacles), or where the gradient is sharply reduced, as at the base of an ICE FALL. Glaciers (and subsequently GLACIAL TROUGHS) with irregular long-profiles are marked by alternating sections of compressing and EXTENDING FLOW. Compressing flow tends to cause a thickening of the ice and an increased potential for erosion and overdeepening , and it is associated with the deep bowl-shape component of the CIRQUE.

concave slope A slope with a progressively declining steepness in a downslope direction. Concavity may be the result of DEPOSITION, but is more usually of erosional origin. Basal concavities are characteristic of arid and semi-arid regions, where they comprise a gently sloping rock surface (see PEDIMENT) with a thin and discontinuous cover of ALLUVIUM. However, concavities also occur in tropical humid and SAVANNA regions, in humid temperate lands, and even under PERIGLACIAL conditions. Over a period of time, basal concavities appear to be extended headwards, at the expense of steeper slopes from which they are separated by a sharp break of slope. [*f* STANDARD HILLSLOPE]

concentration See CENTRALIZATION.

concentric zone model Burgess derived this model of CITY structure from observations made of Chicago during the early 1920s. The model stresses distance from the city centre, together with its associated BID-RENT CURVES, as the major determinant of city structure. Thus the city is seen as comprising five concentric zones that differ in terms of their functions and social attributes, and are delimited at varying distances from the city centre. Burgess also suggested that, as the city grows, there is an outward displacement of the zones, with each zone gradually extending into the next outer zone.

Burgess was a sociologist and possibly for this reason he was interested in the social and residential structure of the city, particularly in the relationships between groups of people and different areas of the city (what

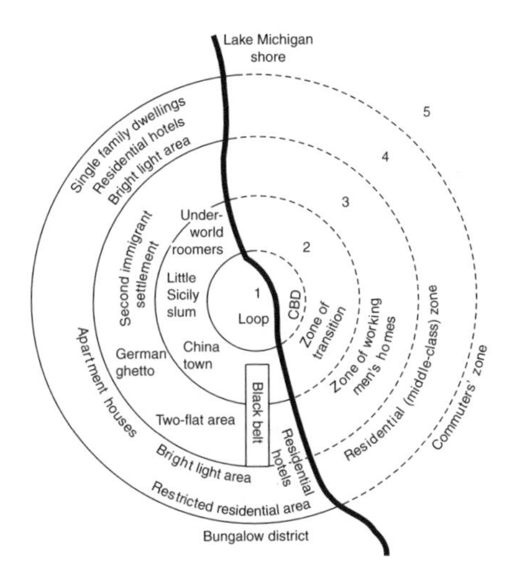

Burgess's concentric zone model based on Chicago

he called *urban ecology*). The main process operating in this part of the model was the tendency for people living in a particular zone to invade, and eventually succeed to, the next outer zone (see INVASION and SUCCESSION). The maintenance of this dynamic system was attributed to the continuing growth of the city's population as a result of persistent immigration to the centre. [*f*]

concept An abstract idea concerning a particular object or situation, e.g. that distance exerts a frictional drag on movement (see DISTANCE DECAY) or that the entrepreneur does not always seek to maximize profits (see SATISFICER CONCEPT).

concordant coast A coastline in which the natural 'grain' of the country, represented by hill-ridges and valleys (whose directions are usually determined by geological structure) runs parallel to the coast. Where such a coastline is affected by subsidence, or a substantial rise of sea-level (as in the POSTGLACIAL period), the ridges give rise to elongated islands separated by sounds (drowned valleys), as on the Adriatic coast of Dalmatia (hence the term *dalmatian coast*). Concordant coasts are also sometimes referred to as PACIFIC-TYPE COASTS.

condensation The formation of droplets of water or ice from water vapour that is cooled below its DEW-POINT (see DEW). This cooling mechanism most commonly involves vertical uplift, for example at a FRONTAL DEPRESSION or as a result of CONVECTION, but it can involve the horizontal transfer of air over a cooler surface (*advection*). Condensation usually occurs within the ATMOSPHERE, giving rise to CLOUD, MIST and FOG. On the ground surface, condensation will result in the formation of dew or HOAR FROST, the latter when dew-point is below 0°C.

conditional instability The condition of an air mass in which the ENVIRONMENTAL LAPSE-RATE is less than the DRY ADIABATIC LAPSE-RATE, but greater than the SATURATED ADIABATIC LAPSE-RATE. A pocket of air, if forced to rise (e.g. up a mountain slope) will remain cooler than the surrounding ATMOSPHERE and thus stable, so long as it is unsaturated. However, if in the process the air is cooled to DEW-POINT and CONDENSATION begins, releasing latent heat, the air pocket may become unstable. It will then continue to rise freely, without forced ascent, and further condensation will give rise to cloud and rain. [*f*]

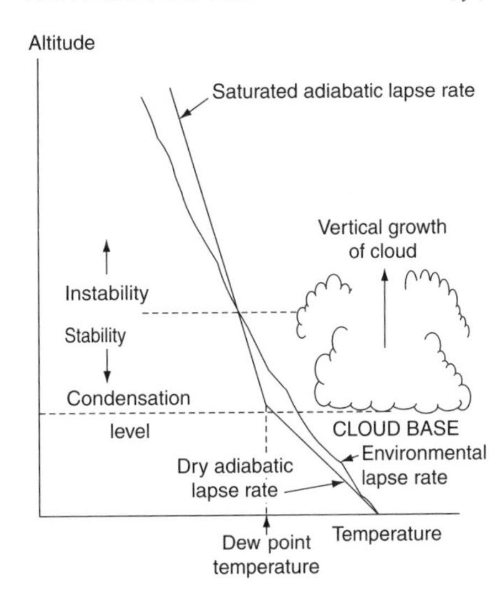

Conditional instability

confidence interval See CONFIDENCE LIMITS.

confidence levels These indicate the degree of confidence that can be placed in the results of a statistical test (e.g. testing a HYPOTHESIS), taking into account the nature and limitations of the data used in the test. They refer to levels of PROBABILITY that the conclusion reached is the correct one. Cf SIGNIFICANCE LEVEL.

confidence limits The proportion of times a given outcome may be expected to occur by chance in statistical analysis. A confidence limit of 95% implies that the given outcome will only be expected to occur by chance 1 in 20 times; a confidence limit of 99% reduces the chance to 1 in 100 times. The range between confidence limits is known as the *confidence interval.* See also CHI-SQUARED TEST, SIGNIFICANCE LEVEL.

congelifraction Sometimes used to describe FREEZE-THAW WEATHERING.

congeliturbation Sometimes used to describe the results of frost action in the ACTIVE LAYER, leading to the formation of PATTERNED GROUND and helping to generate SOLIFLUCTION.

conglomerate (i) A type of SEDIMENTARY rock consisting of rounded PEBBLES that have been cemented together, often by a sandy matrix (they most commonly represent a former beach environment). (ii) In human geography, the term is be applied to a large business organization, usually comprising a holding company and a group of subsidiary companies engaged in a range of business activities. See TNC.

coniferous forest See BOREAL FOREST.

connectivity The degree to which a NETWORK is internally connected; how well its NODES are linked together. It may be assessed by a number of different measures, such as the *alpha index, the beta index* and the *cyclomatic number.* It is thought that a correlation exists between the degree of connectivity shown by a nation's transport network and its level of economic DEVELOPMENT, so that the former may be used to provide a good indication of the latter.

consequent stream A stream whose course follows, or is 'consequent upon', the original slope of the land surface – for example, the direction of DIP of an inclined layer of rock or the limbs of a newly formed ANTICLINE. With time, the initial consequent stream pattern will be modified by the growth of SUBSEQUENT STREAMS and the process of RIVER CAPTURE. [*f* DRAINAGE PATTERNS]

conservation (i) The protection (and possible enhancement) of old buildings, urban areas (see PRESERVATION, URBAN CONSERVATION), historic sites and monuments, wild animals and plants, habitats, etc., because of a growing awareness and appreciation of their intrinsic value and AMENITY, and because of the threat posed by modern destructive influences (e.g. pollution, RESOURCE exploitation, etc.). See ENDANGERED SPECIES. (ii) The wise use of resources for the greatest good of the greatest number; in particular, reducing the rate of consumption of *non-renewable resources* (see SUSTAINABLE DEVELOPMENT). (iii) The introduction of management and/or PLANNING programmes that seek to improve the quality of natural and human ENVIRONMENTS, particularly the latter.

constructive plate boundary See PLATE TECTONICS.

constructive wave A type of wave whose effect is to build up, or aggrade, the BEACH profile. Constructive waves are of relatively low frequency (6–8 per minute) and low amplitude, and they are most commonly associated with gently sloping beaches. When the wave breaks (or surges), there is a strong SWASH, capable of carrying SHINGLE and SAND far up the beach, and a relatively weak BACKWASH, reduced by PERCOLATION into the beach,

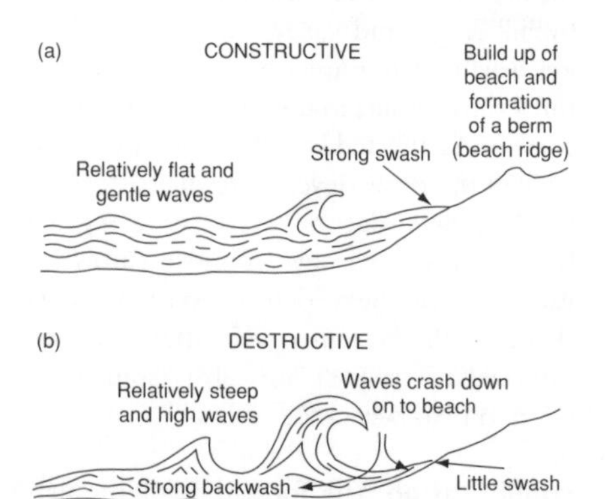

(a) Constructive and (b) destructive waves

and less capable of transporting the material back to its original position. Constructive waves thus contribute to the formation of beach ridges and BERMS. Over time, constructive waves will increase the beach angle, thereby encouraging the waves to adopt a more destructive nature (see DESTRUCTIVE WAVE), with reduced swash and increased backwash. This is an excellent example of NEGATIVE FEEDBACK in geography. [*f*]

consumer goods A range of goods in the form in which they will reach domestic consumers, from foodstuffs to furniture, clothes to cosmetics, tobacco to televisions, etc. A distinction is sometimes drawn between consumer *durables* and *non-durables*. The latter term embraces mainly subsistence goods (food and drink) that are literally consumed, while the former term includes more lasting items (electrical goods, carpets), which because of their inherent durability are purchased much less frequently.

contagious diffusion A form of SPATIAL DIFFUSION occurring where the pattern of spread is outward from a source region. It is well illustrated by the spread of contagious diseases and the diffusion of those other phenomena (e.g. the growth of a city's BUILT-UP AREA or the passing on of news and new ideas) that rely on touch or direct contact for their transmission. The process is strongly influenced by distance, because nearby individuals or areas have a much higher probability of contact than do remote individuals or areas. Ct HIERARCHIC DIFFUSION.

containerization The movement of goods in large standardized metal boxes that are capable of being carried by three different modes of transport (ship, road and rail). Containerization of cargo means an easier and more efficient transfer of goods between transport modes, as well as enabling goods movement to benefit from the particular economies associated with those modes. Thus road transport provides a door-to-door service feeding to and from railway depots; rail operates a speedy service moving the containers to and from port terminals; while sea transport provides cheap, long-haul carriage to other import or export terminals. Containerization has meant reduction both in the number of people employed in cargo-handling activities, since these have become very mechanized, and in the significance of TRANSHIPMENT points as industrial locations. It has also meant the development of new and large container-carrying ships, as well as the installation of extensive and highly automated transport terminals.

continental climate The climate typically associated with continental interiors (for example, those of N America and the Eurasian land-mass). The principal characteristics are: relative aridity, owing to distance from maritime influences (though convective rainfall occurs, mainly in early summer); great seasonal extremes of temperature; wide diurnal ranges of temperature; contrasting atmospheric pressure conditions in summer (when heating of the land surface induces low pressure) and winter (when cooling results in high pressure); sharp seasonal changes between summer and winter, with weakly developed intermediate seasons of spring and autumn. *Continentality* to a large extent results from the differing thermal properties of land and water. SOILS and rock have a relatively low specific heat, and can thus warm and cool rapidly. Water, by contrast, has a high specific heat; it is slow to warm, but is more conservative of heat.

continental drift The hypothesis, presented most convincingly by Wegener in 1915, that the continental land-masses have undergone important shifts of position, notably in the post-Carboniferous period. The continents were interpreted structurally as rigid blocks floating in a more or less fluid substratum and as able to 'drift', possibly as a result of drag by CONVECTION currents within the Earth's core owing to the accumulation of radioactive heat. It is envisaged that the continents were once united in one 'supercontinent' (PANGAEA), comprising LAURASIA to the north and GONDWANALAND to the south. During the Mesozoic era fragmentation began, and individual land-masses became increasingly separated. A wide range of evidence has been cited in support of continental drift: the 'fit' of the continental outlines (as on the eastern and western margins of the Atlantic); the glaciation in Carboniferous-Permian times of parts of S America, S Africa and India; and the matching of geological structures (for example, the Caledonian and Hercynian folds of eastern N America and W Europe) across oceans.

The hypothesis of continental drift has been strongly contested, and by the middle of the 20th century was regarded as untenable, largely because the 'driving force' was unknown and, in the view of some, the process was theoretically impossible. The whole concept of continental drift has now been subsumed by, and modified to form the more recent theory of, PLATE TECTONICS (see also PALAEOMAGNETISM, POLAR WANDERING CURVE).

continental ice sheet A very large ICE SHEET, of continental dimensions, as found today in Antarctica. In Antarctica the mean thickness of the ice is 2000–2500 m, but masses of rock project as NUNATAKS. The ice may be extremely cold (with basal temperatures at some points as low as −30°C), and ABLATION rates are low except at the periphery (where CALVING of ice directly into the sea constitutes a major loss). The ACCUMULATION ZONE occupies a very large area of the ice sheet, although annual inputs of snow are limited due to the aridity of the climate, and the flow of ice is extremely slow.

continental shelf The submerged, gently sloping margins of a continent. The shelf terminates at a pronounced BREAK OF SLOPE (at a depth of 120–360 m), beyond which the ocean depth increases suddenly (the more steeply inclined sea-floor here is called the CONTINENTAL SLOPE). The continental shelf is well developed at some points (for example, around the British Isles, where it extends over 300 km westwards from Land's End), but absent at others (for example, along the Pacific coast of N America).

continental slope The seaward extension of the CONTINENTAL SHELF into the deep oceans where it adjoins the ABYSSAL floor.

continuous variable See VARIABLE.

continuum A smooth, unbroken sequence or gradation. The term is used in central place studies (see CENTRAL PLACE THEORY) to describe a central place system that does not show a clear hierarchical structure (see CENTRAL PLACE HIERARCHY). Rather than being grouped or clustered into distinct CLASSES, the individual central places are spaced out, so that each one occupies a unique position along the axis showing central place status. See also RURAL–URBAN CONTINUUM. [*f*]

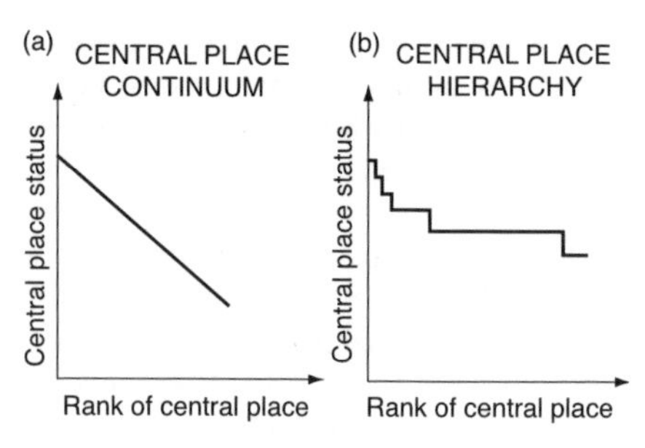

(a) Continuum (b) hierarchy

contraception See BIRTH CONTROL.

conurbation The extensive BUILT-UP AREA formed by the coalescence of once separate, and mainly URBAN settlements. The growth of conurbations was a striking feature of URBANIZATION in the 20th century. They represent large-scale concentrations of population, economic activities and services. Some geographers distinguish between *uninuclear* and *polynuclear* conurbations. The former are produced by outward growth from a single nucleus (usually a major CITY in its own right) engulfing largely small TOWNS and VILLAGES located around it (e.g. the Greater London conurbation), while the latter result from RIBBON DEVELOPMENT filling the interstices between a network of closely spaced towns (e.g. N Staffordshire and W Midland conurbations). Since 1961, eight conurbations have been officially recognized in the compilation of the British CENSUS. It should be noted, however, that the term METROPOLITAN AREA is increasingly used in preference to conurbation.

convection A term describing vertical movements of air within the ATMOSPHERE, most commonly involving the rising of air (convection current) warmed by a heated ground surface. When an air pocket is warmed, it expands and rises due to a reduction in its density. Air is drawn in to replace this rising air and a circular type of air movement is propagated, known as a CONVECTION CELL. Convection is associated with atmospheric INSTABILITY, which causes the rising and cooling of air pockets, CONDENSATION and CLOUD formation, *convectional rainfall*, and phenomena such as THUNDERSTORMS. Convective processes are very important in tropical regions, where diurnal heating of the ground, associated with the prevailing high humidities, frequently leads to heavy convectional rainfall in the late afternoon. Heat can also be transferred through liquids or partial-liquids in the form of convection currents (see PLATE TECTONICS).

convection cell A circular motion of air movement involving the transfer of heat, driven by strong convective updraughts. As an air pocket is forced to rise by intense heating, air will be drawn in at the ground surface to replace it, and a circular flow of air may result. Convection cells operate at a range of different scales. See HADLEY CELL; and LAND BREEZE, SEA BREEZE. [*f*]

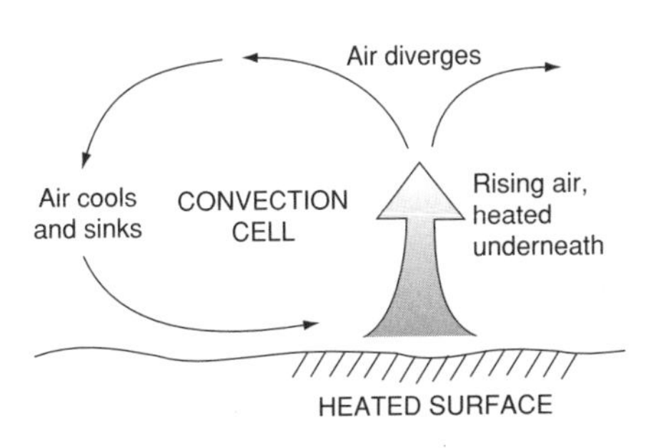

Convection cell

convenience goods Despite the fact that it has a wide current usage, this term is somewhat ill-defined. What is generally implied are the following meanings: CONSUMER GOODS, essentially of a *non-durable* nature, that are mainly in daily demand; those disposable goods that might be seen as the outgrowth of the contemporary *throwaway society*; and goods that are processed and packaged for speedy and convenient consumption (e.g. canned and frozen foods).

convex slope A slope element with a progressively increasing steepness in a downslope direction. Occasionally the whole of a slope profile will assume a convex form, though more usually the convexity is formed only on the upper part of the slope profile. Convex slopes can be structurally determined (as on GRANITE hills, where curvilinear *sheet joints* are subjected to EXFOLIATION), but are commonly the product of WEATHERING and debris TRANSPORT. They are usually underlain by solid rock, over which a layer of SOIL and REGOLITH is slowly being removed by creep and allied processes. Convex slopes are said to be typical of certain rock types (notably LIMESTONE and CHALK) and those climates (e.g. humid temperate and humid tropical) that favour the more slowly acting MASS MOVEMENTS. [*f* STANDARD HILLSLOPE]

coombe A term most commonly used when referring to the short, steep-sided valleys, often without a stream at the bottom, that are found on the scarp slopes of CUESTAS in the chalk country of SE England.

cooperative Originally an association of farmers set up to act as purchasing and distribution agents of seeds, feedstuffs and fertilizers, and as grading and selling agents for farm produce (e.g. creameries in Denmark, fruit-handling cooperatives in California, wine cooperatives in France). The system allows farmers to operate their own holdings as individuals, yet enjoy the advantages of bulk purchasing, grading and standardization, and of large-scale contract marketing. In Britain, the term has been applied to the manufacturing context when a group of workers (*workers' cooperative*) takes over the company by which they were employed, in order to obviate its closure (sometimes referred to as a *management buy-out*). In the THIRD WORLD, cooperatives have been set up to generally foster DEVELOPMENT. For example, in W Africa they have been formed to encourage the introduction of mechanized cultivation to peasant holdings, to modernize traditional craft industries, and to promote trade (through produce marketing and credit provision).

coordinates Numbers used to locate a point on a GRAPH relative to the *x* and *y* axes, or a point on a map relative to LONGITUDE and

LATITUDE. On a graph, the coordinate measured from the *y* axis parallel to the *x* axis is called the *abscissa*, and the other is called the *ordinate*.

core (i) In geology, the term 'core' refers to the iron and nickel-rich inner part of the Earth, often subdivided into a liquid outer core and a solid inner core. (ii) The term is also used in models and theories of *regional science* concerned with the unequal distribution of development to denote a favoured area (it might be a city or a whole region), in which there is considerable CENTRALIZATION of resources, economic wealth, productivity, labour, innovation, political power, etc. (see CORE-PERIPHERY MODEL). The emergence of a core is almost inevitably by means of a BACKWASH EFFECT (see POLARIZATION) acting upon, and to the detriment of, areas elsewhere (see PERIPHERY).

core-periphery model A spatial model of economic DEVELOPMENT based on the observation that it is rarely evenly distributed, be it on a regional, national or global scale. There is a tendency for growth to become concentrated at favoured locations (CORE) which, in their turn, leave in their wake areas of stagnation or decline (PERIPHERY). In his model, Friedmann (1966) recognized four stages in the growth of the SPACE ECONOMY, each reflecting a change in the relationship between the core and the periphery. (i) The pre-industrial society shows a system of local, and largely undifferentiated, cores, each serving a small regional ENCLAVE. (ii) One of the particularly favoured cores develops into a strong core, to which move ENTREPRENEURS and LABOUR. The national ECONOMY becomes focused on a single metropolitan region; its associated BACKWASH creates a large periphery. (iii) The simple core-periphery structure is gradually transformed into a multinuclear structure, as favourable parts of the periphery are developed. Secondary cores form as a result of SPREAD EFFECTS, thereby reducing the periphery on a national scale to smaller intrametropolitan peripheries. (iv) The intrametropolitan peripheries are gradually absorbed into the metropolitan economies. Here, local and national backwash and spread effects seem to be generally in balance. A functional interdependent system of cities emerges, characterized by national integration, efficiency in location and maximum growth potential.

In this model four types of area may be distinguished: *core regions* – urban-industrial concentrations with high levels of technology, capital, labour and high growth rates; *upward transition regions* – increasingly influenced by core regions and characterized by immigration, intensive use of resources and constant economic growth; *resource frontier regions* – part of the periphery, and typified by new settlement and the exploitation of newly discovered resources; *downward transition regions* – characterized by stagnant or declining economies, due to the exhaustion of primary resources or abandonment of industrial complexes. Both upward-transition and resource frontier regions may become secondary cores. [*f*]

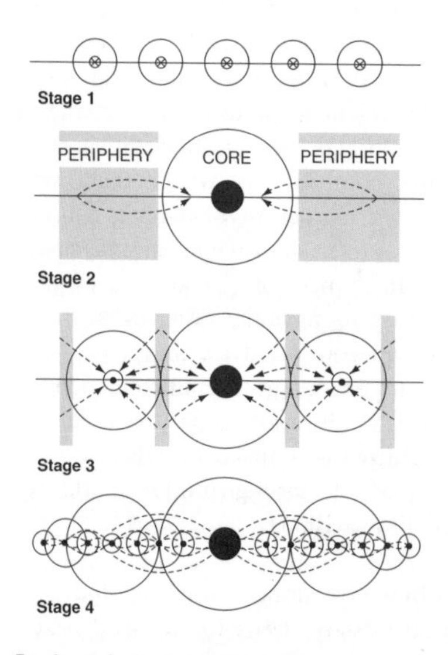

Friedmann's development model

Coriolis force Any object moving at the Earth's surface is subjected to a *deflecting force* due to the Earth's rotation. This deflecting force, known as the Coriolis force, is always to the right in the Northern Hemisphere, and to the left in the Southern Hemisphere. The Coriolis force, which increases with distance from the Equator (where it exerts no influence), explains why winds do not travel in a straight line from areas of high pressure (ANTICYCLONES) to areas of low pressure (DEPRESSIONS), but instead form a circulatory pattern. In the Northern Hemisphere, winds circulate around an anticyclone in a clockwise direction, and around a depression in an anticlockwise direction. The opposite is the case in the Southern Hemisphere. At the ground surface, friction also has a significant influence on wind direction, causing it to cross the isobars at an angle rather than circulating roughly parallel to the isobars as it does higher up in the atmosphere (see GEOSTROPHIC WIND).

corrasion The purely mechanical EROSION of rock surfaces by the impact of debris being transported by streams. The process is most effective if the stream LOAD comprises hard, coarse and angular fragments. The term is usually employed in fluvial geomorphology, though a comparable process is associated with the TRANSPORT of rock debris by other media (ice, waves, wind). See ABRASION.

correlation The degree of relationship between pairs of VARIABLES. When two variables increase or decrease together, the relationship is known as *positive correlation. Negative correlation* occurs when the relationship is inverse, so that one variable increases as the other decreases. *Partial correlation* is the relationship between two variables when a third variable, which is related to both, is controlled. *Multiple correlation* refers to the relationship between two or more INDEPENDENT VARIABLES and one DEPENDENT VARIABLE. See also REGRESSION ANALYSIS.

correlation coefficient An index or measure giving a precise value to the linear relationship (CORRELATION) between two or more variables. Values range from +1.0 (perfect *positive correlation*) to −1.0 (perfect *negative correlation*). If the association of the variables is RANDOM, the coefficient value will be 0 or nearly so.

corridor (i) A strip of territory of one STATE interrupting the territory of another to give it access to the coast: e.g. the Polish Corridor to the Baltic Sea, which existed prior to the Second World War, or the former Israeli corridor to Eilat on the Red Sea. (ii) A prescribed international air route over a country. (iii) A LINEAR PATTERN of URBAN development encouraged as, for example, along valleys and major routeways (see AXIAL BELT).

corrie See CIRQUE.

corrosion The purely chemical EROSION of rock surfaces by flowing water, as in LIMESTONE, which is attacked by carbon dioxide dissolved in streams. It is often difficult to make a clear distinction between corrosion and CHEMICAL WEATHERING. At times of low DISCHARGE, parts of stream CHANNELS become exposed to atmospheric weathering, the products of which are subsequently removed by the stream at high discharge. Weakening of the rock structure by corrosion also assists other erosive processes.

cost-benefit analysis A method, used widely in PLANNING practice, of objectively comparing alternative proposals by quantifying, largely in financial terms, the total of *costs* (disadvantages) and of *benefits* (advantages) that will accrue with each alternative. Cost-benefit analysis involves four basic stages: (i) defining the possible alternatives; (ii) identifying the costs and benefits likely to be associated with each alternative; (iii) measuring those costs and benefits; (iv) on the basis of the relative levels of costs and benefits selecting that alternative offering the greatest *net benefit* (the greatest margin between benefits over costs). Cost-benefit analysis might be employed, for example, in deciding which one of a number of different routes might be followed by a new motorway or in choosing the site for a major international airport.

cost-space convergence Developments in TRANSPORT and COMMUNICATIONS, particularly during the second half of the 20th century, that make locations more equal in terms of ACCESSIBILITY. As a result, the cost differences between locations have been reduced.

cost structure The cost structure of an enterprise indicates the relative cost of the various INPUTS needed either to produce a given OUTPUT or to operate the enterprise over a given period. These inputs include such items as the costs of LABOUR, ENERGY and RAW MATERIALS. It is important to appreciate that cost structures vary with the scale of operation, from activity to activity and from place to place, within the context of the same activity. Because of the last of these, cost structure can have an important influence on the location of economic activities, as well as serving to identify those items that carry particular weight both in the overall PRODUCTION COSTS and in the general location equation.

cost surface The uneven, three-dimensional surface created by spatial variations in PRODUCTION COSTS. The areas of low 'relief' occurring on that surface can be regarded as LEAST-COST LOCATIONS.

cottage industry A system of production in which craftsmen, aided by their families, work at home either on their own account or on behalf of an ENTREPRENEUR, who delivers the RAW MATERIALS and collects the finished goods. It was a system followed in British industry up to the Industrial Revolution, and is widely practised in the THIRD WORLD. Ct FACTORY SYSTEM.

counterurbanization A process of DECENTRALIZATION involving the movement of people and employment from major cities to either small SETTLEMENTS and RURAL areas located just beyond the CITY margins or to more distant, smaller cities and TOWNS. The process was first observed in the USA during the early 1970s (see SUNBELT). The results of the 1981 Census first revealed the operation of the process in Britain, in that all the metropolitan counties (particularly their inner-city areas) lost population during the preceding intercensal period. High rates of growth were experienced in the medium-sized, freestanding towns and cities of S England (referred to as *Sunrise England*).

The term is perhaps rather misleading in that the changes have not in any way reduced the overall degree of URBANIZATION at a national level. The process has led merely to a redistribution of urban growth, 'taking it away' from the largest cities and 'giving' it to small- and medium-sized towns. Indeed, far from being 'anti-urban' it is promoting a wider dissemination of urbanization. The process seems to have been driven by: a general loosening of the location of manufacturing and services (no longer having to be tied to major cities or central-city areas); a growing reaction to 'big city living' and people's preference for the 'better quality of life' perceived to be offered by small towns and the countryside. See also RURAL TURNAROUND.

[*f* SUBURBANIZATION]

crag-and-tail A glacial landform developed where a glacier or ICE SHEET overrides a mass of hard rock (the *crag*), which protects softer rocks in its lee; these form a tapered, gently sloping ridge (the *tail*). A famous example is at Edinburgh Castle.

crater A rounded, funnel-shaped depression, usually at the summit of a VOLCANO, marking the exit of LAVA, cinders and ASH. Very large craters may result from violent explosions or massive subsidence (see CALDERA). In dormant or extinct volcanoes the crater is often occupied by a lake (hence *crater lake*).

craton A modern geological term for a rigid block of ancient pre-Cambrian rocks, previously referred to as a SHIELD.

crevasse A deep fissure in the surface of a glacier, extending to a maximum depth of

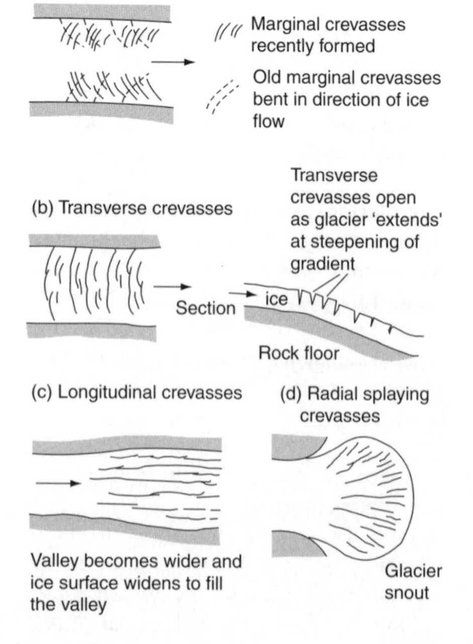

Crevasse

30–40 m and resulting from tensional forces. For example, along the glacier margins, flow is reduced by friction with the valley wall but the centre-line ice velocity is not affected and flows relatively quickly. The resultant tensile stresses cause *marginal crevasses* that are oriented at approximately 45° to the glacier edge and upglacier towards the glacier centre-line. Once initiated, such crevasses may be 'rotated' by glacier flow, so that they 'point' downglacier. Where the glacier is free to expand laterally (near the snout, or when leaving a confined section) *longitudinal crevasses* form; and at the glacier snout itself *radial crevasses* (otherwise termed *splaying crevasses*) are commonly developed. Other factors causing tensional crevasses are irregularities in the glacier bed. For example, where the glacier rides over a rock-step, tension at the ice surface leads to the formation of *transverse crevasses*, which may run across the glacier from one side to the other. [*f*]

critical temperature A temperature of specific importance in terms of vegetation. For example, freezing point, since many plants are vulnerable to frost, or 6°C, since for most plants active growth cannot occur below this temperature.

cross-section (i) The profile revealed when a section is taken through a feature or landscape, usually at right angles to its longest axis – as, for example, across a valley or a CUESTA. [*f* ASYMMETRICAL VALLEY; CUESTA] (ii) The term can also refer to a sample that is representative of, or typical of, the whole. See SAMPLING.

crude birth rate See BIRTH RATE.

crumb structure See SOIL STRUCTURE.

crystalline rock A type of metamorphic or IGNEOUS ROCK whose constituent minerals are of crystalline form, e.g. GRANITE, GNEISS and SCHIST. Crystalline rocks are mechanically strong and are able to resist physical EROSION and WEATHERING. However, individual minerals may be susceptible to CHEMICAL WEATHERING, which in time leads to GRANULAR DISINTEGRATION.

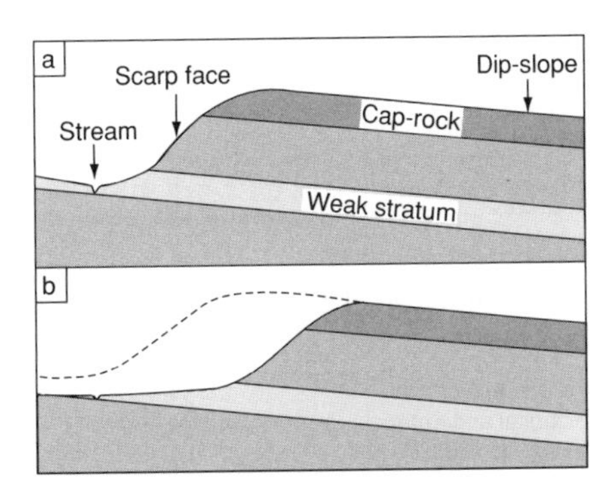

Cross-section of a cuesta showing (a) cap rock and (b) the uniclinal shifting of a scarp-foot stream

cuesta An asymmetrical upland feature usually associated with gently dipping sedimentary rocks, for example CHALK, and comprising a steep scarp slope and a longer, gentler dip slope. The term ESCARPMENT is sometimes used synonymously, but strictly speaking, in this context, it applies to the scarp slope only. [*f*]

cultural geography Study of the spatial aspects of human CULTURE. This includes the distribution of cultural traits (e.g. language and religion), the SPATIAL DIFFUSION of different cultures, and the varying nature of the CULTURAL LANDSCAPE.

cultural landscape The landscape produced by human occupation, where the physical ENVIRONMENT provides the HABITAT and CULTURE modifies it. Key elements of the cultural landscape include architecture, SETTLEMENT traditions, field systems, etc.

culture In geographical studies, the term 'culture' embraces a wide range of human attributes and artefacts – from beliefs, ideas and values, through customs and behaviour, to tools and works of art. Culture is recognized as being a powerful influence on the way in which people respond to, and utilize, the ENVIRONMENT (see CULTURAL LANDSCAPE). Equally, it is held that aspects of culture lie deeply rooted in the environment.

cumulative causation A crucial part of the model of regional development devised by Myrdal (1957). He suggested that REGIONAL

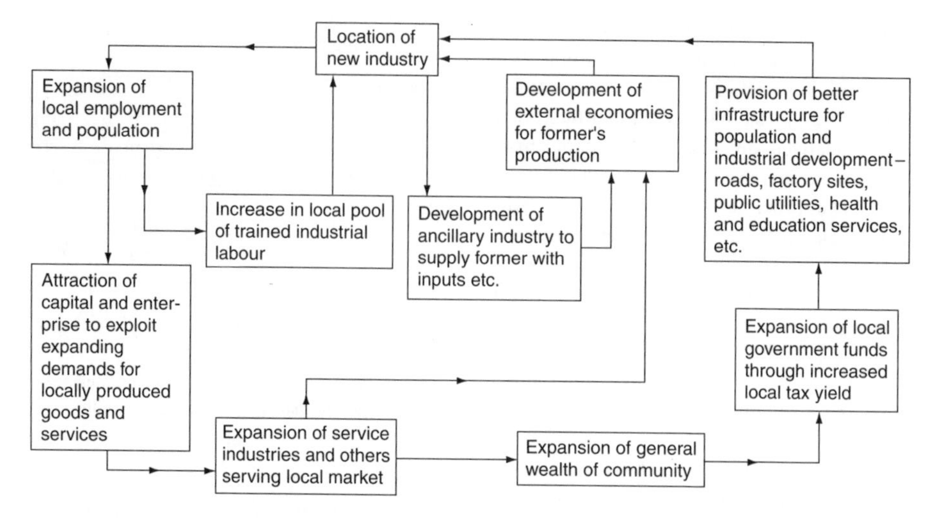

Myrdal's model of cumulative causation

IMBALANCE exists within all countries and that economic forces tend to increase rather than diminish this imbalance. Some form of INITIAL ADVANTAGE possessed by one region sets in motion a process whereby that region becomes progressively more successful (often at the expense of other regions). This is mainly because it will attract innovation, higher investment, better services, more labour, etc. Thus the rise of, and sustained leadership of, CORE regions (e.g. SE England, SE Brazil) may be explained by this process. In the accompanying figure, the trigger to cumulative causation is shown to be the setting up of a new industry (note the beneficial repercussions of this one development). [*f*]

cumulative frequency The FREQUENCY DISTRIBUTION of a set of grouped data is first converted into percentage form with the number of values in each group being expressed as a percentage of the total number of values in the distribution. These percentage figures are then summed successively, normally working from the lowest to the highest group. Thus, where x% of the population is living in cities with populations between 25,000 and 50,000, y% in cities between 50,000 and 10,0000 and z% in cities with populations greater than 100,000, the cumulative frequency would run x, $x + y$, $x + y + z$. The cumulative percentage frequency for a given group indicates the proportion of values lying below the upper limit of that group. This may then be plotted on the vertical axis of a *cumulative frequency graph*, against the range of values on the horizontal axis. Sometimes referred to as *percentage cumulative frequencies*, such frequency distributions are often used for the comparison of variables originally measured on different scales.

cumulonimbus cloud See ANVIL CLOUD.

cumulus cloud A large and usually isolated cloud with considerable vertical extent. The upper surface of the cloud is dome-shaped, with many individual 'protuberances', and a nearly horizontal base. Cumulus clouds result from CONVECTION currents, and develop under conditions of limited atmospheric INSTABILITY. *Fair-weather cumulus clouds* are typical of warm summer days under anticyclonic conditions, when solar radiation heats the ground surface, and air pockets in contact with the ground warm and rise to limited heights. Such clouds tend to disappear in the evening as the ground cools and atmospheric STABILITY is restored. [*f* CLOUD]

current bedding A type of geological structure, usually in SANDS and SANDSTONES, where very thin layers (*laminae*) of SEDIMENT have been formed at a marked angle to the overall stratification. Current bedding is typical of deltaic deposits – for example, where a glacial meltwater stream has entered a lake. A similar structure may result from the deposition of sand grains by wind action on the lee face of a DUNE. Current bedding is also referred to as *cross-bedding.*

cuspate delta A DELTA that projects only a limited distance into a sea or lake, owing to the fact that the river-borne SEDIMENTS are actively redistributed by wave action and longshore currents along the coast on either side of the river mouth.

cuspate foreland An accumulation of BEACH deposits (SAND and SHINGLE) shaped by CONSTRUCTIVE WAVES from two different directions. Dungeness in Kent is a notable example, formed largely during the post-Roman period. It is believed that initially a large SPIT grew across the mouth of a bay (now occupied by Romney Marsh) under the influence of waves from the SW. With the passage of time, and through the addition of beach ridges at some points on its seaward face, and EROSION at others, the spit became reoriented to face towards the SW. However, simultaneously, its far end began to be refashioned by waves from the E (approaching through the Straits of Dover), and numerous individual beach ridges were added here, thus accentuating the triangular shape of the foreland. [*f*]

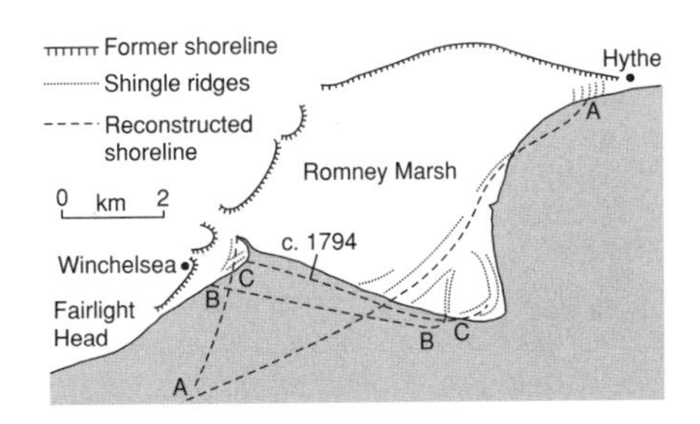

Stages in the growth of Dungeness, a cuspate foreland

cusps See BEACH.

cut-off In a strongly meandering river, the 'neck' between adjacent MEANDERS may be progressively reduced by bank EROSION. Eventually, the point will be reached at which the river breaks through the neck, to form the cut-off, and the meander itself will become abandoned. Banks of SEDIMENT at either end of the latter will result in the formation of an OXBOW LAKE and further sedimentation will convert the lake into a marshy depression.

cwm See CIRQUE.

cycle of erosion The concept, first fully developed by the American geographer Davis in the late 19th century (and adapted subsequently by other geomorphologists), that landscapes develop progressively through time from *initial forms*, the result of earth movements and uplift of land-masses, to *ultimate forms.* The latter occur when long-continued DENUDATION has reduced the initial forms to a near-level surface of EROSION, referred to by Davis as a PENEPLAIN – though other terms for this feature are now used, such as PLANATION surface and PEDIPLAIN. The emphasis in the cycle concept is thus on the *sequential development* of landforms, through the stages of *youth*, *maturity* and *old age*, each of which is associated with specific forms and processes. In modern geomorphology there is much criticism of the cycle concept on the grounds that it: relies too much on generalization; pays little attention to the detailed study of processes, and thus has little application to practical aspects of landform study; is unrealistic in the sense that it assumes very long periods of structural and climatic stability that are unlikely to have existed.

cycle of poverty The idea that poverty and DEPRIVATION are transmitted from one generation to the next, thus creating a self-perpetuating system or VICIOUS CIRCLE. The

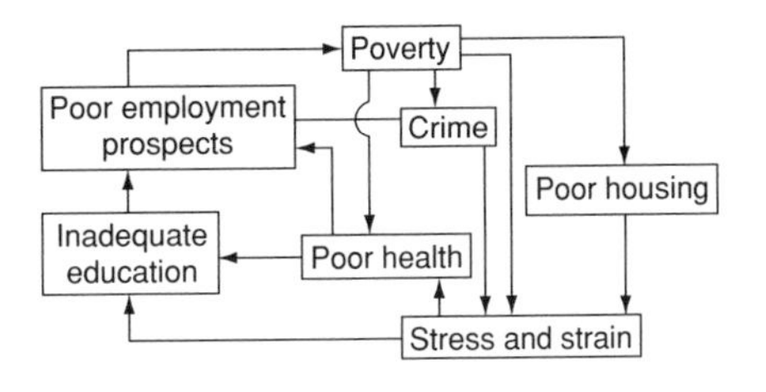

The cycle of poverty

children of poor parents may receive little parental support and may be forced to attend inadequate schools. As a result, they leave school at the earliest possible opportunity and with few qualifications. This, in its turn, means that they have difficulties in finding work and can therefore only expect to earn low wages for doing rather menial tasks. Thus they tend to remain 'trapped' in a cycle of poverty (the *poverty trap*), being largely unable to improve their lot. The same sort of vicious circle may be seen at a regional scale, as for example in rural areas of LEDCS and in peripheral parts of MEDCS (see PERIPHERY). [*f*]

The terms 'cycle of poverty' and 'poverty trap' are also used to describe the plight of the poorly paid who, on receiving a small rise in wages or on being taken out of the income tax bracket, find that they are no longer eligible for those state benefits provided for low-paid households. In this sort of situation, the small gain in wages may be more than outweighed by the loss of benefits, to the extent that financial circumstances of the household actually deteriorate.

cyclone An area of low pressure, with a more or less circular pattern of isobars. The term is sometimes applied to mid-latitude FRONTAL DEPRESSIONS, but is now increasingly restricted to tropical depressions of the HURRICANE type. Such systems are usually quite small, with a diameter of about 650 km. Atmospheric pressure is very low (less than 950 mb), and winds may exceed 120 km hr^{-1} (hurricane force). Cyclones result from intense surface heating (often involving a sea area with a surface temperature of 27°C or more), and are associated with powerful convective uplift, giving cumulonimbus clouds (see ANVIL CLOUD) reaching heights of 12,000 m or more, which release torrential rainfall. However, in the centre of the storm there is an area of calm (the *eye*) where descending air causes the temperature to rise, clouds to become broken and rainfall to cease. Cyclones depend for their development on an initial supply of heat and moisture, and are sustained by the further release of LATENT HEAT through massive CONDENSATION. When the storm moves over land where the supply of moisture is cut off, it quickly decays. Cyclones generally move along westward tracks, either to the N or S of the Equator, but then curve northwards along the eastern margins of continents to become components of the westerly wind systems of the mid-latitudes.

D

dairy farming The rearing of cattle for milk, either for direct sale to consumers or for processing into butter, cheese, etc. Pigs and poultry are often associated as side-lines in that they are fed on the by-products of such milk processing. Formerly, dairy farming was widely practised around towns and cities, but given the development of refrigerated transport and UHT processing, liquid milk can be carried over long distances and kept for long periods, thus encouraging specialization in favoured areas.

dalmatian coast See CONCORDANT COAST.

dam See BARRAGE.

dasymetric technique A method used in the mapping of distributions whereby symbols are placed on a MAP to show the occurrence of a given phenomenon. For example, the technique is frequently used in plotting the distribution of population in a given area using dots (each representing a given number of people) or proportional symbols, located on the map in such a way as to give a reasonably accurate impression of the actual distribution. Ct CHOROPLETH MAP. [*f* PIE DIAGRAM]

data set See POPULATION (ii).

dead cliff A sea-cliff that is no longer being actively eroded by waves, often resulting from a relative fall in sea-level, where it may be fronted by a RAISED BEACH. A dead cliff can also result from significant DEPOSITION of beach material at its foot, either by marine processes or by human intervention. Despite no longer being under direct attack from the sea, dead cliffs are affected by SUBAERIAL processes to give a relatively gentle, often vegetated slope (see BEVELLED CLIFF).

dead ice Ice, forming part of a glacier or ICE SHEET, that is no longer flowing. Dead ice is found in a variety of situations. For example, when an ALPINE GLACIER undergoes surging as a result of increased PRECIPITATION in the ACCUMULATION ZONE, the snout will advance considerably. However, when the impetus of the surge has ended, the lowermost part of the glacier will be abandoned as a dead ice mass, slowly melting away *in situ* and becoming covered by ABLATION moraine. When dead ice melts, it characteristically leaves behind hummocky ground, often with features of GLACIO-FLUVIAL depositions, e.g. KAMES and KETTLE HOLES.

death rate The average number of deaths per 1000 inhabitants in a given population. It is not a very refined measure of mortality, because it does not take into account the overall age–sex structure of the population. A high death rate in a 'young' population is a significantly different situation from that where the same rate prevails in an 'older' population. Ct BIRTH RATE.

debris Scattered and broken fragments often resulting from the decay and disintegration of rocks by weathering processes. Such material occupying slopes is termed SCREE or TALUS.

debt In geography this refers mainly to what is often called *Third World debt*. It is debt owed by LEDS to either MEDCS or international AID organizations. Third World debt increased enormously during the last quarter of the 20th century. The root of the problem lies in the fact that much of the aid was 'given' in the form of loans. Not only do loans have to be repaid, but they also attract interest – this too has to be paid. Debt repayment is only possible if the economy of the receiving country flourishes. Unfortunately, this has rarely happened. This, plus rising interest rates, has meant countries falling more and more behind in their repayments. Providing further loans to pay off debt arrears only makes the crisis worse as countries simply fall deeper into debt (now referred to as the *debt burden*). They enter a downward VICIOUS CIRCLE known as the *debt spiral*. Debt increases DEPENDENCE on the donor nation; it makes the rich richer and the poor poorer, thus widening the DEVELOPMENT GAP. Recently, some MEDCs have agreed in principle to *debt relief*, either to clear or reduce the international debts of the poorest countries. As yet, however, there are few signs that anything has been done. In the meantime, many of the world's LEDCs are trapped as victims of this 'new slavery' created by international debt.

decalcification A soil-forming process, involving the removal of calcium minerals by infiltrating rainwater. It affects SOILS in areas where precipitation is greater than evaporation over much of the year. In humid temperate forest regions decalcification and humifaction (the incorporation of humus within the soil) give rise to brown earth.

decentralization The operation of CENTRIFUGAL FORCES leading to outward movement

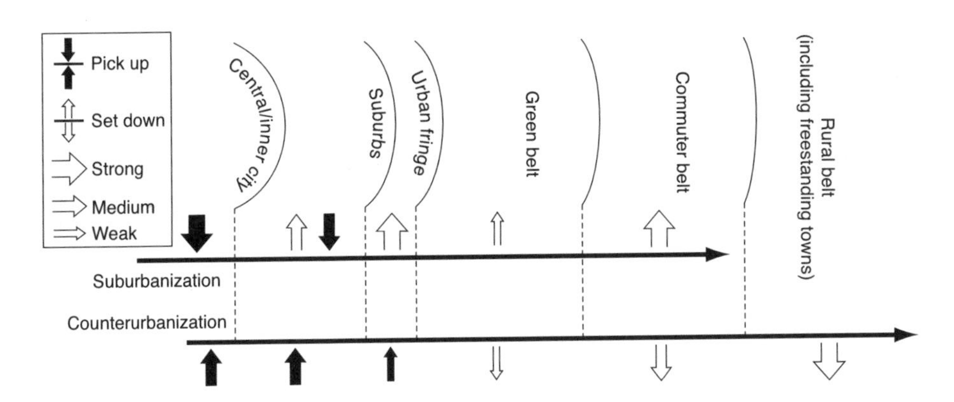

Decentralization

(e.g. of people, economic activities) from established centres. This is illustrated in the CORE-PERIPHERY MODEL, which shows a TRICKLE DOWN of growth from core to periphery (see also SPREAD EFFECT). It is an integral part of SUBURBANIZATION and COUNTERURBANIZATION, and involves 'picking up' population and employment from the inner areas of cities and 'setting down' either in the SUBURBS or in smaller URBAN centres and RURAL areas. Such movements are often a voluntary response to the negative EXTERNALITIES of large cities (especially of their older areas) and the positive externalities that are perceived in the new locations. Planned decentralization from the INNER CITY was part of the OVERSPILL programme in the UK during the 1950s and 1960s. Ct CENTRALIZATION. [*f*]

deciduous forest A type of forest comprising trees that experience annual leaf fall (usually in the autumn or 'fall'). These trees are mainly broad-leafed (such as oak, beech, ash, birch and maple), but some conifers in middle and high latitudes (e.g. larch) have the deciduous habit. Deciduous trees are better adapted than conifers to competing in the climatic conditions of areas such as W Europe. In particular, they are well adjusted to the longer growing season, with six months or more of mean temperatures in excess of 6°C, and to the annually well-distributed rainfall (in the range 750–1000 mm), which provides a winter surplus compensating for any summer deficit caused by EVAPOTRANSPIRATION exceeding PRECIPITATION. The canopy of the forest gives only partial shade in summer (and very little in winter and spring), so that an undergrowth of saplings, shrubs, flowers and grasses is able to thrive. The SOILS beneath summer deciduous forest trees are richer than those of the BOREAL FOREST owing to annual increments of leaves supplying *mull humus*. Most 'natural' deciduous forest has been destroyed or modified by people.

decision-making The process whereby alternative courses of action are evaluated and a decision taken. An important feature of investigations of decision-making is the recognition that location decisions are seldom, if ever, *optimal* (see OPTIMIZER CONCEPT) in the sense of maximizing profits and minimizing resources used. Similarly, consumer behaviour and decision-making hardly ever accord with the assumptions made, for example, in CENTRAL PLACE THEORY. Because of this, two alternative concepts haven put forward: the SATISFICER CONCEPT and the concept of SUBOPTIMAL LOCATION. Attempts to generalize about decision-making are persistently frustrated by inescapable facts: the immense diversity of decision-makers; their varying powers of PERCEPTION; their different circumstances; their varying access to relevant information; differences in their ability to handle that information and to evaluate alternative courses of action. See also BEHAVIOURAL MATRIX.

[*f* PLACE UTILITY]

declining region See DEPRESSED REGION, LAGGING REGION.

decomposers Organisms (particularly bacteria and fungi) that feed on plant and animal remains, breaking them down and in the process releasing energy and nutrients. Decomposers play a major role in maintaining soil fertility, by aiding the decay of leaves and other plant material at the soil surface, and the incorporation of the resultant products within the soil. See NUTRIENT CYCLE.

deconcentration See DECENTRALIZATION.

deep weathering The production of a thick REGOLITH (sometimes referred to as *saprolite*) by prolonged and/or intense CHEMICAL WEATHERING. Deep weathering is especially associated with areas of low RELIEF in humid tropical environments, where warmth and humidity favour processes such as HYDROLYSIS, OXIDATION and CARBONATION, but where TRANSPORT is restricted by the gentle slopes and binding effect of the dense vegetation. It has been suggested that deep weathering was partly responsible for the formation of granite TORS. [*f* TOR]

deferred junction A type of river confluence in which the junction of a tributary with a main stream on a FLOOD PLAIN is delayed by the presence of a LEVÉE. The tributary may thus be forced to run parallel to the main stream for a considerable distance downstream. An extreme example of a deferred junction is the R Yazoo, which flows alongside the Mississippi for 280 km before actually joining it.

deficiency disease A disease caused where some element vital to human health and survival is missing. Most commonly, the deficiency occurs when the DIET lacks a specific mineral, protein or vitamin. For example, *rickets* (a bone softness occurring in children) is caused by a deficiency of Vitamin D. See MALNUTRITION.

deflation The removal of fine material by wind action, particularly in deserts and along coasts. It is often regarded as a form of 'wind EROSION', but in fact it involves only the transport of loosened particles, mainly SAND. Deflation is usually a small-scale and ineffective process, though when operative over a very large area its total effect can be significant. Resultant landforms are usually small hollows and BLOWOUTS.

deforestation The complete felling and clearance of forested land. An action having a range of serious environmental and economic consequences. [*f*]

deglaciation The reduction in size of a glacier or ICE SHEET, resulting from a negative MASS BALANCE and leading to the exposure of the previously ice-covered surface. Deglaciation involves both retreat of the ice margins (as ABLATION exceeds forward movement) and downwasting of the ice surface. Since it releases large quantities of meltwater, various landforms associated with lakes and fluvial activity are formed, together with RECESSIONAL MORAINES. See GLACIO-FLUVIAL.

deglomeration See DECENTRALIZATION; ct AGGLOMERATION.

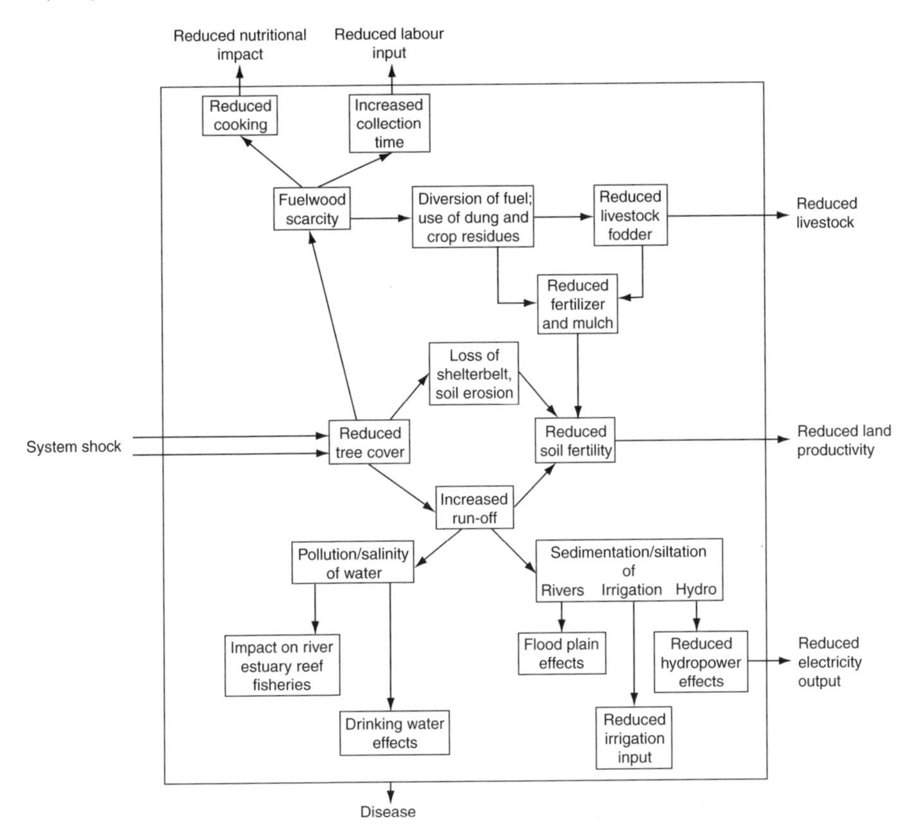

Deforestation and its impacts

degradation A term used loosely to describe the wearing down of the land surface by denudational processes (hence, *degraded slopes* or *degraded beach profiles*). More strictly, the term refers to EROSION carried out to restore or maintain the condition of GRADE. For example, a reduction in SEDIMENT in a stream will cause it to have excess energy and it will erode its bed. This will cause a reduction in channel slope (degradation), thus reducing stream energy to the point at which stream CAPACITY is just sufficient for the reduced LOAD.

degrees of freedom The number of items in a *data set* to which arbitrary values can be given, so that in a data set of n values, with a known mean value of x, there are $(n - 1)$ degrees of freedom, i.e. one less than the total number of items in the data set. It is a measure used particularly in the analysis and comparison of samples. [*f* CHI-SQUARED TEST]

de-industrialization The decline (relative or absolute) in the importance of manufacturing experienced by most MEDCs since the mid-1960s, as measured by its contribution to GDP and its percentage of the total labour force. It is explained by declining productivity (perhaps due to old plant, restrictive labour practices, high taxes, a strong currency, etc.) and strong foreign competition, particularly from countries where labour is less costly and unionized. The USA and UK have been among those countries most affected to date, and their areas of heavy (iron and steel, shipbuilding) and traditional (textiles) industry have been hardest hit. The decline of the manufacturing sector is worrying to some governments, because they believe that the TERTIARY SECTORS and QUATERNARY SECTORS are incapable of raising the same level of overseas earnings. A significant amount of the manufacturing lost by MEDCs has been relocated in LEDCs with their attractions of cheaper land and labour, and less stringent controls on pollution. Ct INDUSTRIALIZATION; see also GLOBALIZATION, NIC, RIC.

delta An accumulation of river-borne SEDIMENTS at the mouth of a river (in a sea or lake), formed where the rate of DEPOSITION exceeds the rate of removal by wave action or tidal currents. The principal methods of deposition are: the dumping of BED LOAD; the settling of suspended sediment as river velocity is reduced; the process of *flocculation*, whereby CLAY particles in suspension coagulate on contact with seawater and settle rapidly to the bed. See ARCUATE, BIRD'S-FOOT and CUSPATE DELTAS.

demographic coefficient An index designed to give a measure of future population growth and pressure in any region. It is derived by the formula:

$$C = dR$$

where C is the demographic coefficient, d the density of population and R the NET REPRODUCTION RATE. An alternative version of the formula is

$$C = dT$$

where T represents the rate of NATURAL INCREASE per 1000 inhabitants.

demographic regulation theory This states that every society tends to keep its fertility and mortality in a sort of balance. When the DEATH RATE is high, there is little need for the regulation of the BIRTH RATE. But when mortality is reduced to low levels, then the persistence of high fertility (and therefore of high rates of NATURAL INCREASE) will produce a scale of population growth that may well exceed the CARRYING CAPACITY of the land. This, in turn, will be perceived as threatening to lower general WELL-BEING. In these circumstances, the regulation of fertility (see BIRTH CONTROL) becomes accepted as something needing to be undertaken for the collective good. This theory is important in explaining the DEMOGRAPHIC TRANSITION.

demographic transformation See DEMOGRAPHIC TRANSITION.

demographic transition Sometimes also known as the *demographic transformation* or the *population development model*. A model representing changing levels of fertility and mortality over time, their changing balances and their net effect on rates of population

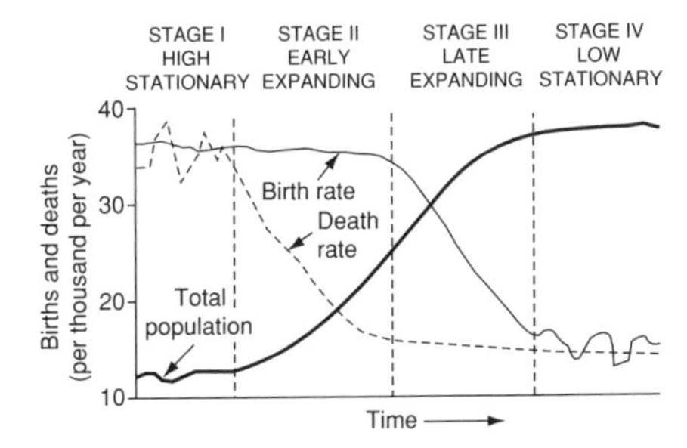

Stages of the demographic transition

growth. These demographic changes are linked with the broad process of DEVELOPMENT over a sequence of four stages.

1 *The high stationary stage* – fertility and mortality levels are high and subject to shorter fluctuations. Deaths due to natural checks such as famine, disease and war are the most significant influence on population growth, which tends to be relatively small. This stage is associated with largely undeveloped societies, relying on primitive technology and minimal subsistence.
2 *The early expanding stage* – population begins to grow at an accelerating rate as a result of the BIRTH RATE being sustained at a high level and of the DEATH RATE falling quite dramatically in response to the introduction of modern medicine, better diet and more sanitary living conditions. During this phase, economic developments might include the emergence of COMMERCIAL AGRICULTURE and the initiation of INDUSTRIALIZATION.
3 *The late expanding stage* – the rate of growth begins to slacken off as the death rate stabilizes at a low level and the birth rate declines (old traditions and taboos weaken and more people practice contraception). During this stage, society becomes highly urbanized and industrialized.
4 *The low stationary stage* – this occurs when both fertility and mortality levels are low (but with the birth rate more prone to fluctuations) and population growth is minimal, if at all. Society at this stage enjoys considerable economic wealth (much of it derived from the TERTIARY SECTORS and QUATERNARY SECTORS) and a high standard of living.

The model is a broad generalization based on the experience of MEDCS. It does not mean that all LEDCS will necessarily follow exactly the same pathway. Differences in culture, economy and technology may cause some deviation. Those countries that do may make the transition may do so more quickly than today's MEDCs; others more slowly. See DEMOGRAPHIC REGULATION THEORY. [*f*]

demography The study of human POPULATION, particularly its VITAL STATISTICS.

denationalization See PRIVATIZATION.

dendritic drainage A common type of drainage pattern, associated with areas of horizontal or gently dipping strata. Dendritic drainage comprises a multitude of small tree-like branch *streams* that unite, usually at an acute angle, to give larger streams and, eventually, one major *trunk stream.* Overall, the pattern resembles that of the branches of a large tree. The actual closeness of the pattern (see DRAINAGE DENSITY) will vary greatly, depending on rock permeability, and the amount and nature of PRECIPITATION. It is also likely to vary through time. [*f* DRAINAGE PATTERNS]

dendrochronology The science of reconstructing past climatic changes, using the evidence of tree rings, each of which reflects (in its width) the amount of growth as determined by the temperature and precipitation of the year in which it was formed. Samples from living trees (obtained from cores of wood extracted from the tree trunk) and long-dead trees can be assembled to give a climatic record over a period of several thousand years. Individual tree species of great longevity, such as the Bristlecone Pine of the Sierra Nevada in California, can provide a record of up to 4000 or more years.

density gradient A term used in URBAN GEOGRAPHY to describe spatial variations in population density within URBAN areas. It has been claimed that in the MEDC city, population densities generally lessen with increasing distance from the centre in a negative exponential manner (see ALONSO MODEL, EXPONENTIAL GROWTH RATE), while in LEDC cities density

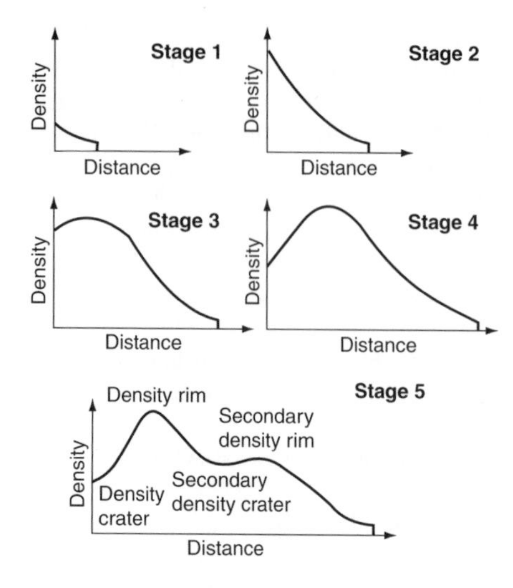

Stages in the evolution of urban population density gradients

gradients are relatively constant. Critics of the former generalization have pointed out that relatively low densities occur in the centre due to the displacement of residence from the CBD, and that the highest densities occur outside the CBD, thus forming an inner-city *density rim* around the central *density crater*. Outward from this rim, the density gradient is accepted as assuming a negative exponential form. It is suggested that urban density gradients change as the city grows (as shown in the accompanying figure). Two developments are particularly important: (i) the inner-city density rim is lowered by DECENTRALIZATION; (ii) a secondary density rim develops as planning controls stop the outward spread of the BUILT-UP AREA (see GREEN BELT) and encourage in-filling in existing SUBURBS.

In a similar way, the existence of SHANTY TOWNS around the margins of LEDC cities helps to raise population densities. See DISTANCE DECAY. [*f*]

denudation A term referring to all those processes of WEATHERING, TRANSPORT and EROSION that are responsible for the lowering and shaping of the physical landscape. The term *denudation chronology* refers to the study of the evolution of the landscape to its present form over long periods of time, as indicated by past 'denudational events' such as the formation of EROSION SURFACES, marine platforms, RIVER TERRACES, etc., and associated deposits.

dependence The condition in which a nation, region or group of people is only able to survive and progress by reliance on support provided by another. This support can take various forms, including trade, AID, political protection and defence. Dependence at a national level, particularly in the THIRD WORLD, has been the outcome of COLONIALISM. Traditional societies were originally reasonably self-contained – it was the arrival of European colonists that converted them into dependent countries.

dependency ratio The number of children (aged under 15) and of old people (aged 65 and over) expressed as a ratio of the number of adults (aged between 15 and 64), as illustrated in the accompanying tables. Basically it indicates the number of people

Dependency ratios in the UK (1961–2011)

Year	Under 15 years of age	Over 64 years of age	Dependency ratio
1961	25	12	0.59
1996	21	16	0.59
2011	18	20	0.61

Dependency ratios in selected LEDCs

LEDC	Under 15 years of age	Over 64 years of age	Dependency ratio
Brazil	35	4	0.64
Ethiopia	45	2	0.89
Haiti	40	3	0.75
Saudi Arabia	45	2	0.92

that the economically active population has to support. For MEDCs the ratio tends to be understated because quite large numbers of young adults remain in full-time education and therefore are part of the dependent population. There are important differences in the dependency ratios of LEDCs and MEDCs. In the former, dependency mainly involves the young, whereas in MEDCs it is the elderly. In both scenarios, and where ratios are high, there is always the challenge of whether those at work can provide the necessary support.

dependent place In CENTRAL PLACE THEORY, a SETTLEMENT of whatever size dependent on the goods and services provided by a particular central place. The dependent places of a high-order central place will inevitably include lower-order central places, because the latter do not provide the full range of CENTRAL PLACE FUNCTIONS performed by the former. [*f* CENTRAL PLACE HIERARCHY]

dependent territory See DEPENDENCE.

dependent variable A range of values which, in CORRELATION and REGRESSION ANALYSIS, are assumed to be related in some way to another INDEPENDENT VARIABLE. If data concerning levels of SO_2 in the ATMOSPHERE are obtained from a number of sites and an attempt is made to relate these figures to data about the burning of fossil fuels, then those consumption values would constitute the *independent variable* and the SO_2 data the dependent value (i.e. assumed to be related to fuel consumption). In the construction of a GRAPH to show the relationship between the two variables; the ordinate or y axis would be used to plot the SO_2 data and the abscissa or x axis the fuel consumption figures. The precise relationship between the two variables may also be established by statistical methods (e.g. CORRELATION COEFFICIENT, RANK CORRELATION).

depopulation A decline in the number of people living in an area, more usually brought about by net MIGRATION loss rather than by excessively high mortality. Such losses from country areas may be referred to as *rural depopulation.*

deposition Most commonly, the laying down of SEDIMENTS produced by the WEATHERING and EROSION of land-masses. In addition, deposition can involve chemical PRECIPITATION in oceans and lakes (forming evaporite deposits or LIMESTONE), and plant growth and decay. Deposition can take place in a variety of environments (fluvial, marine, aeolian, etc.) and is largely controlled by the size of particles being carried and the velocity of TRANSPORT (see HJULSTROM CURVE). Larger particles are deposited at higher velocities in relatively high-energy environments (e.g. on a storm beach), whereas finer particles are deposited at lower velocities in low-energy environments (e.g. in a sheltered lagoon).

depreciation Reduction in the original value of an asset due to use and/or obsolescence, as suffered by, say, a car or a washing machine.

depressed region A region that, in economic terms, is performing less well than the national average, and often increasingly so. A region characterized by economic distress and decline (hence the often-used alternative designation *declining region*), as indicated by symptoms such as: relatively high levels of unemployment; outflow of CAPITAL; a marked contraction of a basic industry on which the area has long depended (e.g. the shipbuilding industry in NE England); closure of FIRMS, often linked to the basic industry; emigration of labour. Once started, a sort of reverse MULTIPLIER EFFECT or *downward spiral* (see VICIOUS CIRCLE) is created, prompting gradual deterioration in the region's INFRASTRUCTURE and public services. This, in turn, will create further unemployment, encourage more out-migration and reduce investment confidence. In most MEDCs, such regions receive various forms of government assistance (see DEVELOPMENT AREA, GOVERNMENT INTERVENTION, GROWTH POLE). Often this is simply for the political expedient of being seen to be doing something to counteract a decline that, in practice, is extremely difficult to reverse. In the CORE-PERIPHERY MODEL, depressed regions would be regarded as *downward transition regions.* See also LAGGING REGION.
[*f* CYCLE OF POVERTY]

depression See FRONTAL DEPRESSION.

depression storage PRECIPITATION held temporarily on the ground surface commonly in the form of puddles. It represents a 'store' in the DRAINAGE BASIN HYDROLOGICAL CYCLE.

deprivation A concept applying to both people and areas, based on the notion of disadvantage relative to other people and other areas. Deprivation relates particularly to the three realms of housing, employment and services. For example, there are some sections of society deprived of the opportunity: to live in decent housing; to earn an adequate wage; to have proper access to various services (see WELFARE). Thus deprivation has a number of different facets (see TERRITORIAL SOCIAL INDICATORS), and these tend to be interrelated. for example, lack of a job or of an adequate income prevents access to proper housing; hence the frequently used term, *multiple deprivation.* equally, deprivation is a spatial phenomenon in that there are whole areas tending to miss out – e.g. DEPRESSED REGIONS and the inner city (see also TERRITORIAL JUSTICE, ETHNIC CLEANSING).

deprivation cycle DEPRIVATION is frequently self-perpetuating and takes the form of a DOWNWARD SPIRAL. It affects both people (see CYCLE OF POVERTY) and areas (see VICIOUS CIRCLE).

deranged drainage A disordered and somewhat chaotic pattern of drainage, characterized by numerous short streams and basins of INTERNAL DRAINAGE (occupied by lakes, marshes and bogs). Deranged drainage is found in those parts of Scandinavia and E Canada where selective ice sheet EROSION and DEPOSITION have produced a highly irregular land surface, with numerous low hills separated by shallow depressions or elongated furrows. [*f* DRAINAGE PATTERNS]

deregulation Freeing businesses and firms from government laws and regulations. This is regarded as a vital but controversial part of CAPITALISM. On the one hand, it promises more competition and less public spending; on the other, it removes worker protection and may result in a cutting back of services. In the case of the deregulation of public transport in the UK, for example, it is feared that companies will only run services to profitable destinations at profitable times of the day. See PRIVATIZATION.

derelict land Land formerly used, but now abandoned, unproductive and in need of RECLAMATION; e.g. the tip-heaps of exhausted mineral workings, abandoned factory premises and old docklands. Cf BROWNFIELD SITE.

descriptive statistics STATISTICS designed to simplify data to a more manageable form and used to describe the form of a FREQUENCY DISTRIBUTION; e.g. ARITHMETIC MEAN, MEDIAN, MODE, STANDARD DEVIATION. See also CENTRAL TENDENCY; ct INFERENTIAL STATISTICS.

descriptive technique See DESCRIPTIVE STATISTICS.

desert An arid region that experiences very low annual rainfall and, as a result, has little or no vegetation. Most of the world's deserts lie between 15–30°N and S of the Equator, corresponding with the sub-tropical belt of high pressure. Here the air is generally sinking, hence the lack of convective activity leading to cloud formation and the lack of rain. With an absence of cloud, diurnal temperature ranges are very high; during the day temperatures will often exceed 30°C, whereas at night frosts may occur, especially in the winter. While the term 'desert' is most often associated with hot deserts, it also applies to regions of low rainfall in polar areas, e.g. large parts of Antarctica.

desert pavement A stony desert surface, formed where SAND particles have been removed by DEFLATION, leaving a closely packed layer of *pebbles* (*lag gravel*), which are themselves often faceted by wind ABRASION (see HAMMADA). The pebbles may be cemented together by minerals drawn upwards to the surface in solution and precipitated as a result of evaporation.

desert varnish A precipitated layer (usually of iron minerals or manganese oxide) on the surface of rocks in hot deserts. The minerals forming desert varnish are drawn to the

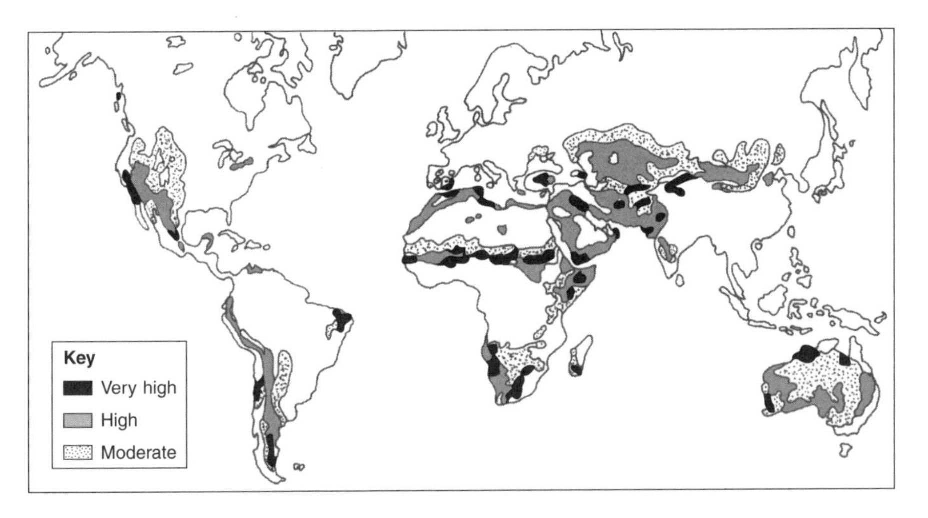

The global distribution of desertification

surface by capillary action, and deposited as a thin skin as evaporation removes moisture.

desertification The spread of DESERT conditions into former areas of semi-arid bush, STEPPE grassland and even woodland. For example, in recent decades, the southern margins of the Sahara have been advancing at an alarming rate, particularly during and since the great Sahel droughts of the early 1970s. While desertification can in part be linked to 'natural' climatic changes, it is increasingly accepted that the activities of people (for example, OVERGRAZING of domestic animals, overcultivation and the large-scale destruction of woodland for firewood) is a major contributory factor, or possibly even the main cause of desertification. When land becomes desertified, it is no longer able to support its people. Food shortages result and people are forced to migrate to other regions, abandoning their land and their livelihoods. When the infrequent rains do arrive, often in the form of downpours, serious SOIL EROSION can result as the land has been stripped of its protective vegetation layer. River channels become silted and the flood hazard increases. Desertification is now regarded as one of the greatest and most urgent problems confronting many tropical countries. [*f*]

desilication A soil-forming process, involving the removal of silica from the upper to the lower SOIL HORIZONS, or out of the SOIL PROFILE altogether, by organic SOLUTIONS. It occurs in all humid regions, but is most effective in tropical RAINFOREST environments, where it contributes to the formation of LATOSOLS.

desire line In transport studies, a straight line drawn on a map joining two points between which there is a desire or reason to travel, though not necessarily the actual route to be followed. On a desire-line diagram, one line is drawn for each separate movement, contiguous to the next, so that the overall thickness of the lines gives a visual indication of the total number of desired journeys.

deskilling The process whereby jobs are broken down into a series of separate, simple tasks, with each task demanding less skill of the operative than the original job. The replacement of moderately skilled workers by labour-saving machinery. It has been associated with the move in manufacturing towards ASSEMBLY-LINE PRODUCTION, AUTOMATION and MASS PRODUCTION.

destructive plate boundary See PLATE TECTONICS.

destructive wave A type of wave that has the effect of eroding, or 'combing down', BEACHES. Destructive waves tend to occur

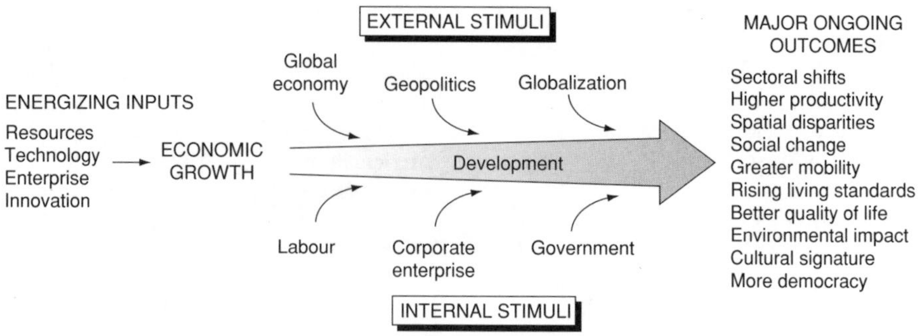

EXTERNAL STIMULI
Global economy
Geopolitics
Globalization
ENERGIZING INPUTS
Resources
Technology
Enterprise
Innovation
ECONOMIC GROWTH
Development
Labour
Corporate enterprise
Government
INTERNAL STIMULI
MAJOR ONGOING OUTCOMES
Sectoral shifts
Higher productivity
Spatial disparities
Social change
Greater mobility
Rising living standards
Better quality of life
Environmental impact
Cultural signature
More democracy

Development

during stormy conditions, are 'steep' in form, and break at a high frequency (13–15 per minute). There is, at the break-point, an almost vertical plunging motion, which generates little SWASH and thus weak TRANSPORT of material up the beach. The BACKWASH is relatively more powerful, and effectively transports SEDIMENT back down the beach face, resulting in a net loss of material. Most beaches are subjected to the alternating action of constructive and destructive waves. The former tend to be more active in the summer, and result in substantial beach ACCRETION, while the latter remove – temporarily – beach material during the winter. Beaches thus normally experience an annual cycle of growth and decay.

[*f* CONSTRUCTIVE WAVE]

detour index The shortest distance between two points expressed as a percentage of the most direct route between them. Thus the lower the index value, the more the 'direct' route deviates from the straight line. Sometimes referred to as the *index of circuity*.

developed country (world) See MEDC, THIRD WORLD and the NORTH.

developing country (world) See LEDC, THIRD WORLD and the *South*.

development In human geography this refers to the state of a particular society and the processes of change experienced within it. Development is generally regarded as involving some sort of progress in four main directions: ECONOMIC GROWTH; advances in the use of technology; improving WELFARE; MODERNIZATION. These dimensions are widely used for distinguishing between the FIRST WORLD and the THIRD WORLD, and between MEDCs and LEDCs. The meaning of development shifted considerably during the second half of the 20th century. Originally, it usually meant economic development or economic growth. Nowadays the view of development is altogether much broader, involving the whole of society and embracing cultural and social as well as economic and technological change. See DEVELOPMENT INDICATORS. [*f*]

development area A DEPRESSED REGION recognized by the British government as being in need of special assistance under the current regional planning policy; e.g. S Wales, NE England, the Highlands and Islands of Scotland. Such areas receive a variety of financial and infrastructural assistance, principally to encourage private investment, create new employment opportunities and to improve social conditions. Ct INTERMEDIATE AREA.

development gap The difference in terms of standard of living and WELL-BEING between the world's richest and poorest countries (between the MEDCs and the LEDCs). Despite some efforts to the contrary (see AID), the gap continues to widening (see DEBT) particularly as GLOBALIZATION puts more and more power and wealth in the hands of the strongest nations.

development indicators Statistical measures used by the WORLD BANK to monitor the DEVELOPMENT process. The measures relate principally to such economic aspects as

INDUSTRIALIZATION, production, consumption, investment and trade (see ECONOMIC INDICATORS). However, also included are indices of population (BIRTH RATE, DEATH RATE, LIFE EXPECTANCY), health and education. See HUMAN DEVELOPMENT INDEX.

development-stage model A sequential model recognizing six stages in the development of a region or country, each being marked by distinctive sectoral characteristics.

1 The PRIMARY SECTOR is all-dominant, with an emphasis on self-sufficiency.
2 Increased specialization within the primary sector accompanied by rising levels of production and of interregional/international trade.
3 The development of a SECONDARY SECTOR involving particularly the processing of selected primary products, and thus creating a narrow manufacturing base.
4 Diversification within the secondary sector encouraged by the proliferation of INDUSTRIAL LINKAGES, by rising incomes and, therefore, higher levels of consumer spending.
5 Expansion of the TERTIARY SECTOR in response to the opportunity to export capital and services to less advanced areas, and in response to greatly increased consumer spending.
6 The emergence of a QUATERNARY SECTOR, as the region or country specializes in the production and refinement of new ideas and processes for export.

The development-stage model is also known as the *Clarke-Fisher sector theory.* Cf STAGES OF ECONOMIC GROWTH MODEL. [*f*]

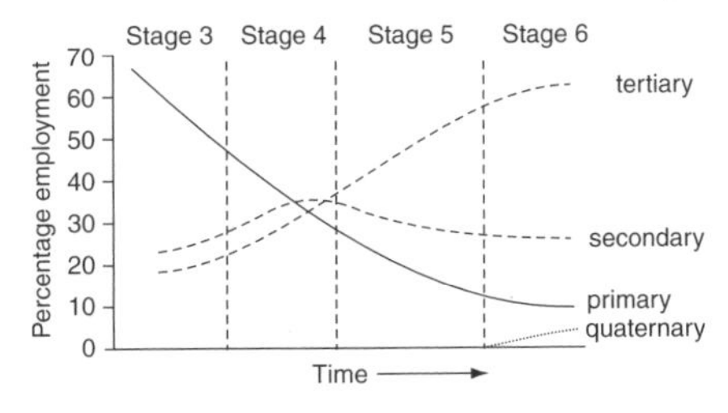

Development-stage model

devolution The process by which the central government of a STATE allows areas within its boundaries a degree of political AUTONOMY. It has been used to satisfy demands for independence made by militant MINORITY groups, as for example in Belgium where the Flemish and Walloons form what are virtually two separate states. The Basques, Catalonians and Andalusians in Spain have their separatist movements pressing for devolution while, in Britain, the Scots and the Welsh have their own parliamentary assemblies.

dew Water droplets formed on the ground, or on the surfaces of grass and leaves, by the direct CONDENSATION of atmospheric moisture (though that found on growing vegetation may be derived partly from transpired moisture). Dew is possible wherever warm moist air passes over a cool surface, but is mainly the result of nocturnal cooling of the ground on a still, cloudless night. As the ground temperature is rapidly lowered, a thin layer of air in contact with it is cooled to its DEW-POINT and moisture within the air is condensed to form water droplets. In deserts, where nocturnal cooling is very intense, dew fall may make a significant contribution to the total PRECIPITATION, and in some parts of the world (e.g. Chile) it is harvested using dense wire mesh and collecting vessels to help supply fresh water to local communities. See also RELATIVE HUMIDITY.

dew-point The temperature at which air becomes saturated, i.e. its RELATIVE HUMIDITY reaches 100%. Once dew-point has been reached, the process of CONDENSATION will commence, forming DEW, FOG and CLOUD.

diet The kind and amount of food consumed for cultural, medical or personal reasons.

differential erosion The selective EROSION, by processes such as river and wave action, of rocks of widely varying resistance (e.g. bands of LIMESTONE and CLAY) or along lines of weakness such as JOINTS and FAULTS. The effects of differential erosion are usually most evident on cliffed coasts, where headlands are formed by hard igneous intrusions (see IGNEOUS ROCK), metamorphic rocks, and SANDSTONES and limestones, and bays are developed in weaker SHALES and clays. Wave action also attacks faults, joints and BEDDING PLANES, to form minor inlets, caves and GEOS.

diffluence (glacial diffluence) The breaching of WATERSHEDS by distributary ice flows from a VALLEY GLACIER. In a valley system occupied by glaciers, the junction of smaller glaciers with a larger 'trunk' glacier may result in a 'congestion' of ice. The glacier surface may build up to the level at which it 'spills' into an adjoining valley system, often by way of a pre-existing COL.

diffusion See SPATIAL DIFFUSION.

diffusion curve The representation on a GRAPH of the SPATIAL DIFFUSION of an INNOVATION from its origins to its general use. Typically, the curve is S-shaped, showing an initial acceleration and then a slowing down in the spread and adoption. [*f* URBANIZATION CURVE]

diffusion lag See HIERARCHICAL DIFFUSION.

dilatation The process whereby rock joints are developed by the spontaneous expansion of rock masses when confining pressure is reduced. For example, a rock such as GRANITE or GNEISS, formed deep within the Earth's crust, will be increased in strength by the great pressure to which it is subjected. However, if overlying strata are removed by DENUDATION, the rock will experience elastic expansion. It is possible that dilatation may also occur extensively in glaciated regions, partly because as ice erodes solid rock it leads to the replacement of a substance with a high density by one with a low density, and partly as a result of the melting of large thicknesses of ice that have been pressing downwards.

dip Most commonly used to describe the maximum angle of inclination of a layer of SEDIMENTARY ROCK, it can, however, also apply in other geological contexts – for example, to the plunge of a SILL or a FAULT. It is possible to express the dip of a rock in terms of its angle of dip (with 90° being vertical) and its orientation, being the direction that it plunges. A line drawn at right angles to the dip is called the STRIKE. [*f*]

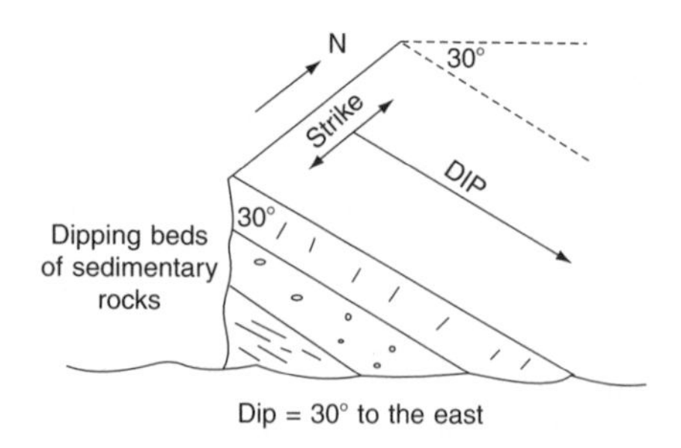

Dip and strike

dirt cone A conical or elongated mound of ice, covered by a layer of debris usually 1–3 cm in thickness, developed on the ABLATION ZONE of a glacier. Dirt cones result from differential ABLATION and occur wherever patches of sandy or gritty debris form on the glacier surface. The debris protects the underlying ice from melting, by reducing ablation. A debris-covered mound will thus grow, while the sliding of the debris over the slopes of the mound will gradually produce the cone form.

discharge The flow of a river, either in total or at a particular point (gauging station) along its course, expressed in terms of the volume of water passing in a unit of time. Discharge is expressed as cubic metres per second ('cumecs'). The formula for calculating discharge is: CHANNEL cross-sectional area x mean velocity. (The latter is determined from measurements of actual velocity, by a current meter, at a number of points distributed evenly through the stream cross-section.) In practice, once the necessary measurements and calculations have been made, the discharge of a river at a gauging station can be related to the *height* of the water surface (the *stage* of the river), by way of a graphical plot showing the relationship between discharge and level (see HYDROGRAPH).

discordant Descriptive of geomorphological features that 'cut across', or are discordant to, the structural grain of an area. For example, *discordant drainage* comprises streams that cut across a series of anticlinal and synclinal axes (see AXIS), as in the case of the rivers Avon, Test and Itchen in S England. A *discordant coastline* is oriented transversely

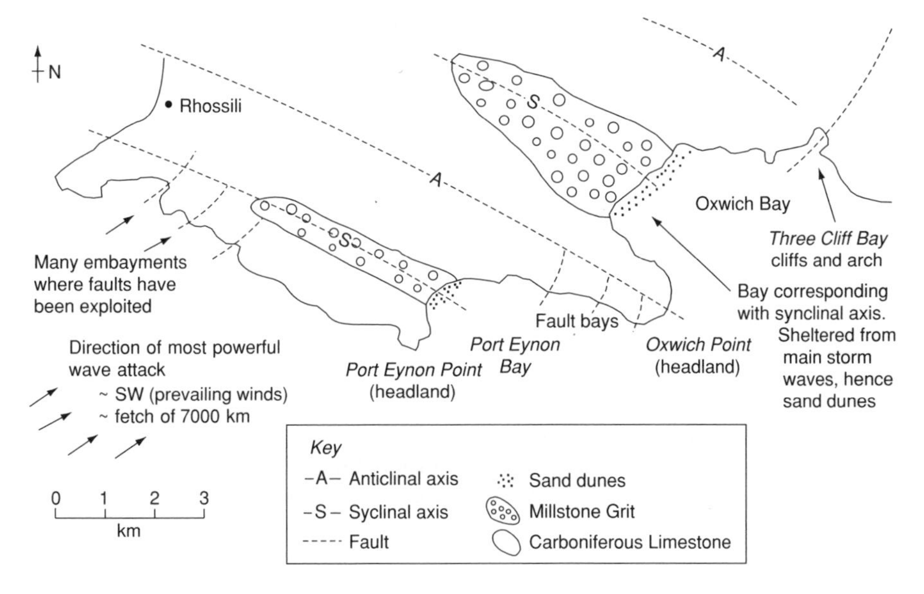

South Gower coast: a discordant coastline

to the dominant structures, as in the Gower coast in S Wales where the resistant Carboniferous limestone and less resistant Millstone grit have been folded into a series of east-west ANTICLINES, and DIFFERENTIAL EROSION has produced a series of synclinal lowlands (bays), separated by anticlinal ridges (headlands). Discordant coastlines are sometimes referred to as ATLANTIC-TYPE COASTS. [*f*]

discovery-depletion cycle See EXPLOITATION CYCLE.

discrete data See HISTOGRAM.

discrete variable See VARIABLE.

discrimination The act of making a distinction between people or groups of people in a manner that is likely to be unfair or unjust. Discrimination based on PREJUDICE and that acts against minority ETHNIC GROUPS is common in many countries. It generally results in the restriction of opportunities available to those people or groups, particularly in such matters as employment and housing (see APARTHEID). Equally insidious is discrimination based on wealth (or rather the lack of it), age, sex and SOCIAL CLASS. See SELECTIVE MIGRATION.

diseconomies of scale These occur where an increase in the scale of production is accompanied by an increase in production costs or where an increase in the size of a firm is accompanied by a decrease in profitability. The latter occurs because the more a company grows, the higher become the costs of communication, decision-making and generally running the business. Ct ECONOMIES OF SCALE.

disequilibria See SPATIAL DISEQUILIBRIA.

dispersal See DECENTRALIZATION.

dispersed settlement A pattern of rural SETTLEMENT, with isolated farms or cottages not grouped into VILLAGES and HAMLETS, as in much of the Celtic west of Britain. Here the dispersion may be attributed to the persistence for many centuries of the *gavelkind* inheritance system and to the effects of parliamentary enclosure of the uplands in the 18th and 19th centuries. In other areas, it is thought that dispersed settlement is a relatively recent phenomenon, resulting from the break-up of formerly nucleated settlements. Certainly, in areas of traditionally NUCLEATED SETTLEMENT, enclosure of open common fields, the loosening of community ties and the demands of modern efficient farming have all encouraged a certain degree of *settlement dispersal.*

dispersion diagram A diagram showing the distribution (dispersion) of a number of values measured over a period of time. Dispersion diagrams are particularly useful in the study of climatic VARIABLES such as PRECIPITATION and temperature. For example, in a *rainfall dispersion diagram* there is a vertical column for each month of the year, on which a dot is placed against the scale used (rainfall measured in mm) for each individual month's rainfall recorded over the whole of the study period. From the diagram, the MEDIAN (middle) and QUARTILE (at 25% and 75% levels of occurrence) can easily be read off. See RAINFALL RELIABILLTY. [*f*]

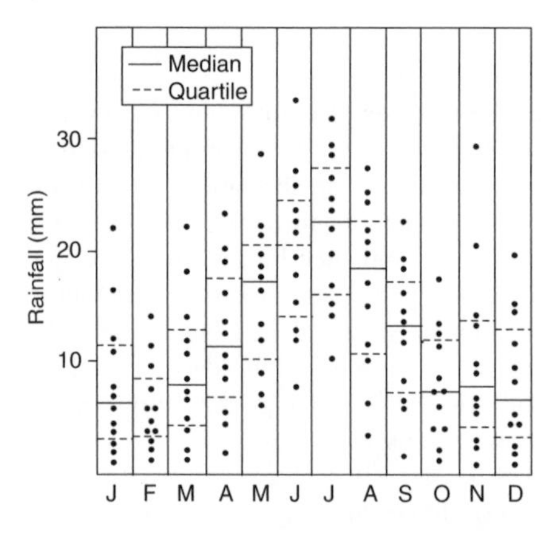

Dispersion diagram for monthly rainfall over a 12-year period

disposable income The amount of personal income that is left after all direct taxes (e.g. income tax) and national insurance payments have been deducted from gross income. For both the individual and the economy as a whole, this gives a measure of the amount of income available for expenditure on consumption, and for investment and saving.

distance decay Sometimes referred to as the *distance lapse rate*, based on the general principle that while everything is related to everything else, near things are more related than distant things. In other words, the amount of interaction between two places (SPATIAL INTERACTION) or objects decreases as the distance between them increases. The rate of interaction decline with distance will vary enormously, depending on the particular objects or places and on the particular functional context. Thus *distance-decay curves* can assume many different forms, but are generally of an exponential nature. The principle of distance decay underlies many of the classic models of SPATIAL STRUCTURE, e.g. CENTRAL PLACE THEORY, SPATIAL DIFFUSION, BID-RENT THEORY, DENSITY GRADIENTS. In GRAVITY MODELS, distance is assumed to exert a negative exponential effect on interaction.

distributary A newly created stream channel resulting from the division of a larger channel, which does not rejoin the main channel. Distributaries are very characteristic of DELTAS and ALLUVIAL FANS, where low gradients and huge sediment loads promote the splitting of channels.

distribution In geography, usually taken to be synonymous with SPATIAL DISTRIBUTION.

distribution costs The costs of delivering a product to a consumer (freight charges, insurance, storage, etc.). Ct PROCUREMENT COSTS.

diversification The process whereby there is a broadening of economic activity. This might involve a farmer widening his range of crops, a manufacturing FIRM adding to its lines of production or a city or region encouraging the growth of new activities. By this broadening, the economic risks attached to specialization are reduced.

diversified expansion One of three basic ways by which an enterprise may expand (ct HORIZONTAL EXPANSION, VERTICAL EXPANSION). This occurs when a FIRM undertakes production of a new commodity, without ceasing production of any of its existing lines. Expansion by *diversification* is undertaken for a number of reasons: to spread risk; to compensate for seasonal or cyclical fluctuations in the demand for those commodities already produced; to seize a chance opportunity to break into a new line of production; to exploit a recent INNOVATION. Expansion will often take place by merger, whereby the expanding company simply acquires companies already operating in those fields in which diversification is being sought.

division of labour The specialization of workers in particular parts of the production process. With specialization, the basic aptitudes of the individual worker can be better exploited; in addition, frequent repetition of the same task is likely to increase speed and skill, while time is saved not having to switch from one operation to another. Possibly one of the best examples of the division of labour in practice is to be found in the motor vehicle assembly plant. Cf INTERNATIONAL DIVISION OF LABOUR.

doldrums The equatorial belt of low winds or calms, lying between the trade-wind belts occurring between 5° and 30° N and S of the Equator. The doldrums constitute a zone of constantly high temperatures, high humidity and generally low pressure (see INTER-TROPICAL CONVERGENCE ZONE). However, there are no steep pressure gradients, and it is estimated that calm conditions exist for approximately one-third of the year.

dolerite A dark-coloured, dense, medium-grained basic IGNEOUS ROCK, usually occurring in the form of small-scale DYKES and SILLS within SEDIMENTARY ROCKS.

doline A closed depression found in LIMESTONE terrain. Dolines may have rocky margins or be bounded by smooth, soil-covered slopes, and their floors are often littered by limestone blocks or covered by thick accumulations of TERRA ROSA. Two main methods of formation have been suggested: (i) *solution dolines* result from slow downward development by SOLUTION processes that are concentrated beneath a SOIL mantle, without significant disturbance of the BEDROCK; *collapse dolines* (often more rocky in appearance) form where the roof of a subsurface cavern, formed at no great depth by subsurface solution, has caved in.

dominant wave The most powerful wave, capable of moving the greatest quantities and largest sizes of BEACH material, occurring on a particular stretch of coastline. Factors influencing dominant waves are the FETCH, and the direction of approach of gale-force winds. In S England the maximum fetch (from the SW) coincides with the direction of most strong winds, so that the dominant waves arrive from a SW direction. It has been noted that both beach and bay forms tend to become aligned to the pattern of dominant waves. However, the latter are often modified by WAVE REFRACTION as they approach the shore, with the result that the relationship between coastal features and the direction of dominant wave approach is locally much modified (as in the case of the recurved ends of large SPITS).

dormitory settlement A large residential settlement lying within the COMMUTING area of a TOWN OR CITY, i.e. functioning as a residential base for people who work elsewhere. Hence also *dormitory suburb.*

downthrow (throw) The change of level of rock strata on either side of a FAULT, expressed in terms of the amount of vertical displacement of a rock layer on the lowered (*downthrow*) side of the fault, in relation to the same layer on the raised (*upthrow*) side.

downtown An American term for the CBD.

downward spiral See VICIOUS CIRCLE.

downward transition region See CORE-PERIPHERY MODEL.

draa The largest category of desert sand dune, having a wave length of 300–500 m and a height of 20–450 m. Many are believed to be the product of SEIFS that have amalgamated over a long period, perhaps under wind conditions that were different from those of the present.

drainage basin An area of land drained by a river and its tributaries, bounded by a WATERSHED. A drainage basin can be viewed as an example of a largely self-contained geomorphological and/or hydrological system (see DRAINAGE BASIN HYDROLOGICAL CYCLE). The relationship between the component parts of the system can be analysed in terms of STREAM

ORDER. Such methods constitute the basis of *drainage basin morphometry* (measurement and analysis of the form of drainage basins). It is widely believed that the drainage basin provides the most convenient and logical unit for studies in FLUVIAL GEOMORPHOLOGY, since there is a close relationship between the relief and shape of a basin; its climate, vegetation and underlying geology; its stream network (see DRAINAGE DENSITY); its pattern of discharge, and geomorphological processes operative within it.

drainage basin hydrological cycle The open system (see CENTRAL SYSTEMS THEORY) describing inputs, stores, links and outputs of water within a drainage basin. The drainage basin HYDROLOGICAL CYCLE is of great importance to hydrologists and planners concerned with water supply, FLOODING and low flows. Drainage basin hydrology is also fundamental in understanding slope processes and the formation of certain types of soil – for example, GLEY SOILS. [*f*]

drainage density A statistical expression for the density of stream channels in a drainage basin. It is usually stated in terms of the length of stream CHANNEL in km per unit area of drainage basin (km²). Values for drainage density vary widely, from less than 5 km per km² (in permeable rocks such as LIMESTONE) to approaching 500 km per km² (in BADLANDS). One of the main controls of drainage density is rock type. For example, permeable rocks have a much lower density than impermeable rocks, however, it is interesting to calculate DRY VALLEY densities in, say, CHALK areas as a measure of past climatic regimes.

drainage patterns The spatial arrangement of stream networks. It is possible to identify several common drainage patterns (see DENDRITIC DRAINAGE, DERANGED DRAINAGE, RADIAL DRAINAGE, TRELLISED DRAINAGE) that are linked to drainage basin characteristics, such as geological structure, relief, vegetation cover and erosional history. Drainage patterns can also be quantified using such measures as DRAINAGE DENSITY and STREAM ORDER. [*f*]

drift A term most commonly applied to glacial deposits comprising unstratified TILL and stratified OUTWASH sands and gravels.

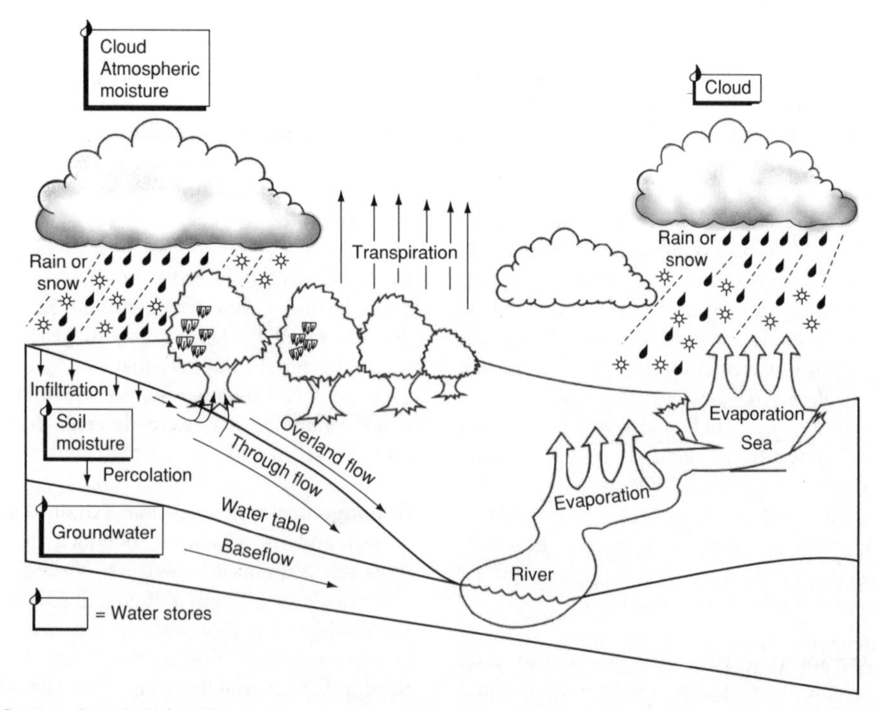

Drainage basin hydrological cycle

	IDEAL PATTERN
TRELLISED Typical of alternating resistant and non-resistant rocks	
RECTANGULAR Frequently minor joint/ fault patterns in underlying bedrock	
PARALLEL Typical of steep relief, especially where there is a lack of vegetation	
DENDRITIC Occurs in well-adjusted, gently sloping basins with a fairly uniform rock type	
DERANGED Where the original pattern has been altered by glaciation, for example	
ANNULAR A curving, broadly concentric pattern, where rivers exploit weaker bands of rock around a dome feature	

Drainage patterns

Drift is a superficial deposit that overlies the solid geology across large parts of E England.

drought A continuous period of dry weather, which may last for months or even years (e.g. the great Sahel droughts in Africa in the early 1970s). Within more humid climates, droughts are more infrequent and of shorter duration. It is now possible to detect the prevalence of drought conditions through the analysis of SATELLITE IMAGES.

drumlin An elongated 'streamlined' mound of TILL deposited and shaped beneath an ICE SHEET. The long AXIS of the drumlin lies parallel

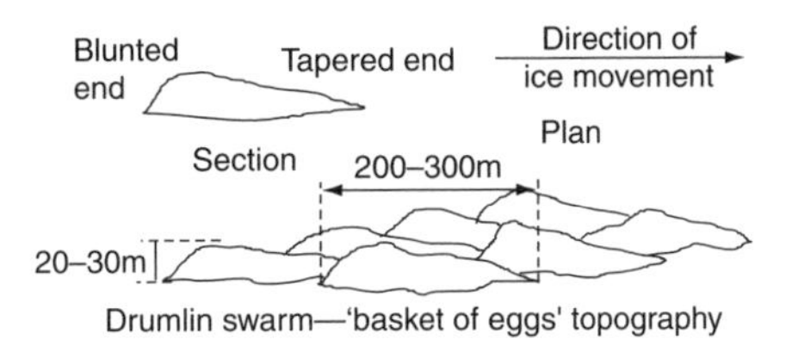

Drumlin

to the direction of ice flow enabling the landform to be used as an indicator of past glacier movement. Within the drumlin, although the till is non-stratified, individual stones are oriented parallel to ice movement, giving a well-developed *tip fabric*. In long-profile the drumlin is asymmetrical; the steeper slope, forming the 'blunt end' or *stoss slope*, faces upglacier, while the gentler slope, forming the 'tail' or *lee slope*, extends downglacier. Drumlins are usually quite small (20–30 m in height, and a few hundred metres in length), and often occur in swarms. The processes of formation include lodgement of debris as it melts out of the basal ice layers, reshaping of previously deposited GROUND MORAINE, and accumulation of till around BEDROCK obstacles (*rock drumlins*). [*f*]

dry adiabatic lapse-rate The rate of temperature change of a pocket of unsaturated but not perfectly dry air. A pocket of air that rises spontaneously under conditions of atmospheric instability, or is forced to rise up

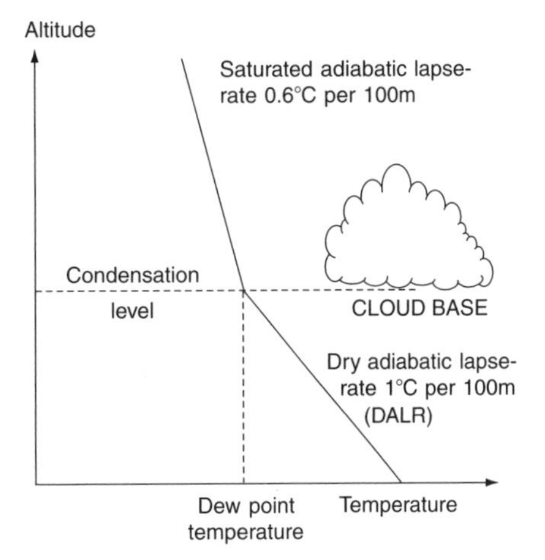

Dry adiabatic lapse-rate

a mountainside or above a frontal surface cools at a steady rate of 1°C for every 100 m of ascent. As the air pocket encounters progressively decreasing pressure, it expands and cools without the exchange of heat with the surrounding air. Unsaturated air that sinks – for example, on the leeward side of a mountain range (see CHINOOK and FÖHN WIND) – warms up (due to compression) at the same rate. Ct SATURATED ADIABATIC LAPSE-RATE. [*f*]

dry farming AGRICULTURE in a semi-arid area, without the help of IRRIGATION, made possible by conserving moisture through mulching, the maintenance of fine tilth and the utilization of two years' rain for one crop.

dry valley A valley formerly occupied by a permanent stream, but which is now – except under abnormal weather conditions – dry. Dry valleys are especially characteristic of LIMESTONE and CHALK terrains, but may occur in many other types of rock, though usually in smaller numbers. They have been attributed to many causes and different conditions. For example, under past PERIGLACIAL conditions, the development of PERMAFROST impeded PERCOLATION in normally permeable rock, so that meltwater from winter snowfall ran over the surface, forming streams capable of eroding valleys. See also COOMBE, MELTWATER EROSION.

dual economy (i) A regional or national ECONOMY that appears to consist of two separate parts or two different systems of production and exchange. Well exemplified in former colonial countries where a Western and largely export-oriented economy operated alongside a native, essentially subsistence economy. (ii) The term is increasingly used with reference to the two-part industrial structure beginning to emerge in MEDCS, made up of *centre firms* (largely TNCS) and *periphery firms* (small, relatively simple organizations), often functionally subordinate to, and dependent upon, the centre firms. Japan has a clear-cut dual economy.

dual job market A highly simplified view of the LABOUR market that sees all employment as falling broadly into two types: jobs that are highly paid, secure, and offer fringe benefits and good promotion prospects; jobs that are poorly paid, insecure and offer little training or chance of advancement.

dumping The sale of goods abroad at less than their average cost; a way of disposing of surplus production. In recent decades, Japan has been accused of dumping electrical goods, motor vehicles and microchips on European and N American markets.

dune A low hill or ridge of SAND, resulting from DEPOSITION by wind, along coasts or in deserts. The accumulation of the sand is often aided by vegetation (for example, *marram grass* at the coast) or rocky obstacles. Dunes are rarely stable landforms (unless deliberately anchored by a grass sward or pine trees), and are liable to migration (by AEOLIAN transport of sand grains from the windward to the leeward slopes) and severe EROSION (see BLOWOUT). In many areas of Britain (e.g. the coast of S Wales and the Culbin Sands of E Scotland) dune sands have extended inland to overwhelm agricultural land and even settlements. Much dune migration of this type occurred during the 13th and 14th centuries, which were evidently a time of great storminess in W Europe. (See also BARCHAN, DRAA, ERG, SEIF).

durables See CONSUMER GOODS.

duricrust A hard layer, often several metres in thickness, resulting from the cementation of SOIL and REGOLITH particles either at or beneath the land surface. Some appear to result from the PRECIPITATION of minerals drawn upwards by capillary action, while others are due to the accumulation of leached minerals at some depth in the SOIL PROFILE. The principal cements include calcium carbonate, as in *calcrete*, a formation found in drier regions; silica, as in *silcrete*; and iron, as in *ferricrete*, widely developed in SAVANNA regions.

dust bowl A semi-arid area, from which the SOIL is being or has been removed by the wind, especially where vegetation cover has been destroyed by overcultivation or OVERGRAZING. The term was widely used in the 1930s with reference to large areas of SW

USA, which suffered severe SOIL EROSION, with strong winds raising huge *dust storms*.

dyke A relatively narrow band of intrusive IGNEOUS ROCK that cuts across the BEDDING PLANES of the SEDIMENTARY ROCK into which it has been intruded. Dykes frequently occur in large numbers, running parallel to each other forming *dyke swarms*. [*f* LACCOLITH]

dynamic equilibrium A concept developed by modern geomorphologists in which it is envisaged that a condition of balance can exist in nature – for example, between the rate of production of debris on a slope by WEATHERING, and its rate of removal by transportational processes such as SOIL CREEP and RAINWASH. The concept of dynamic equilibrium is applicable in the systems approach to physical geography (see GENERAL SYSTEMS THEORY). Many landforms – and indeed ECOSYSTEMS – can be viewed as open systems, in which INPUTS and OUTPUTS of energy and mass become balanced. For example, a glacier is sustained by inputs of snow, but reduced by outputs of meltwater. When all the inputs and outputs are 'equal', the glacier is in a state of EQUILIBRIUM; thus it will not advance or retreat. If, however, the climate becomes colder and snowfall increases, the glacier cannot remain in equilibrium and input will exceed output. It will therefore increase in size, and the snout will advance into a lower and warmer area. This will cause increased ABLATION, which will offset increased accumulation to restore equilibrium.

E

earth flow A type of MASS MOVEMENT in which loose slope materials become saturated with water and flow at moderate to very rapid speeds. It may occur at the toe of a LANDSLIDE, owing to the concentration there of soil water. Where the flow affects mainly CLAY-sized particles with a high moisture content, an earth flow becomes a *mud flow*.

earth movement A disturbance within the Earth's crust as a result of compressive or tensional forces. Earth movements lead to the initiation and development of fold and FAULT structures, as well as crustal uplift and depression. The actual occurrence of earth movements is indicated by EARTHQUAKES and slight earth tremors.

earth pillar A sharply pointed pinnacle, developed in a mass of broken rock or debris subjected to intense EROSION by RAINWASH and small rivulets. Earth pillars are particularly striking in some parts of the Alps, such as the Pennine Alps and the Tyrol, where old LATERAL MORAINE deposits containing BOULDERS have been deeply gullied following glacier recession. As EROSION proceeds, the largest boulders are exposed and then act as 'umbrellas', sheltering the finer material beneath from rain-splash erosion. The boulders are thus left increasingly upstanding as 'caps' to pillars as the surrounding material is washed away.

Earth Summit The name given to the conference held in Rio de Janeiro in 1992 which was attended by world leaders and conservationists concerned about major environmental issues such as GLOBAL WARMING, the OZONE LAYER, DEFORESTATION and the rapid depletion of *non-renewable* RESOURCES. If these problems are to be overcome, it is widely believed that governments should pursue policies of SUSTAINABLE DEVELOPMENT.

earthquake A sudden and brief period of intense ground shaking. Earthquakes are usually associated with volcanic activity or movement along a FAULT. The point of origin of an earthquake is called the *focus*. The point on the ground surface immediately above the focus is called the *epicentre*. When an earthquake occurs, shock waves (seismic waves) travel outwards like ripples in a pond. There are three types of wave: (i) *primary* or *P waves*, which are the fastest; (ii) *secondary* or *S waves*; (iii) *long* or *L waves*, which despite being the slowest are responsible for causing the greatest destruction because they travel through the crust only.

Earthquakes are recorded by *seismographs* and the time-interval between the arrival of P and S waves at a seismic station can allow calculation of distance to the earthquake focus and time of occurrence. The most powerful earthquakes are associated with

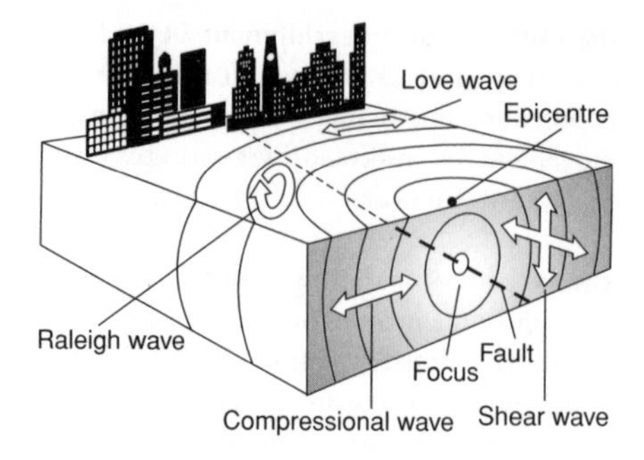

Features of an earthquake

major fault-lines – for example, the San Andreas 'megashear', which gave rise to the great San Francisco earthquake of 1906 and the severe Turkish earthquake on the Anatolian Fault in 1999. The *magnitude* of earthquakes (a term for the total energy released by an earthquake) is measured by the open-ended Richter Scale, where each point on the scale represents a x30 increase in the amount of energy release. The effects of an earthquake can be measured using the Modified Mercalli Scale, which has 12 gradations indicated by Roman numerals (see AFTERSHOCK, BENIOFF ZONE).

Earthquakes represent a serious hazard to human activity accounting for 47% of all fatalities by natural disasters in the last millennium and 35% of all economic costs. The most severe earthquake of the last millennium occurred at Shaanxi, China, causing 830,000 people to lose their lives. Increasingly it is believed that inappropriate building construction and the lack of emergency preparedness are partly to blame for the sheer scale of many disasters. [*f*]

ecological balance A 'steady state', achieved by an ECOSYSTEM in which inputs have come to equal outputs; this is also referred to as *homeostasis.* An example of ecological balance is CLIMATIC CLIMAX VEGETATION, in which the 'natural' vegetation has become perfectly adjusted to the prevailing climatic conditions. However, it is now widely accepted that ecological balance is more of a theoretical concept, since controlling factors such as climate and soil are always undergoing change (see CLIMATIC CHANGE).

ecology The scientific study of the mutual relationships of plants and animals to their ENVIRONMENT (a distinction is sometimes made, for convenience, between *plant ecology* and *animal ecology*). An ecologist is concerned particularly with the biological processes at work within the ECOSYSTEM, including the ways in which organisms gain energy and matter both from the physical environment and each other, and in turn release that energy and matter back to the environment.

economic bloc See TRADE BLOC.

economic climate See ECONOMIC ENVIRONMENT.

economic distance The distance a commodity may travel before the costs of transport exceed its value. The economic distance of a small product will exceed that of a bulky one of the same value, because more units of the former can be transported for the same cost.

economic environment A term referring to the external conditions in which FIRMS operate and which include such elements as interest rates, market buoyancy, unemployment levels, taxes, government subsidies, etc. Basically, the economic environment covers conditions over which the firm has no real control (cf EXTERNALITIES), but to which it must respond. The alternative terms *economic climate* and *business climate* are widely used.

economic geography That aspect of geography dealing with the distribution of economic activities, and with the factors and processes affecting them.

economic globalization The processes that are causing the economies of the world to move closer together and to become more integrated. These include TRADE, FOREIGN OVERSEAS INVESTMENT and AID, and are reinforced by the growth of TNCs and regional economic BLOCS, as well as advances in TRANSPORT, COMMUNICATIONS and information handling. See GLOBALIZATION.

economic growth A vital aspect of DEVELOPMENT, involving rising levels of material production and consumption, and frequently changes in the *sectoral balance* of the economy (e.g. the SECONDARY SECTOR and the TERTIARY SECTOR gain in importance) (see DEVELOPMENT-STAGE MODEL). The progress of economic growth may be monitored by a range of measures, e.g. income, GDP and GNP per capita, industrial production, levels of service provision, energy supply and consumption, trade, capital flows, etc. As such, these measures indicate that economic growth is, in itself, a multi-faceted process of change, the precise course and character of which will vary from place to place, depending on the nature and quality of RESOURCES (physical and human) and on opportunities (perceived and realized). [*f* DEVELOPMENT]

economic indicators STATISTICS that are sensitive to changes in the state of industry, trade and commerce within the national or regional ECONOMY. In the UK, economic indicators include the statistics of unemployment and unfilled job vacancies, gold reserves, bank advances, output of steel and motor vehicles, BALANCE OF TRADE, etc. Cf DEVELOPMENT INDICATORS.

economic overhead capital This refers to CAPITAL invested in INFRASTRUCTURE. It is widely regarded to be a vital stimulant of DEVELOPMENT. For example, investment of such capital in the construction of the M4 motorway was crucial to the economic revival of S Wales, for there was little prospect of attracting new employment until the ACCESSIBILITY of the region had been improved.

economic planning The identification of future economic needs, and the organization and deployment of scarce RESOURCES in the most efficient way to satisfy those needs. In this sense, economic planning can be undertaken by the individual, FIRM, local authority, regional and national governments. In capitalist societies, economic planning tends to take the form of varying degrees of GOVERNMENT INTERVENTION, particularly with respect to regional economic development (e.g. correcting REGIONAL IMBALANCE), and controlling interest rates and taxes.

economic rent The difference between the total revenue received from the sale of a commodity and the total costs of production and transport. As used in AGRICULTURAL GEOGRAPHY, economic rent is the surplus return or profit resulting from using land for one type of production rather than another. The revenue received is determined by the market price for that commodity and this, in turn, is a response to the supply and demand situation. PRODUCTION COSTS are assumed to be fixed (i.e. they do not vary from place to place), while TRANSPORT COSTS increase with distance from the market. The greater the transport costs, the smaller the difference between revenue and total costs, and therefore the smaller the economic rent. Thus economic rent decreases with distance from the market, but the *economic rent gradient* will vary according to the type of agricultural production. In the accompanying figure it can be seen that the transport costs for vegetables and dairying are high and, as a result, the rent gradients are steeply inclined. This concept of economic rent (sometimes also referred to as *locational rent*) is an integral part of VON THUNEN'S MODEL. [*f*]

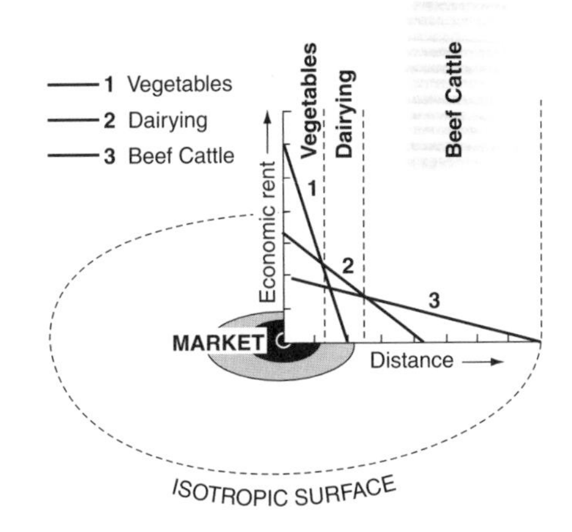

Economic rent and farming patterns

economic sector A major structural division of an ECONOMY. Most commonly, four functional divisions are recognized (see PRIMARY, SECONDARY, TERTIARY and QUATERNARY SECTORS). However, it is possible to recognize two sectors on the basis of ownership (see PRIVATE SECTOR, PUBLIC SECTOR). See DEVELOPMENT-STAGE MODE.

economic welfare See WELFARE.

economies of scale The cost advantages that are obtained from concentrating production into larger-scale processing units. A distinction is drawn between *internal* and *external economies of scale*. The former occur as a result of the expansion of the individual FIRM, and independently of changes in the size of the other firms in the industry. They are economies derived by the firm from its own efforts. For example, increasing the scale of its activities permits greater specialization and a better DIVISION OF LABOUR among workers. This in turn will lead to a saving in costs.

External economies of scale occur where a firm takes advantage of developments within the industry as a whole. This results in a lowering of its own costs. For example, an expansion of the whole industry may lead to a significant increase in the demand for a particular component; the price of it then falls (because of internal economies of scale in its manufacture), thereby presenting a saving to all firms using that component. Also, the individual firm is likely to benefit from technological improvements resulting from R&D undertaken by other firms. Ct DISECONOMIES OF SCALE.

economy (i) The whole economic structure and function of a country, region or society, particularly the exploitation of RESOURCES and the production, distribution and consumption of goods and services. (ii) A potential saving or cost advantage (see ECONOMIES OF SCALE, EXTERNAL ECONOMIES).

ecosystem An organic community of plants and animals viewed within its physical environment or HABITAT, e.g. a freshwater pond, a mixed woodland, or a hedge. An ecosystem can be described as a 'complex of interacting phenomena', within which there are many complicated and often subtle relationships (between climate and vegetation, vegetation and SOILS, animals and vegetation, and so on). It is possible for an ecosystem to become *stable*, with the various components achieving a condition of balance or equilibrium. However, it is also possible for ecosystems to become seriously disturbed, or unstable, as a result of natural catastrophes – for example, a major volcanic ERUPTION or interference by people. [*f* TROPHIC LEVEL]

ecotourism A form of TOURISM that seeks to minimize the environmental impacts of visitor pressure and the construction of tourist INFRASTRUCTURE. It is a reaction to the huge environmental, social and cultural damage that has been caused by mass tourism and package holidays. There have been some successful pioneering ventures, particularly in tropical RAINFOREST areas of Peru and Costa Rica, to arrive at a sustainable mode of tourism that is essentially environmentally friendly. For some, an important aspect of ecotourism is that it should be owned and run by local people and the profits retained in the local area rather than 'leaking' to a foreign tour company.

edaphic A term describing the SOIL conditions that influence the growth of plants and other organisms. These include soil texture and acidity, the presence or absence of vital minerals (including trace elements) and the SOIL MOISTURE content. Plant geographers recognize that, while major vegetation regions are climatically determined, within those regions local variations in vegetation (PLANT COMMUNITIES) often reflect the influence of edaphic factors.

edge A term used in GRAPH THEORY to describe the link between two vertices (NODES) of a TOPOLOGICAL MAP. The edges (sometimes also referred to as *links*) of a transport NETWORK would be the routes linking SETTLEMENTS. See also NETWORK ANALYSIS.

edge city The product of urban processes (mainly DECENTRALIZATION) leading to parts of the SUBURBS becoming more city-like through the AGGLOMERATION of offices, factories and large shopping complexes at favoured, accessible locations. They are features of post-suburban N America, but are beginning to appear in W Europe. [*f* GLOBAL CITY]

effective precipitation The amount of PRECIPITATION that enters the SOIL and is available for plant growth. In other words, the balance of the precipitation remaining after losses from evaporation, the rate of which is largely controlled by temperature.

EIA (environmental impact assessment) A process of detailed research undertaken to assess the likely environmental effects of a development project (such as the building of a large oil terminal on a tidal estuary). The assessment would attempt to set against the economic advantages of such a project the adverse consequences in environmental destruction or deterioration (including reduction in the natural beauty, or aesthetic quality, of a landscape, as well as the impact on wildlife resources, such as habitats for wintering flocks of birds or populations of rare animals). It is, of course, very difficult to evaluate aesthetic against economic considerations in a type of profit-and-loss account. In the USA, environmental impact statements have been required by law for many years. In the UK, such assessments have been

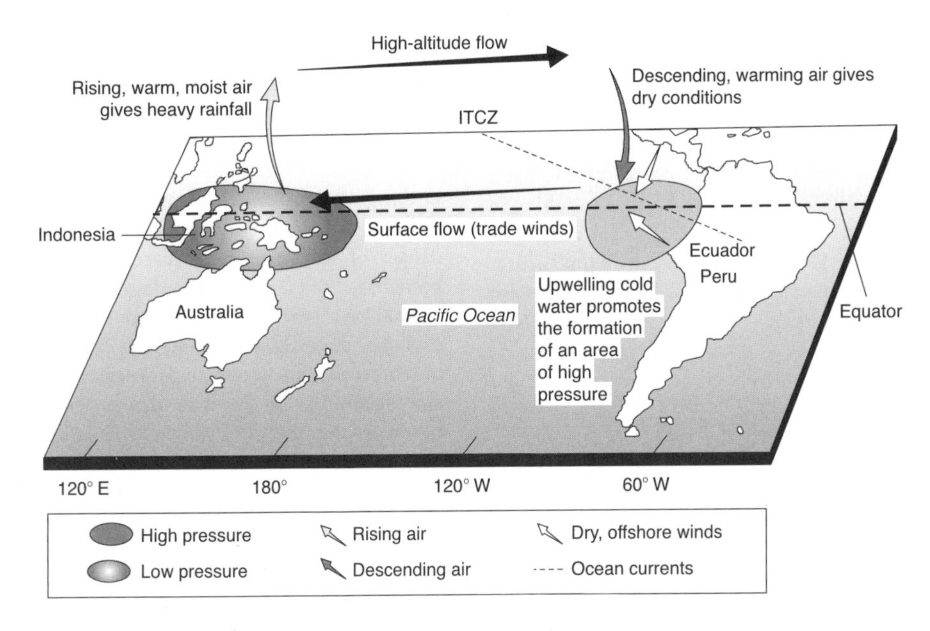

(a) Normal (Walker) circulation

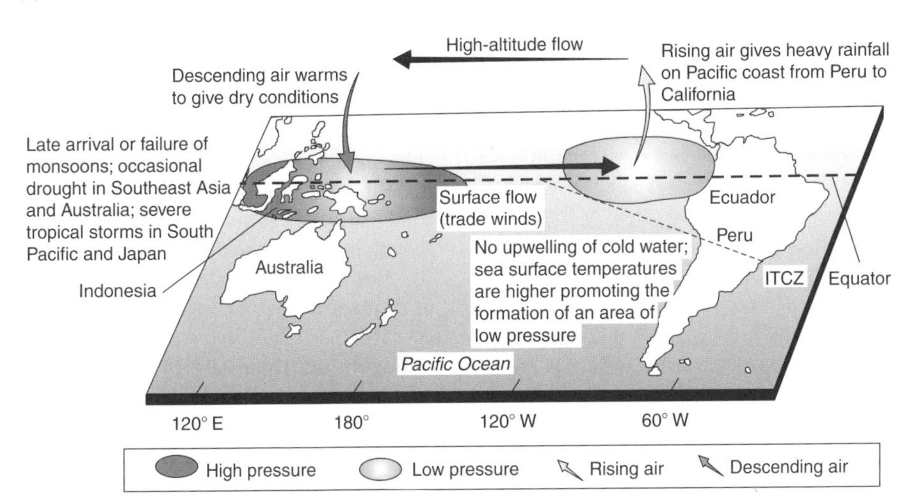

(b) El Niño event

made voluntarily in some fields (i.e. by oil companies concerned with exploration); but since July 1988, EU directives have made assessments obligatory for certain types of development.

El Niño A pronounced warming of the E Pacific Ocean, by as much as 10°C, following the large-scale transference of warm water from the W Pacific Ocean eastwards (rather than the normal westward flow, via the South Equatorial Current). In effect, the cold Peru Current becomes temporarily replaced by a warm ocean current. One direct effect is a major reduction in marine plankton, which thrive in cold water, and a depletion in fish numbers (e.g. anchovies in Peruvian waters). More indirectly, the tropical atmospheric circulation is seriously disturbed, resulting in major climatic anomalies (for example, during 1997/98 there were widespread fires in Indonesia, floods and landslides in Peru and a serious drought in Papua New Guinea). El Niño 'events' are thought to occur at intervals of about seven or eight years. [*f*]

elbow of capture See RIVER CAPTURE.

electricity Power generated in dynamos, in which energy is derived from either hydraulic turbines, as in the case of *hydro-electricity* (see HYDRO-ELECTRIC POWER), or steam turbines, as in the case of *thermal electricity*. For the latter, heat is obtained from: coal, especially substandard varieties such as lignite and brown coal; peat, as in Ireland; *nuclear power*; *geothermal power* from subterranean sources. Tidal (see BARRAGE) and WIND POWER have been harnessed in a few areas.

eluviation In general usage, the washing-down of soil constituents through a SOIL PROFILE by the process of LEACHING (involving the dissolving of nutrients) and MECHANICAL DOWNWASH (involving solids such as HUMUS or CLAYS). Eluviation is promoted by high precipitation and acidic vegetation, and is associated with the formation of PODSOLS. Materials may be re-deposited to form an illuviated lower horizon (see ILLUVIATION).

emergent coast A coastline that has risen, or is still rising, relative to sea-level. Emergence can be due to vertical uplift of the land – for example, as a result of isostatic recovery (see ISOSTAYSY) following the melting of a large ICE SHEET, or an actual fall in sea-level, e.g. at the onset of a glacial advance (see EUSTATIC). Typical features of an emergent coast are abandoned cliff-lines, RAISED BEACHES, and extensive coastal plains representing the former sea-floor.

emigration The act of leaving an area and settling in another. The term is most commonly used when referring to the movement of people across national frontiers. Ct IMMIGRATION.

endangered species An increasing number of plant and animal species in danger of extinction as a direct result of human activities, e.g. by the destruction of natural HABITATS through clearance and drainage, the excessive commercial exploitation of NATURAL RESOURCES, etc. More than 1000 animal species are listed in the *Red Book* produced by the *International Union for Conservation of Nature and Nature Reserves*, while over 25,000 species of plant are on the endangered list. This important issue, with its resulting loss of BIODIVERSITY, requires a much greater responsibility on the part of governments and TNCs. It also needs internationally agreed CONSERVATION programmes and a fundamental reappraisal of the relationship between the human species and the natural world.

endemic species A species of plant or animal that is restricted to a particular locality (commonly an island) due to a unique combination of factors (climate, soil) and geographical isolation.

end-moraine See TERMINAL MORAINE.

endogenetic Those factors and processes in landform development arising from *within* the Earth (ct EXOGENETIC). For example, endogenetic factors affecting slope formation include rock type, geological structure (e.g. JOINTS, BEDDING PLANES, and angles of DIP) and GROUNDWATER hydrology.

Landforms are the product of a complex interplay between endogenetic and exogenetic processes.

energy The capacity of a material or of radiation to do work. *Energy resources* are generally thought to encompass ELECTRICITY, gas, steam and nuclear power, together with fuels such as coal, oil and timber. See also RESOURCE.

energy crises The so-called *oil crises* date from 1973 and 1979–80 when the Libyan government, followed by the rest of OPEC, challenged the world oil-production and oil-pricing policies, which until then had been laid down by the international oil companies. In many Middle Eastern countries, the assets and operations of those companies were confiscated and nationalized and, although oil was never in short supply, those countries cleverly engineered the situation in such a way as to cause the price of crude oil to rise in a spectacular manner. Indeed, between 1971 and 1981 the world price of oil increased by nearly 20 times, and oil became an influential weapon in the politics of international relationships. The effects of the crisis were most keenly felt by those countries dependent on imported oil supplies, particularly those THIRD WORLD countries with few RESOURCES to offer in exchange. It also badly affected Japan, which derived over 75% of its total energy requirement from Middle Eastern oil. It encouraged the international oil companies to undertake exploration for oil in those parts of the world deemed to be politically stable (i.e. free from the risk of confiscation), but that are environmentally difficult (e.g. Alaska, the North Sea). It reawakened interest in the more traditional fuels, such as coal, as well as prompting R&D into possible ALTERNATIVE ENERGY based on renewable resources (e.g. wind, wave, tidal and solar power). Furthermore, it persuaded a number of countries to embark on industrial restructuring.

Engels' law A generalization that the proportion of income spent on food tends to decline as income increases.

englacial Contained within a glacier or ICE SHEET. *Englacial debris* is derived from rock fragments that have fallen into crevasses or, more usually, on to the glacier surface in the ACCUMULATION ZONE where, owing to subsequent burial by winter snow, they become part of the glacier's internal structure. Near the glacier snout, some englacial debris is revealed again at the surface as the ice is lowered by ABLATION.

enterprise (i) A business concern or FIRM. (ii) A readiness or ability to take an initiative or a risk in the context of business; recognized by some as a FACTOR OF PRODUCTION.

enterprise culture A belief in the benefits to society of profit-making and the setting up of small businesses (see CAPITALISM). It was an integral part of *Thatcherism.*

enterprise zone One of a number of planning measures introduced in the UK in the 1980s and 1990s in an attempt to arrest INNER-CITY DECLINE and to stimulate the economic regeneration of inner city and derelict areas by public-sector investment in manufacturing and commerce. Nearly 30 zones were established, including some well-known sites such as London Docklands, Salford and the Lower Swansea Valley. In 1989 the government decided to end the scheme, since the zones had done little to increase employment.

entrepôt See FREEPORT.

entrepreneur One who initiates and undertakes an enterprise or business; a business organizer whose aim is profit. The main responsibilities are risk-bearing, deciding what goods or services to produce and at what scale, and marketing those goods or services. *Entrepreneurial skill* is a crucial quality in the successful operation of the capitalist economy (see CAPITALISM).

environment The surroundings within which people, animals and plants live. It is a complex thing with many interacting factors. It is also dynamic and constantly changing.

environmental hazard See NATURAL (ENVIRONMENTAL) HAZARD.

environmental impact assessment See EIA.

environmental lapse-rate The actual decrease of temperature with altitude in the earth's ATMOSPHERE above a particular place at a specific time. The environmental lapse-rate (ELR) has a mean value of 0.6°C per 100 m. It changes with the passage of different air masses. For example, polar maritime air warmed from below in its passage southwards tends to have a 'steep' ELR over Great Britain, while tropical maritime air cooled from below as it moves northwards has a 'gentle' ELR. ELR also varies diurnally. During the daytime, as the ground is heated by solar radiation, it steepens, thereby increasing the likelihood of atmospheric instability. At night, it is reduced with the cooling of the ground and the overlying air, thus favouring stability. While temperature generally decreases with altitude, there may be occasions where it remains constant (isothermal lapse-rate) or actually increases to form a temperature inversion.
[*f* CONDITIONAL INSTABILITY]

environmental perception The way in which an individual or group of people regards its ENVIRONMENT.

environmental pollution The disturbance of the natural environment resulting directly and indirectly from human activities. The causes or sources of pollution are many, ranging from the atmospheric emissions associated with the burning of fossil fuels to the application of FERTILIZERS and pesticides, from the discharge of industrial and domestic effluent into rivers and other water bodies to the generation of intolerable noise levels. The outcomes of pollution are equally diverse, ranging from the creation of ACID RAIN to rising levels of mercury in the sea, from the breaking or disruption of FOOD CHAINS to the hazarding of human health. In recent times, there has been a growing public awareness in many countries of the true scale of environmental pollution, and programmes of environmental protection are becoming commonplace although they are not always easy to police. See ACCEPTABLE DOSE LIMIT, CONSERVATION.

environmentally sensitive area. See ESA.

ephemeral stream (river) A stream or river that flows intermittently or seasonally. A feature of DESERTS and, although short-lived, can have very high discharges.

epicentre The point on the surface of the Earth lying immediately above the focus of an EARTHQUAKE. The latter usually occurs at a depth of 0–50 km; however, 'deep focus' quakes, with a depth of origin greater than 250 km, have been identified. The world distribution of epicentres shows a marked concentration around the margins of the Pacific Ocean, along the centre-line of the Atlantic, and through the Mediterranean basin into Turkey, Iran and beyond. In other words, along plate margins (see PLATE TECTONICS). Generally, the greatest destruction associated with an earthquake occurs close to the epicentre. There are, however, exceptions. In the 1985 earthquake that damaged large parts of Mexico City and killed 7000 people, the epicentre was hundreds of kilometres away. The disaster occurred because the city had been built on unconsolidated lake bed sediments that became unstable when affected by the earthquake's shock waves.
[*f* EARTHQUAKE]

epidemiology The study of the spread of diseases; an important field of MEDICAL GEOGRAPHY and involving some of the concepts and ideas of SPATIAL DIFFUSION.

epiphyte See LIANA.

EPZ (export-processing zone) A relatively small, clearly defined area within a country in which the aim is to attract export-oriented industries by special concessions such as: exemption from taxes and duties, particularly with regard to the import of raw materials and parts used in the production of goods for export; provision of the necessary physical infrastructure; no restrictions on foreign ownership. There are now over 200 EPZs in operation, most of them set up after 1971. Notable clusters occur in Central America and the Caribbean; Singapore is in effect one large export-processing zone, as was Hong Kong before it was returned to China.

equatorial forest See RAINFOREST.

equilibrium A condition of balance which, when established, tends to perpetuate itself unless controlling conditions change markedly (see GENERAL SYSTEMS THEORY). The balance may involve only *materials*, as on a BEACH where the deposition of SEDIMENTS by CONSTRUCTIVE WAVES and LONGSHORE DRIFT equals the removal of sediments by DESTRUCTIVE WAVES and longshore drift, so that the beach size is maintained over a period of time. However, ENERGY is also involved. For example, a river in a state of equilibrium (see GRADE) will adjust its slope to provide exactly the velocity, and thus energy, needed to transport the sediment LOAD. The relationship between equilibrium and form can be demonstrated in slope study. *Equilibrium slopes* are those in which the steepness is such that TRANSPORTATION processes are able to remove weathered material as rapidly as it is produced. See also DYNAMIC EQUILIBRIUM.

equilibrium (firn) line The boundary line between the ACCUMULATION ZONE of a glacier and its ABLATION ZONE. It marks the level at which, over the year, there is no net increase or decrease in the mass of the ice. In other words, at the equilibrium line annual accumulation exactly equals annual melting. The position of the equilibrium line on a particular glacier may vary over a period of years. After several seasons of heavy winter snowfall, giving a 'positive mass budget', it will become lower in altitude. By contrast, a series of warm summers, increasing ABLATION and resulting in a 'negative mass budget', will lead to an increase in the *equilibrium line altitude* (ELA). At present, the ELA of most glaciers in the Alps lies at approximately 2900 m above sea-level. [*f* MASS BALANCE]

equilibrium price See SUPPLY AND DEMAND CURVES.

equinox The time of year when day and night are of approximately the same duration throughout the Earth. This occurs on two dates (21 March and 22 September), when the Sun is directly overhead at midday at the Equator. In the Northern Hemisphere, 21 March is known as the spring, or vernal, equinox, and 22 September as the autumnal equinox. See SOLSTICE.

erg A very large area of SAND DUNES specifically within the Sahara desert. Ergs have in the past been regarded as the product of wind TRANSPORT and DEPOSITION on a massive scale. However, the Saharan ergs occupy low-lying basins, surrounded by REG and HAMMADA from which sand has been removed by running water (stream floods and SHEET FLOODS) during more humid conditions than at present, and deposited in the basins. The dunes of the ergs represent the reworking *in situ* of ALLUVIUM since the Sahara became more arid.

erosion The sculpturing action of, for example, running river water, sliding glacial ice, breaking waves, and wind armed with rock fragments (see ABRASION and CORRASION). Erosion is thus largely a physical process and involves the removal or transportation of material (usually rock) – it is this element of transportation that distinguishes it from the process of WEATHERING. However, in some circumstances *chemical erosion* can occur, as on LIMESTONE coasts and in limestone streams.

erosion surface An extensive near-level surface formed by erosive processes acting over a long period of time (see PLANATION, PENEPLAIN, PEDIPLAIN). Erosion surfaces are sometimes well preserved, and constitute striking elements in the landscape, as in the great pediplains eroded across the ancient pre-Cambrian rocks of parts of tropical Africa.

erratic A rock fragment that has been transported by a glacier or ICE SHEET, and deposited (sometimes far from its source) in an area of unrelated geology.

eruption The extrusion or emission of solid, liquid or gaseous materials from a volcanic vent. This may be a CENTRAL ERUPTION or a FISSURE ERUPTION, and may take the form of outflows of LAVA or explosive activity, involving pyroclastic material (see PYROCLASTIC FLOW) and possibly ignited gases (see NUÉE ARDENT).

escarpment A steep slope at the margins of an upland (for example, a PLATEAU edge or a CUESTA). The term is often abbreviated to *scarp* (hence *scarp face* or *scarp slope*).

[*f* CUESTA]

ESA (environmentally sensitive area) An area of fragile or scarce HABITAT within the EU that has been designated for special protection. British examples include the Avon Valley in S England, N Kent and the Brecklands of East Anglia.

esker A sinuous (winding) ridge of SILT, SAND and GRAVEL, laid down by meltwater in a SUBGLACIAL tunnel oriented approximately at right angles to the ice front. An esker represents the former course of a meandering meltwater stream beneath the ice. Esker ridges are sometimes small, but, in the case of former continental ICE SHEETS, may be very large features, 1 km or more in width and several tens of metres in height. Such large eskers usually comprise several individual ANASTOMOSING ridges, and extend for hundreds of km across the landscape, as in parts of Finland. Esker sediments display the bedding characteristics of GLACIO-FLUVIAL formations, comprising stratified deposits of sand and gravel.

essential services See PUBLIC UTILITIES.

estuary The mouth of a river, where the channel broadens out into the sea and in which the TIDE flows and ebbs (e.g. the Thames estuary and the Severn estuary). Most estuaries represent the lower parts of former river valleys that have been drowned by the POSTGLACIAL rise of sea-level.

ethnic cleansing A term coined during the break-up of what until 1990 was Yugoslavia. It refers to attempts made by the three main ethnic groups in the population – the Serbs, Croats and Moslems – to untangle their respective populations in areas where they formerly intermingled in varying proportions (most notably in Bosnia-Hercegovina and Kosovo). Besides seeking territorial gain, the aim has been to redraw the political map in terms of homogeneous ethnic areas (i.e. areas that have been 'cleansed' of minorities). The whole process has been undertaken with great force, immense destruction, considerable bloodshed and a blatant disregard for human rights. Huge numbers of people have become homeless refugees and have been forced to suffer the most appalling DEPRIVATION.

ethnic group A group of people united by a common characteristic or set of characteristics related to race, nationality, language, religion or some other aspects of culture. The term also often implies that the group constitutes a MINORITY element in some larger population, as for example the Asian and West Indian groups in Britain. In many instances, the distinctiveness of an ethnic group is reinforced by other secondary characteristics, such as their general social status, occupations, affluence or poverty, and their residential concentration in particular areas. The acquisition of such secondary traits tends to inhibit the ASSIMILATION process and to exacerbate concentration rather than dispersal. See GHETTO.

EU (European Union) Originally referred to as the *European Common Market* or *European Economic Community* and later the *European Community*. It was formed in 1958, following agreements embodied in the Treaty of Rome (1957). Its membership today comprises Austria, Belgium, Denmark, Eire, Finland, France, Germany, Greece, Italy, Luxembourg, The Netherlands, Portugal, Spain, Sweden and the UK. The principal aims include: to secure freedom of movement for persons, goods and capital between member countries; to create a customs union with common external TARIFFS; to make available development AID for THIRD WORLD countries and for LAGGING REGIONS within the community; to formulate a CAP (Common Agricultural Policy); to establish a zone of monetary stability. At present, there are other countries interested in joining. These include Cyprus, Turkey and former communist countries in eastern Europe. It is the wish of some present members that economic union will lead eventually to political union – a 'United States of Europe'.

eustasy The theory that worldwide changes

of sea-level (*eustatic sea-level changes*) can occur, owing to an addition to, or a reduction of, the amount of water in the oceans. In the recent geological past, most eustatic changes have been connected with glaciation. During a glacial period, water on land is stored as snow and ice, and less water flows into the oceans, causing a fall in the ocean level. When the climate warms up again, the snow and ice melts, and ocean levels rise. The effect of these oceanic changes in the sea-level is complicated by *isostatic recovery*, which is the response of the land to changes in glacial mass.

eutrophication The process of nutrient enrichment that ultimately leads to the reduction of oxygen in streams and lakes, and the consequent death of fish, molluscs and other species. This impacts on the FOOD CHAIN by starving waterfowl (swans, coots and ducks). It is primarily caused by the excessive leaching of nutrients (e.g. from agricultural FERTILIZERS or slurry), which promotes rapid algae growth ('bloom') and results in the water becoming starved of oxygen. *Eutrophic lakes* are common in low-lying, gently undulating agricultural landscapes.

evapotranspiration The loss of moisture at the Earth's surface by direct evaporation from water bodies and the SOIL plus TRANSPIRATION from growing plants. Evapotranspiration cannot be measured directly, but can be derived indirectly from the *moisture balance equation*: precipitation = run-off + evapotranspiration + changes in soil moisture storage. *Potential evapotranspiration* is the maximum possible that can occur from a soil that is kept continually moist by IRRIGATION. It can be measured directly, using an 'evapotranspirometer', which records PERCOLATION; hence potential evapotranspiration is derived from PRECIPITATION minus percolation. Evapotranspiration is a loss, or output, from the DRAINAGE BASIN HYDROLOGICAL CYCLE.

exclusive economic zone See LAW OF THE SEA, TERRITORIAL WATERS.

exfoliation An effect of WEATHERING in which layers or sheets of rock peel away from an exposed rock surface. It is especially characteristic of INSELBERGS in the tropics, where the landform is shaped by the successive splitting of curvilinear sheets of rock.

exhumation The uncovering of surfaces buried beneath REGOLITH or younger overlying deposits. For example, granite TORS are thought to have been exposed when the overlying regolith was removed by fluvial action and MASS MOVEMENT.

exogenetic Those factors and processes in landform development arising from outside the Earth (ct ENDOGENETIC). For example, exogenetic factors affecting slope formation are dominantly climatic (temperature, PRECIPITATION) or influenced by climate (vegetation, SOILS). These factors, in conjunction with rock type and structure (the principal endogenetic factor), influence slope-shaping processes such as WEATHERING, surface wash, seepages, THROUGHFLOW and various kinds of mass TRANSPORT of weathered debris.

expanded town Specifically, a British town enlarged under the provisions of either the Town Development Act (1952) or the Housing and Town Development Act, Scotland (1957), principally for the purpose of accommodating OVERSPILL from some large city. One of the earliest schemes was the expansion of Swindon, initiated in 1954 (when its population was 69,000) and involving the movement of some 10,000 Londoners and an unspecified number of employers. Cf NEW TOWN.

expansion diffusion One of two broad types of SPATIAL DIFFUSION (ct RELOCATION

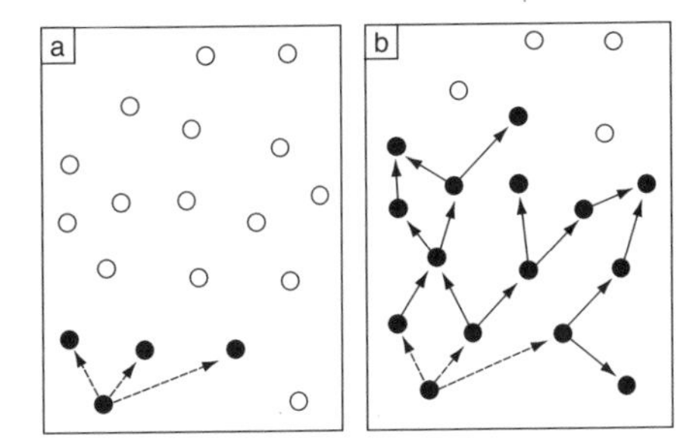

Expansion diffusion: (a) the initial stage and (b) a later stage

DIFFUSION). Here the thing being diffused remains, and is often intensified, in the originating region. For example, the communication of an idea by one person to others who do not know about it and who, in turn, pass it on means that the total number of 'knowers' increases through time as well as in space, and the rate of diffusion tends to accelerate. The spread of a new farming technique through an agricultural area would be likely to involve expansion diffusion. Cf HIERARCHICAL DIFFUSION, CONTAGIOUS DIFFUSION. [*f*]

exploitation cycle A generalized sequence that depicts the exploitation of a *non-renewable resource.* Following discovery, the resource undergoes an increasing rate of exploitation; this is encouraged as the discovery of new RESERVES moves ahead of demand. Following peak production, the rate of discovery of new reserves begins to fall and proven reserves become progressively depleted. Although consumption declines as supply diminishes, exploitation eventually reaches the point where the resource becomes exhausted. Also referred to as the *discovery-depletion cycle.* [*f*]

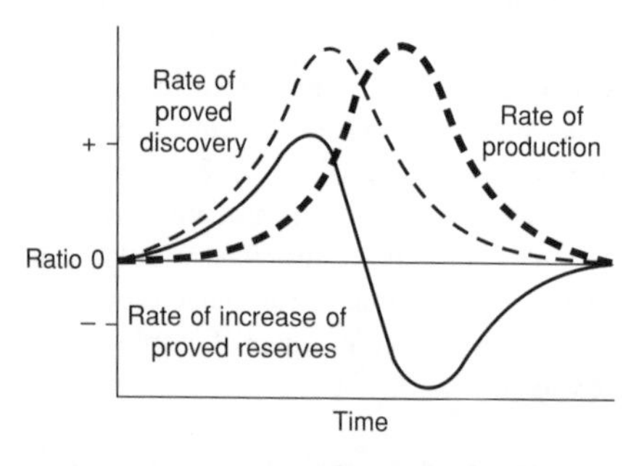

Rates of change in the discovery, production and reserves of a non-renewable resource

exponential growth rate A growth rate that is constant over time and in which the number increases as a constant proportion of the number at a previous time; i.e. growth that is geometric rather than arithmetic. When plotted on normal arithmetic graph paper, an exponential rate of growth will appear as a rising curve, but on semi-logarithmic paper it will appear as a straight line. Cf LOGNORMAL DISTRIBUTION.

export-processing zone See EPZ.

export quota See QUOTA.

exports Items transported out of a country for sale abroad as part of its TRADE (ct IMPORTS). These are *visible exports.* Also part of exports are *invisible* earnings, which comprise payments received for services, transport and loan interest, together with revenue received from foreign investments and tourism. [*f* BALANCE OF TRADE]

export substitution The process whereby a FIRM or country produces goods and services in the consuming country rather than exporting them. Much of Japanese OFFSHORE INVESTMENT in branch plants has been to do with export substitution. The reason has been to reduce TRADE FRICTION with those countries where the BALANCE OF TRADE tips heavily in favour of Japan (e.g. in the UK and USA). A product made by a Japanese firm in the consuming country does not count as a Japanese export. Ct IMPORT SUBSTITUTION.

extended family A group of people living close to one another and comprising not only the *nuclear family* of parents and children, but also blood relatives and relatives by marriage. A social characteristic more commonly encountered in RURAL societies and in the THIRD WORLD.

extending flow A type of glacier flow, in which there is a stretching and thinning of glacial ice, often in response to an increased bedrock gradient and subsequent increase in glacial velocity. As the ice becomes thinner, so there is a reduction in the erosive potential of the glacier. It is the opposite of COMPRESSING FLOW, which involves a thickening of the ice often in response to a decrease in bedrock gradient. Extending flow is responsible for the development of transverse tensional CREVASSES, typically associated with breaks of slope, e.g. where a glacier descends from a CIRQUE basin over a rock lip into the valley below.

extensification This occurs where the inputs of an economic activity are reduced. In the majority of cases, but not all, this might be expected to lead to a fall in output. It is currently being advocated in the context of agriculture in the EU (see CAP). The need to reduce food surpluses is one powerful argument in favour of extensification (see SETASIDE). Another is that a reduced application of FERTILIZERS and pesticides would benefit CONSERVATION of the environment. Ct INTENSIFICATION.

extensive agriculture A type of AGRICULTURE characterized by relatively low levels of INPUT and OUTPUT per unit area of farmland. Farming in relatively large holdings, usually highly mechanized, employing little labour, with relatively low yields per unit area, but with large total yield and a high yield per worker; e.g. as practised in the wheatlands of N America. Extensive agriculture tends to be associated with relatively low rural population densities (in part a reflection of farm size and mechanization) and to be carried out in areas relatively distant from markets. Although showing some different characteristics, extensive agriculture is also practised in the THIRD WORLD (e.g. SHIFTING CULTIVATION).

external diseconomies See DISECONOMIES OF SCALE.

external economies Cost advantages or potential savings obtained from sources outside the individual firm (see AGGLOMERATION ECONOMIES, LOCALIZATION ECONOMIES, COMMUNICATIONS ECONOMIES). They may be derived in a variety of ways, as for example from: the local availability of workers having skills relevant to the activity in question; R&D facilities; ancillary activities providing equipment and specialized services; well-developed LINKAGES. As such, the economies help to reduce *operating costs* and are most commonly reaped where there is a localized concentration of firms involved in the same field of activity. See also ECONOMIES OF SCALE, REGIONAL SPECIALIZATION.

external economies of scale See ECONOMIES OF SCALE.

externalities The impact of one person's (or organization's) actions on another, and over which the person (or organization) affected has no direct control. Such externalities can be either *positive*, where the repercussions are beneficial, or *negative*, where the impact is adverse and creates costs. For example, the construction of a bypass will create positive externalities for the residents of the bypassed village in that they will benefit from the reduction in through traffic. This environmental improvement, in its turn, might be mirrored in an increase in house prices as the perception of the village as a residential location becomes more favourable. Equally, that same bypass might create negative externalities for shopkeepers, who suffer as a result of the reduction in the number of potential customers passing through the village.

extractive industry The exploitation of non-renewable RESOURCES, e.g. mining and quarrying. Some would extend the definition to include the exploitation of certain *renewable resources*, e.g. fishing and forestry.

extrapolation Used in forecasting and based on the principle that the recent past is a good guide to the near future. Once a trend has been established on the basis of past values, it can be extended into the future either mathematically or by eye.

extrusive (volcanic) rock A type of IGNEOUS ROCK formed from the extrusion of LAVA on to the Earth's surface. Extrusive rock is cooled very rapidly (by contrast with INTRUSIVE ROCK), and is therefore fine-grained (as in BASALT) or even glassy (as in obsidian).

eye of a storm The central area of a tropical CYCLONE (hurricane) characterized by descending air, calm conditions and little rainfall.

fabric effect The way in which elements of the TOWNSCAPE inherited from the past affect present-day LAND USE. For example, old buildings may not lend themselves to new uses, and thus their conservation will constrain land-use change (see URBAN CONSERVATION). Similarly, the preservation of old and narrow street systems may impede the movement of traffic.

factor A cause or control in the sense of one VARIABLE contributing to variations in another set of observations – for example, in climate, factors include latitude, altitude, distribution of land and sea, ocean currents, influence of RELIEF barriers, etc.

factor analysis A statistical procedure that measures the apparent interrelationships (*covariance*) of selected VARIABLES. In geographical research such variables are usually recorded either at a series of different locations (e.g. investigating the characteristics of a set of SOIL samples taken at different places) or over a network of sub-areas (e.g. investigating the correlation between POPULATION DENSITY, SOCIO-ECONOMIC STATUS and housing conditions in the wards of a town).

factor cost What the producer receives for the sale of a product. This is not synonymous with *market price*, but refers to the net amount received after indirect taxes or similar charges have been paid.

factors of production The basic elements of the production process; those things that are necessary before production can begin. They are generally regarded as being threefold: LABOUR, land and CAPITAL. Some would recognize ENTERPRISE as a fourth factor. Labour requirements vary according to the type of production, while land is fundamental in a variety of ways, e.g. as a source of RAW MATERIALS, providing industrial sites, etc. See also CAPITAL.

factory farming A system of livestock farming in which animals are reared intensively under cover and in conditions of restricted mobility. Poultry (see BATTERY FARMING), pigs and calves are most often reared in this way. Factory farming is aimed at increasing output, reducing costs and producing a standardized food product, most often sold through supermarket chains. Cf AGRIBUSINESS.

factory system The system of employment and production established during the Industrial Revolution when, for the first time, work on a large scale was conducted under supervision in factories. Ct COTTAGE INDUSTRY.

falling limb See HYDROGRAPH.

family life cycle See LIFE CYCLE.

family planning See BIRTH CONTROL.

famine A scarcity of food, leading to MALNUTRITION and STARVATION, provoked primarily by some failure in the food production process, such as: by flooding (as in Bangladesh); by persistent DROUGHT (as in the Sahel); by the ravages of diseases and pests (such as locusts); by political unrest (as in Somalia). Overpopulated areas tend to be especially vulnerable to such failures. Famine may also be precipitated by insufficiently comprehensive AID programmes, which may be effective in reducing levels of mortality, but are found wanting when it comes to producing the extra food required by the resulting increase in population. See MALTHUS' THEORY OF POPULATION GROWTH.

FAO See UNO.

farm consolidation A vital part of agricultural reform aimed at overcoming the inefficiency and diseconomies of FARM FRAGMENTATION. It involves the amalgamation and regrouping of both fields and farm holdings into generally larger units.

farm diversification A scheme introduced in the UK in 1988 with a view to reducing agricultural surpluses (see CAP), maintaining farm income and reducing rural DEPOPULATION. Farmers receive grants for developing alternative sources of income from such things as golf courses, riding and visitor centres, farm shops and PYO crops.

farm fragmentation This occurs where the fields of an agricultural holding are not contiguous and so do not form a single continuous unit. Fragmentation may have a variety of causes, such as equal-inheritance practices, the commercial consolidation of non-contiguous farms, the enclosure of strips in former open-field systems, piecemeal land reclamation, etc. It is generally regarded as being contrary to farming efficiency.

fault A fracture in a rock, induced by either tensional or compressive forces, in which there is displacement along a fault-line. A fault can be distinguished from a JOINT in that, with the latter, no displacement occurs. The side of the fault that has experienced a relative fall is called the *downthrow side*; the other side is termed the *upthrow side*. Among the main types of fault are: *normal faults*, developed by tension and involving lowering of the rocks on the side towards which the fault-plane is dipping; *reversed faults*, resulting from compression, which has the effect of raising the rocks on the side towards which the fault-plane is dipping; *tear faults* (otherwise known as *transform faults*), in which the movement of the rocks on either side of a near-vertical fault-plane is almost entirely horizontal and parallel to the fault. Faulting is directly responsible for the production of certain landforms (see FAULT SCARP and RIFT VALLEY), and also forms lines of weakness that can, for example, be exploited by erosional processes to form valleys. [*f*]

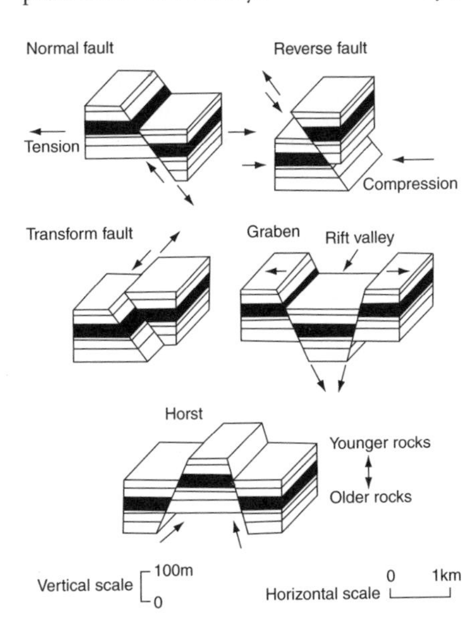

Types of fault structure

fault block An area bounded by FAULTS that has either been upraised (see BLOCK MOUNTAIN and HORST) or lowered (see GRABEN) in relation to surrounding areas. [*f* FAULT]

fault-line scarp A steep slope, coincident with a fault-line but not actually the product of the original faulting movement. One important result of faulting is to bring unresistant rocks against resistant rocks. Where the former are lowered or removed altogether by EROSION a fault-line scarp is formed. [*f*]

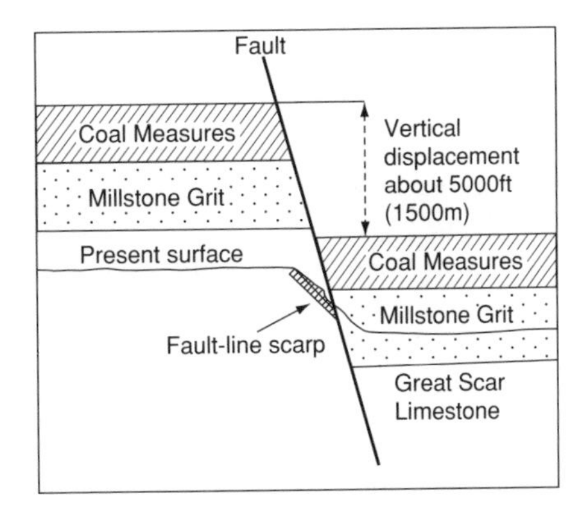

Giggleswick fault-line scarp, North Yorkshire

fault scarp The most basic landform produced by normal faulting. It is a steep slope coinciding with the line of a FAULT, and equal in height to the throw of the fault. The development of a fault scarp will also affect drainage development. For example, HANGING VALLEYS will be formed where streams downcutting on the upthrow side cannot keep pace with downfaulting; waterfalls may also occur. [*f* FAULT]

feedback See NEGATIVE FEEDBACK, POSITIVE FEEDBACK.

felsenmeer A 'sea of stones' or *block field*, comprising large numbers of angular BOULDERS produced by intense FREEZE-THAW WEATHERING acting on JOINTS in the rock. They are characteristic of many present-day PERIGLACIAL regions and also occur on high PLATEAU surfaces and mountain tops in temperate regions.

ferralitic soil See LATOSOL.

Ferrel cell This mid-latitude circulation cell involves the surface movement of air towards the Poles from the sub-tropical high pressure zone at approximately 30°N and S. The poleward limb of the cell converges with polar air travelling Equatorwards at a boundary called the POLAR FRONT (approximately 60°N and S) where it is forced to rise. Compensatory flow at altitude towards the Equator completes the Ferrel cell. In recent years, the Ferrel cell has been regarded as being too simple a model for the highly complex atmospheric circulation in the middle latitudes. Other models have been proposed. See ROSSBY WAVES. [*f* HADLEY CELL]

fertility ratio The number of young children in the population of an area related to the number of women of child-bearing age.

$$\text{fertility ratio} = \frac{\text{no. of children under 5 years}}{\text{no. of women aged 15 to 50}} \times 1000$$

The ratio provides a valuable indication of future population trends; the higher the ratio, the greater the expected increase in population.

fertilizer A substance of animal (dung, dried blood, bone meal), vegetable (compost, sea weed) or chemical (artificial) origin, added to the SOIL to ensure the supply of necessary elements for plant growth. The three main requirements are nitrogen, phosphorus and potassium.

fetch The distance of open sea over which wind blows to generate waves. Fetch is a major factor controlling the height and energy of waves, and hence their ability to modify coastal landforms by EROSION, TRANSPORT and DEPOSITION. The direction of maximum fetch, which helps to determine the approach of the DOMINANT WAVES, is particularly important within enclosed or partially enclosed seas. For example, in the North Sea the greatest fetch is towards the N and NE, and the least towards the E and SE. The resultant generation of dominant waves determines LONGSHORE DRIFT of SEDIMENT (N to S in the North Sea) and also the orientation and form of BEACHES and bays.

field capacity The state of the SOIL when all 'gravity' water has been drained away, usually over a period of several days (or even weeks) after the cessation of rainfall. The remaining 'capillary' water, held to individual soil particles by surface tension, is sufficient to provide for the needs of growing plants. However, in a major DROUGHT this will be reduced by EVAPOTRANSPIRATION to levels well below field capacity, and plant wilting will occur unless IRRIGATION is practised.

fieldwork The act of leaving an office or classroom and, at first hand, observing or collecting information about a particular area, group of people, issue or problem. Analysis and the drawing of conclusions are likely to take place back at base.

fill-in migration When a person moves from a town to a large city as part of a MIGRATION CHAIN, a sort of vacuum is created in the town. This draws in, and is filled by, a person moving from the surrounding rural area. Cf STEPWISE MIGRATION.

filtering The process by which ageing housing is thought to become occupied by lower groups in the social HIERARCHY. It is assumed that housing deteriorates with age and that, the poorer the social group, the less it is able to demand adequate housing standards. In this way, and with the passage of time, housing is seen as being handed or filtered *downwards*, while social groups, because they come to occupy property vacated by a higher group, may be thought of as filtering *upwards*. A basic flaw in the concept is that it implies that only the rich build new houses. Clearly this is not so, as the new housing market caters for a wide spectrum of social classes. A further weakness lies in the fact that, in many cities, the rich often move into older property in areas deemed to be 'good' INNER-CITY locations. See GENTRIFICATION.

finger (ribbon) lake An elongated lake occupying a formerly glaciated valley. Finger lakes are characteristic of glacial troughs in mountain regions (e.g. the English Lake District), where selective overdeepening has produced basins in solid rock that have become flooded in the POSTGLACIAL period.

finite Used in the context of *non-renewable*

RESOURCES to emphasize the fact that, once they are used, they are gone and cannot be replaced.

firm A unit of management operating under a trade name and organized to engage in economic activity (be it mining, manufacturing, or selling goods and services). A firm may be a sole proprietorship, a private or limited liability company or a state-owned enterprise.

firn Literally meaning 'last year's snow', it is snow that has lain for a year without melting. Firn represents the balance of the winter snowfall on the ACCUMULATION ZONE of a glacier after the removal of the uppermost layers by summer ABLATION. The remaining snow then becomes buried and compressed by the next winter's snowfall. In this way the glacier is built up by successive increments of firn, which undergo gradual modification, involving recrystallization and the exclusion of contained air bubbles. Thus the density of the firn increases, the ice crystals grow larger, and eventually it is transformed into true glacier ice. The rate of change varies with local conditions (e.g. the thickness of winter snowfall, the amount of thawing and refreezing during the summer season, prevailing temperatures, etc.) and is very much more rapid in temperate than polar glaciers. The firn line is the lowermost limit of firn (see EQUILIBRIUM (FIRN) LINE) on a glacier and represents the boundary between the accumulation zone and the ABLATION ZONE.

First World This comprises the so-called 'advanced', 'industrialized' and 'developed' countries of N America, W Europe, Japan, Australia and New Zealand. They are characterized by MARKET ECONOMIES. As with other global terms (e.g. the NORTH, MEDC, the DEVELOPED WORLD), there are problems posed by certain groups of countries. For example, are the NICS and the E European states that were formerly part of the SECOND WORLD to be included or not? Like the others, First World is at best a useful but crude shorthand term. Ct THIRD WORLD.

fiscal policy That part of a government's policy to do with raising revenue (through taxes) and levels of public spending (as on education and health).

fish farming See AQUACULTURE.

fissure eruption A linear volcanic ERUPTION in which fluid LAVA emerges in large quantities and with little explosive activity, along extensive lines of crustal weakness – FAULTS, fractures or major JOINT lines. Over a period of time successive lava flows build up, sometimes to thicknesses of hundreds of metres, and 'inundate' the pre-existing RELIEF features, such as valleys and INTERFLUVES. The latter are replaced by extensive PLATEAUS (usually comprising BASALT).

fixed costs (i) Costs that do not vary with the volume of production, i.e. costs not affected by ECONOMIES OF SCALE (e.g. *overhead costs* such as rent, investment in plant and machinery). (ii) Costs that are constant in space and therefore have no influence on COMPARATIVE ADVANTAGE (e.g. financial capital or the costs of labour and materials where nationally agreed rates apply). Ct VARIABLE COSTS.

fixed-k hierarchy See K-VALUE.

fjard (fiard) A coastal inlet produced by the 'drowning' of an undulating landscape of shallow valleys and low INTERFLUVES. A typical fjard coastline is that of SE Sweden and SW Finland, where the glaciated rocky lowland has been submerged by the POSTGLACIAL rise of sea-level (though the coast is now emerging quite rapidly, owing to isostatic recovery). Fjards are very subdued features, with shallow depths of water and gentle margins, and contrast strongly with the spectacular FJORDS in which glacial EROSION has been more concentrated and intense.

fjord (fiord) A deeply glaciated valley in a coastal region that has been partially flooded by the sea in POSTGLACIAL times. This inundation may be partly due to a EUSTATIC rise of sea-level, but is mainly the result of Pleistocene glaciers eroding well below the level of the sea even at that time. Fjords are well developed in Norway (whence the term is derived), but equivalent landforms are found in British Columbia, Alaska, Greenland, southern Chile and the South Island of New Zealand. Fjords display all the usual features of glacial troughs (U-shaped cross-section, hanging valleys, etc.) though on an exaggerated scale. See GLACIAL TROUGH; ct FJARD.

flagship development A project that is intended to improve an area's image and prestige or, in a sense, 'to lead the way'. The London Docklands project has proved to be a successful flagship development, but not so the nearby Millennium Dome.

flash flood A short-lived FLOOD, characterized by: a very rapid onset (or 'steeply rising hydrograph'); a brief period of peak flow; a relatively slow decline in river DISCHARGE ('recession period'). Flash floods are particularly associated with DESERTS, where rain is often in the form of short (infrequent) intense showers, the ground surface is 'baked' and IMPERMEABLE and there is little vegetation to impede the flow of water. They are also characteristic of mountainous environments where the steep gradients, possibly combined with DEFORESTATION, transmit water rapidly. In December 1999 the N coast of Venezuela suffered severe flash flooding following torrential rainfall; some 50,000 people were thought to have been killed by the floods and the associated mudslides. Flash floods perform an important role in transporting SEDIMENTS (in suspension and as BED LOAD) over relatively short distances; however, as discharge is reduced and sediment concentration increases there is a rapid change from TRANSPORT to DEPOSITION in a downstream direction.

flint A very hard, dark grey or black concretion found commonly in the uppermost divisions of CHALK. When subjected to hard impacts flints break with conchoidal fractures, to produce extremely sharp edges. As a result flints were widely used in the PALAEOLITHIC and NEOLITHIC PERIODS for the construction of weapons (axe-heads, arrowheads) and tools. When exposed at the surface by long-continued SOLUTION of the chalk, flints become increasingly weathered and, with clay impurities, form the superficial deposit known as CLAY-WITH-FLINTS.

flocculation See DELTA.

flood A period of high DISCHARGE of a river, resulting from conditions such as: heavy PRECIPITATION; intense melting of snow and ice; the breaching of natural barriers (such as ice dams); the collapse of artificial BARRAGES. River floods have been defined as 'events' of such magnitude that the channels cannot accommodate the peak discharge. In other words, a flood is a flow in excess of the channel capacity, and results in inundation of low-lying flat land adjacent to the channel (see FLOOD PLAIN). Floods may occur seasonally (as on the Blue Nile) or at more irregular intervals depending on the occurrence of individual high-intensity rainstorms, such as that producing the catastrophic flood at Lynmouth, N Devon, England, on 18 August 1952).

One important aspect is the prediction of river floods, in order to aid safe engineering construction of bridges, river embankments, etc. and to avoid the siting of houses in areas subject to serious flood hazards. From a study of past records, an attempt is made to determine the *recurrence interval* of floods of particular dimensions. In simple terms, the largest flood that occurred during the past 50-year period is likely to be matched by a corresponding flood during the next 50 years. In reality prediction is likely to be far more complicated for a variety of reasons. The available period of study may not embrace the truly exceptional flood (the '1000-year flood'); but it is conceivable that such a flood could occur at any time in the near future. Another factor is the modification of river CATCHMENTS by people's actions in DEFORESTATION, AGRICULTURE, land drainage, URBANIZATION, etc., which may considerably alter the 'probability' of floods of a particular size.

The term flood is also used in a wider sense to refer to the inundation of land by other than river water – for example, as a result of high lake levels resulting from exceptionally high precipitation, or abnormally high sea-levels, as during the 'storm surges' in E England in 1953.

flood plain That part of a valley floor over which a river spreads during seasonal or short-term FLOODS. During such events the velocity of flow is less than that within the river channel, and in the relatively slack water over the plain suspended SEDIMENT slowly settles out. As a result the flood plain is slowly built up by increments of SAND, SILT and CLAY (see ALLUVIUM). The flood plain is also modified by shifts of the river course (in the development and migration of MEANDERS), and flood plain deposits also comprise material from extended POINT-BARS. By definition, flood plains are areas of gentle RELIEF, giving an impression in the field of almost perfect flatness. However, in detail they are diversified by marshy depressions marking abandoned channels and OXBOWS, and along the river margins localized DEPOSITION may give rise to LEVÉES.

flow chart A diagram in which a sequence of interlinked topics, events or items is presented to show the development or evolution of some theme, objective or product.
[*f* STAGES OF ECONOMIC GROWTH MODEL]

flow-line map A map showing movement, as of freight, passengers, shipping and people. A line indicates the general direction of the routeway concerned, while the quantitative indication of traffic is given by the thickness of that line.

flow production An AUTOMATION system used in manufacturing where stages in the making of a particular product are juxtaposed and linked by some form of conveyor belt. Because it minimizes production time, it is reckoned to be cost-effective.

fluted moraine Linear glacial landforms (oriented parallel to the valley sides) composed of TILL, probably resulting from the moulding of GROUND MORAINE by moving ice.

fluvial A term applied to the processes (EROSION, TRANSPORT, DEPOSITION) and outcomes of rivers and streams.

fluvio-glacial See GLACIO-FLUVIAL.

focus The point of origin of an EARTHQUAKE within the Earth's crust. See also EPICENTRE.
[*f* EARTHQUAKE]

fog A weather phenomenon resulting from CONDENSATION, near the ground, of atmospheric water vapour. In effect, fog can be thought of as 'cloud' at ground level. Although the basic mechanism of fog formation is cooling of the ATMOSPHERE, and a resultant increase in relative humidity towards 100%, several different types of fog are recognized (see ADVECTION FOG, RADIATION FOG and STEAM FOG). The main result of fog formation is the reduction in visibility.

Fog is said to exist when visibility is reduced to less than 1 km (where visibility is impaired but greater than 1 km, this is defined as *mist*). In terms of reduced visibility there are several different categories of fog: *dense fog* (visibility less than 50 m); *thick fog* (less than 200 m); *fog* (less than 500 m); *moderate fog* (less than 1000 m). Fog is often formed in valley bottoms where cold air sinks and collects during calm anticyclonic conditions and where additional moisture is available from rivers. It can be a major road

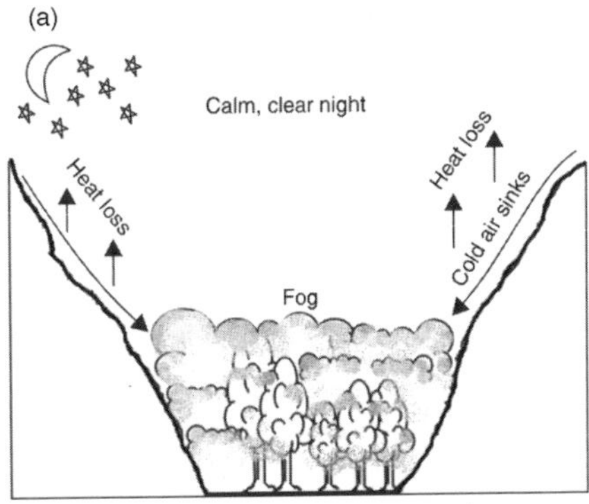

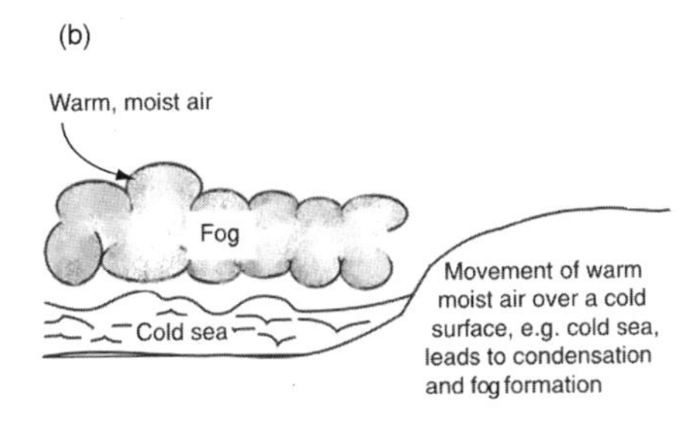

(a) Radiation fog and (b) advection fog

safety hazard particularly where motorways cut across valleys. [*f*]

föhn wind A very warm, dry wind that descends mountain slopes (the term originates from the European Alps, although the phenomenon occurs throughout the world – see CHINOOK), causing rapid melting of snow and triggering AVALANCHES in spring. It is particularly well developed when warm, moist air is drawn towards a mountain range from an area of relative high pressure. The air is cooled, on rising over the mountain barrier, at the SATURATED ADIABATIC LAPSE-RATE, and much of the contained moisture is lost by CONDENSATION and PRECIPITATION. On the leeward side, the drier air sinks and warms at the DRY ADIABATIC LAPSE-RATE. There is thus a net increase in air temperature by as much as 15–20°C as the air crosses a mountain range. [*f*]

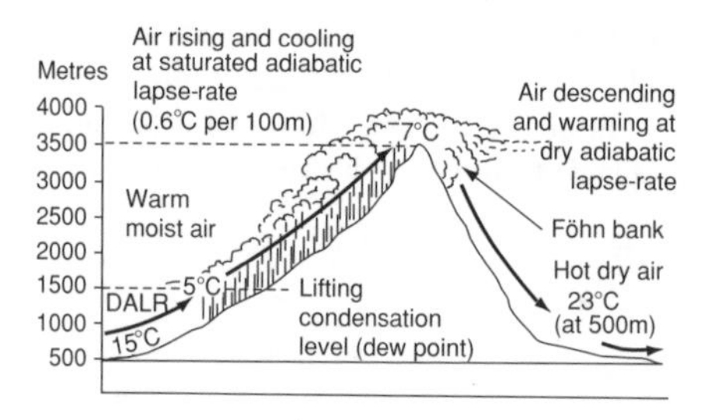

Föhn wind

fold A contortion of rock layers by intense pressures within the Earth's crust (see PLATE TECTONICS). Folding is most commonly associated with convergent plates (destructive margins) where the stresses are compressional. Rocks may be upfolded to form an ANTICLINE or downfolded to form a SYNCLINE. When the pressures are extreme, the rocks may become faulted. [*f* ANTICLINE]

fold mountain A mountain (or more usually range of mountains, e.g. the Alps, Himalayas and Rockies) resulting from the bending and crumpling of stratified SEDIMENTS, by compressive forces in the Earth's crust. It is commonly associated with PLATE TECTONIC action. A period of mountain building is termed an OROGENY. VOLCANOES and EARTHQUAKES are often associated with YOUNG FOLD MOUNTAINS, where they are on an active plate margin. The severe crumpling causes folding and faulting, and the intense heat and pressure generated can lead to the formation of metamorphic rocks. See also OLD FOLD MOUNTAINS.

food chain A series of organisms with interrelated feeding habits, each organism serving as food for the next in the chain. The chain usually begins with organisms that produce vegetal matter, by way of photosynthesis involving sunlight (solar energy). Thus grasses, herbs and shrubs provide food for grazing and browsing animals *(herbivores* such as gazelle, zebra and wildebeest), which are in turn preyed upon by *carnivores* (leopard, lion and hyena). Again, plants and seeds are consumed by small and medium-sized birds (sparrows, pigeons), which are preyed on by birds of prey, e.g. sparrow hawks and peregrine falcons. In effect, there is a continual, but ever-decreasing, transfer of energy through the food chain. See AUTOTROPH, BIOLOGICAL CONTROL, HETEROTROPH, TROPHIC LEVEL. [*f* TROPHIC LEVEL]

food production chain The sequence of linked stages that starts with the agricultural inputs to farm production, and ends with food distribution and consumption. It is the central thread running through the overarching *food supply system.* [*f*]

footloose industries Industries that do not seem either to be tied to any special kind of location or to have any overriding locational requirements. FIRMS involved in the same footloose industry (e.g. light engineering) are frequently to be found in different types of location. It is claimed that the development of grid systems for the distribution of energy and power has tended to make industry as a whole rather more footloose.

foreign (overseas) investment Undertaken by FIRMS as they grow, in order to extend their business interests and profits by increasing production, enlarging markets or diversifying. There are three main forms: (i) setting up branch plants and subsidiary companies (as by the Japanese company Sony in the UK); (ii) entering into joint ventures with foreign companies (as between Rover

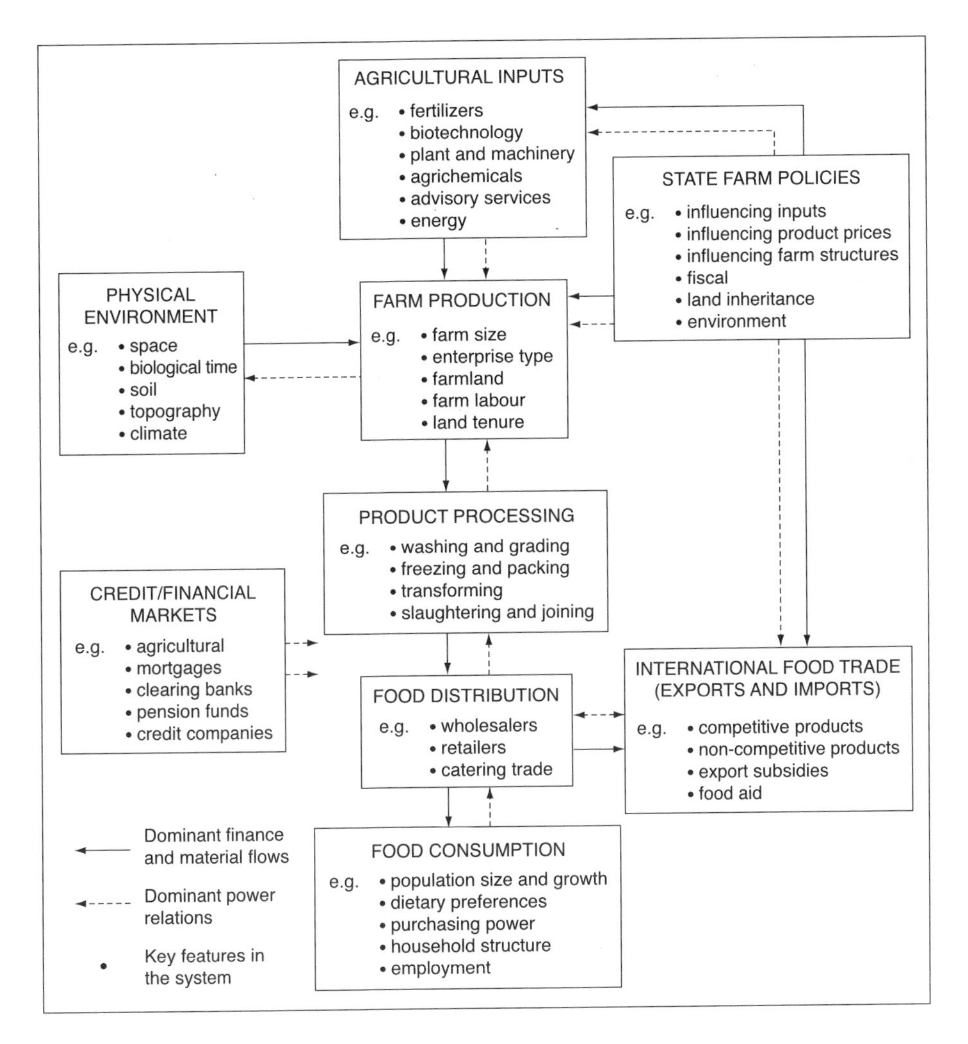

Food production chain

and Honda); (iii) acquiring equity (shares) in foreign companies. *Inward investment* refers to the above as viewed from the 'receiving' country. Foreign investment is very much part of GLOBALIZATION and the 'business' of TNCS.

foreland (i) In PHYSICAL GEOGRAPHY, the term is most commonly used to describe a coastal promontory of erosional or depositional origin (see CUSPATE FORELAND). (ii) In ECONOMIC GEOGRAPHY, the term refers to the seaward trading areas of a PORT that are connected to it by shipping routes. Ct HINTERLAND.

foreshore A term loosely applied to the part of a coast lying between the lowest low-water line and the average high-water mark. Ct BACKSHORE.

forest A continuous and extensive tract of woodland, usually of commercial value. In Britain, the term is also used to describe a royal hunting ground, outside common law and subject to forest law, e.g. the New Forest. See AFFORESTATION.

forest park An area of forestry which, in addition to yielding timber, is used for RECREATION. In times of growing personal leisure, the recreational value of forests is being increasingly recognized, be it for picnicking, walking, pony-trekking or natural history. The opening up of a forest for recreation need not adversely affect the supply of timber.

Forestry Commission Set up in 1919 to rectify the large-scale DEFORESTATION of Britain before and during the First World War and to build up the nation's stock of timber for the future. Much of the 1.2 million ha of land held by the Commission is given over to coniferous PLANTATIONS. Although very successful in replenishing timber stocks, over the years its work has been criticized on a number of counts. For example, the wholesale planting of quick-growing, but 'exotic' conifers has meant altering the traditional appearance of the RURAL landscape (a particularly contentious issue in the English Lake District). The Commission is also active in fields other than just economic forestry. For example, it is responsible for the management of the New Forest, where its time and resources are also directed towards the care of deciduous woodlands, the maintenance of grazing for large numbers of livestock and the provision of recreational infrastructure (car parks, picnic areas, camping sites, etc.). The recently established Forest Enterprise is a commercial trading arm of the Forestry Commission that manages the forestry estates on a multi-use basis.

forward linkage See LINKAGE.

fossil fuel Combustible material made from the fossilized remains of plants and animals; e.g. peat, lignite, coal, oil and natural gas. These fuels are prime examples of *non-renewable* RESOURCES. See PRIMARY ENERGY.

Fourth World A collective term sometimes used to identify the poorest and *least-developed countries* (LLDCs) of the THIRD WORLD.

free face A steep and largely bare rock face from which weathered debris falls, slides or is washed by rain as quickly as it is released from bedrock. Free faces are widely developed on coasts (*sea cliffs*) and in glaciated valleys where slopes are oversteepened by glacial EROSION. They are also formed in river valleys where lateral CORRASION undercuts the slope base (*river cliffs*) or where particularly hard bands of rock outcrop. Over a period of time, free faces will undergo parallel retreat, as a result of unimpeded WEATHERING over the whole rock surface. However, unless removed by basal TRANSPORT, the weathered material will accumulate in increasing quantities, thus masking the lower part – and perhaps in time the whole – of the free face.

[*f* STANDARD HILLSLOPE]

free market economy An ECONOMY unfettered by any sort of GOVERNMENT INTERVENTION; the type of economy associated with CAPITALISM.

freeport An area set aside, usually at the coast, for economic activities where costs are saved because the area is exempted from the taxes that apply elsewhere in the country. Freeports have proved attractive to manufacturing as raw materials can be imported and finished goods exported without incurring any duty or tax payments. Singapore is perhaps the global leader in terms of making use of the freeport device. The benefits are the creation of jobs and the increased consumer spending associated with those jobs. Sometimes known as a *free-trade zone*, as in China. See EPZ.

free trade area Formed by a group of countries agreeing to free the trade between themselves of all restrictions, such as import and export duties and QUOTAS; e.g. the EU, NAFTA and LAFTA. It is customary for member countries to be allowed to trade with 'outsiders' on their own terms. Equally, the creation of these TRADING BLOCS can easily lead to trade wars. See PROTECTIONISM.

freeze-thaw weathering (frost shattering) The disintegration of rocks as a result of ice crystal growth in BEDDING PLANES, JOINTS and pores when ground temperatures fall below 0°C. It is regarded as the most pure form of physical WEATHERING, and occurs widely in PERIGLACIAL and glacial environments, in mid-latitude regions (especially mountains) in winter, and possibly in some hot DESERTS. The process is determined by the expansion of water by approximately 10% as it changes from liquid state to solid ice. The process tends to be self-reinforcing in a jointed rock, since with the first frost cycle (of freezing and melting) individual joints may be widened by 10%, thus allowing the entry of more meltwater and a correspondingly greater expansion with the next cycle, and so on.

There are differences of opinion as to the climatic conditions producing the most effective freeze-thaw weathering. It has been shown that *diurnal* frost cycles may effect only shallow penetration of the rock, with little disintegration resulting. Some authorities argue that it is only the *annual* frost cycle (winter freezing, summer thaw) that is of real significance. It is also believed that other processes, such as DILATATION, and the opening of joints by CHEMICAL WEATHERING assist freeze-thaw weathering. See also BLOCK DISINTEGRATION, GRANULAR DISINTEGRATION.

freight rates The cost of transporting a commodity over a given distance. Freight rates vary according to: the mode of transport; the type or quantity of the commodity to be moved. Looking at the accompanying figure, it is apparent that the simple assumption that TRANSPORT COSTS vary proportionally with distance rarely holds (*a*). The inclusion of TERMINAL COSTS in the calculation of freight rates means that average transport costs per kilometre decrease or show a *tapering* as the length of haul increases (*b*). It should be remembered, however, that there are some freight-rate structures that operate irrespective of distance, as for example postage (*c*). There are also instances where stepped tariffs apply, with a uniform rate being adopted within a given zone (*d*). [*f*]

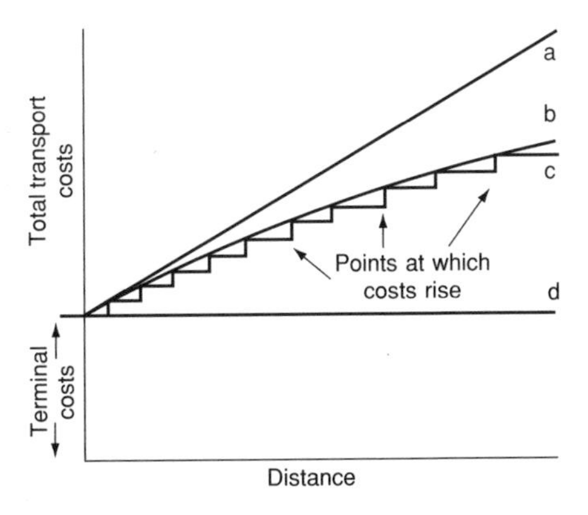

Effect of distance on total transport costs (a) Linear relationship, total cost increase proportionally with distance (b) Curvilinear relationship, total costs not increasing proportionally with distance (i.e. tapering) (c) Stepped relationship, costs do not increase proportionally with distance and they increase in stages (d) No increase of costs with distance

frequency distribution The frequency distribution of a VARIABLE refers to the number of occurrences of different values. Such frequency data must be measured either on a continuous scale (e.g. stream flow velocity) or grouped into classes (e.g. the number of SETTLEMENTS in a given country falling into each of a specified series of size classes). Plotting such data on a GRAPH produces either a frequency curve, in the case of continuous data, or a HISTOGRAM in the case of grouped data. There are many types of frequency curve; they include normal, skewed and bimodal. These terms describe the general 'shape' of the distribution. [*f*]

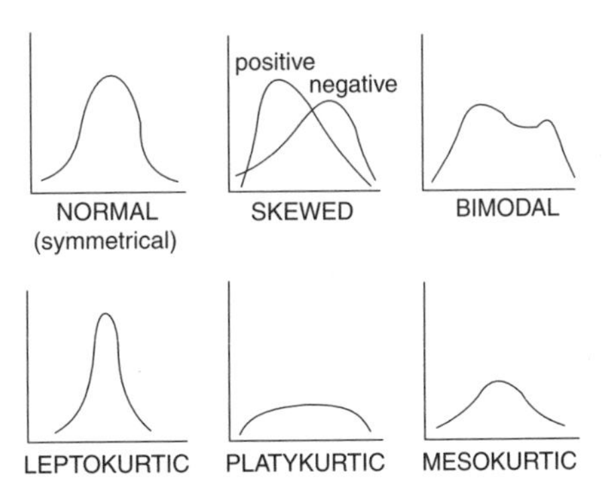

Frequency distributions

friction of distance A basic geographical concept concerning the impediment to movement that occurs because places, objects or people are spatially separated. The greater the separation, the greater is the friction of distance and the greater are the costs incurred in overcoming that distance by means of TRANSPORT, telecommunications, etc. Cf DISTANCE DECAY.

Friedmann See CORE-PERIPHERY MODEL.

fringing reef A platform of coral, attached to the coastline and extending seawards for a distance of a few hundred metres. The surface of the reef is highly irregular, with much broken coral and many large hollows resulting from selective SOLUTION; however, an inner lagoon is absent or weakly developed (ct BARRIER REEF). The outer edge of the reef is clearly defined, and drops steeply into deep water. Its continuity is broken at points where rivers enter the sea or where ESTUARIES have resulted from a relative rise of sea-level.

front A narrow zone of transition, dividing two air masses of differing temperature and humidity characteristics, intersecting the Earth's surface (see COLD FRONT, WARM FRONT, OCCLUDED FRONT, POLAR FRONT). Fronts are most clearly developed in areas where air masses converge, as in mid-latitude FRONTAL DEPRESSIONS. Here the change from one type of air to another is sufficient to be represented conveniently by a line on a weather map.

[*f* FRONTAL DEPRESSION]

frontal depression An area of low atmospheric pressure, developed along the POLAR FRONT and characterized by WARM and COLD FRONTS, and a WARM SECTOR. Depressions form most readily over the oceans in mid-latitudes and track eastwards, bringing clouds and rain to the western margins of continents. At a front, warm moist air is forced to rise. As it does so, it cools, leading to CONDENSATION of water vapour, cloud formation and rainfall. At the warm front (of relatively gentle gradient), uplift is less rapid but extends over a

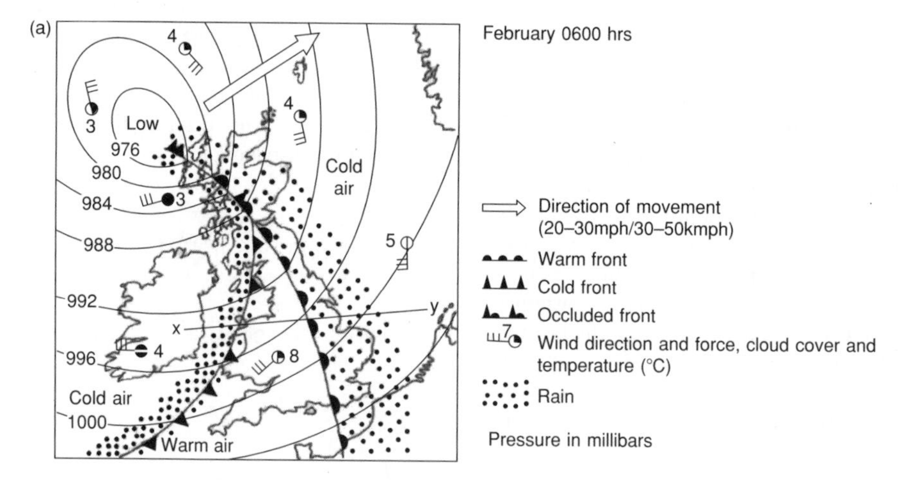

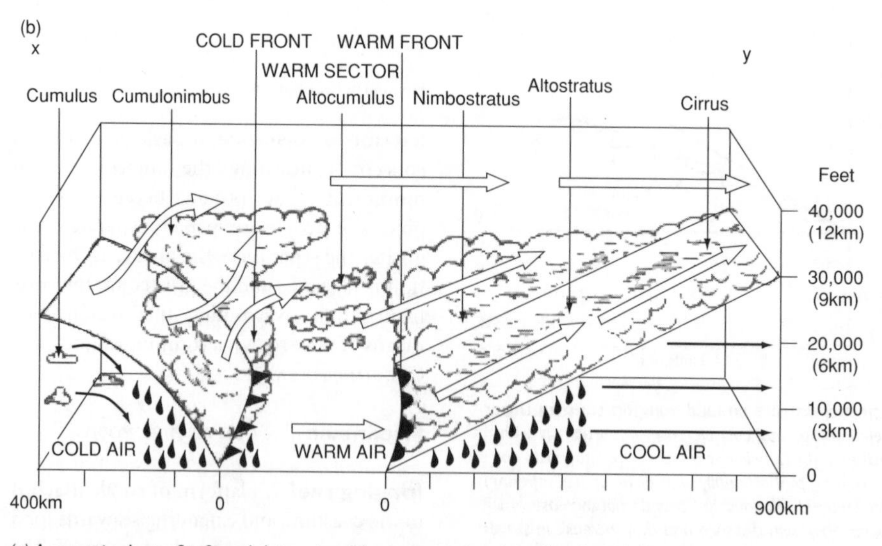

(a) A synoptic chart of a frontal depression (b) a cross section through a frontal depression

very wide area; the resultant rainfall is steady and may continue for several hours. At the cold front (of relatively steep gradient), uplift is more rapid and affects a much smaller area; the resultant rainfall is more intense, short-lived and usually comprises heavy showers. [*f*]

frontier (i) A narrow zone that 'fronts' or faces a neighbouring country; otherwise called the BORDER or march. It should be distinguished from BOUNDARY, which is the actual line of demarcation between the two countries. (ii) A term applied particularly in the history of N America to a thinly populated pioneer zone (see PIONEER SETTLEMENT) on the margins of more settled lands and representing a stage in the westward expansion of population and settlement. The Canadian Northlands and Alaska are popularly referred to today as 'the last great frontier', a description also applied to Siberia, the Australian 'outback', and other harsh and largely unsettled environments. See WILDERNESS.

frost creep See SOLIFLUCTION.

frost heave The upward movement of rock or soil particles as a result of the pressures generated by the formation of ice segregations in the ground. Most often there is also a degree of lateral displacement (termed frost thrust) such that the term *frost expansion* may be the most appropriate to use. Frost heave takes several forms; for example, the formation of ice crystals (especially NEEDLE ICE) will uplift soil particles; and larger ice bodies may raise large stones. One of the principal results of frost heave is the vertical sorting of debris, with the coarse particles being raised to the surface above underlying fines.

frost hollow A low-lying depression or valley into which cold air drains at night, particularly under the stable atmospheric conditions associated with ANTICYCLONES. On the adjacent slopes, nocturnal radiation of heat causes the ground to become cold. The overlying air is in turn cooled and, being relatively dense, it flows downhill to accumulate in the depression or valley. Ground temperatures on the slopes may remain above freezing, but in the hollow they may fall well below it, so that the ground surface and vegetation become white with frost. Indeed, temperatures within a frost hollow may fall to a very low level, particularly in a valley where the flow of air along the floor is impeded by a road or railway embankment. Fruit growers in particular seek to avoid frost hollows, since the frost will cause serious damage should it occur during blossom time.

frost shattering See FREEZE-THAW WEATHERING.

fuelwood Wood used for cooking and heating, particularly in the THIRD WORLD. Besides being a highly time-consuming task, the collection of fuelwood can and does cause much environmental damage, particularly in SAVANNA areas. Felling of trees and shrubs can easily lead to SOIL EROSION.

fumarole A small volcanic vent in the Earth's crust from which steam rather than acid gases or carbon dioxide is emitted. Associated with volcanic landscapes that are dormant or declining.

fungicide A chemical used to control or eradicate a fungal infection; for example, of plants in AGRICULTURE and HORTICULTURE.

G

G8 A grouping of the world's leading economies – Canada, France, Germany, Italy, Japan, Russia, the UK and the USA – that has assumed a collective role in running the GLOBAL ECONOMY. For many years it was known as G7, that was until 1998 when Russia was invited to join. The aim of the group is to use its considerable political muscle and economic power to bring about closer economic cooperation at an international level. One specific action has been to lend money to the IMF to support that organization's activities.

gabions Wire cages filled with rocks and commonly stacked together to produce a wall-like structure to protect vulnerable cliffs or river banks from erosion. They are common forms of slope protection and afford a buffer to the HYDRAULIC ACTION of water. In time, vegetation may colonize the surface, thereby making them less visually obtrusive.

Gaia hypothesis A concept in which the Earth is viewed as a complex organism capable of regulating and organizing its many component parts and processes. It is particularly useful in understanding the complex adjustments that maintain the planet's long-term climatic equilibrium.

gap town A town situated in, or near, or commanding, a gap in a ridge – e.g. Guildford and Dorking in the UK's North Downs. Gap towns frequently owe their origin not only to their ability to command routes through the gap along the valley floor, but also to their commanding position with respect to routes crossing the gap from high ground on either side of it. Ancient routes often ran transverse to the gap, not through it.

garden city A name coined by Howard (1898) to describe a carefully and wholly planned TOWN, designed to maintain something of an open RURAL character, with relatively low housing densities, and with particular attention paid to the location of INDUSTRY, services and AMENITIES. The intention was to combine the advantage of town life with the attractions of living in a healthy rural environment. Howard produced a model of a garden city with the declared aim of creating 'cities in gardens, and gardens in cities'. The ideal city size was deemed to be about 30,000 inhabitants. Howard and his followers (known as the *Garden City Movement*) attempted to demonstrate the feasibility of the model by undertaking two garden-city ventures in Hertfordshire: one at Letchworth (started 1903) and the other at Welwyn (started 1920). The application of the same design ideas to a suburban context was demonstrated by the construction of Hampstead Garden Suburb (started 1907). These experiments pioneered ideas of urban design later applied in the British NEW TOWNS and housing estates of the postwar era.

garigue (garrigue) A degenerate form of scrub woodland associated with thin, poor and dry SOILS, and especially characteristic of LIMESTONE areas around the Mediterranean. The vegetation is low-growing, patchy and comprises many aromatic plants (sage, lavender and thyme). In the driest areas, even true desert plants (such as prickly pear and aloe) are found.

gatekeepers In HUMAN GEOGRAPHY the term refers to professional people (estate agents, bankers, building society managers) who have the power to allocate scarce RESOURCES, particularly housing, among competing people. Since, for most people, buying a house requires obtaining credit from a bank or building society, the branch managers of these institutions are put in a position of power. They control, as it were, the 'gate' to home ownership, closing it to those who are deemed to have a poor credit rating. In the USA estate agents have become gatekeepers by directing Black buyers to particular residential areas and away from others. The landlords of privately rented accommodation are also able to perform the same role of DISCRIMINATION. Cf URBAN MANAGERS.

GATT (General Agreement on Tariffs and Trade) This was set up in 1947 by the world's leading nations after the Second World War to promote free trade by removing tariffs and other types of trade barrier. The original 23 signatory nations have now grown to nearly 100. Since the mid-1980s, much of its business has been to resolve the TRADE FRICTION between the EU and the USA over agricultural subsidies. A basic weakness of GATT is that each member state reserves the right to PROTECTIONISM when it believes that its trading interests are being adversely affected. Recently, it has been superseded by the World Trade Organization (WTO).

gavelkind See LAND REFORM.

GDP (gross domestic product) The total value of goods and services produced by the ECONOMY of a country over a specified period, normally a year. In terms of international comparisons, it is more appropriate to express GDP in per capita terms. Percentage contribution to GDP is a widely used method of monitoring trends in the sectoral balance of the economy, i.e. assessing the

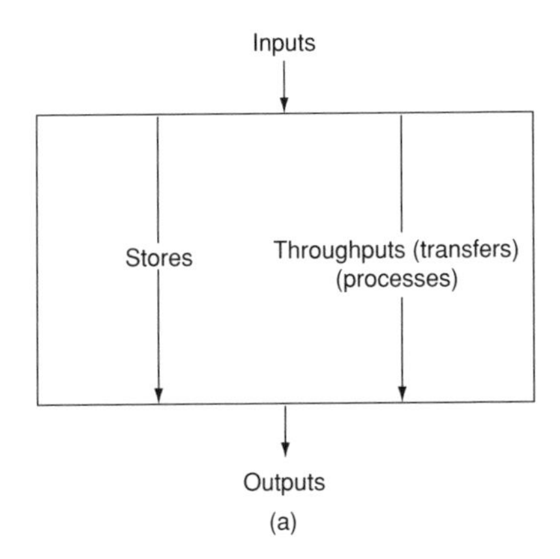

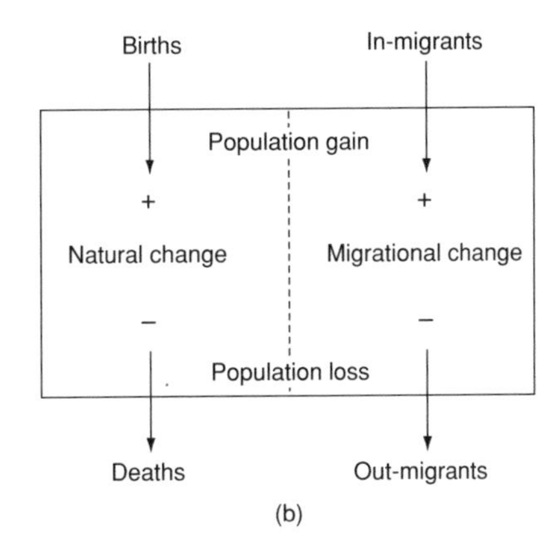

General systems theory: (a) an open system and (b) a systems view of population

changing contributions made by the four ECONOMIC SECTORS to the national economic effort. [*f* DEVELOPMENT-STAGE MODEL]

gender The distinction between gender and sex is increasingly being made. While the male and female sexes are biological facts, gender is shaped by society and culture. It is to do with the roles that society expects men and women to play. The traditional stereotype was for men to be the 'bread-winners' and for women to be child-bearers and home-keepers. That view is still held in many parts of the world and reinforced by religions such as Islam. However, in other areas, gender roles are being changed, as evidenced by more women working outside the home and the advent of 'house-fathers'.

General Agreement on Tariffs and Trade See GATT.

general systems theory A *system* is a set of interrelated objects comprising components (stores) and processes (links). Systems theory aims to demonstrate the nature and complexity of these interrelationships, and thus to show the multivariate nature of phenomena. Two main types of system are recognized: *closed systems*, in which there are no inputs or outputs of materials or energy across the system 'boundaries'; *open systems*, in which such inputs and outputs do occur, and indeed are necessary for the continuance of the system. A slope, for example, may be viewed as an open system, with inputs of solar energy, PRECIPITATION and WEATHERING products, and outputs of water and SEDIMENTS at the slope base. Open systems display two important attributes: They attain a condition of balance, or *equilibrium*, when inputs equal outputs; landforms that achieve this state may undergo no change of form with time, and thus are *time independent*. When the condition of balance is disturbed (e.g. by increased input of materials), the system will undergo *self-regulation* and change its form to restore equilibrium. This is usually achieved by NEGATIVE FEEDBACK, whereby the system adapts itself in such a way that the effects of the initial change are countered. Thus, where more sediment enters a river, some will be deposited to steepen the channel gradient; thus stream velocity will be increased, so that the enlarged load can be transported without further DEPOSITION. Ct POSITIVE FEEDBACK. [*f*]

genetically modified food See GM FOOD.

gentrification A process occurring in INNER-CITY areas whereby old, substandard housing is bought, modernized and occupied by middle-class and wealthy families. The process is well demonstrated in inner London districts such as Chelsea, Fulham and Islington, which have become much sought-after and expensive residential locations. Gentrification is probably triggered by the survival of once-elegant, but rundown housing and by locational advantages such as ready access to central city employment and services. It is also probably helped by the availability of improvement grants. Once started, the process is no doubt sustained by the perceived social prestige derived from living in such fashionable areas. Gentrification represents an interesting reversal of the normal FILTERING process, in that it involves a social upgrading of obsolescent residential areas.

geo A very narrow, steep-sided inlet in a cliffed coastline. Geos are the result of wave EROSION along a clearly defined line of weakness such as a FAULT, major JOINT or BEDDING PLANE. They may develop initially as caves that are extended some tens of metres into the CLIFF face; eventually, however, the roof will collapse to form the inlet. Geos are particularly characteristic of SANDSTONE and LIMESTONE cliffs.

geographical inertia See INDUSTRIAL INERTIA.

geographic information system See GIS.

geography The content, scope and emphases of geography have undergone considerable change over the past 50 years, as paradigm has succeeded paradigm, and it is highly unlikely that any one definition of the subject would satisfy everyone. Most are agreed that it comprises study of the Earth's surface as the home of the human race. But how much geography is the science of SPATIAL DISTRIBUTIONS and SPATIAL RELATIONSHIPS, how far it is concerned with the interaction between people and their physical ENVIRONMENT, and to what extent study of the REGION is the focus of the subject – these are all matters for debate. Recently, the focus seems to have shifted towards a more issues-based approach, ranging in scale from to the local to the global, from POLLUTION to poverty (see CYCLE OF POVERTY), The fact that geography is located at the interface between the natural and social sciences adds to the difficulty of arriving at a universal definition.

geological (stratiphical) column The division of the geological time-scale into eras, which are then subdivided into periods as shown in the table.

Era	*Period*
Quaternary	Holocene Pleistocene
Tertiary (Cenozoic)	Pliocene Miocene Oligocene Eocene
Secondary (Mesozoic)	Cretaceous Jurassic Triassic
Primary (Palaeozoic)	Permian Carboniferous Devonian Silurian Ordovician Cambrian Pre-Cambrian

The geological column

geometric mean This average is used when the values in a data set show a geometric or exponential progression (see EXPONENTIAL GROWTH RATE). It is found by calculating the nth root of the product of all the values. Ct ARITHMETIC MEAN.

geomorphology The study of landforms.

geopolitics That aspect of political geography that emphasizes the geographical relationships of STATES.

geostationary satellite A satellite maintaining a fixed position over a point on the Earth's surface. Thus it can only monitor that belt of the Earth that passes under it during the completion of an orbit.

geostrophic wind The theoretical movement of air parallel to the isobars, well above the surface of the Earth and unaffected by the friction that causes surface winds to cross the isobars obliquely. The geostrophic wind is the consequence of two opposing forces acting on each other: the PRESSURE GRADIENT operating from high to low pressure, and the CORIOLIS FORCE operating in the opposite direction. [*f*]

geosyncline A major downfold in the Earth's crust, containing a vast thickness of SEDIMENTS. The latter appear to have accumulated at approximately the same rate as

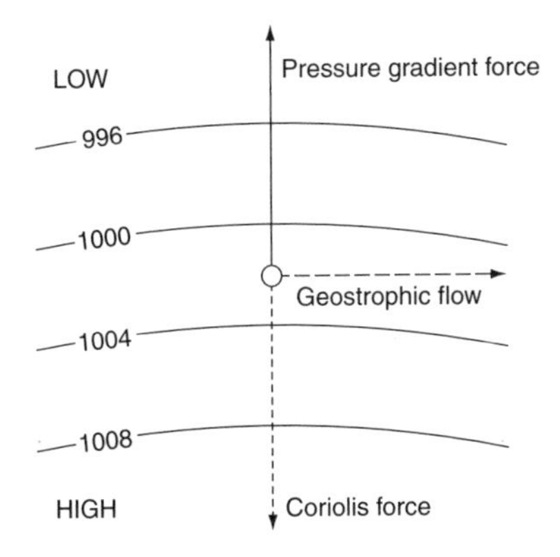

Geostrophic wind

downsinking of the geosynclinal floor has occurred. After a long period of accumulation, compressive forces result in large-scale folding of the geosynclinal sediments and the formation of FOLD MOUNTAINS. See also YOUNG FOLD MOUNTAINS, OLD FOLD MOUNTAINS, PLATE TECTONICS.

geothermal heat The heat energy derived from hot rocks within the Earth's crust. Geothermal heat escapes to the surface by way of hot water or steam, or is slowly conducted upwards through the rocks themselves. Geothermal heat is sufficient to cause some basal melting of ICE SHEETS and glaciers, and thus contributes to their sliding over BEDROCK. Geothermal heat can be used in the generation of ELECTRICITY – for example, in Iceland, where geothermal heat supplies the country with almost all its energy. See GEYSER.

geyser An intermittent fountain of hot water and steam, spurting from a small hole in the Earth's crust. The water is heated up, geothermally, in reservoirs beneath the surface. These are connected to the outlet by a narrow pipe. Temperatures within the reservoirs and lower pipe are raised to above 100°C and the heated water begins to expand. The water in the upper pipe is lifted towards the surface and, suddenly, as pressure is 'released' in the lower pipe, part of the water is suddenly converted into superheated steam. As the latter expands it ejects the water in the pipe into the air, often in spectacular fashion.

ghetto Originally the term was used to denote that part of a TOWN or CITY reserved for the Jewish community, but nowadays it refers to any residential area that is largely occupied by one ethnic or cultural MINORITY group. The emergence of a ghetto represents a degree of SEGREGATION brought about partly because members of a minority group wish to live together and partly in response to discriminatory pressure by the host community. While many ghettos are INNER-CITY areas of poor housing, they are not exclusively so. Where ethnic groups have been able to move up the social scale, they have moved away from the SLUMS and become concentrated in areas of better residential quality, thus forming what have been called *gilded ghettos*. Although the ghetto is relocated, it is preserved – it might even expand. Equally, a ghetto may gradually dissipate as a result of ASSIMILATION. Members of the ethnic group will still loosely associate in what are termed *urban villages*. [*f*]

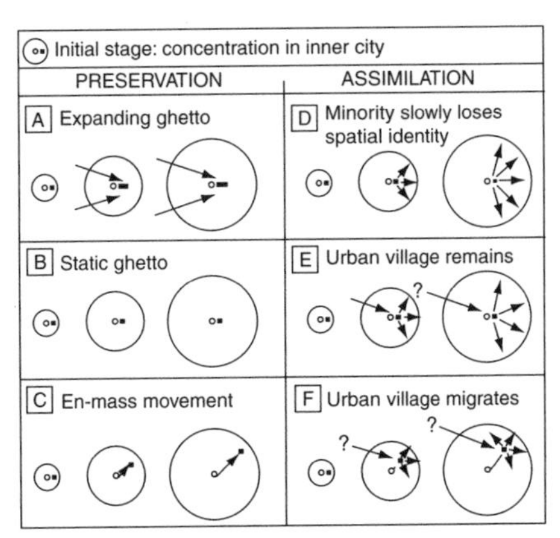

Models of ghetto evolution

ghost town A term used to describe a once-flourishing settlement, now completely abandoned or inhabited by few people. Many former mining settlements, particularly those associated with the great gold rushes of 19th century N America and Australia, are thus aptly described.

gilded ghetto See GHETTO.

Gini coefficient A statistical measure of the degree of correspondence between two sets of percentage frequencies. It is derived by the formula

$$G = \Sigma (X_i - Y_i),$$

where X_i and Y_i represent the two sets of percentage frequencies. The value of G can range from 0 to 100, with a value of 0 indicating exact correspondence between the two frequencies. The Gini coefficient is often used in conjunction with the LORENZ CURVE as a means of assessing the degree to which a given distribution of data differs from a uniform distribution. It is sometimes referred to as the *index of concentration.*

GIS (geographic information system) A store of geographical data held in digital form on a computer. The capacity of the modern computer is such that vast amounts of information can easily be stored, updated and analysed. The growth of GIS is closely related to REMOTE SENSING, which is constantly providing new information about the surface of the Earth and other planets.

glacial diversion of drainage The modification of the preglacial drainage pattern resulting from the activities of an ICE SHEET or glacier. Such diversions may be temporary (lasting during the glacial episode only) or – where the diverted river channels are deeply incised into rock – may become permanent. The mechanisms of drainage diversion vary. The formation of ice-marginal lakes (PROGLACIAL LAKE) may lead to the formation of OVERFLOW CHANNELS. However, it is now believed that many so-called overflow channels were actually produced by SUBGLACIAL streams of meltwater, flowing at great hydrostatic pressure beneath decaying ice sheets and capable of cutting deep channels across old WATERSHEDS. Permanent drainage diversion may also be associated with GLACIAL WATERSHED BREACHING. [*f*]

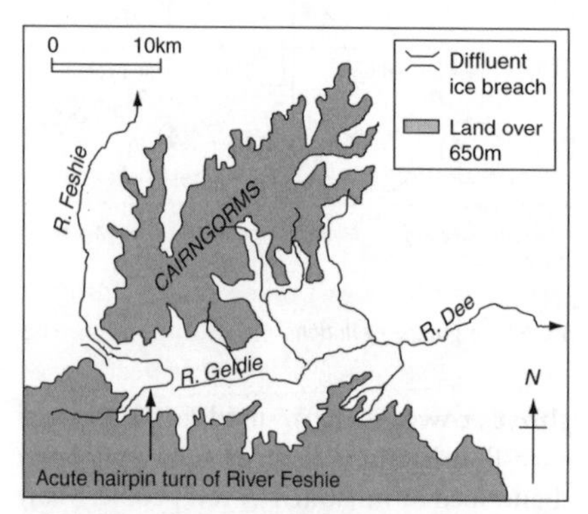

Glacial diversion of River Feshie, Cairngorms

glacial eustatism (sometimes **glacio-eustatism**) A worldwide change of sea-level resulting from the development and decline of ICE SHEETS, notably during the PLEISTOCENE period. The mechanism involved is the transfer of large volumes of seawater on to the continental land-masses by PRECIPITATION in the form of snow and ice, and their subsequent return to the oceans as the ice sheets are dissipated. Many RAISED BEACHES and coastal platforms have been explained in terms of glacial eustatism. It has been estimated that melting of the world's remaining ice sheets and ICE CAPS (mainly in Antarctica and Greenland) would cause a further eustatic rise of sea-level by some 50 m or so. See GLOBAL WARMING.

glacial lake A body of meltwater impounded between a glacier and the valley wall.

glacial outburst (*jökulhlaup*) A short-lived but sometimes catastrophic flood resulting from the sudden release of meltwater stored within or on the surface of a glacier or ICE SHEET, or from a lake dammed up against the ice margin. Particularly large outbursts (known as *jökulhlaups)* are frequent in Iceland, where large quantities of meltwater result from normal ABLATION and the escape of GEOTHERMAL HEAT beneath the ice in this volcanically active region. One famous example is that of the periodic evacuation (every 10 years or so) of the large meltwater lake adjacent to the volcano Grimsvotn, beneath the Vatnajokull ICE CAP. Water builds up and then escapes, via an ice tunnel 40 km in length, to causes spectacular flooding on the SANDUR plain beyond the ice cap margin.

glacial stairway The irregular long-profile of a glaciated valley, resulting from the formation of alternate rock basins and ROCK STEPS. Some glacial stairways are geologically determined; the basins result from glacial overdeepening involving ABRASION and plucking of the weaker rocks, while the steps are developed from harder, more massive rock outcrops. Glacial stairways are thought to be related to the nature of GLACIER FLOW. Along large valleys there are frequently found alternating sections of COMPRESSING and EXTENDING FLOW, perhaps resulting from small initial breaks in the valley long-profile. The sections of compressing flow produce glacial overdeepening, while extending flow is associated with reduced EROSION. Hence the initial irregularity of the profile becomes ever more exaggerated, and a typical glacial stairway is formed. [*f* ROCK STEP]

glacial trough A deep U-shaped valley resulting from the overdeepening processes associated with valley glaciers.

glacial watershed breaching A process that occurs in its simplest form when the outlet of ice from a group of CIRQUES and/or a valley is impeded by a larger glacier. The ice is forced to build up in level, and may eventually escape across a pre-existing WATERSHED, following a line of weakness (such as a preglacial COL). The latter may be so deepened by intense glacial EROSION that it provides a route for POSTGLACIAL rivers. See GLACIAL DIVERSION OF DRAINAGE.

glacier budget See MASS BALANCE.

glacier flour Suspended SEDIMENT (mainly fine SILT) carried by meltwater streams issuing from beneath glaciers. Where the suspended sediment concentration is very high, the flowing water takes on a milky, opaque appearance. The glacier flour itself is the product of intense ABRASION at the glacier base. There is usually a seasonal rhythm to the discharge of the sediment. During winter (when meltwater flows are minimal) the products of EROSION build up beneath the ice; during summer (when meltwater is abundant) the sediment is flushed out.

glacier flow The complex processes by which ice is transferred from the ACCUMULATION ZONE of the glacier, through the ABLATION ZONE, to the snout. There are three important mechanisms. (i) *Basal sliding* – where ice temperatures are at, or close to, 0°C a layer of basal meltwater readily forms between ice and BEDROCK, thus acting as a lubricant and aiding the sliding process. In many temperate glaciers, basal sliding constitutes between 50% and 90% of total glacier movement. However, in 'cold-based' or polar ice (where ice temperatures may be as low as 10–20°C below freezing) basal sliding may be totally absent. (ii) *Internal deformation* – ice crystals under pressure are affected by recrystallization and continual slight movements relative to each other. In this way the ice as a whole appears to take on 'plastic' qualities. Internal deformation is relatively more important, as a component of total glacier flow, in 'cold' than in 'warm' ice. (iii) *Faulting* within glaciers – this may occur where the glacier is strongly compressed, as at the base of steep ice falls or at the snout where the glacier is overriding TERMINAL MORAINE. At such points THRUST FAULTS may develop, with the ice lying above the fault advancing over the less mobile ice beneath. See also EXTENDING FLOW, COMPRESSING FLOW.

glacier karst The highly irregular glacial RELIEF sometimes formed by the differential melting (*differential ablation*) of large stagnant ice masses. An uneven cover of surface morainic debris results in varied surface ABLATION and the formation of numerous surface depressions, broadly equivalent to the DOLINES of true KARST.

glacier table A very common landform on the surface of glaciers, resulting from the presence of large BOULDERS that protect the ice from the Sun's rays. As the surrounding bare ice is rapidly lowered by ABLATION, the boulder is left upstanding on a pedestal of ice.

glacio-fluvial Descriptive of phenomena (both erosional and depositional) resulting from the action of meltwater streams associated with glaciers and ICE SHEETS. Some important glacio-fluvial landforms include SUBGLACIAL stream channels and valleys, spillways from ice-impounded lakes, and a wide range of depositional features formed beneath, at the margins of, and beyond the ice (such as ESKERS, KAMES, kame terraces, and OUTWASH PLAINS and fans. The latter comprise stratified SEDIMENTS, by contrast with the unstratified sediments (notably TILL) resulting from direct glacial DEPOSITION.

gley soil A soil developed under conditions of intermittent waterlogging, resulting in impeded SOIL drainage (as in valley bottoms or where a pan has formed in the B-HORIZON). In an *anaerobic environment* (i.e. one not subjected to aeration) oxidation of ferric compounds is restricted, and the soil develops a bluish-grey mottled appearance owing to the presence of ferrous compounds. Where pockets of oxygen are present, oxidation may occur to form red or orange blotches. Gley soils tend to be sticky, compact and display no recognizable soil structure. [*f*]

Wet season water table
Peat accumulation
Mottled soil layer affected by oxidation and reduction
Dry season water table
Saturated anaerobic layer affected by reduction
Blue-grey sticky clay
Parent material

Gley soil profile

global city One of the world's leading cities – a major NODE in the complex economic networks being produced by ECONOMIC GLOBALIZATION. The influence of global cities (e.g. London, New York, Tokyo) is linked to their provision of financial and producer services. Most of the leading TNCS have their headquarters there. Global cities have impressive central areas that symbolize their immense power, status and wealth. They have been described as the 'command capitals' of the GLOBAL ECONOMY. Because of their scale, they are heavy consumers of RESOURCES and large producers of POLLUTION. In their national contexts, global cities create huge degrees of spatial disparity and inequality. Each represents a massive POLARIZATION of investment and employment. Many are truly PRIMATE CITIES. Cf GLOBAL VILLAGE. [*f*]

global economy The evolving macroeconomic system that increasingly links together all the countries of the world. It is largely to do with the worldwide exploitation of RESOURCES, and the production and marketing of goods and services.

globalization Any process of change operating at a world scale and having worldwide effects. It can be physical (e.g. sea-level changes) or human (e.g. economic DEVELOPMENT) or both (e.g. GLOBAL WARMING). The processes are essentially geographical because they have the potential to affect

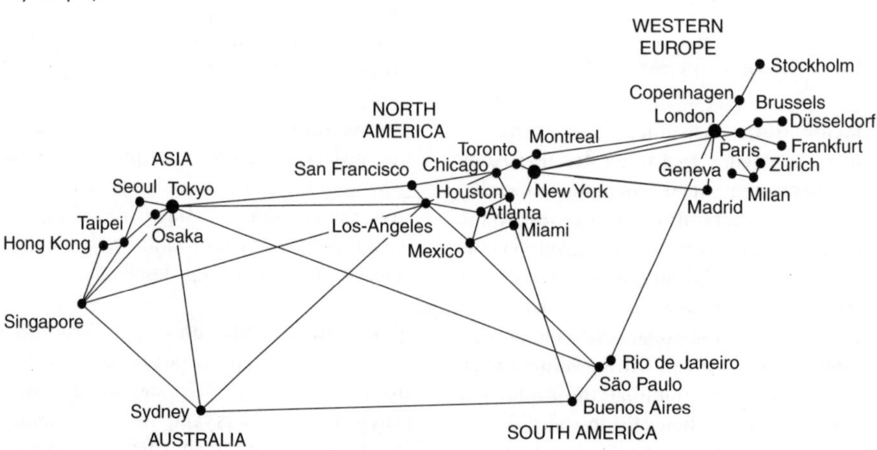

The network of global cities

much of the Earth's environment and people. They are also geographical because they can lead to different outcomes in different parts of the world. It is wrong to think that globalization is only to do with the growth of the GLOBAL ECONOMY. See ECONOMIC GLOBALIZATION.

global shift This term is usually confined to the redistribution of manufacturing at a global scale. Following the Industrial Revolution, much of the world's production became concentrated in N America and W Europe. In the 1970s and 1980s, however, TNCS began to move manufacturing to the NICS and favoured locations in the THIRD WORLD. This shift was driven by the search for cheaper and less unionized labour, cheaper land and fewer controls on pollution. It was made possible by advances in transport that lowered the costs of moving raw materials and finished products over large distances. The shifting continues to this day. In its wake, the former strongholds of manufacturing have either undergone DE-INDUSTRIALIZATION or industrial RESTRUCTURING. Their economies have also experienced another 'shift' as they have become more dependent on the TERTIARY and QUATERNARY SECTORS.

global triad see NAFTA.

global village A term used to convey the idea that people around the world are being drawn closer together into one global community. Advances in TRANSPORT have reduced the time and cost involved in overcoming the FRICTION OF DISTANCE. Advances in COMMUNICATIONS are also playing a major part through their ability to keep distant locations in constant and instantaneous touch. The term GLOBAL CITY is also used to describe the same development, but in addition it underlines the important point that rising URBANIZATION is an integral part of this 'coming together'. The emergence of the global village and the global city are important outcomes of GLOBALIZATION.

global warming The currently held view that the Earth's temperatures are experiencing a significant, if slow, rise, as indicated by: a wide range of climate data; evidence such as the current recession of many glaciers and ice caps, and of scientific indicators in bores from ice caps. Using computer-generated models, it has been predicted that by the middle of the next century, global temperatures may rise on average by 5°C or more (mainly in polar regions), with consequent displacement of the Earth's climatic belts and serious implications for grain-producing areas such as the Prairies of N America. Another likely effect will be a rise in sea-level of 2–3 m or more, with disastrous consequences for many low-lying regions (indeed, many Pacific islands would be totally inundated). Many scientists believe that global warming is a human-induced phenomenon, resulting from the build-up in the atmosphere of excessive quantities of carbon dioxide (from the use of fossil fuels, and as a result of the destruction of RAINFORESTS in Amazonia and other tropical regions). This enhances the natural GREENHOUSE EFFECT in the Earth's atmosphere. However, there is much dispute about the extent, causes and likely impact of global warming. An optimistic view would be that the Earth as a natural system is, in the long run, capable of adjusting to these changes (see GAIA HYPOTHESIS). In the short term at least, most governments and coastal authorities are taking the threat seriously as they plan for the future.

GM (genetically modified) food An outcome of genetic engineering undertaken to increase the productivity, disease-resistance and quality of crops and livestock. The scientific techniques involved either transfer genes from one organism to another (*transgenic*) or change genetic materials within an organism. Genetic modification is the latest chapter in a long history of human endeavour to improve food production. Experimentation with livestock has been going on longer than with crops. However, it is the latter that is currently the cause of greater concern. Most transgenic crops have been bred to incorporate greater herbicide tolerance or to increase resistance to pests. If crops do breed with wild near-relatives, then it is possible that these traits may be passed on to wild plants and cause major ecological disruption. Of course, the only way to find out is by controlled field trials. But there are some who feel that even these should not be undertaken because of the risks to wildlife. It is interesting that public concern about GM food centres on production and the environment rather than on consumption and what it might do to people!

gneiss A coarse-grained crystalline metamorphic (see METAMORPHISM) rock resulting from intense heat and pressure.

GNP (gross national product) This is the GDP of a country plus all the income earned by residents from investment abroad, but less the income earned in the country by foreigners abroad. As with GDP, the measure is of greatest value when it is expressed in per capita terms. It is widely used, as when making international comparisons of economic growth.

Gondwanaland The name given to a former large land-mass, comprising the southern continents of S America, Africa, S Asia, Australasia and Antarctica. During the Mesozoic era, this 'supercontinent' began to fragment, and the various components drifted during the Tertiary era to their present-day locations (see CONTINENTAL DRIFT, PLATE TECTONICS).

goods and services See PUBLIC GOODS AND SERVICES.

government intervention A term currently used with reference to capitalist economies (see CAPITALISM) when government acts in order to rectify what are regarded as deficiencies and defects created by, or associated with, normal FREE MARKET forces, hence creating a MIXED ECONOMY. Intervention can be made in all sectors of the economy from agriculture (see CAP) through to the provision of SOCIAL SERVICES. Intervention can also take diverse forms, from the granting of subsidies to encourage particular types of production to the imposition of import QUOTAS, from regional AID programmes (see ASSISTED AREA, GROWTH POLE) to the creation of ENTERPRISE ZONES in INNER-CITY areas. It might be argued that it is the degree of government intervention that differentiates capitalist from socialist states, it being partial in the former and comprehensive in the latter. See also BALANCED GROWTH.

graben A type of geological structure in which a downfaulted mass lies between two parallel FAULTS. Initially, a graben will give rise to a RIFT VALLEY. [*f* FAULT]

grade A concept of EQUILIBRIUM (see GENERAL SYSTEMS THEORY) applied in particular to landforms such as streams and valley-side slopes. The term was used at the end of the 19th century by writers such as Gilbert and Davis. In Davis's interpretation, graded river profiles and slopes were identified by their smoothness. A graded long-profile was described as having a smooth concave profile. A more modern view is that a graded river is: an open system in a state of equilibrium, constantly adjusting its slope as SEDIMENT load and DISCHARGE change over time; in an average condition of balance, when viewed over a period of years. In *graded slopes*, the equilibrium is between the rate of production of sediment by WEATHERING, and its removal by SOIL CREEP and wash processes. It can be diagnosed by its smoothness, an even SOIL cover of no great depth, and a gradient that is adjusted to provide the necessary TRANSPORT rate.

graded time This refers to the time period over which a landform (such as a slope or river CHANNEL) remains in a graded condition (see GRADE).

granite A coarse-grained, acidic IGNEOUS ROCK usually occurring in the form of large intrusive masses (see BATHOLITH). Deep weathering of granite, involving the exploitation of closely spaced JOINTS, has led to the formation of TORS. In tropical-humid regions granite is particularly prone to rapid CHEMICAL WEATHERING, due to high temperatures and humidity (see REGOLITH, GRUSS).

granular disintegration The breakdown, by WEATHERING processes, of a rock into its constituent minerals or groups of minerals. It is often attributed to purely physical mechanisms. However, it is more likely to result from selective CHEMICAL WEATHERING, for example, the oxidation of an 'iron' cement binding together quartz particles in a SANDSTONE, or the HYDROLYSIS of feldspar minerals in GRANITE. Granular disintegration is therefore best defined as a *physical* process that commonly has a *chemical* cause.

graph In its simplest form, a graph is a diagram to locate the position of a given VARIATE with respect to any two VARIABLES represented by two axes. On the vertical or *y* axis is plotted the DEPENDENT VARIABLE, while the INDEPENDENT VARIABLE is represented by the horizontal or *x* axis (see REGRESSION ANALYSIS). In some cases, each of the points that locate the variates may be joined by a line to form a curve, which gives an immediate visual impression of the nature of the relationship between the two variables under investigation. In other cases, such as in a SCATTER DIAGRAM, the plotted points may show a tendency to cluster rather than form a linear arrangement.

Many modifications may be made to the *simple graph* just described in order to show more complicated relationships. For example, various scales other than arithmetic ones may be employed on the axes (lognormal, probability), while a *compound graph* allows the graphical comparison of two or more dependent variables. The term is also used in a wider sense to denote any form of symbolic diagram, e.g. PIE DIAGRAM, TOPOLOGICAL MAP. See also TRIANGULAR GRAPH.

[*f* LOGARITHMIC SCALE]

graph theory That part of mathematics concerned with the study of simple TOPOLOGICAL MAPS (alternatively known as GRAPHS) drawn to represent various geographical NETWORKS. See also ACCESSIBILITY, CONNECTIVITY.

graphicacy The communication through the medium of MAPS and diagrams of spatial relationships that cannot successfully be communicated by words or mathematical formulae. It is the particular province of the geographer and ranks alongside the other basic skills of literacy and numeracy. Graphicacy may also be used to represent time relationships (e.g. rates of change) as well as conceptual abstractions. Cf CARTOGRAPHY.

gravel A deposit of rounded, sub-angular or angular stones, with diameters approximately within the range 2–60 mm. A *river gravel* usually comprises well-rounded stones, resulting from active rolling, sliding and ATTRITION on the river bed. However, a *solifluction gravel* usually consists of more angular material, produced by FREEZE-THAW WEATHERING and transported over a limited distance.

gravity anomaly A term often used in the context of PLATE TECTONICS to describe an exceptionally high (positive) or low (negative) value of gravity compared to that which would be expected. For example, at a constructive plate margin, where new crust is being formed (such as the mid-Atlantic Ridge), a positive anomaly exists. This is explained by the fact that new, dense rock material is being brought to the surface. At a destructive plate margin, where dense rock is being subducted, a relatively low value – a negative anomaly – is often recorded.

gravity model The application of Newton's Law of Universal Gravitation to a variety of different situations arising in human geography where movement is involved, as in MIGRATION, shopping, traffic and trade. In the case of migration, the formula underlying the model is:

$$M_{ij} = g \frac{P_i P_j}{d_{ij}^2}$$

This indicates that the volume of migration between two places (i and j) is directly proportional to the product of their populations and inversely proportional to their distance apart, raised an exponent. In the case of shopping, the model might be used to predict the likelihood of a shopper going to one particular shopping centre rather than another. In this application, the notion of distance between the shopper's home and accessible shopping centres might be measured in terms of intervening opportunities (see INTERVENING OPPORTUNITY THEORY) rather than linear distance. In other words, the likelihood of a shopper going to a particular centre is reduced proportionally by the number of alternative centres that are located nearer to the shopper's home than the centre in question.

green belt A girdle of land, designated by

PLANNING and encircling a TOWN or CITY with a view to preventing the further outward spread of its BUILT-UP AREA. Possibly one of the best-known green belts is the one around London, which was given official recognition in the Town and Country Planning Act (1948). This has always been a controversial aspect of British postwar planning. It has undoubtedly prevented the further spread of London's contiguous built-up area and helped to firm up the edge of the CONURBATION. At the same time, however, it has to be noted that the continuing ECONOMIC GROWTH of London has prompted the spawning of new and detached SUBURBS for London-bound commuters on the outer side of the green belt. In this sense, then, the introduction of the green belt might be seen as contributing an increase of some 30 km to the length of the JOURNEY-TO-WORK for many commuters. In this respect, it was unfortunate that the physical restraint of London by the green belt was not matched by effective restraint of the metropolitan economy. Other 'unforeseen' consequences include the fact that the villages and small towns located within the green belt itself, because of the protection that the legislation gives them, have become highly desirable and much sought-after residential locations. Powerful bidding for housing in these DORMITORY SETTLEMENTS by middle-class families has meant not only the inflation of house prices, but the pressuring-out of lower-income households, thus disrupting traditional social structures and balances. Some have suggested that, rather than corseting London within a green belt, it might have been better to have defined *green wedges* between the principal growth corridors reaching out from London along major transport routes.

Green movement The word 'Green' was adopted in the 1970s by environmentalist organizations (e.g. Greenpeace, Friends of the Earth) when defining what modern society and individuals needed to do in order to protect the ENVIRONMENT and make DEVELOPMENT sustainable. Since then, a political movement has emerged that, for example, is well represented in the European Parliament. During the 1990s environmentally friendly organizations and causes sprang up everywhere. Many people now favour Green trust funds, where companies that do not harm the environment are backed. Green consumers have a growing impact on the preparation and packaging of goods, and are prepared, if necessary, to pay more for Green products. The RECYCLING of household waste now receives widespread support. In short, the Green movement involves a diverse range of actions designed to spare the planet.

Green Revolution A term used to describe the repercussions and problems associated with the introduction, since the late 1950s, of new strains of cereals (rice and wheat) that are high yielding. These have been developed for the benefit of the THIRD WORLD (particularly S Asia, the Philippines and parts of Latin America) in the hope of overcoming food shortages. In order to reap the benefits of these new strains, nitrogenous FERTILIZERS are needed in great quantities, as is a careful management of water supply, often through IRRIGATION schemes. The costs of maintaining adequate supplies of both of these may be considerable. In addition, supplies of fertilizer have repeatedly fallen below demand. There have also been difficulties in persuading farmers to cooperate in the introduction of new strains where systems of SHARE CROPPING prevail. Even where the strains have been successfully introduced, there have been some unwanted KNOCK-ON EFFECTS. For example, the release of labour from the land has helped to swell rural-to-urban MIGRATION. Self-sufficiency in cereals has tended to disrupt traditional patterns of trade, to the detriment of former suppliers, while the overall rise in food production appears to have been accompanied by a significant rise in the price of food, as well as by a further increase in population. In the words of one expert, 'through the Green Revolution low-energy, self-provisioning, labour-intensive AGRICULTURE is being transformed into high-energy, capital-intensive farming dependent upon a wide range of industrialized inputs'. Because of this, and contrary to intention, the Green Revolution has increased, rather than decreased, the dependence of the LEDCs on the MEDCs. In short, much of the early optimism attached to the Green Revolution has evaporated.

greenfield site Literally, a field or plot of land that has not been subject to any significant non-agricultural development; therefore usually located in a RURAL area. The construction of out-of-town shopping centres (see HYPERMARKET) almost inevitably involves the somewhat controversial development of accessible greenfield sites. Indeed, the gathering momentum of urban DECENTRALIZATLON is putting greater pressure on such sites and their development by other activities, such as offices and industry. Ct BROWNFIELD SITE.

greenhouse effect The warming of the Earth's ATMOSPHERE by the absorption of long-wave radiation from the Earth (note that *short-wave* radiation from the Sun can readily penetrate the atmosphere). It is an entirely natural process without which it would be too cold for life to exist on Earth. There is, however, growing concern that the 'natural' greenhouse effect in the Earth's atmosphere is being increased by factors such as pollution and large-scale DEFORESTATION, which have increased the concentration of the so-called 'greenhouse (absorptive) gases', such as carbon dioxide, and methane. It has been argued that this could produce a long-term (but spatially very uneven) rise in temperature, with disastrous consequences, including the melting of polar ice, changes in the atmospheric and oceanic circulation patterns, increased storminess and increased DESERTIFICATION. See GLOBAL WARMING. [*f*]

grid A uniform pattern (usually of squares, but it could be of equilateral triangles or hexagons) that is superimposed on a surface on to which data have been plotted, in order to carry out a statistical or SPATIAL ANALYSIS of those data. Grids are frequently used in SAMPLING. See also NATIONAL GRID.

grike (gryke) A deep groove in a LIMESTONE pavement (ct CLINT) resulting from solutional processes (see CARBONATION) acting along a JOINT.

grit A coarse SANDSTONE, usually comprising grains of variable size, which accumulated under marine deltaic conditions. The Millstone Grit, a geological formation of Carboniferous age, particularly well developed in the Pennines of N England, includes alternating grits, sandstones and SHALES. These have been differentially eroded into a series of PLATEAUS or CUESTAS and vales, with the harder grits forming prominent rocky ESCARPMENTS.

gross accessibility index See SHIMBEL INDEX.

gross domestic product See GDP.

gross national product See GNP.

gross reproduction rate A ratio obtained by relating the number of female babies (i.e. potential mothers) to the number of women of child-bearing age, and used in population studies as an indication of future trends. Cf NET REPRODUCTION RATE, FERTILITY RATIO.

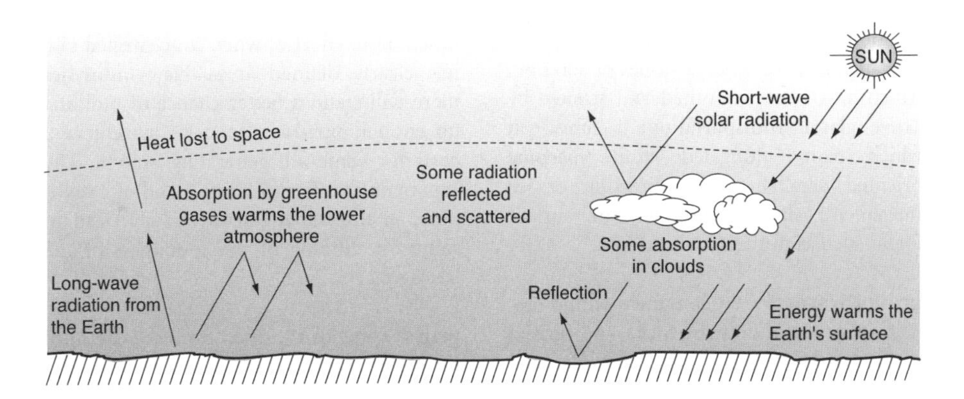

Greenhouse effect

ground frost A phenomenon whereby the temperature of the ground surface is lowered to or a little below 0°C, while that of the overlying ATMOSPHERE remains above freezing point. Although ground frosts are liable to occur throughout winter in Britain, they are especially characteristic of autumn and spring. They develop on clear, still nights when nocturnal radiation of heat from the ground surface is unimpeded.

ground ice The ice that is formed within the SOIL, REGOLITH and rock by intense freezing under PERIGLACIAL conditions. It is an important constituent of PERMAFROST; it also forms seasonally in the ACTIVE LAYER overlying the permafrost, where it performs a geomorphological role in disturbing and moving SEDIMENTS (see FROST HEAVE, SOLIFLUCTION). The melting of ground ice formations, owing to climatic change or disturbance by man, leads to the development of THERMOKARST and can inadvertently lead to building subsidence and other engineering problems in periglacial regions.

ground moraine Debris that, having been transported at the base of a glacier or ICE SHEET, is then left smeared on the BEDROCK. The debris may have been transported beneath the ice or in its lowers layers, to be freed by melting due to heat friction from sliding or the escape of GEOTHERMAL HEAT. Where an ice sheet is involved, a thick layer of LODGEMENT TILL will gradually build up. It is believed that the extensive TILL deposits, or BOULDER CLAYS, of areas such as East Anglia comprise the ground moraine of PLEISTOCENE ice sheets. The features of such tills are: an absence of stratification or sorting of the constituent particles; the presence of large stones or BOULDERS, which have been smoothed and striated by active glacial transportation; a 'fabric' in which many elongated stones become oriented parallel to the direction of ice movement; the development of surface forms such as flutings and DRUMLINS.

groundwater Water contained within SOIL, REGOLITH or the underlying rocks, and derived mainly from PERCOLATION of rainwater and meltwater. Groundwater is contained within pores (as in a porous SANDSTONE) or JOINTS and BEDDING PLANES (as in CHALK or LIMESTONE). Where these are interconnected, lateral movement of groundwater is possible, and it may return to the surface by way of seepages and SPRINGS. Groundwater may occur in sufficient quantities to form an underground reservoir or AQUIFER from which water can be withdrawn by wells and borings. See also PHREATIC WATER, VADOSE WATER and WATER TABLE.

Group of Eight See G8.

growing season That part of the year when temperatures are sufficiently high to sustain plant growth (usually taken to be above 6°C). In practice, the growing season is defined not merely by the minimum growth temperature, but also by the occurrence of damaging night frosts; for example, the growing of cotton requires 200 days between the last 'killing frost' of spring and the first of autumn.

growth pole This consists of expanding industries that are concentrated in a particular area and which, being by nature PROPULSIVE INDUSTRIES, set off a chain reaction of minor expansions throughout the HINTERLAND of the concentration. The idea, first formulated by the French regional economist Perroux, has since been incorporated into the regional development policies of several MEDCS. It involves the deliberate selection of one or a few growth poles within a DEPRESSED REGION. Into these, new investment is concentrated rather than being thinly spread over the whole region. The justification for such a strategy is that public expenditure will be much more effective when concentrated in a few clearly defined areas. New industries there will stand a better chance of building up enough EXTERNAL ECONOMIES to achieve a basis for some self-generating growth. The experiences of Fos (S France) and of growth poles in the Mezzogiorno (S Italy) lead to some questioning of the soundness of the basic concept.

gruss A deposit formed from the chemical rotting of GRANITE.

guest worker A person who migrates to

another country for full-time employment, but who does not intend (or is not permitted) to settle permanently in that country. The receiving countries are usually those that suffer labour shortages, particularly in unskilled, poorly paid jobs. Examples of such temporary MIGRATION include the movements of Turkish men into former W Germany, of Algerians into France and of workers from Botswana into S Africa.

gully erosion A type of EROSION, of SOIL or soft rock, resulting from concentrated RUN-OFF (ct SHEET EROSION), and creating numerous deep gashes in the hillslope. Gully erosion is frequently induced by human activity, particularly in tropical areas where rainfall intensity is high. Gullying may result simply from the removal of vegetation or cultivation (particularly where land is ploughed up and down a slope), but can also be caused in many minor ways. For example, cattle tracks associated with compacted and IMPERMEABLE soil can become sites for gullies. However, gully erosion is also a 'geological' form of erosion in many BADLANDS, where impermeable rocks (CLAYS and SHALES), poor natural vegetation cover and occasional heavy downpours produce a close network of stream courses and periods of rapid downcutting.

Gutenberg channel A layer of less rigid (plastic) material, at a depth of 100–200 km beneath the Earth's surface, in which the speed of EARTHQUAKE waves is reduced. It is thought to coincide with the ASTHENOSPHERE and is thought to be the layer on which the rigid lithospheric plates move.

Gutenberg discontinuity The boundary between the solid mantle and liquid outer core of the Earth, at a depth of some 2900 km. At this boundary, seismic 'S' waves stop being transmitted (they only pass through solid matter) and 'P' waves slow down.

guyot A flat-topped hill rising from the deep ocean floor but not breaching the sea surface. Many guyots appear to be submarine volcanic peaks, up to 4000 m in 'height'. Having originated at active plate margins, they have subsequently been planed across by marine erosion as the plate has spread away (conveyor belt-like) from the MAGMA source (see SEA-FLOOR SPREADING).

H

habitat A term used in ECOLOGY to describe the specific ENVIRONMENT of plants and animals, in which they are able to live, feed and reproduce.

Hadley cell A feature of the Earth's general circulation in low latitudes, postulated by Hadley in 1735. Intense warming of the surface in the Equatorial zone causes large-scale uplift of air. This is compensated by a low-level flow towards the Equator from the sub-tropical ANTICYCLONE at about 30°N. This is itself formed partly from the descent of air that has moved at a high level away from the Equatorial zone. The surface winds in the resultant 'cell' are deflected by the Earth's rotation (see CORIOLIS FORCE) to result in the NE Trades. A comparable cell is identifiable in the Southern Hemisphere. The Earth's general circulation comprises, in addition to the Hadley cells, two other cells in each hemisphere: a northerly low-level flow from the sub-tropical high towards the POLAR FRONT

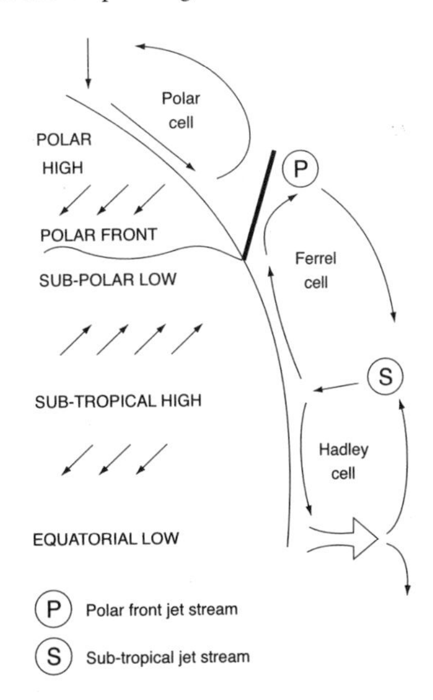

Simple tricellular model of atmospheric circulation

(with a compensating return flow at a high level); and a southerly flow from the polar anticyclone (and a compensating return flow at high altitudes). In more recent times this simple model of the Earth's general circulation has been modified to take account of JET STREAMS and horizontal patterns of air movement (such as migrating high- and low-pressure systems near the surface). [*f*]

hail (or **hailstone)** A pellet of ice, usually formed under conditions of extreme atmospheric INSTABILITY (e.g. in a thunder cloud), comprising concentric ACCRETIONS of opaque and clear ice. When a raindrop is carried upwards into a cooler environment by a rapidly ascending draught of air it will freeze. Supercooled droplets in the surrounding air, the result of condensation, will freeze on impact with the embryo hailstone, forming a layer of opaque ice. Other supercooled moisture will at times accumulate as a 'wet layer' on the hailstone; this will subsequently freeze to give a layer of clear ice. The banded structure of hailstones reflects the continual 'recycling' of hailstones, involving perhaps up-and-down movements in a thunder cloud. Hailstones normally have a diameter of up to 5 mm or so, but sometimes grow to the size of a golf ball, or even larger, and can cause considerable damage to property (e.g. cars) and fields of crops.

halophyte A plant that thrives in a salt-impregnated SOIL, as in an ESTUARY or on a coastal SALT MARSH, or in the presence of a salty spray (as on the lower parts of sea-cliffs). On salt marshes halophytes perform the important task of trapping SEDIMENT when inundated at high tide, thus causing the marsh to grow upwards over time. In this respect they can be considered a PIONEER SPECIES, the first stage in a VEGETATION SUCCESSION.

hamlet A small RURAL settlement (often no more than a cluster of a few houses), too small to be called a VILLAGE and usually lacking a church.

hammada (ramada) A stone pavement, comprising a relatively thin layer of angular or rounded fragments, in a hot DESERT. The concentration of surface fragments results from: the DEFLATION of SAND particles; the selective removal of fines by surface wash processes during infrequent, episodic storms; the upward migration of stones by wetting and drying of the surface soil layers.

hanging valley A tributary valley that joins a main valley by way of a sharp break of slope, often marked by a WATERFALL. The most frequent cause of hanging valleys is glacial EROSION. A large glacier in a main valley will cause overdeepening in excess of that achieved by a smaller glacier in the tributary valley. When the glaciers have melted, the tributary valley is left 'hanging' for as much as hundreds of metres above the main valley floor. When a river returns to the tributary valley, it cuts into the break of slope and, in time, the waterfall will be replaced by a series of rapids within a steep-sided gorge. An ALLUVIAL FAN may be formed where the tributary river joins the main river valley, and the rapid reduction in gradient causes a drop in the river's velocity leading to DEPOSITION.

harbour A haven or anchorage for shipping involving a stretch of sheltered water, close to the shore and protected from the open sea by natural (e.g. a SPIT or headland) or artificial (e.g. a breakwater or jetty) means. RIAS, with their deep and secluded channels, probably provide some of the finest *natural harbours*, such as those in Sydney, Australia, and Milford Haven, Wales. See MARINA.

hardness scale The hardness of rock minerals as expressed by Mohs' Scale of Hardness, with grades ranging from 1 (very soft – talc) to 10 (extremely hard – diamond).

hardpan A cemented layer within the SOIL (usually in the B-HORIZON) resulting from ILLUVIATION and the PRECIPITATION of leached minerals from the A-HORIZON. The hardpan may hinder plant growth, impede soil drainage and, when exposed at the surface by the removal of the topsoil, render cultivation difficult.

harmattan A very dry NW wind, blowing from the Sahara Desert towards the Gulf of

Guinea in W Africa. It is dominant from November–January, when the high-pressure cell over the desert is most strongly developed. The harmattan is characterized by extremely low RELATIVE HUMIDITIES (sometimes of less than 10% at midday) and cool temperatures. The wind is often dust-laden, up to heights of 3000–5000 m, and visibility may be reduced to 300 m or less. It offers a period of relief from the prevailing humid heat of W Africa. However, its effects are often far from benevolent. The onset of the wind is marked by an increase in respiratory infections, and germs borne by the dust particles appear to be responsible for outbreaks of cerebral spinal meningitis. The fire hazard, as a result of the extreme desiccation, can be serious.

hazard See NATURAL (ENVIRONMENTAL) HAZARD.

hazard assessment See NATURAL (ENVIRONMENTAL) HAZARD.

hazard perception See NATURAL (ENVIRONMENTAL) HAZARD.

HDI (human development index) Used as a measure of DEVELOPMENT in a country and for making international comparisons. The index takes into account three variables that are given equal weighting: income per capita; adult literacy; life expectancy. The HDI takes the highest and lowest values recorded for each variable by the countries of the world. The interval between them is given a value of 1 and the value of each country is then scored on a scale of 0 to 1 (from worst to best). The HDI is the average score of the three variables and so, too, is expressed as a value between 0 and 1. The wealthiest MEDCS have an index approaching 0.999 and the poorest LEDCS range down to less than 0.300. Cf PQLI.

head A deposit of poorly sorted stones and BOULDERS found at the foot of slopes and sea-cliffs, and resulting from past PERIGLACLAL activity. The debris, frequently angular, is released by FREEZE-THAW WEATHERING, and is then soliflucted down the slope or cliff-face to accumulate as a basal deposit, often several metres in thickness.

headward erosion A process frequently associated with steepened sections of a river's long-profile, whereby erosion takes place in an up-valley direction. One common example of headward erosion is the upstream recession of WATERFALLS. Headward erosion is also common at SPRINGS (e.g. in CHALK and LIMESTONE) where it is more specifically referred to as spring-sapping.

heat balance The average condition of balance in the ATMOSPHERE between incoming solar radiation and outgoing heat losses due to reflection or re-radiation from the surface. The heat balance thus prevents, at least in the short term, significant rises or falls in atmospheric temperatures. However, the heat balance is not maintained, in the terms stated, at every point on the earth. For example, at or near to the Equator more heat

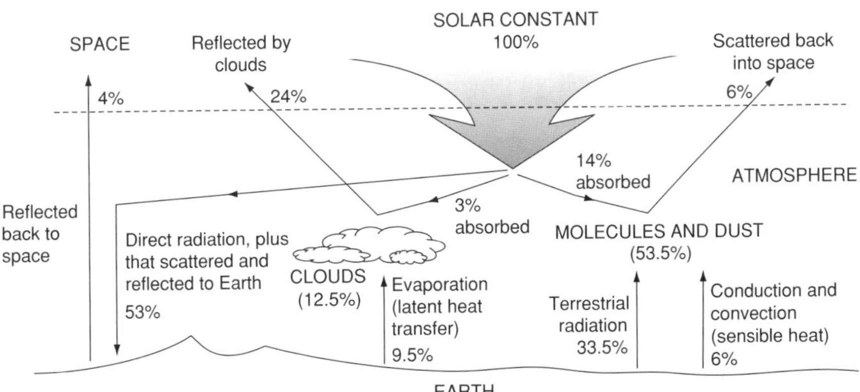

The Earth's heat balance

is gained from solar radiation than is lost by re-radiation, whereas in higher latitudes more heat is lost than is gained. Thus EQUILIBRIUM is maintained by the lateral transfer of heat, courtesy of migrating air masses and ocean currents that carry heat from warm to cold latitudes. [*f*]

heat island The phenomenon whereby temperatures within a CITY or large TOWN are often significantly higher than those of surrounding RURAL areas. The heat island effect is greatest at night, when city temperatures may be 6–8°C higher than those in rural areas, as warmth stored by URBAN surfaces during the previous day is slowly released. There are various factors involved, such as: the release of heat by domestic and industrial fuel consumption; the high capacity for heat absorption of many urban surfaces, such as brick walls and asphalt roads; the reduced consumption of heat by processes such as evaporation, owing to the loss of rainwater by rapid drainage and the resultant 'arid' environment; atmospheric pollution, which reduces losses of long-wave radiation from the ground at night. It has been shown that the centre of London has a mean annual temperature of 11.0°C, compared with 10.3°C for the SUBURBS and 9.6°C in rural areas. At 9pm on 26 February 1988, the centre of Oxford recorded a temperature of 4°C compared to about 0.5°C in the countryside just 7 km away. [*f*]

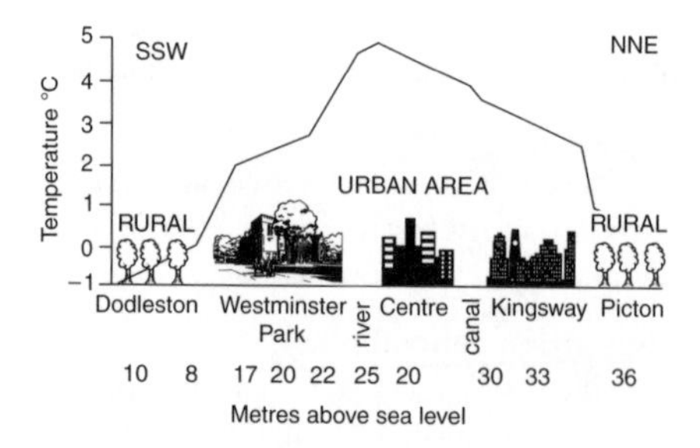

Urban heat island in Chester, England

heath (heathland) An area of largely treeless country, dominated by heathers, gorse and various grasses able to thrive on poor acidic SOILS (see PODSOL). True heaths (e.g. those of the New Forest and Dorset in S England) are confined mainly between 50° and 60°N, and are usually below 200 m above sea-level (ct MOORS). Heaths have been formed by the degeneration of former FOREST areas, as a result of the widespread burning of trees by Neolithic man, for cultivation and the pasturing of animals. This has been accompanied by soil impoverishment (by active LEACHING, especially in sandy areas), acidification and the formation typically of a thin surface peat layer. At the present time, heaths, which provide a unique environment for certain plants and animals, are being seriously reduced in extent by URBANIZATION, AFFORESTATION and RECLAMATION.

heavy industry BASIC INDUSTRY of national economic importance and in which large quantities of materials are handled or processed; e.g. coalmining, iron and steel production, shipbuilding, petrochemicals. Ct LIGHT INDUSTRY.

helicoidal (helical) flow A type of flow associated with sinuous, and in particular meandering, river channels. It is thought that the main current of the river will tend to flow in a straight line, thus impinging on the outer bank where the channel bends. As a result, a slight head of water will be built up, leading to a compensatory return flow across the channel. Since the maximum velocity in a river (the *thalweg*) is just beneath the water surface, this compensatory flow will be close to the channel floor, giving a circular motion when viewed in cross-section. This secondary flow is superimposed on the main downstream flow of the river, resulting in a type of spiral motion, or helicoidal flow. It is possible that helicoidal flow plays some part in the transference of SEDIMENT, eroded from the outer banks of MEANDERS, to the inside banks where it is deposited. [*f*]

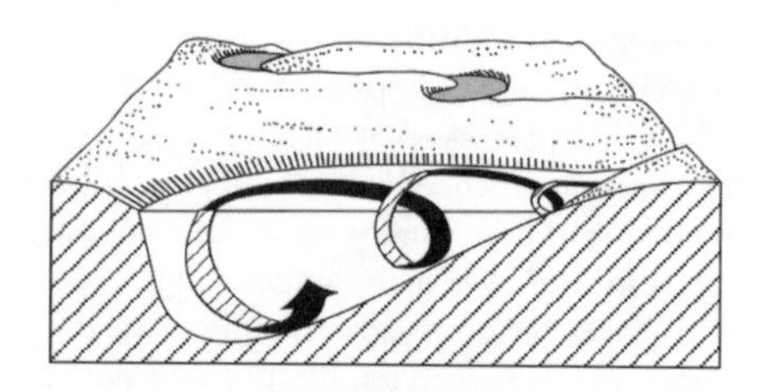

Helicoidal flow in a meandering stream

HEP (hydro-electric power) Electric power generated in turbines, the motive energy for which is derived from moving water. In some cases, HEP is generated at natural WATERFALLS (e.g. Niagara), in others it involves building a dam across a valley (e.g. Grand Coulee Dam, USA), while it can also be achieved by piping water down a mountainside. In all three instances, the critical requirement is a good head of water maintained at a constant rate of supply.

herbivore An animal that derives most of its food from eating plants. Herbivores represent the second TROPHIC LEVEL in the FOOD CHAIN.

heritage Anything inherited from the past. In geographical studies this includes both natural and human features. See HERITAGE TOURISM.

Heritage Coast A designation made in Britain since 1970 in order to ensure that stretches of coast of high scenic quality are protected from development and are subject to effective public management.

heritage tourism The marketing of landscapes, particularly their unusual scenic qualities and historic relics (e.g. stately homes, archaeological sites, old urban areas, etc.) as tourist attractions. A form of TOURISM that seeks to exploit the past and the unique as RESOURCES. Sometimes referred to as the *heritage industry.*

heterotroph An organism incapable of self-feeding, which derives its necessary supply of sugars and starches from AUTOTROPHS, e.g. a cow that eats grasses. Heterotrophs represent the second and third TROPHIC LEVELS in the FOOD CHAIN.

hierarchical diffusion A pattern and process of SPATIAL DIFFUSION (a sub-type of EXPANSION DIFFUSION) characterized by 'leapfrogging', whereby the diffusing phenomenon (be it an idea or an innovation) tends to leap over many intervening people and places (ct CONTAGIOUS DIFFUSION). In this instance, simple geographic distance is not always the strongest influence on the diffusion process. Instead, hierarchical diffusion recognizes that large places or important people tend to get the news first, subsequently transmitting it to others lower down the HIERARCHY. It occurs because in the diffusion of many things space is relative, depending on the nature of the COMMUNICATIONS network. Big cities, for example, linked by very strong information flows, are actually 'closer' than they are in a simple geographic space. The diffusion of clothing fashions provides a good example: new fashions are launched in the major fashion centres of London, New York and Paris, and first diffuse internationally to other capitals, before percolating down through the respective national settlement hierarchies. The frequently observed contrast between fashion in METROPOLITAN areas and in small country TOWNS is indicative of the *diffusion lag* that characterizes hierarchical diffusion. [*f*]

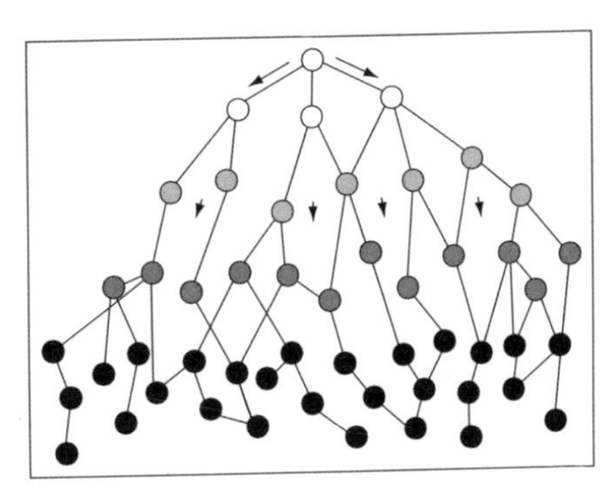

The process of hierarchical diffusion

hierarchic sampling See NESTED SAMPLING.

hierarchy A vertical class system made up of a series of distinct levels or orders (see CENTRAL PLACE HIERARCHY, URBAN HIERARCHY). Ct CONTINUUM; see also CLASSIFICATION.

high-order central place A CENTRAL PLACE providing *high-order goods and services*; i.e. goods and services with high THRESHOLD and RANGE values. A central place enjoying a high ranking or status within the CENTRAL PLACE HIERARCHY, such a status being more likely to be associated with a CITY rather than with a TOWN. Ct LOW-ORDER GOODS AND SERVICES.

high-technology industry Frequently referred to as 'high-tech industry', an industrial activity involving advanced technology and spearheading the fifth KONDRATLEFF CYCLE. The term covers quite a diversity of activities, such as the manufacture of semi-conductors, computers, microchips and industrial robots; the development of new metals and ceramics; telecommunication and information technology; biotechnology and fibre optics. A vital input to all high-tech industry is R&D, often undertaken at nearby universities, technical colleges and other public institutions. Thus, access to R&D establishments can constitute a significant location factor for these new industries, as does proximity to an airport in order to ensure speedy access to national and international markets. See also SUNBELT. Ct LOW TECHNOLOGY.

hill farming General farming carried out in upland areas, but in Britain the term is strictly applied, for government grant purposes, to a specific category of farming. It is particularly concerned with sheep-rearing and the raising of beef cattle.

hinterland (i) The area from which a PORT derives its exports, and within which it distributes its imports (ct FORELAND). (ii) The area from which a CENTRAL PLACE draws its customers, and throughout which it distributes goods and services. Alternative terms used in the context of CENTRAL PLACE THEORY include MARKET AREA, SPHERE OF INFLUENCE, *tributary area*, UMLAND and *urban field*.

histogram A graphical technique used for showing the FREQUENCY DISTRIBUTION of non-continuous or *discrete data*, i.e. data that are grouped into CLASSES (see CLASS INTERVAL). The class boundaries are marked off on the *x* axis. Vertical columns are then drawn proportional in length to the number of occurrences in each class – in other words, the vertical or *y* axis is used as the frequency axis. [*f* BINOMIAL DISTRIBUTION]

historical geography The geography of the past; the academic 'borderland' between geography and history.

Hjulstrom curve A GRAPH, based on the researches of Hjulstrom in 1935 into the movement of SEDIMENT by streams. A minimum velocity is required for the 'entrainment' (picking up) of particles of a given size, resting loosely on the bed or banks of a stream CHANNEL; this is the *critical erosion velocity*. The velocity at which particles are deposited is called the *critical deposition velocity*. Hjulstrom found that much lower EROSION velocities are needed to move SAND particles than either SILT (finer) or GRAVEL (coarser), and that erosion velocities for very fine particles (CLAY) need to be surprisingly high, owing to the fact that clay particles stick to each other, and form a smooth bed. In simple terms, the Hjulstrom curve shows that a channel in sand is 'unresistant', while one in gravel, silt or clay is 'resistant'. One result is that sandy channels are abnormally wide, owing to the rapidity of bank erosion. The situation is rather less complex with the critical deposition velocity, where, as velocity falls, so successively smaller particles are deposited. In between the two critical velocities, particles will be held in suspension and will be transported. [*f*]

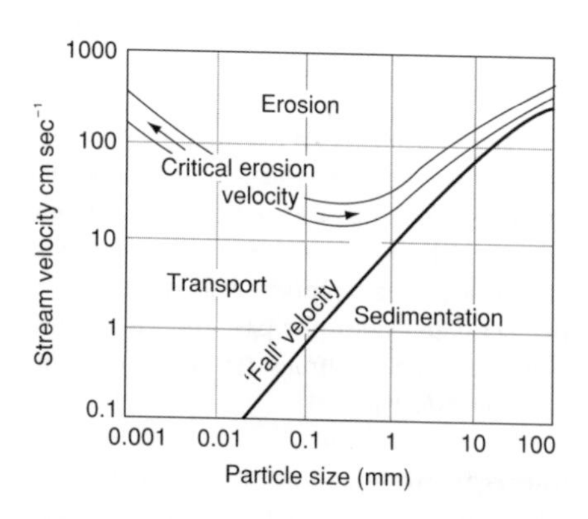

Hjulstrom curve

hoar frost The most common type of 'white' frost observable on a winter's morning, following a clear night and the rapid radiation of heat from the Earth's surface. As the latter is cooled below the DEW-POINT of the air lying immediately above it, CONDENSATION of water vapour directly on to the surface (whether SOIL or vegetation) occurs. If the temperature of condensation is below 0°C, ice is formed immediately. Alternatively, moisture condensed at above 0°C will become frozen if the ground temperature subsequently falls below 0°C. Ct RIME.

hog's back (hog-back) A narrow, broadly symmetrical ridge developed in a steeply dipping or vertical hard rock strata.

hollow block See DESERT VARNISH.

honeypots A term frequently used in the context of RECREATION and TOURISM to denote places of special interest or appeal that are highly popular with visitors and that tend to become very congested at peak times, e.g. Land's End, the Lake District, the Tower of London (see HERITAGE TOURISM). When it comes to planning for recreation and tourism, honeypot policies have often been adopted (especially in national parks) that encourage the development of selected honeypots with the aim of: ensuring that proper provision is made at these sites for the accommodation of visitor traffic (providing car parks, picnic sites, toilets, etc.); diverting traffic away from rather more 'sensitive' and unspoilt areas that could not possibly cope with large numbers of visitors without seriously detracting from their intrinsic appeal.

horizontal expansion One of three ways in which an enterprise may expand (ct DIVERSIFIED EXPANSION, VERTICAL EXPANSION), involving the *takeover* by a FIRM of other, presumably rival, companies in the same business. Such absorption of other firms provides the opportunity to increase the scale of production or operation, and thus for the enlarged enterprise to benefit from ECONOMIES OF SCALE.

horn See PYRAMIDAL PEAK.

horse latitudes The sub-tropical, high-pressure belts of the ATMOSPHERE, about 30°–35°N and 30°–35°S (though interrupted by the pattern of distribution of land and sea). These are zones of calm and descending air, from which air masses move poleward and Equatorward (see HADLEY CELL). The horse latitudes are said to derive their name from the throwing overboard of horses in transport from Europe to N America if the ships' passage was delayed by calms.

horst An upraised FAULT block, with sharply defined and sometimes parallel marginal faults. Initially, a horst is associated with the formation of a BLOCK MOUNTAIN; however, the latter may eventually be destroyed by continued DENUDATION, though the horst will persist as a geological structure. A horst is thus the converse of a GRABEN. Examples of horsts are the Black Forest and Harz Mountains in SW Germany and the Vosges in E France. [*f* FAULT]

horticulture Originally, the cultivation of a garden, but the term is now used more widely to include the intensive cultivation of vegetables, fruit and flowers. It includes MARKET GARDENING, *nursery gardening* and *glasshouse cultivation.*

hot spot An area where, beneath the Earth's crust, strongly localized rising currents of MAGMA known as *plumes* occur. The magma approaches the base of the crust, and then spreads horizontally in all directions in the subcrustal zone. Where the crust above hot spots is weak, volcanic activity occurs, as in the Hawaiian islands. Here magma succeeds in penetrating the Pacific Ocean Plate. One theory is that hot spots remain 'stationary', while overlying plates undergo lateral movement (see PLATE TECTONICS). As a result, a line of VOLCANOES, decreasing in age towards the present-day location of the hot spot, will be formed over a long period of geological time. Volcanic chains formed in this way should be contrasted with those developed along major fractures in the Earth's crust, such as fault-lines, and constructive and destructive plate margins.

hot spring Otherwise referred to as *thermal springs*, hot springs occur in areas of vulcanicity, particularly during the declining phase of activity. Hot water (from 20°C to boiling point) emerges more or less continuously, by contrast with the spasmodic ejection associated with GEYSERS.

housing association A non-profitmaking organization set up in the UK to provide rented accommodation. During the late 1980s and early 1990s such organizations gradually replaced local authorities as providers of this type of housing, particularly for those in need (such as the poor, the single-parent family, the elderly and those with disabilities). Since then, however, housing associations have also become providers of housing for owner-occupiers – for example, by renovating derelict properties and selling homes that have been repossessed by banks and building societies.

housing class A group of people occupying the same broad type of residential accommodation and experiencing the same sort of residential conditions. Seven housing classes have been recognized in Britain: (i) outright owners of large houses in desirable areas; (ii) mortgage payers who 'own' their houses; (iii) council tenants in public (i.e. council) housing; (iv) council tenants in slum houses awaiting demolition; (v) tenants of private house owners, usually in the INNER CITY; (vi) house owners who must take lodgers in order to meet loan repayments; (vii) lodgers in rooms. Access to these different housing classes depends, to a large extent, on income, but is also controlled by GATEKEEPERS and URBAN MANAGERS.

Howard See GARDEN CITY.

Hoyt See SECTOR MODEL.

hum A residual hill of LIMESTONE, in a late-stage KARST landscape. See also MOGOTE.

human development index See HDI.

human ecology The application of the concepts of ECOLOGY to investigation of the relationships between people and their ENVIRONMENT. It is to be seen in Burgess's CONCENTRIC ZONE MODEL of city structure. See also INVASION AND SUCCESSION.

human geography That part of GEOGRAPHY concerned with the spatial analysis of the human population. It encompasses such aspects of the population as numbers, composition, economic and social activities, and SETTLEMENT, with each aspect tending to generate its own branch of human geography (see POPULATION GEOGRAPHY, CULTURAL GEOGRAPHY, ECONOMIC GEOGRAPHY, SOCIAL GEOGRAPHY, URBAN GEOGRAPHY).

human rights The UN Universal Declaration of Human Rights (1948) states that all people have the right to: life, liberty and education; freedom of movement, religion, association and information; a nationality; equality before the law.

humus Organic material within the SOIL derived from leaves, twigs, roots, dead organisms, animal excreta, etc., which has gradually been broken down by chemical decay and the activities of micro-organisms into a finely divided, dark-brown gelatinous substance. Within the soil layers (particularly the A-HORIZON) the humus enters into a complex chemical relationship with CLAY minerals, forming the so-called CLAY-HUMUS COMPLEX or *colloid.* In cool, moist climates, organic matter may accumulate at the soil surface as a *raw humus layer.* In coniferous forests and heathlands, plant litter is composed mainly of compounds that decay very slowly, giving so-called *mor humus.* This is strongly acidic, discourages soil organisms, and is associated with slow *humification* (the decay and incorporation of humus into the soil). By contrast, temperate deciduous woodlands, with extensive autumn leaf fall, provide *mull humus.* This is more readily decomposed and is rich in soil organisms, especially earthworms, which help to mix the humus and soil particles; it is also associated with base-rich soils.

hurricane A very powerful tropical storm, characterized by winds of extreme velocity (in excess of 120 km/hr) and capable of causing widespread damage on land, as well as constituting a serious hazard to shipping. Hurricanes initially form mainly in latitudes 5–10°N and S over the western sections of the Atlantic (particularly the West Indies and Gulf of Mexico region), the Indian and Pacific Oceans. They are more frequent in the Northern Hemisphere (annual frequency 50–60) than in the Southern Hemisphere (annual frequency 20–30). The hurricane season in the N Atlantic lasts from July to October, and occurs when the equatorial low-pressure zone is displaced northwards. Each individual hurricane normally lasts from two to three days. The storms are generated over warm ocean waters (with surface temperatures above 27°C), and the energy 'driving' them is derived from the large-scale CONDENSATION of moist air and the resultant release of large amounts of latent heat. For further details of pressure conditions, winds, weather and track, see CYCLONE.

hydration A process of CHEMICAL WEATHERING whereby certain rock minerals take up (absorb) water. In the process changes in volume occur, physical stresses are set up within the rock and, as a result, *physical* disintegration can occur, usually in the form of small-scale flaking from the rock surface.

hydraulic geometry A study of the changing geometry of a stream CHANNEL, expressed in parameters such as width, depth and slope, and in relation to variations in DISCHARGE.

hydraulic ratio The ratio between the cross-sectional area of a river CHANNEL (d x w) and the length of the *wetted perimeter* (the line of contact between the water and the channel). Hence R = A/P, where R is the hydraulic radius, A is cross-sectional area, and P is the wetted perimeter. Hydraulic radius is a measure of the *efficiency* of a stream, referring to the proportional losses of energy by friction between the flowing water and the channel bed and banks, as compared with the losses within the water. Large values for R are associated with streams with large DISCHARGES, and with cross-sections that are approximately semi-circular. Conversely, low values (and thus reduced efficiency) are given by small streams with a considerable width:depth ratio.

hydro-electric power See HEP.

hydrograph A GRAPH on which variations of river discharge (in m^3/s) are plotted against time. Hydrographs are often characterized by prominent peaks, representing increased channel flow following periods of heavy rain or the melting of a thick snow cover. In form, these peaks are markedly asymmetrical, with a steep *rising limb* (representing a rapid increase in the amount of water reaching the channel via OVERLAND FLOW), and a gentler, *concave falling limb* (representing a decrease in the amount of water reaching the channel). In the case of a hydrograph peak generated by a rainstorm, the time interval between the peak rainfall intensity and the maximum river flow, is known as the *lag-time*. A short lag-time is usually associated with rapid water transfer involving overland flow. There are a number of variables that determine the effectiveness of overland flow, including SOIL type, underlying rock, the antecedent soil moisture conditions (reflecting previous rainfalls), vegetation cover and the steepness of basin slopes. Peaks on the hydrograph are superimposed on a more steady river flow, of smaller volume, derived from BASE FLOW. This results from the slow but continuous entry of GROUNDWATER, via springs and seepages, into the river CHANNEL. Sometimes, hydrograph peaks are composite, with an early and prominent peak due to overland flow (*quickflow*), and a later, more protracted but subdued peak representing water that has travelled to the channel more slowly via THROUGHFLOW routes. [*f*]

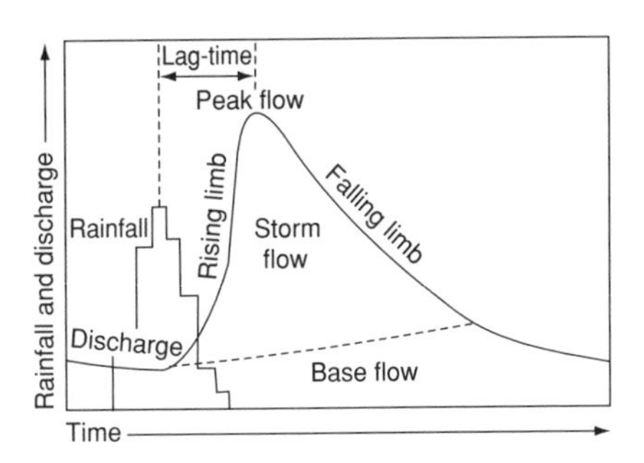

Hydrograph showing a run-off peak following a rainstorm

hydrological cycle The unending transference of water from the oceans to the land (via the ATMOSPHERE), and vice versa (via rivers). In broad terms, the hydrological cycle involves the following sequence. Water is transferred from the surface to the atmosphere by evaporation from sea, lake and land surfaces, and *transpiration* from growing vegetation. It is then transferred within the atmosphere, both vertically (by *convection*) and laterally (by *winds*) where it cools and condenses. PRECIPITATION then transfers the water back to the surface where some of it

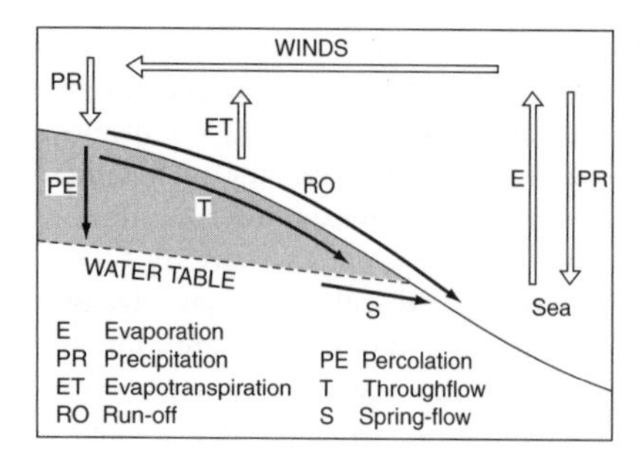

The hydrological cycle

lands directly in the oceans and the rest falls on land, to flow as surface run-off or percolate to supply GROUNDWATER. This in turn re-emerges by way of SPRINGS to augment river flow. In reality, the hydrological cycle is extremely complex. [*f*]

hydrology The scientific study of water, both surface and underground, including its properties, distribution, movement and use by people.

hydrolysis A complex process of CHEMICAL WEATHERING, involving a reaction between water and a rock mineral (ct HYDRATION). The process is involved in the WEATHERING of GRANITE.

hydrophyte A plant that grows in water or SOIL that is saturated by water. Good examples are papyrus and mangrove, a small tree with stilt like roots that rapidly colonizes bare or submerged mud along rivers and the fringes of ESTUARIES in the tropics.

hydrosphere The total 'free' water at the Earth's surface, in either liquid or solid form. The most important component of the hydrosphere, by far, is oceanic water (volume 1350 x 106 km^3), but significant amounts are contained within lakes and rivers (0.2 x 106 km^3), and ICE SHEETS and glaciers (26 x 106 km^3). Additionally, 7 x 106 km^3 is stored as GROUNDWATER within the LITHOSPHERE and 0.01 x 106 km^3 as water vapour within the ATMOSPHERE.

hypabyssal A type of INTRUSIVE IGNEOUS ROCK formed at intermediate depths within the Earth's crust, usually in the form of DYKES or SILLS.

hypermarket Large RETAILING development (over 4645 m^2) of floor space) catering mainly for car-owning customers, usually strategically located at points of good ACCESSIBILITY in a regional road NETWORK. The term is increasingly misused to denote any sizeable retailing development. Most often, a hypermarket accommodates a range of retail outlets under one roof, and provides large car parks and services such as cafés and restaurants, petrol stations, banks and launderettes. Many have been built in N America and in France on GREENFIELD SITES, but in Britain planning policies have been much 'less permissive' and as a result, such developments tend to be found in URBAN (e.g. Brent Cross, N London) and suburban (e.g. Asda, Southampton) areas. Ct RETAIL PARK, SUPERMARKET, SUPERSTORE.

hypothesis A proposition to be proved or disproved by reference to facts, e.g. that the volume of MIGRATION is inversely proportional to the distance travelled. Once proven, the hypothesis graduates to the status of *theory* or LAW. See NULL HYPOTHESIS.

I

ice age A geological period of widespread glacial activity, when continental ICE SHEETS developed and glacial EROSION, TRANSPORT and DEPOSITION operated on a massive scale. The most recent ice age was that of the PLEISTOCENE period. This began some 2 million years ago. Although a major withdrawal of the ice occurred some 10,000 years BP, it is unlikely that the Pleistocene ice age has actually terminated. Within the Pleistocene itself, many important fluctuations of climate took place, giving alternating glacial and INTERGLACIAL periods. It was formerly believed that there were four main glacial periods, but more recent evidence from ocean-floor cores has indicated the possibility of up to 20 glacial periods (each of up to 100,000 years in duration), separated by brief interglacials of some 10,000 years.

ice cap A continuous area of land ice that is less extensive than an ICE SHEET. Two types have been identified. *Mountain ice caps* occupy upland PLATEAUS, and attain thicknesses in excess of 500 m. At their margins rapidly flowing *outlet glaciers* cause considerable dissection of the plateau edges. *Lowland ice caps* are found in areas of gentle RELIEF at lower elevations in the High Arctic.

ice fall A very steep section of a valley glacier, comprising numerous deep CREVASSES and, at the top of the ice fall, broken ice masses and pinnacles (*seracs*). Ice falls occur at the head of a glacial trough, where tributary glaciers from CIRQUE basins converge, or where the glacier passes over a bar of resistant rock. Flow velocities on ice falls are high, possibly in the order of 1000 to 2000 m/yr.

ice segregation See GROUND ICE.

ice sheet A very extensive, continuous area of land ice that may, at its margins, grade into sea ice. Only two ice sheets exist at present, those of Antarctica (covering some 11.5 x 10^6 km^3) and Greenland (1.7 x 10^6 km^3). However, in the recent past continental ice sheets also covered much of NW Europe and N America (the *Pleistocene ice sheets*). The ice sheets of Antarctica and Greenland are characterized by great thicknesses of ice (2000 m on average in the former), very low temperatures (with the result that even the basal ice is locally at -30°C), and relatively low rates of accumulation and ABLATION (and consequently low velocities).

ice wedge A tapering mass of GROUND ICE, formed in a fissure several metres in depth under PERIGLACIAL conditions. Where there is intense cold, the SOIL OR REGOLITH will contract to form a polygonal network of deep cracks (*fissure polygons*). In summer these may become occupied by meltwater, which in the following winter will freeze, expand and enlarge the fissures laterally, giving rise to *ice-wedge polygons.*

iceberg A large mass of floating ice that has broken away from an ICE SHEET or glacier terminating in water. The most important source of icebergs is the Ross Ice Shelf in Antarctica. As the ice, moving out from the Antarctic ice sheet crosses the sea bed, it at first remains grounded; however, the sea eventually becomes deep enough to float the 300–400 m layer of ice, leading to the breaking away of many tabular ice masses beyond the 'grounding line'. The icebergs then drift northwards, at a rate of several km a day, and can penetrate as far as 60°S in the Pacific Ocean. In the Northern Hemisphere many icebergs are derived from the CALVING of ice at glacier snouts – for example, along the coast of Alaska.

igneous rock A type of rock formed by the solidification of MAGMA, either within the Earth's crust (INTRUSIVE ROCK) or at the surface (EXTRUSIVE (VOLCANIC) ROCK). Igneous rocks vary according to: chemical composition (whether the magma is silica rich, as in GRANITE, or basic, as in BASALT, dolerite and gabbro); the rate of cooling, which is rapid at the surface, giving fine-crystalled or glassy rocks such as basalt, and much slower at great depths, giving coarse-crystalled rocks such as granite or gabbro). Some igneous rocks develop from SEDIMENTARY ROCKS that have undergone an intermediate stage of powerful METAMORPHISM. The latter process may be so extreme that magma is 'recreated', as in the emplacement of large granite BATHOLITHS.

illuviation The deposition of soil matter (organic HUMUS, CLAY-size particles and solutes) in an illuvial B-HORIZON, having been removed by ELUVIATION from the A-HORIZON. Concentrations of clay particles may result in the formation of a claypan, and concentrations of iron may form an ironpan (see HARDPAN).

IMF (International Monetary Fund) This was established in 1944 as an agency of the UNO. It exists to promote international monetary cooperation, the expansion of world trade (cf GATT, WTO) and exchange stability. It also aims to eliminate foreign exchange restrictions. The IMF is not a bank in the sense that, strictly speaking, it does not lend money, but it does help member countries to acquire foreign currency. See G8.

immature soil A 'young' SOIL in which

soil-forming processes have had little effect. The mineral content of immature soils is high, the HUMUS content is low, and there is little or no development of SOIL HORIZONS. See AZONAL SOIL, RENDZINA.

immigration The act of moving into an area with the intention of settling in it. The term is most commonly used when referring to the movement of people across national FRONTIERS. Ct EMIGRATION.

imperfect competition A market situation in which neither absolute MONOPOLY nor PERFECT COMPETITION prevails; a situation closest to real life in most circumstances. It is characterized by the ability of sellers to influence demand by product branding and advertising, by restricting the entry of competition through restrictive practices, by the existence of UNCERTAINTY and the absence, to varying degrees, of price competition.

imperialism The policy of making an empire whereby a powerful STATE develops a relationship (e.g. through TRADE, military protection, technical AID) with other and dependent territories, as formerly between Britain and her colonies and between the Former Soviet Union (FSU) and her dependent countries in E Europe (cf NEOCOLONIALISM). Imperialism and COLONIALISM are sometimes taken as being synonymous.

impermeable A term describing rocks or superficial deposits that do not allow water to pass through them. This is due either to the lack of interconnection between pores, or the absence of JOINTS and BEDDING PLANES (as in some types of MASSIVE rock, including dolomitic LIMESTONES). Impermeable rocks are characterized by supporting surface drainage, and rates of fluvial EROSION are relatively high, although absolute amounts of erosion will depend on other factors, such as rock type.

impervious A term that is sometimes regarded as synonymous with IMPERMEABLE, but is also used to denote rocks that do not possess JOINT planes or fissures along which water movement can occur (ct PERMEABLE). In other words a rock may be non-porous and impermeable (such as an individual block of GRANITE), and also impervious because there are no interconnected cracks (as in a massive gabbro). Ct PERVIOUS.

import A good or service that is brought into a country from another in the context of international TRADE. Ct EXPORT.

import control Government-created devices (see QUOTAS, TARIFFS) designed to limit the amount of imports entering a country and so ensure a more favourable BALANCE OF TRADE.

import penetration The degree to which a country is dependent upon imports. For example, the UK has a high import penetration in motor vehicles in that many of the vehicles sold there are made overseas.

import quota See QUOTA.

import substitution The process whereby a country seeks to produce for itself goods and services that it once imported. Hence *import substitution industries* (ISIs). Ct EXPORT SUBSTITUTION.

improved land An area of land that has been made more productive, as by drainage, RECLAMATION or the application of FERTILIZERS. Use of the term is usually restricted to the context of AGRICULTURE.

improvement See URBAN RENEWAL.

incised meander A general term for a meandering river that has cut down into its bed as a result of REJUVENATION, resulting in the meanders being well below the level of the adjacent land. If rejuvenation is relatively slow, the meander will continue to erode laterally as it downcuts so forming a 'new' valley that is asymmetrical in profile – this is termed an *ingrown meander*. Good examples are found along the Seine valley between Paris and Le Havre. By contrast, if downcutting is rapid, a narrower valley with symmetrical sides is more likely to form. This is known as an *entrenched meander*. Excellent examples are found in the valley of the Wear at Durham. [*f*]

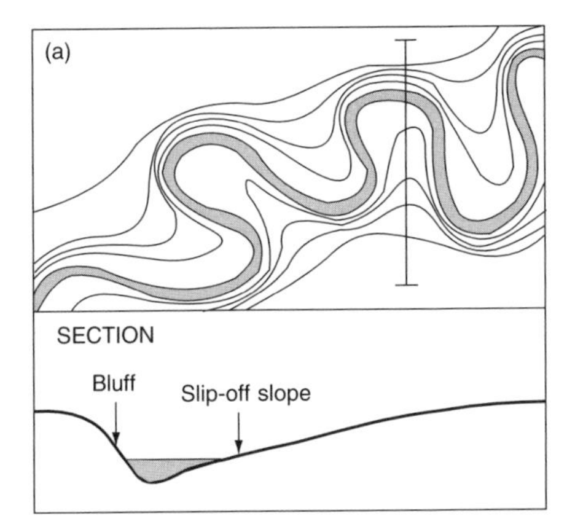

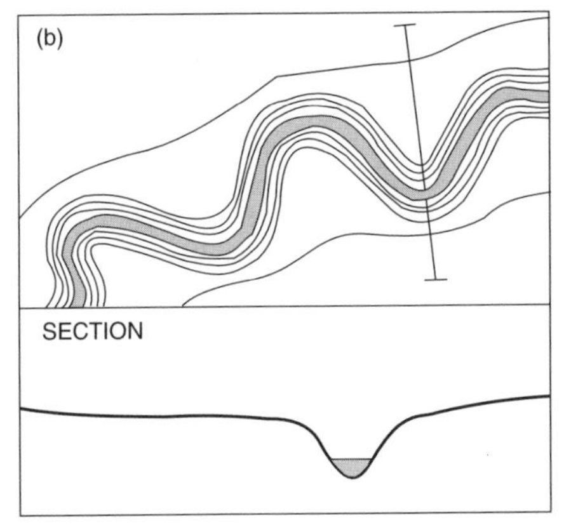

(a) Ingrown meanders and (b) intrenched meanders

inconsequent A drainage pattern that shows no clear relationship to the existing geological structure and pattern of rock outcrops. See DISCORDANT drainage.

incubator hypothesis See SEEDBED GROWTH.

incumbent upgrading A type of residential change whereby old and substandard housing is voluntarily improved by low-income people. In contrast to GENTRIFICATION, incumbent upgrading does not involve any degree of social upgrading – in effect, it halts the FILTERING process.

independent variable See DEPENDENT VARIABLE.

index numbers Numbers that show the change in a variable over time. The value of the variable in a given base year is expressed as 100. Values for subsequent years are then expressed as a percentage of the base-year value.

index of change See INDEX OF DISSIMILARITY.

index of circuity See DETOUR INDEX.

index of circularity See SHAPE INDEX.

index of dissimilarity A technique used to measure the differences between any two sets of paired percentages, e.g. comparing the percentages of the working population engaged in each of the four economic sectors at the beginning and the end of a 10-year period. The formula for the index of dissimilarity is:

either $\Sigma\,(a - b)$, where a is $> b$,
or $\Sigma\,(b - a)$, where b is $> a$,

where a and b are the paired percentage values to be compared. If both sets of percentages add up to exactly 100, then both formulae will give exactly the same result. The index will range from 100 (maximum dissimilarity) to 0 (complete similarity). It is sometimes referred to as an *index of change* or the *locational coefficient.*

The technique may also be used to measure spatial concentration. Given the percentage of the national population living in each regional subdivision and the percentage of the national area occupied by each region, and using the same formulae as above, the higher the value of the index, the greater the dissimilarity between the two sets of figures; therefore, the greater the concentration of population in a particular region or regions. Cf LOCATION QUOTIENT.

index of residential segregation See INDICES OF SEGREGATION.

index of similarity This is a more general measure of relative concentration or dispersal than the LOCATION QUOTIENT. It compares two data sets, as for instance the percentages of a city's retired population resident in different districts might be compared with the percentages of the city's total population resident in those same districts. The formula for the index of similarity is:

$$I = \frac{\Sigma d}{100}$$

where *d* is the difference between the paired data sets. Either the positive or negative differences can be used since both should be equal. The lower the index, the greater the dissimilarity between the two data sets and therefore, in the example given above, the greater the concentration of retired people in a particular district or districts. Cf INDEX OF DISSIMILARITY.

indicator species A species of plant or animal that has particular environmental requirements, and can thus be used as an indicator of the condition or quality of an environment. These requirements may relate to climatic or soil characteristics, or to the non-pollution of the environment. An example of an indicator species is the peregrine falcon, a predator of pigeons, doves and other bird species, which experienced a serious decline in Britain from 700 breeding pairs before 1939 to probably fewer than 300 by the early 1960s. This fall was associated with the deaths of adult birds, and also a failure to breed successfully owing to thin eggshells, infertile eggs or the inability to lay eggs at all. Research indicated that the cause was environmental pollution, involving the use of insecticides such as DDT and other chlorinated hydrocarbons, which were absorbed by the peregrines through the prey species. Since the banning of the use of DDT, and with the aid of rigid protection, peregrine numbers have recovered spectacularly to well over 1000 pairs, thus indicating a return to 'cleaner' environmental conditions.

indices of segregation The term given to a range of different statistical measures employed in SOCIAL GEOGRAPHY to measure the degree of social stratification and residential segregation within a given population. Of these, four have proved to be particularly widely used: (i) the INDEX OF DISSIMILARITY, used where the focus of investigation is on the percentage distribution of two specific groups; (ii) the *Index of Residential Segregation*, which indicates the percentage difference between the distribution of one group and the distribution of the rest of the population (index values range from 0 to 100 – the higher the value, the greater the segregation of that particular group); (iii) the LOCATION QUOTIENT and INDEX OF SIMILARITY may also be used to assess the relative concentration of a particular group within a given area; (iv) the LORENZ CURVE, which provides a graphical method for showing segregation.

industrial archaeology (heritage) The study of the remains of past industrial activity – as, for example, of old mills, defunct mines and smelting works, disused canals and railway lines. At present, there is much interest in preserving such relics and promoting them as part of HERITAGE TOURISM. The Ironbridge Gorge Museum is a good example, being especially important as the birthplace of the Industrial Revolution.

industrial complex A large concentration of manufacturing in a relatively small area. Firms located in such a complex benefit from AGGLOMERATION ECONOMIES, particularly so far as INFRASTRUCTURE is concerned. Such benefits include good TRANSPORT and COMMUNICATIONS, and access to a wide range of support services.

industrial estate See TRADING ESTATE.

industrial geography The study of the spatial arrangement and organization of industrial activity. As a sub-branch of ECONOMIC GEOGRAPHY, it is principally concerned with SECONDARY SECTOR or manufacturing activity. See WEBER'S THEORY OF INDUSTRIAL LOCATION.

industrial inertia The tendency for an industry to remain located in an area when the reasons for it first locating there are no longer significant or no longer exist. Examples are when sources of fuel, ore or other RAW MATERIALS become exhausted in one locality, or when they can be obtained more economically elsewhere. Inertia is sustained in a number of ways, as for example by FIXED COSTS (capital investment in infrastructure), by the existence of tight LINKAGES with other activities located in the area, or by the persistence of a pool of specifically skilled labour. Sometimes referred to as *locational inertia.*

industrial linkage See LINKAGE.

industrial location theory An aspect of INDUSTRIAL GEOGRAPHY that came to prominence in the 1960s, possibly stimulated by the much earlier theory of industrial location propounded by Weber (see WEBER'S THEORY OF INDUSTRIAL LOCATION). The work of Weber and other theorists, such as Hoover, Isard and Lösch, typically make assumptions that are hardly tenable. For this reason, industrial location theory currently receives little support, being superseded by suboptimal (see SATISFICER CONCEPT), behavioural and structural (see STRUCTURALISM) approaches.

industrialization The process whereby industrial activity (particularly manufacturing) assumes a greater importance in the ECONOMY of a country or region. It is widely thought of as a basic dimension of DEVELOPMENT. While it undoubtedly creates 'benefits' in the form of economic wealth and higher standards of living, it is necessary to bear in mind some of its costs, such as ENVIRONMENTAL POLLUTION, ecological damage, exhaustion of *non-renewable* RESOURCES and excessive reliance on vast supplies of energy. For much of the 20th century, it was believed that industrialization offered the best path for LEDCS to follow in order to stimulate the DEVELOPMENT process. Industrialization is now beginning to take place in some of those countries, as TNCS keen to exploit the vast supply of cheap labour set up branch factories (see GLOBAL ECONOMY, ECONOMIC GLOBALIZATION). This may look promising, but it remains to be seen whether this form of industrialization will bring lasting benefits to the people of those countries. There is real concern that it will lead to an unacceptable degree of economic and political DEPENDENCE on 'foreign' countries and companies (see NEOCOLONIALISM).

industry In its widest sense, any work or activity undertaken for gain and that promotes employment (e.g. AGRICULTURE, MANUFACTURING, RETAILING, TOURISM). Sometimes the term is used very specifically to denote simply the manufacture of goods (e.g. as traditionally in INDUSTRIAL GEOGRAPHY). In Britain, the government makes the distinction between MANUFACTURING INDUSTRY and *basic industry*, the latter comprising mining, quarrying, PUBLIC UTILITIES, transport and communications, agriculture, fishing and forestry. Other possible distinctions include the differentiation of LIGHT INDUSTRY and HEAVY INDUSTRY, or of EXTRACTIVE INDUSTRY, MANUFACTURING INDUSTRY and SERVICE INDUSTRY.

infant mortality The average number of deaths of infants under 1 year of age per 1000 live births. High infant mortality is often indicative of limited medical services, malnutrition and retarded DEVELOPMENT.

inferential statistics Statistics used to make statements about a carefully defined POPULATION from a properly selected random sample of that population (see RANDOM SAMPLING); statistical techniques used to test how far samples represent the whole (see SAMPLING). To this end, inferential statistics rely on PROBABILITY theory and on the application of SIGNIFICANCE TESTS. Sometimes referred to as *inductive statistics*, they may also be used in the testing of hypotheses. Ct DESCRIPTIVE STATISTICS.

infiltration The movement of water, derived from rainfall or melting snow, into the SOIL. The rate of infiltration, called the *infiltration capacity*, depends on several factors, such as soil POROSITY, the degree of compaction of the soil surface, the presence of plant roots, and the degree to which SOIL MOISTURE is already present (the antecedent condition of the soil). When rain occurs, infiltration is at first rapid. However, as the empty voids within the soil become filled with water, the rate of entry will slow down to equal the amounts lost at the base of the soil, or by THROUGHFLOW. When rain falls at a rate exceeding infiltration capacity, OVERLAND FLOW (*infiltration-excess overland flow*) will take place. However, this is likely only in extreme storm conditions or in poorly vegetated semi-arid regions.

inflation A process combining rising prices and devaluing money. A condition in which the volume of purchasing power in a given territory is constantly running ahead of the OUTPUT of goods and services, with the result that as incomes and prices rise, so the value of money falls.

inflationary spiral An upward movement of prices that is partly the result, and partly the cause, of increases in wages and salaries, dividends, interest rates, etc.

informal sector The part of the ECONOMY that comprises a range of activities. In MEDCS, these are mostly of a service kind and are undertaken by individuals. There is little or no official control over them. It includes activities such as domestic cleaning, child-minding, bar-tending and various forms of casual labouring, where payments are made, but often not declared, thereby avoiding payment of income tax and national insurance contributions. It is popularly referred to as the *black economy*, because of its undisclosed and clandestine nature. In LEDCS, the informal sector shows the following recurring features: a preponderance of self-employed people and small family enterprises; a high incidence of children in the workforce helping to supplement family incomes; the production of low-quality goods through the recycling of waste and use of low-grade raw materials; the marketing of goods and services mainly undertaken on the street.

information technology See IT.

infra-red radiation Invisible electromagnetic waves that are perceived as heat (see INSOLATION). Because of their wave lengths, they are able to penetrate cloud and darkness. Infra-red photographic devices are used in REMOTE SENSING and are able to distinguish features of the Earth's surface regardless of light or atmospheric conditions.

infrastructure The AMENITIES and services that are basic to most types of human activity. These include the provision of roads, power supplies, communications, water and sewage disposal systems. Some would define the above examples as *physical infrastructure*, as distinct from *social infrastructure* that includes such things as schools, hospitals and prisons. Both are provided at a community, regional or national level mainly by public funds, to which individual users contribute largely through the levy of RATES, taxes and standing charges. See ECONOMIC OVERHEAD CAPITAL.

initial advantage The advantage that accrues to a FIRM, CITY, REGION or country as a result of it being first – be it in the introduction of a new product, the exploitation of a new RESOURCE, the commercial application of new technology or in the opening up of a new market. Once the initial advantage has been seized, then the process of CUMULATIVE CAUSATION sets in, reinforcing the benefits to be gained.

inlier An outcrop of older rock, surrounded by younger rocks, that has been revealed by the localized removal of the younger rocks by EROSION. Inliers are commonly associated with ANTICLINES and domed structures. Erosion of these will expose older rocks brought up along the fold AXIS (see BREACHED ANTICLINE).

inner city That part of the BUILT-UP AREA close to the city centre, considerable areas of which are characterized by old and obsolescent housing, by MULTI-FAMILY OCCUPATION of dwellings, by poor commercial and SOCIAL SERVICES, and by the presence of industrial and commercial enterprises that contribute to the URBAN BLIGHT typical of such areas. Inner-city populations tend to comprise three distinct groups: (i) indigenous residents – mainly people in the older age groups; (ii) immigrants (often belonging to minority ETHNIC GROUPS); (iii) transients – people on the move who use the area as a temporary base. The last two groups move in because their financial circumstances oblige them to occupy low-rent accommodation, while the first group of people remain there because of their *immobility* or inability to afford better and more expensive housing elsewhere, and because they need to be close to the city centre (principally to save on transport costs). The inner city broadly coincides with the ZONE OF TRANSITION in the CONCENTRIC ZONE MODEL. In many countries, such areas have been the subject of large-scale URBAN RENEWAL schemes. See also GENTRIFICATION.

inner-city decline The loss of population and employment due to the deterioration in the ENVIRONMENT and services of the INNER CITY. A phenomenon increasingly experienced by large cities to the extent that it is now a major issue for city government and PLANNING. It is doubtful whether the decline can be attributed to one single factor. Rather it appears to have been triggered by a number of interacting causes, such as the general deterioration of the urban fabric that inevitably comes with age, the perception of the good life as being found in the SUBURBS and beyond, increased personal MOBILITY, enabling people to live further from the city centre, the availability of more modern (often cheaper) housing and premises elsewhere, and so on.

Another element in the decline is the fact that the process has its own negative MULTIPLIER EFFECT – it creates its own *downward spiral.* As people and firms move out, the revenue derived by the local authority from taxes also diminishes. This leads to reduced investment in the maintenance of INFRASTRUCTURE and SOCIAL SERVICES, thus contributing to the general deterioration that goes along with the ageing of the urban fabric. This rundown, in its turn, serves only to persuade more people and firms to leave, and such further losses mean diminished support for commercial services already badly hit by the loss of customers ... and so the decline gathers momentum. Clearly, there is a need for something to be done, if not to bring about the revival of the inner city, then at least to halt the accelerating decline. Much has already been attempted, but little has as yet been accomplished. There is no obvious or simple solution, neither are unlimited funds available. Ct GENTRIFICATION. [*f*]

innovation The introduction of new ideas and new ways of doing things; the act of making changes. The spread of innovation in the twin dimensions of space and time is a field of geographical study that rose to prominence in the second half of the 20th century (see SPATIAL DIFFUSION).

innovation wave The SPATIAL DIFFUSION of innovation over time takes a wave-like form. The wave originates at the point of innovation and moves outward, its amplitude

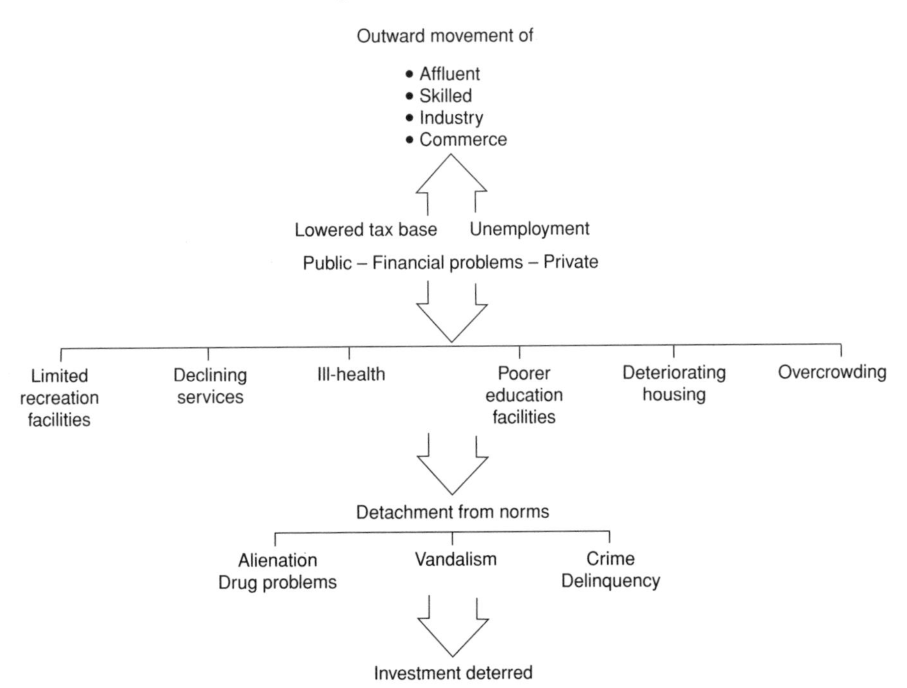

Inner-city decline – the downward spiral

tending to diminish both with distance and the passage of time. This weakening of the wave reflects the probability that the rate of acceptance of the innovation will decline over time and space. The momentum of the wave eventually fizzles out, possibly when it enters 'hostile' territory, comes up against a physical barrier or encounters innovation waves emanating from other centres. Identification of these waves was initially made in an investigation of the spread of agricultural innovations, but the concept can be applied to the spatial diffusion of a wide diversity of 'items', from people to fashions, from disease to videos, from news to riots. See DIFFUSION CURVE. [*f*]

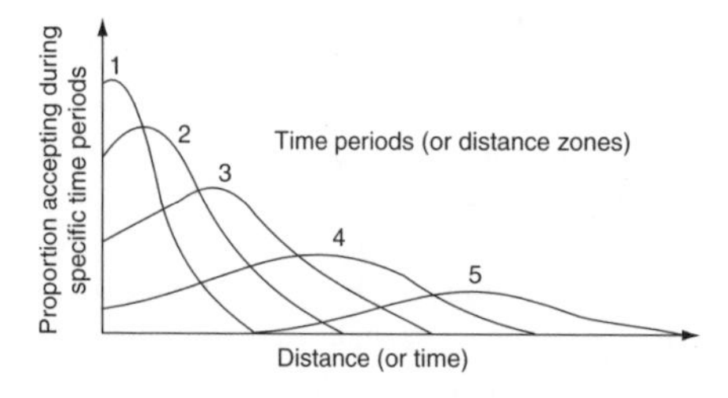

Waves of innovation losing strength with distance from the source area

input (i) In ECONOMIC GEOGRAPHY, the term is sometimes applied to the FACTORS OF PRODUCTION. (ii) In computing, input normally refers to that information or data transferred from external storage to internal processing. (iii) Material or energy that enters a system. See GENERAL SYSTEMS THEORY.

inselberg Literally an 'island mountain', an inselberg is an isolated residual hill standing above an extensive plain of EROSION (see PEDIPLAIN). Most inselbergs are steep-sided, and many are dome-like (e.g. the BORNHARDTS resulting from large-scale EXFOLIATION in granitic rocks). At the base of an inselberg there is frequently an abrupt change of slope, beyond which a gentle concave slope (PEDIMENT) leads down to the plain. Inselbergs are characteristic of late-stage SAVANNA landscapes, but are also found in humid tropical regions, DESERTS and even in temperate latitudes.

insolation The heat energy from the sun that reaches the Earth's surface in the form of short-wave ultraviolet rays (9%), visible light (45%) and infra-red rays (46%). The amount of solar energy received at the outer limit of the ATMOSPHERE (the solar constant) is depleted by reflection and 'scattering', as a result of clouds, dust and water vapour in the air. However, approximately 47% penetrates to the ground. Some of this is immediately reflected back (the ALBEDO effect), but most is absorbed and heats up the surface (most effectively on the land, and less so on the sea). This heat is subsequently lost by long-wave infra-red radiation, which warms the atmosphere (particularly in the presence of cloud cover, which impedes the escape of radiant heat), and conduction to the overlying air. The amount of insolation received at particular points is greatly influenced by latitude. Over the year, insolation at the Equator is 2.5 times that at the poles. However, in mid-summer (when daylight at the poles lasts for 24 hours) the solar energy received at the poles is slightly greater than that at the Equator. Yet, in mid-winter, when there is 24-hour darkness, the poles register a net loss of energy. [*f*]

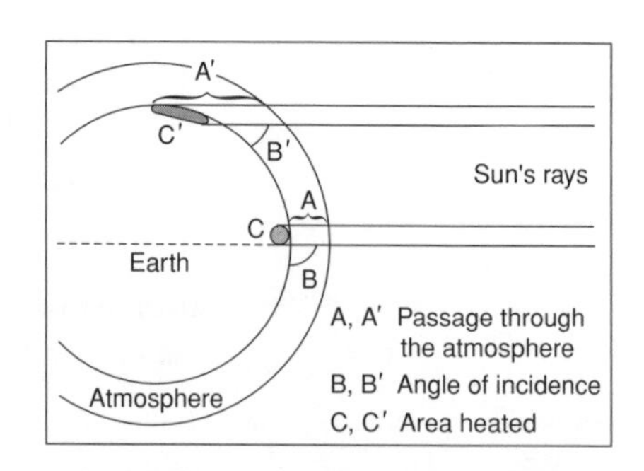

The effectiveness of insolation

insolation weathering The cracking of a rock surface owing to alternate heating and cooling, which may lead to EXFOLIATION. It has been suggested that dark-coloured rocks (which absorb solar energy most effectively), and heterogeneous crystalline rocks (comprising crystals with different colours and coefficients of expansion) are most prone to insolation weathering and subsequent thermal cracking. However, it is increasingly believed that, even in hot deserts where ranges of temperature from day to night are at a maximum, the physical breakdown of rock surfaces is due more to chemical processes such as HYDRATION and SALT WEATHERING.

instability The condition of the ATMOSPHERE in which, if a parcel of air rises (because it is initially warmer and lighter than the surrounding air), it will continue to rise, owing to the fact that – although cooled adiabatically – it remains warmer than the surrounding air. See ABSOLUTE INSTABILITY, CONDITIONAL INSTABILITY. [*f* STABILITY]

installed capacity The potential capacity of a plant (e.g. of an HEP station) as when fully utilized, as distinct from the capacity actually being used.

integration (i) The act or process by which diverse groups (distinguished on such bases as colour, class and religion) become unified within a community without necessarily losing their individual identities. Ct ASSIMILATION. (ii) Undertaking at the same location or within the same firm the successive stages involved in the production of a particular good. See VERTICAL INTEGRATION.

intensification The process of increasing the INPUTS of an economic activity with the aim of not just raising output, but achieving a higher productivity per unit of input. This might involve a manufacturing firm investing capital in new machinery and equipment, or a farmer deciding to cultivate a CATCH CROP. Ct EXTENSIFICATION.

intensive agriculture A system of AGRICULTURE where there is a relatively high level of INPUTS (capital, labour, etc.) and/or of OUTPUTS per unit area of farmland, e.g. MARKET GARDENING, VITICULTURE.

interaction See SPATIAL INTERACTION.

interception The process by which raindrops are intercepted by plant surfaces (and in particular the leaves of large trees), and thus prevented from falling directly on to the SOIL surface. During prolonged rainstorms, the 'capacity' of the foliage will be exceeded, and water will begin to drip from the CANOPY to the ground (*throughflow*). Some water will also run along branches and down the trunk (*stemflow*). However, water droplets retained by the leaves will eventually be evaporated or absorbed (*interception loss*), thus reducing the effectiveness of the rainfall as a whole. Interception is very important in tropical RAINFORESTS, and helps to reduce the effects of rainsplash erosion on the fragile soils. When interception is reduced, say by DEFORESTATION, the ground becomes more vulnerable to processes of SOIL EROSION and, in addition, water is likely to be transferred more rapidly (for example, by OVERLAND FLOW) so increasing the risk of FLOODING.

interdependence (i) Used in ECONOMIC GEOGRAPHY to denote that what occurs at one location will affect what happens at others. This occurs because of specialization in economic production, trade and LINKAGES. A fall in the market price of a commodity will prompt a reduction in OUTPUT; that fall in production (implying that the commodity becomes scarcer) may eventually serve to restore the price to its former level. (ii) The drawing together of the world's nations into the growing web of the GLOBAL ECONOMY by the process of ECONOMIC GLOBALIZATION. Perhaps best summed up in the expression that no nation today is an 'island'. Cf DEPENDENCE.

interface The zone or point of contact between two different phenomena. For example, the coast is the interface between land and sea; a PORT represents the interface between land and sea transport.

interfluve An area of high ground separating two valleys whose rivers flow into the same drainage basin (see WATERSHED).

interglacial A period of relatively warm climate, separating two glacial periods within the PLEISTOCENE. Recent evidence has suggested that there may have been up to 20 interglacials of relatively short duration (approximately 10,000 years). The climatic and vegetational conditions of interglacials have been reconstructed (e.g. by pollen analysis) from deposits of peat preserved by chance in hollows in the surface of the 'preceding' glacial TILL. Within a glacial period, minor ice withdrawals may occur to give an *interstadial*, which is less marked and of shorter duration than a true interglacial.

interlocking spur One of a series of projecting spurs forming an alternating sequence along the sides of a river valley, commonly near its headwaters in an upland area. The river itself winds around the spur ends, which, when viewed upvalley or downvalley, overlap or 'interlock' with each other. If the valley is subsequently glaciated, the spur ends may be eroded away by the ice to leave behind TRUNCATED SPURS.

intermediate area An element in the British regional development policy, introduced in 1969 with the aim of directing various forms of assistance to areas of slow economic growth (see LAGGING REGION) as opposed to areas of actual decline (see DEPRESSED REGION). Sometimes referred to as *grey areas*, these include parts of NW England, Yorkshire and Humberside, where reasonably low unemployment rates concealed major structural economic problems, high out-migration, low ACTIVITY RATES and poor INFRASTRUCTURE. Ct DEVELOPMENT AREA.

intermediate technology A term used in the context of economic DEVELOPMENT to denote small-scale and labour-intensive industries that use local skills and traditional tools, and that serve local needs. Many LEDCs lack the resources (e.g. CAPITAL, INFRASTRUCTURE, know-how), as well as the adequate domestic markets, to embark on large-scale industrial programmes of the type undertaken in MEDCs. For these reasons, some LEDC governments have introduced and encouraged intermediate technology as an early step in the process of INDUSTRIALIZATION. Such a policy appears to work well, particularly where labour is abundant and capital is in short supply, as for example in China, where it has resulted in the proliferation of COTTAGE INDUSTRIES employing traditional craftsmen. See BRANDT COMMISSION.

internal drainage A drainage system in which individual streams converge on an inland depression or lake – for example, the RIFT VALLEY lakes such as Lakes Nakuru, Naivasha and Magadi in Kenya.

internal economies of scale See ECONOMIES OF SCALE.

international division of labour The spatial separation of different components of industrial production, such as mining, manufacturing, assembly, administration and R&D. Each is located in a different part of the world according to costs and local considerations. It is a common practice among TNCS and very much part of the GLOBALIZATION of the world's economy.

International Monetary Fund See IMF.

international trade See TRADE.

interpolation The insertion of assumed values between measured values. For example, the plotting of isopleths (see ISOPLETH MAP) is frequently undertaken by interpolation; i.e. isopleths (e.g. contours) are drawn at prescribed intervals (10, 20, 30 m) on the basis of values (height above sea level) recorded at a series of dispersed data points (spot heights). [*f*]

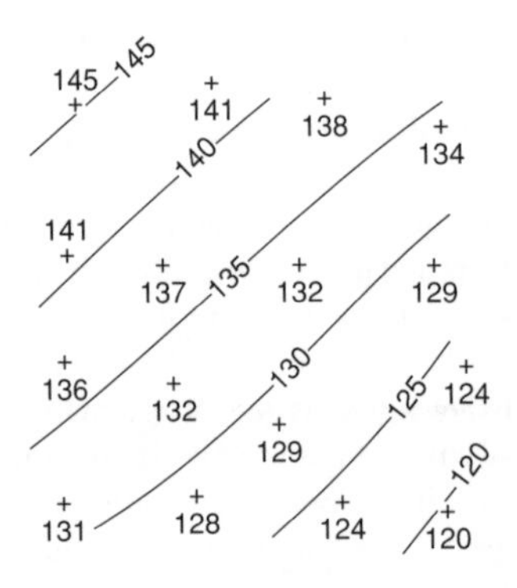

Interpolation of contours on a grid of spot heights

inter-quartile range The difference between the upper and lower QUARTILES of an arrayed set of values; 50% of the values in the distribution will fall within the inter-quartile range. [*f* DISPERSION DIAGRAM]

interstadial See INTERGLACIAL.

inter-tropical convergence zone See ITCZ.

interval data One of three major categories of data. Such data indicate both the order of magnitude and the degree of magnitude, as for example data relating to rainfall, production, population, etc. When it comes to applying SIGNIFICANCE TESTS to interval data, it is imperative to use PARAMETRIC TESTS. Ct NOMINAL DATA, ORDINAL DATA.

intervening opportunity theory This is based on the idea that the amount of migration over a given distance is directly proportional to the number of opportunities at the point of destination, but inversely proportional to the number of opportunities between the point of departure and the destination. *Opportunities* might be defined as vacant houses, employment prospects, social services and similar *pull factors* (see MIGRATION). The theory has been applied subsequently in other contexts, such as RETAILING (the presence of closer, better shopping opportunities will tend to diminish the attractiveness of more distant shopping centres), and often in conjunction with use of the GRAVITY MODEL.

intervention See GOVERNMENT INTERVENTION.

intervention price See CAP.

intrazonal soil A type of SOIL whose formation is related not to general climatic controls (as in ZONAL SOILS), but to particular conditions of rock type or GROUNDWATER. Examples are the SALINE SOILS of poorly drained arid and coastal areas, TERRA ROSA soils on LIMESTONES, and GLEY SOILS in areas of seasonal or permanent waterlogging.

intrusive rock A type of rock formed by the injection of MAGMA into existing rocks, usually along lines of weakness such as BEDDING PLANES or FAULTS, but sometimes involving engulfment of older rocks on a massive scale. Intruded masses vary greatly in form and scale (see BATHOLITH, DYKE, LACCOLITH, LOPOLITH and SILL). Intrusive rocks, which include GRANITE and gabbro, commonly have coarse crystals resulting from relatively slow cooling, unlike EXTRUSIVE (surface) IGNEOUS ROCKS, which tend to be fine-grained or glassy.

invasion and succession An ecological concept (see VEGETATION SUCCESSION) relating to the colonization of areas by plants and animals, which has also been introduced into SOCIAL GEOGRAPHY and URBAN GEOGRAPHY (see HUMAN ECOLOGY). The ideas have been applied to the processes by which areas undergo changes in LAND USE and population. The classic example is provided by the ZONE OF TRANSITION (see also CONCENTRIC ZONE MODEL), where formerly wholly residential districts are progressively invaded by commercial and industrial activities mainly associated with the CBD. Thus residence is gradually succeeded by these intrusive activities. At the same time, outward movement of the more affluent households, in response to the perceived blighting created by commercial and industrial invasion, provides the opportunity for invasion of the area by poorer households and often MINORITY ethnic groups, who progressively occupy the remaining housing. See also INNER-CITY DECLINE. [*f*]

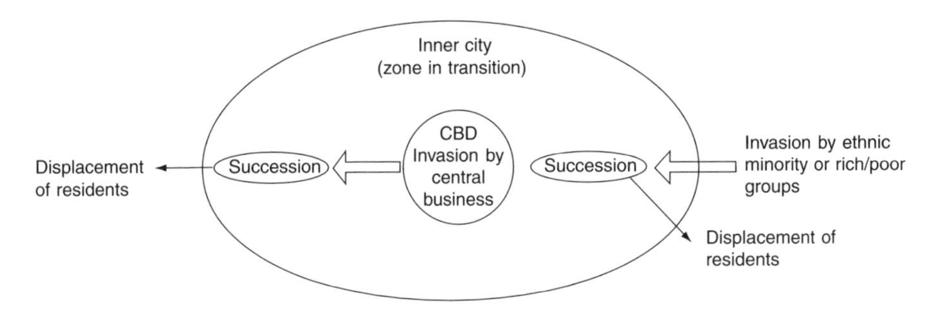

Invasion and succession

inversion of relief The process whereby, for example, an ANTICLINE that once formed a relative upland area, is breached and eroded such that its centre becomes a relative lowland valley. [*f* BREACHED ANTICLINE]

inversion (of temperature) A phenomenon in which temperature increases with height (by contrast with the more usual decrease with height). Inversions frequently affect relatively thin layers of air close to the ground surface, particularly on a clear, still night when heat is radiated rapidly from the Earth and the overlying air is cooled by being in contact with the cold ground (convection). Such inversions are well marked in valleys, where cold air, because of its greater density, drains into the valley bottom (see KATABATIC WIND). Inversions of temperature are associated with the formation of HOAR FROST, MIST and FOG. On a much larger scale, inversions develop where air is subsiding and being warmed adiabatically (as under anticyclonic conditions). Temperature inversions can be associated with pollution incidents, particularly over large cities such as Los Angeles, Mexico City and Athens. The cold sinking air prevents the escape of pollutants from industry and vehicles, and the gases and particulates become concentrated near the ground surface leading to health problems, particularly among the elderly and sick.

investment See FOREIGN INVESTMENT.

invisible trade See BALANCE OF TRADE.

involution A contorted structure formed usually in unconsolidated SEDIMENTS subjected to repeated freeze-thaw cycles in a PERIGLACIAL climate. Some involutions appear to be associated with small surface mounds, such as earth hummocks or *thufurs*, separated by a network of depressions.

inward investment See FOREIGN INVESTMENT.

Iron Age The culture period succeeding the BRONZE AGE, in Europe from *c.* 1500BC, in Britain from about the 6th century BC, characterized by the use of iron weapons and implements. The first main impulse in Britain came in the 5th century BC, with the invasion of Celts of the *Halstatt* culture; in the 3rd century BC came Celts of the more advanced *La Tone* culture in E and S England, and later in the SW. These introduced better weapons, slings, chariots, burial in round barrows, the construction of hill forts and lake villages, and considerable technical improvements in the working of iron, wood and pottery. In the 1st century BC the *Belgae* introduced further improvements.

irrigation The artificial distribution and application of water to the land to stimulate, or make possible, the growth of plants in an otherwise too arid climate. Irrigation may be: (i) *basin* or *flood irrigation*, in which water brought by a river in FLOOD is held on the land in shallow, basin-shaped fields surrounded by banks, e.g. in Egypt, along the banks of the Nile and in the delta; (ii) *perennial* or *'all-year' irrigation*, where water is lifted on to the fields from a low-level river, a well, a tank or small reservoir. Primitive devices have been long used, operated by human or animal power, later by windmills, and recently by steam, petrol or diesel pumps. Perennial irrigation is especially possible where a supply of water can be obtained from mountains, particularly from snow-melt. Modern perennial irrigation involves large BARRAGES or dams to hold back a great volume of water during river floods, which can subsequently be released through aqueducts and canals as required (e.g. the Aswan Dam on the Nile). Many modern barrages are multi-purpose, i.e. are also concerned with HEP production, flood control and navigation.

island arc A long, curving line of islands formed by uplift and volcanic activity. On the outer side of the arc, and running parallel to it, a deep *ocean-floor trench* is usually developed. Island arcs are characteristic of the W Pacific – for example, those running southwards from Honshu, Japan, and the related Bonin, Marianas and W Caroline trenches. They are developed at active plate boundaries (see PLATE TECTONICS), where one plate overrides another. The trench represents the SUBDUCTION ZONE, and the islands are formed by highly folded marine SEDIMENTS that are 'scraped off' the surface of the descending plate by the overriding plate. The volcanic activity is due to re-melting of parts of the descending slab, and extrusion of the resultant MAGMA. [*f*]

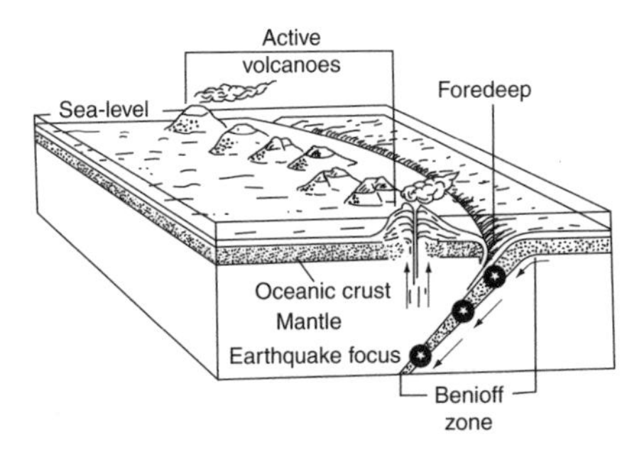

Island arc system

iso- A prefix for lines on an ISOPLETH MAP linking points of equal value or quantity, for example: *isobar* (atmospheric pressure); *isochrone* (time); *isohyet* (PRECIPITATION); *isotherm* (temperature); *isoquant* (production costs). See also ISODAPANE.

isoclinal folding A very intense form of folding related to great compressive forces, in which ANTICLINES and SYNCLINES are so closely 'packed' that the limbs of the folds in cross-section are virtually parallel to each other.

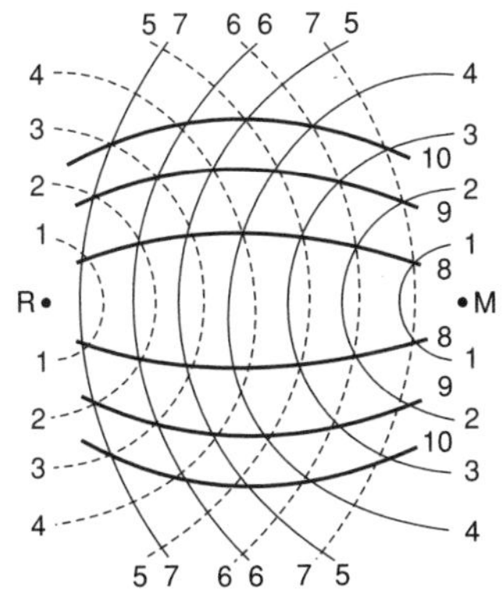

Isodapanes drawn where there is one market, one raw material, no weight-loss and with transport costs proportional to distance

isodapane In INDUSTRIAL LOCATION THEORY, an isodapane is a line connecting points of equal total TRANSPORT COSTS (i.e. the ASSEMBLY COSTS of materials plus the DISTRIBUTION COSTS of products). The OPTIMAL LOCATION is said to occur where the isodapane value is least. In the example shown in the accompanying figure, *isotims* have also been drawn, both for the costs of assembling RAW MATERIALS from source *R* and for the costs of transporting the finished product to market *M*. In all cases, the total cost will be the same as at *X*. [*f*]

isopleth map A MAP involving lines being drawn to join points of equal value, as with a contour map or weather map showing atmospheric pressure. [*f* ISODAPANE]

isostasy A state of balance in the Earth's crust whereby equal mass at depth underlies equal surface mass. Thus where mountains (comprising less dense SIAL) rise high above the average surface level, there must be a compensating 'roof' of sial penetrating deeply into the more dense SIMA. In general, continental land-masses can be viewed as 'rafts' of light sial floating on denser sima.

Isostatic adjustment occurs when, as a result of EROSION, some of the sialic mass is removed; there is then a compensatory uplift to restore EQUILIBRIUM. YOUNG FOLD MOUNTAINS, undergoing rapid erosion because of their strong RELIEF, appear to experience quite marked isostatic uplift. By contrast, oceanic floors, where SEDIMENTS from the land are accumulating in large quantities, undergo isostatic depression. It is argued that such isostatic compensation can occur only by virtue of the lateral transfer of subcrustal mass from areas of subsidence to areas of elevation. Under conditions of continental glaciation, a special form of isostatic adjustment occurs. Crustal depression occurs with the addition of vast weights of ice to the land. Then, when the ice melts, there is a corresponding recoil of the land, often known as *isostatic re-adjustment*. In the UK isostatic re-adjustment has led to a relative fall in sea-level, particularly along parts of the west coast where degraded cliffs and RAISED BEACHES are common.

isotropic surface A theoretically featureless plain showing uniformity with respect to SOILS, climate, ACCESSIBILITY, the distribution of population and purchasing power, etc. An important simplification made in a number of theories, such as CENTRAL PLACE THEORY and VON THUNEN'S MODEL.

IT (information technology) A relatively new field of technology brought about by the fusion of computing and telecommunications, and made possible by the MICROCHIP REVOLUTION. Its main focus is the storing, processing and transmission of data at a global scale, thus enabling faster and more efficient access to a wider range of information. Organizations active in the IT field include British Telecom (BT), IBM and the Meteorological Office, as well as many TNCS. Advances have largely been made in CORE regions, and as a consequence have helped to maintain their superiority and influence. Aspects of IT are also having important geographical effects on everyday living. These include: videoconferencing, which is reducing the need for people to travel to business and professional meetings; homeworking, whereby an increasing number of people are able to work from home and so reduce the amount of COMMUTING; in-home shopping (including banking), which reduces the number of retailing trips. These three examples clearly indicate that IT offers the chance to decentralize and disperse economic activities at a local and regional scale, particularly from URBAN to RURAL areas. At the same time, IT has been an important driving force behind ECONOMIC GLOBALIZATION.

ITCZ (inter-tropical convergence zone) A broad zone of low pressure, migrating northwards and southwards of the Equator with the seasons, towards which tropical air masses converge. The ITCZ is relatively weak over the oceans, but is more clearly defined over land areas such as West Africa, where it lies over the Gulf of Guinea in winter, but moves northwards (for example, over Nigeria) in April and May. Associated convectional storms bring heavy rains, which mark the onset of the wet season. In this region the ITCZ has more of the characteristics of a true FRONT, as it marks the junction of hot dry continental air from the Sahara and cooler humid equatorial air from the south. In general, the ITCZ is discontinuous, with widely separated convective units associated with cloud formation and rain (see HADLEY CELL).

jet stream A high-altitude, very fast-flowing wind (usually between 180–270 km/hour) blowing at approximately 9–15 km above the Earth's surface. A number of jet streams have been identified, but one of the most important is the Polar Front Jet, which is associated with the mid-latitude frontal zone. It appears to play a very important role in the generation and movement of mid-latitude depressions, which are steered eastwards. In winter, the Polar Front Jet migrates northwards in the Northern Hemisphere, and a second *Sub-tropical Jet Stream*, with a more continuous westerly flow, is initiated at approximately 30°N. In summer, an *Easterly Tropical Jet Stream* develops over India and Africa. Jet streams form along lines where the TROPOPAUSE has a steep gradient or is actually 'fractured'. [*f* HADLEY CELL]

joint A narrow crack or fissure in a rock, with no displacement on either side of the fracture (ct FAULT). Joints are developed in several ways: by contraction of IGNEOUS ROCKS during cooling; by stresses caused by earth movements; by DILATATION (pressure release).

joint-block removal The detachment of JOINT-bounded blocks by glacier ice. Joint-block removal (also referred to as *plucking*) is believed to be important in the formation of CIRQUE headwalls, the risers of rock steps, and the downglacier (*stoss*) faces of a ROCHE MOUTONNÉE. It was once thought that the ice froze on to the rock mass and wrenched it away. However, this is now somewhat debatable since the tensile strength of ice is much less than that of rocks and, in this situation, the moving ice would probably fracture first unless the rocks had first been sufficiently weakened by WEATHERING or DILATATION.

jökulhlaup See GLACIAL OUTBURST.

journey-to-work Travel between home and place of work (see COMMUTING). Journey-to-work can be measured in terms of either distance or travel time. Mode of transport is another important criterion used in the analysis of journey-to-work patterns.

justice See TERRITORIAL JUSTICE.

just-in-case production A system of organization used in manufacturing, involving the stockpiling of RAW MATERIALS and components at a factory. This is undertaken 'just in case' there is a rise in the cost of those things or there is a sudden rise in demand for the factory's output. It has been found to be inefficient and wasteful. Ct JUST-IN-TIME PRODUCTION.

just-in-time production This system of organization in manufacturing is designed to minimize the stocks of raw materials and components held by a factory. In this way, it is cost-efficient. In order to work properly, it needs an accurate forecasting of demand and very careful scheduling of flows along the PRODUCTION CHAIN to meet that demand. Ordering systems have to be efficient. Raw materials and components have to be 'defect free' and their delivery totally reliable. The system was pioneered, and is used to great effect, by Japanese manufacturers. It has since been widely used in NICS and, belatedly, in the UK.

K

kame A mound of stratified GRAVELS and SANDS, formed by meltwater from a decaying glacier or ICE SHEET. Kames vary in scale, form and origin. Some are low hillocks rising only a few metres, while others are steep-sided hills 30–50 m in height. Some kames form as fans or DELTAS of SEDIMENT laid down along the ice margin, often where small lakes are impounded. these subsequently form asymmetrical hillocks. others result from the accumulation of sediment in large SINKHOLES formed by ABLATION close to the margins of a stagnant glacier. When the ice finally melts, a complex of kames will result. *Kame terraces* develop between the ice and a valley wall, often within marginal lakes that become filled with sediment. When the ice melts, the ice contact slope forms the steep TERRACE edge. [*f*]

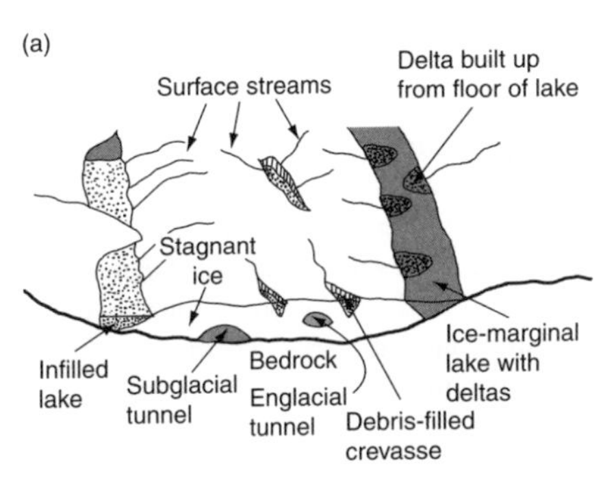

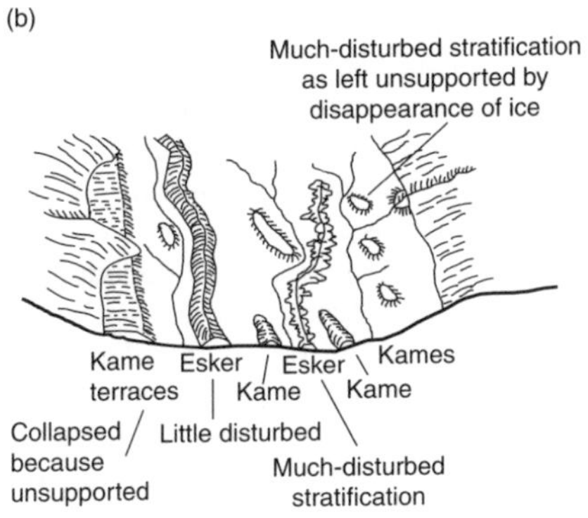

Kames (a) glacial landscape and (b) postglacial landscape

Kant's index of concentration Used in the analysis of RURAL settlement to measure the degree of concentration or dispersion of SETTLEMENT, as reflected in terms of the distance between habitations. The formula used to derive the index is:

$$X = \frac{1}{M} \sqrt{\frac{A}{D}}$$

where X is the interval between two settlements, $1/M$, is the scale of the map used in the investigation, A is the area under scrutiny and D the density of habitation. The lower the values of X, the greater the concentration of settlement.

kaolin A whitish CLAY (*kaolinite*) resulting from the breakdown of feldspar (for example, within GRANITE) either by recent CHEMICAL WEATHERING or hydrothermal processes operating at the time of rock emplacement (otherwise known as *pneumatolysis*). In its refined state used in a wide range of manufacture from china to cosmetics, paper to pharmaceuticals.

karren Furrows produced by solutional processes acting on the surface of LIMESTONE. Some are produced by acidulated rainwater on exposed limestone rocks, while others result from subsoil WEATHERING. Karren vary in scale from shallow grooves running down limestone faces to deeper channels up to 0.5 m deep and several metres in length. See GRIKE.

karst A specific area of LIMESTONE in the former Yugoslavia. The term is now widely used to describe any limestone region characterized by underground drainage and an abundance of surface solutional forms such as DOLINES, UVALAS and POLJES. The process of *karstification* refers to: the gradual disappearance of surface streams underground, by way of SINK HOLES; the formation of underground passages and cavern systems; the lowering of the limestone WATER TABLE, resulting in numerous DRY VALLEYS and depressions; and extensive surface and subsurface SOLUTION, to give small features such as KARREN and larger features such as DOLINES. See also COCKPIT KARST, TOWER KARST.

katabatic wind A local wind, blowing downhill on a cold night. Katabatic winds are due to the flow of dense air, chilled by contact with slopes that have themselves been cooled by nocturnal radiation, especially on cloudless nights. When the cold air ponds-up in valley bottoms, frost and FOG might become a hazard. A type of katabatic wind also occurs when cold air drains from a glacier surface into the valley below.

[*f* ANABATIC WIND]

Kendall's correlation coefficient See RANK CORRELATION.

kettle hole A circular depression, initially filled by meltwater, resulting from the gradual decay of a block of ice buried by overlying SEDIMENTS. Kettles can sometimes be seen forming today in large numbers ahead of Alpine glaciers, from beneath which streams are washing out and depositing large quantities of SAND and GRAVEL, concealing masses of calved ice from the glacier snout. Sometimes individual kettles coalesce to form quite large lakes.

key settlement A term used in RURAL PLANNING to denote a village or small TOWN selected for promotion as a small growth centre. The strategy has been adopted as a way of trying to halt the loss of population from remote rural areas. It is based on the belief that by concentrating (rather than dispersing) investment, and by using that investment to extend and improve the range of CENTRAL PLACE services, to provide housing and to introduce new sources of employment, growth will be engendered in the selected key settlement. Eventually, this growth will rub off on adjacent areas and on nearby settlements not so selected. The key settlement is therefore a sort of mini-GROWTH POLE.

kibbutz An Israeli form of rural SETTLEMENT, planned and organized according to collective principles, and in which land is communally owned (see COLLECTIVE FARMING). The early kibbutzim were wholly agricultural in function, but more recently there has been increasing and successful involvement in industrial production.

kinetic energy The 'free' energy continually being dissipated as heat friction by running water, breaking waves, sliding ice, etc. In streams, kinetic energy is defined by the formula:

$$Ek = \frac{MV^2}{2}$$

where M is the mass of the water and V is mean velocity. Thus a large rapidly flowing river generates more than twice the kinetic energy of an equally large river flowing at half the velocity.

knickpoint The point at which a river long-profile undergoes a marked steepening, as a result of either a resistant rock outcrop or a relative fall of sea-level. The latter type arises because the offshore gradient is usually steeper than that of the lower course of the river. Thus a fall in sea-level will add a relatively steep section to the extended river profile. This will be transmitted upstream as a 'wave of EROSION', often forming a gorge as the river cuts down, eventually to the head of the stream. Many rivers are characterized by a succession of knickpoints, related to the numerous base-level changes of the PLEISTOCENE. Knickpoints are also formed when a river is captured. In effect, the capturing river provides a lowered *local* base-level for the captured stream to work towards.

knock-and-lochan A type of landscape, found particularly in NW Scotland, resulting from the erosion of a relatively low-lying area of hard rock by an ICE SHEET. The latter scours areas and lines of geological weakness (well-jointed rock, FAULTS and SHATTER BELTS), to produce numerous depressions. Following the withdrawal of the ice, these become occupied by small lakes (*lochans*) and peat bogs. Separating the depressions, there are low rounded hills, often with little soil or vegetation (*knocks*), which have been scoured by glacial ABRASION.

knock-on effect Used in a variety of circumstances broadly to indicate the consequences of a particular action, decision or event, e.g. the 'ripple' effects associated with the opening of a new motorway or the wider repercussions of closing a large steel plant. In some instances, the use of the term may be synonymous with MULTIPLIER EFFECT.

Kondratieff (long waves) cycles Kondratieff was a Russian economist who, in 1925, published a paper making the observation that empirical evidence suggested that capitalist economies undergo regular BUSINESS CYCLES of 'growth, boom and bust' of approximately 50 years or so in duration, and that the initiation of each new cycle appears to have been heralded by important innovations in technology. The cyclical sequence he suggested went as follows.

- 1st cycle (*c.* 1780–1842) – initiated in Britain by the first smelting of iron ore by coal and by the mechanization of the textile industry.
- 2nd cycle (1842–97) – the age of steam, railways and Bessemer steel; again a cycle initiated in Britain.
- 3rd cycle (1898–1939) – heralded by electricity, chemicals and motor vehicles. A cycle initiated in the USA and well under way when Kondratieff wrote his paper.

Since then, a 4th cycle has been recognized by economists, spanning from the end of the Second World War to the ENERGY CRISES of the 1970s. Again, the USA took a leading part in important technological developments, including air transport, the development of oil-based industries, and the growth of electrical and electronics industries. It is now thought that a 5th cycle has already started, and that the new technology heralding this cycle relates to such fields as microprocessors, genetic engineering, new metals and energy sources (see HIGH-TECHNOLOGY INDUSTRIES). Arguably it is now the turn of Japan and the NICS to take the innovative lead. [*f*]

koppie (also **kopie, castle kopie**) A small rocky hill in S Africa. However, the term is now widely applied to similar features throughout Africa. Koppies are characteristically composed of JOINT-bounded BOULDERS, often in a state of collapse. They result from the protracted WEATHERING of INSELBERGS, of which they are therefore a degenerate form. In practice it is sometimes difficult to differentiate koppies and TORS. [*f* RUWARE]

kurtosis A measure of the peakedness of a curve, as used in the analysis of FREQUENCY DISTRIBUTIONS.

k-value Used in CENTRAL PLACE THEORY to denote the number of central places of a given order dependent on a central place of the next highest order in the CENTRAL PLACE HIERARCHY (see also DEPENDENT PLACE). Christaller recognized three basically different k-value systems, each involving a different *principle*, and a different spatial relationship between the central place lattice and the network of hexagonal HINTERLANDS. These were: k-value 3 – the MARKET PRINCIPLE; k-value 4 – the TRANSPORT PRINCIPLE; k-value 7 – the ADMINISTRATIVE PRINCIPLE. While Christaller

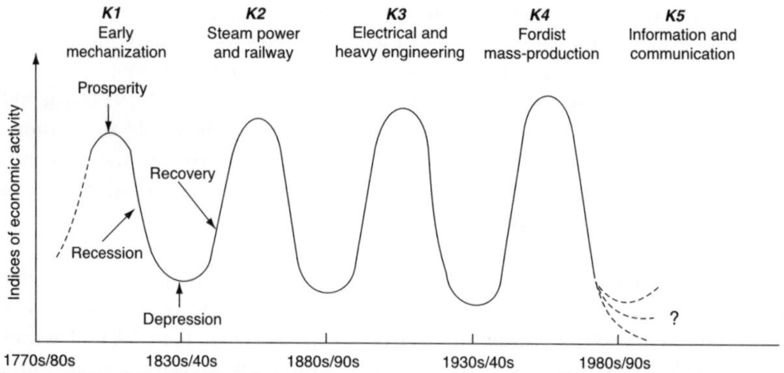

Main 'carrier' branches	Textiles Textile chemicals Textile machinery Iron working/ castings Water power Potteries	Steam engines Steamships Machine tools Iron and steel Railway equipment	Electrical engineering Electrical machinery Cable and wire Heavy engineering/ armaments Steel ships Heavy chemicals Synthetic dyestuffs	Automobiles Trucks Tractors Tanks Aircraft Consumer durables Process plant Synthetic materials Petrochemicals	Computers Electronic capital goods Software Telecommunications Optical fibres Robotics Ceramics Data banks Infomation services
Infrastructure	Trunk canals Turnpike roads	Railways Shipping	Electrical supply and distribution	Highways Airports/ airlines	Digital networks Satellites
Organization of firms and forms of co-operation and competition	Individual entrepreneurs and small firms (<100 employees) competition. Partnership structure facilitates co-operation of technical innovators and financial managers. Local capital and individual wealth	High noon of small-firm competition, but larger firms now employing thousands rather than hundreds.	Emergence of giant firms, cartels, trusts, mergers. Monopoly and oligopoly become typical. Regulation or state ownership of 'natural' monopolies and public utilities.	Oligopolistic competition. Transnational corporations based on direct foreign investment and multiplant locations. Competitive sub-contracting on 'arm's length' basis or vertical integration.	'Networks' of large and small firms based increasingly on computer networks and close co-operation in technology, quality control, training, investment planning and production planning ('just-in-time'), etc.
Geographical focus	Britain, France, Belgium	Britain, France, Belgium, Germany, USA	Germany, USA, Britain, France, Belgium, The Netherlands, Switzerland	USA, Germany, other EEC, Japan, Switzerland, Sweden, other EFTA, Canada, Australia	Japan, USA, Germany, other EU and EFTA, Sweden, Taiwan, Korea, Canada, Australia

Kondratieff cycles

believed in a *fixed-k hierarchy* (with the same k-value applying to all levels in the central place hierarchy), Lösch advocated a *variable-k hierarchy* with k-values varying not only from level to level, but also from place to place within the same level.

[*f* ADMINISTRATIVE PRINCIPLE]

L

labour One of the three main FACTORS OF PRODUCTION (the others being CAPITAL and land). It is fundamental to the operation of all production systems. Labour requirements vary with the nature of the economic activity concerned, many enterprises having very specific needs, be they for highly skilled technicians, clerical staff, unskilled manual workers or machine operatives. The spatial availability of specific types of labour skill can have quite a considerable influence on the location of firms, as can other qualities of the labour force of an area, such as its wage expectations, reliability, adaptability, etc. It is sometimes the case that the accumulation over time of particular labour skills in an area can contribute significantly to INDUSTRIAL INERTIA. See also DUAL JOB MARKET, LABOUR COSTS, MARXISM.

labour costs The total cost of wages and salaries paid to the employees of a FIRM, together with other labour-related expenditures (e.g. contributions to national insurance, pension schemes, etc.). Despite mechanization, ASSEMBLY-LINE PRODUCTION and AUTOMATION, the costs of labour remain generally a very significant element in total production costs. This possibly stems from the pressure for higher wages and salaries brought to bear in MEDCs by organized labour largely through trades unions. In the UK today, the wages and salaries of employees are equivalent to 54% of value added in manufacturing; for the USA the equivalent figure is 30%. While wage and salary levels still vary from activity to activity, it should be noted that as a result of trades union activity and the adoption of nationally agreed rates of pay, there is nowadays relatively little spatial variation in labour costs within the same industry. Despite substantial moves towards sex equality, however, there still remain some differentials between the wages of men and those of women, a fact that continues to persuade some firms to prefer to recruit female labour.

At the international scale, there are still *low-wage areas* to be found, and as such they are attractive to labour-intensive industries and to the TNCs. The removal during the last quarter of the 20th century of electrical and electronic firms from the USA and W Europe to NICs and then to LEDCs is partly a reflection of the continuing and considerable pull of the labour-cost factor on industrial location. See ECONOMIC GLOBALIZATION.

laccolith A concordant type of igneous intrusion, formed by the injection of MAGMA along a BEDDING PLANE and the resultant doming of the overlying strata. Some laccoliths have been partly or wholly exposed, by removal of the weak sedimentary rocks, to give large domed landforms. A *cedar-tree laccolith* is a type formed where magma, rising along a central pipe, is intruded into several BEDDING PLANES within the updomed structure. [*f*]

lacustrine An adjective indicating an association with a present or former lake, and used mainly to denote SEDIMENTS, e.g. a *lacustrine delta* formed by the accumulation of

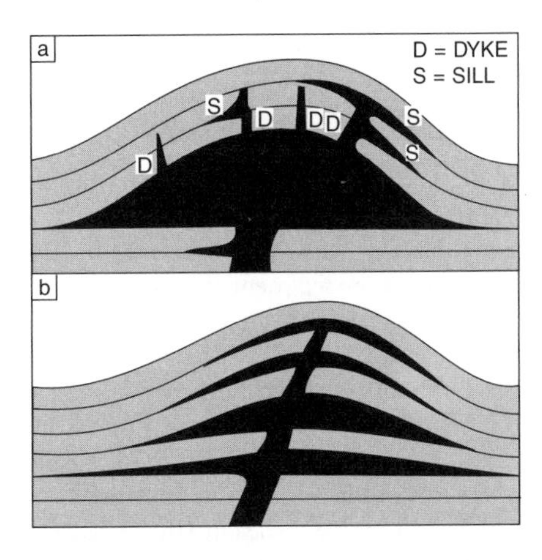

(a) Laccolith with dykes and sills, and (b) cedar-tree laccolith

SANDS and GRAVELS at a point where a stream enters the lake, a *lacustrine terrace*, resulting from the formation of BEACH deposits along the margins of a former lake; and *lacustrine plain*, in which the lake has been filled by in-washed sediments.

LAFTA (Latin American Free Trade Association) Set up in 1961 to promote economic cooperation within Latin America.

lag industry An industry that depends for its growth on the expansion either of other industries (see PROPULSIVE INDUSTRY) or other parts of the economy. The manufacture of motor vehicle accessories, and the extraction of sand and gravel might be cited as examples.

lag-time See HYDROGRAPH.

lagging region A type of DEPRESSED REGION that may possibly be distinguished by the fact that, although it is performing below the national average in terms of economic growth, there is nonetheless some measure of progress. In this respect, the distinction may be made with a *declining region*, where the basic problem is rather more one of absolute decline in productivity and prosperity; if not that, then a widening gap between the region's performance and that of the nation as a whole. Thus a lagging region may be regarded as something less of a *problem region*. In Britain it is lagging regions that have been given INTERMEDIATE AREA status, while

DEVELOPMENT AREA status has been reserved for the even more serious problems of the declining region.

lahar A rapid flow of mud and volcanic sediment, resulting from the overflow of a crater lake or lava-dammed lake, or the saturation of the sediment (e.g. volcanic ash) by prolonged heavy rain. Velocities of the lahar may be very high when hot (up to 100 km/hour), and the flow may travel for 100 km or more. Lahars often overwhelm farms and settlements, sometimes causing severe loss of life in densely populated areas such as Java. In 1985 a deadly lahar, triggered by the eruption of Nevado del Ruiz, wiped out the Colombian town of Armero killing some 22,000 people and burying parts of the town in up to 8 m of mud.

lame duck A term used to describe an economic activity that cannot compete on the open market without the help of GOVERNMENT INTERVENTION. Without it, the activity would go out of business with a consequent loss of jobs and revenue. The whole issue of whether governments should support lame ducks is a controversial one. Some would argue that it is better to use government help to encourage growth activities than to prop up ailing ones. Others, however, say that the loss of one activity can so easily create a much wider and more serious *downward spiral* of decline. This, in turn, would make even greater demands on government funds, as for example in the form of unemployment and other social benefits.

laminar flow A type of flow (e.g. in a river) where the water is transported at low velocities in a smooth straight CHANNEL, with parallel 'layers' of water shearing over one another in such a way that velocity is least next to the bed and greatest near the water surface. In most actual streams laminar flow is replaced by TURBULENT FLOW, particularly where the bed is rough and velocities are relatively high.

land breeze A gentle, cool wind blowing from the land towards the sea at night. As the land is cooled by nocturnal radiation, atmospheric pressure is slightly raised, by contrast with the lower pressure over the still-warm sea. This leads to a compensatory movement of air seawards (ct SEA BREEZE). Land breezes are characteristic of tropical regions, but may occur on a fine summer's day in Britain under calm anticyclonic conditions.

[*f* SEA BREEZE]

landfill site A natural or man-made hole in the ground (e.g. mines, quarries and gravel pits) used for the tipping and disposal of waste. There are potential environmental hazards associated with the practice. For example, there is a risk of contaminating GROUNDWATER supplies. The decomposition of rubbish under ANAEROBIC conditions produces methane and, with it, the risk of explosions.

land reclamation See RECLAMATION.

land reform An alteration to the LAND TENURE arrangements that exist in an area; for example, the breaking up of large estates and the redistribution of the land by sale to existing owner-occupier farmers and former tenants, or the abolition of the custom of equal inheritance (*gavelkind*), which over the centuries has been responsible for acute FARM FRAGMENTATION. In some former colonial territories, land reform has been effected rather more by governments dispossessing large land holders or by the spontaneous occupation of land held by expatriates.

land tenure The system of land ownership and of title to its use; it is particularly significant in the context of AGRICULTURE (and also housing). Land tenure can take a number of different forms. The principal types are: (i) *owner-occupation*, where the user or occupier is the owner; (ii) *tenancy*, where the user or occupier is not the owner and where the user either pays the owner rent or makes payment in the form of labour or SHARE CROPPING; (iii) *use right*, where ownership is not significant and where a group or individual establishes rights by use (as in SHIFTING CULTIVATION); (iv) *institutional*, where land is owned by institutions (companies, etc.) and labour is contracted (as in PLANTATION agriculture); (v) COLLECTIVISM, where the land is owned by the state or the community, and individuals share the produce or revenue from sales (see also

KIBBUTZ). Other types of land tenure include *common law, absentee ownership and state ownership* (i.e. of military areas).

land use The use of land by human activities, not necessarily always for financial profit or gain. A basic distinction may be drawn between *rural* (AGRICULTURE, forestry, RECREATION, etc.) and *urban* (INDUSTRY, commerce, housing, etc.) land use.

land-value surface An uneven 'surface' created by the fact that land values and rents vary from place to place in response to spatial variations in the level of demand and in the quality of land. See also URBAN LAND-VALUE SURFACE, COST SURFACE.

landscape A term used to describe the sum total of the 'appearance' of an area. Thus *physical landscape* refers to the combined effect of the landforms, 'natural' vegetation, SOILS, rivers and lakes, while *human landscape* includes all the modifications made by people (cultivated vegetation, field patterns, COMMUNICATIONS, SETTLEMENTS, open-cast mines and quarries, etc.). See also CULTURAL LANDSCAPE. Ct ENVIRONMENT.

landscape evaluation A method of quantifying the 'quality' of a particular landscape, as an aid to PLANNING, DEVELOPMENT and CONSERVATION. One difficult problem is to derive criteria for evaluation that are not entirely subjective (though one method is for a number of observers, acting independently, to classify views from selected positions in terms such as *unsightly, undistinguished, pleasant, distinguished, superb* and *spectacular*). Marks are assigned to these value judgements (ranging from 0 to 32), and average scores for particular landscapes are obtained.

landslide A type of MASS MOVEMENT in which the material displaced retains its coherence as a single body as it moves over a clearly defined plane of sliding. Landslides are often promoted by large accumulations of soil water from rainfall, SPRINGS or melting snow. This adds to the weight of the sliding mass, and – as PORE WATER PRESSURE increases – reduces friction between constituent particles. The latter is especially important when the sliding mass comprises weathered materials resting on a substratum of CLAY. Landslides can cause significant damage to lines of communication (railways, roads, etc.) and can cause loss of life and injury, for example where low-income SHANTY TOWN housing has been constructed on marginal land comprising steep deforested slopes. Landslides often become NATURAL (ENVIRONMENTAL) HAZARDS.

[*f* MASS MOVEMENT]

lapse-rate The rate of temperature change with altitude, either in a stationary column of air (see ENVIRONMENTAL LAPSE-RATE) or in a rising pocket of air (see ADIABATIC).

latent heat The heat energy expended in changing the state (phase) of a body without raising its temperature, expressed in calories per gram. Thus the *latent heat of evaporation* is that required to convert water into water vapour; this heat is then released again if, subsequently, the water vapour is condensed into water droplets (the *latent heat of condensation*). The large-scale release of latent heat is important in maintaining or accelerating rising currents of air in the Earth's atmosphere. See CYCLONE, SATURATED ADIABATIC LAPSE-RATE, THUNDERSTORM.

lateral dune A minor SAND DUNE forming along the flanks of a larger DUNE that has accumulated around a large rock obstacle.

lateral erosion EROSION performed by a stream experiencing lateral (sideways) migration, as in the development of MEANDERS or braided CHANNELS. The stream bank is actively undercut, usually at a point where a thread of high-velocity flow (the *thalweg*) impinges on it (ct VERTICAL EROSION). In some cases, lateral erosion may affect a valley-side slope, and over a very long period may lead to the formation of an extensive surface of lateral PLANATION (*panplain*), surmounted by steep-sided residual hills (remnants of the former INTERFLUVES). It is widely believed that lateral planation, by SHEET FLOODS and stream floods, may be important in the evolution of DESERT scenery (see PEDIMENT).

lateral moraine A deposit of unsorted, angular material, most commonly derived from WEATHERING and collapse of the slope above the ice, which has accumulated along the margins of a valley glacier. The debris may be in transit on the ice, forming a *supraglacial lateral moraine*, or may accumulate as a *lateral dump moraine* between the glacier and the valley wall.

laterite A term used in two different senses. First, to denote tropical SOILS resulting from *lateritization* (see LATOSOL) and, second, to describe a hardened layer of 'ironstone' resulting from the PRECIPITATION of iron minerals (see DURICRUST), either at the surface or in the subsoil, resulting in a concrete-like layer or a series of massive concretions.

latifundia Large landed estates found in Italy, Spain and parts of Latin America.

Latin American Free Trade Association See LAFTA.

latitude The angular distance of a point along a *meridian* (line of longitude), either N or S of the Equator (0°). The latitude of London is 51°30'N and of Washington 39°20'N. A *parallel of latitude* is a circle drawn around the Earth, parallel to the Equator, so that every point along it has the same latitude. Parallels of latitude intersect all meridians at right angles. See LONGITUDE.

latosol A major SOIL type associated with the humid tropics, and characterized by red, reddish-brown or yellow colouring. The A-HORIZON is weakly developed, comprising a thin layer of quartz particles. There is little HUMUS here, as plant litter is decayed very rapidly by bacterial activity. The B-HORIZON is usually thick, and comprises CLAY and SAND (the product of powerful CHEMICAL WEATHERING), together with sesquioxides of iron (giving red coloration) and aluminium (yellow). In general, latosols are not fertile, though an illusion of fertility is created by the profuse forest growth, which is achieved by the very rapid RECYCLING of a limited supply of plant nutrients. [*f*]

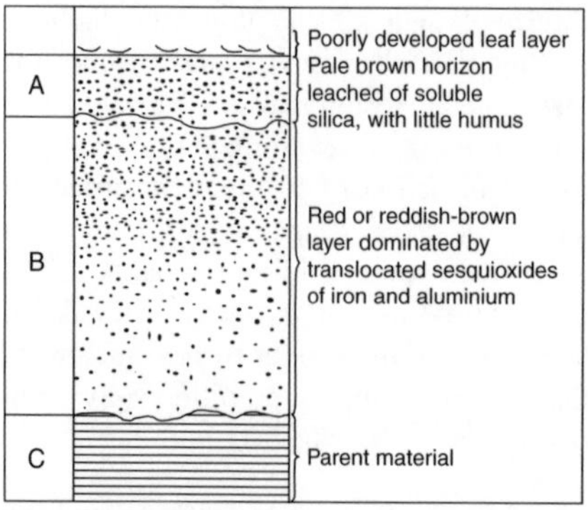

A latosol soil profile

lattice In geography, the term is used where there is a regular spatial arrangement of some phenomenon (e.g. SETTLEMENT). In CENTRAL PLACE THEORY, Christaller assumed that central places would be arranged on the ISOTROPIC SURFACE according to a lattice of equilateral triangles, with each central place equidistant from six neighbours. [*f* CENTRAL PLACE THEORY]

Laurasia The ancient northern 'supercontinent', comprising N America and Greenland, Europe and much of Asia, which was separated from the southern continent of GONDWANALAND by the former Tethys Sea (see CONTINENTAL DRIFT). Laurasia was fragmented by the westward drift of N America, and the opening up of the N Atlantic, during and since the Cretaceous period.

lava Molten MAGMA from the Earth's interior that has been extruded on to the Earth's surface by volcanic activity (see ACID LAVA, BASIC LAVA, ERUPTION, EXTRUSIVE ROCK).

law A theory or hypothesis that has been proven by empirical evidence. In geography, such laws often refer to cause-and-effect relationships – see e.g. BUYS BALLOT'S LAW, ENGELS' LAW.

law of minimum effort See LEAST EFFORT.

law of retail gravitation A law formulated by Reilly and based on the GRAVITY MODEL, which states that 'two centres (*a* and *b*) attract trade (*T*) from an intermediate place in direct proportion to the size (*P*) of the centres and in inverse proportion to the square of the distances (*D*) from these two centres to the intermediate place'. So the proportion of the retail trade from an intermediate place attracted by each centre is found by the formula

$$\frac{T_a}{T_b} = \frac{P_a}{P_b} \times \frac{D_b^2}{D_a^2}$$

A version of this formula may be used to predict the boundary between the MARKET AREAS of two centres (see BREAKING-POINT THEORY).

Law of the Sea The name given to a UN Convention held in Geneva in 1958 to clarify the rights of STATES to exploit underwater RESOURCES occurring on the CONTINENTAL SHELF. It resolved that the right to exploit minerals from the ocean floor should be allowed offshore to a water depth of 200 m. A UN conference, bearing the same title, took place in 1977 and decreed that the earlier *Continental Shelf Convention* should be replaced by the 200 nautical mile (370 km) *Exclusive Economic Zone* that refers to fish as well as to sea-bed mineral resources. See TERRITORIAL WATERS.

LDC (less developed country) See LEDC, THIRD WORLD.

leaching The removal of dissolved chemicals, particularly bases, by organic solutions known as *leachates*, which are often derived from the decay of surface litter. Leaching is most commonly associated with well-drained, sandy soils and a climate where precipitation exceeds evaporation. In removing the bases, leaching results in an eluviated A-HORIZON that becomes increasingly acidic and generally less fertile. Leaching is a vital process in the formation of LATOSOLS and PODSOLS. Ct ELUVIATION.

lead industry See PROPULSIVE INDUSTRY.

least-cost location The basic notion underlying WEBER'S THEORY OF INDUSTRIAL LOCATION that each FIRM seeks to occupy the site that involves the lowest level of costs. Often taken to be synonymous with OPTIMAL LOCATION.

least effort This lies at the root of the idea that where some form of movement is concerned and where there are a number of alternative options, people will tend to choose the option that involves the least effort, as measured in terms of cost, energy, time, etc. For example, the principle of least effort assumes that people will tend to shop at the nearest appropriate retail outlet (a basic premise of CENTRAL PLACE THEORY). The whole idea is closely related to the concepts of DISTANCE DECAY and FRICTION OF DISTANCE. The principle is sometimes referred to as the *law of minimum effort.*

least squares A statistical method used to discover the best-fit *regression line* (see REGRESSION ANALYSIS) for two variables plotted on a GRAPH. The least squares method determines the gradient of that line and its axial interception. Basically, the method places the regression line between the points representing the two sets of observations in such a way that the sum of the squares of the separate distances between the line and each point (i.e. RESIDUALS) is kept at a minimum value. The method will, in fact, yield two regression lines, one being measured and plotted against the *x* axis and the other against the *y* axis. [*f*]

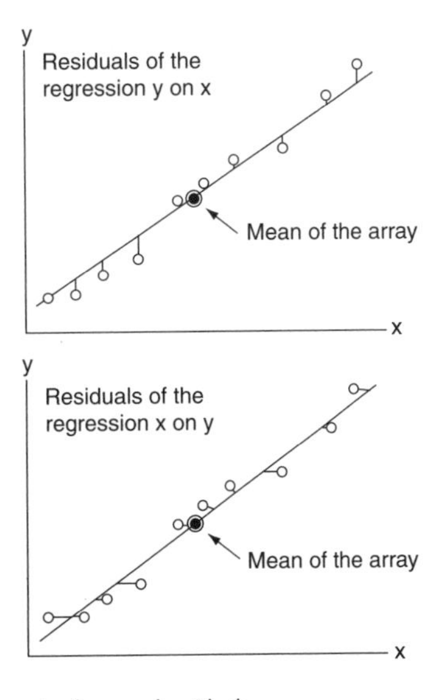

Regression lines and residuals

LEDC (less economically developed country) This term is one of a number used when referring to the countries of the THIRD WORLD (cf DEVELOPING COUNTRY, *less developed country, low-income country*). Although a useful shorthand term, it is nonetheless a crude one that fails to acknowledge the large differences in levels of economic DEVELOPMENT that occur within the Third World, as for example between RICS and LLDCS. Ct MEDC.

leisure Time free from employment. With the advent in MEDCS of shorter working hours and early retirement, leisure time is increasing, allowing greater participation in a range of possible recreational activities. At the same time, however, this increased leisure is generating a greatly enlarged demand for the provision of recreational amenities, such as football pitches, tennis courts, golf courses, country parks, swimming pools and marinas. Aspects of leisure that are of particular concern to the geographer include the ACCESSIBILITY of leisure AMENITIES and the degree of accord between the distribution of those amenities and the distribution of leisure demand. Cf RECREATION.

levée A containing bank of SEDIMENT formed along the edge of a river CHANNEL occupying a FLOOD PLAIN. When flooding occurs and the river overtops its banks to inundate the plain, there is widespread sedimentation. However, the greatest amounts of DEPOSITION occur in a concentrated zone along the junction between the rapidly flowing water in the channel and the slacker water over the flood plain. In this way, the levées build up over time to a height that is greater than that of the flood plain. This increases the potential scale of inundation should the levées be topped or breached. To reduce the threat of flooding, natural levées are often strengthened and raised artificially.

level of living The degree to which the needs and wants of a community are satisfied; the factual circumstances of WELL-BEING. See NEED-SATISFACTION CURVE.

liana A woody climbing plant that entwines itself around large trees, particularly in tropical RAINFORESTS. Unlike *epiphytes* (which germinate and grow on the branches of the trees), lianas have their roots in the soil. Lianas commence their growth in the deep shade at the forest floor, but then grow rapidly upwards towards the light. Often they pass from one tree to another, achieving a length of 60 m, binding together the forest structure, and preventing individual trees from toppling, even when the latter are broken off at the base.

lichenometry A method of dating rock surfaces, using the size and occurrence of lichens growing on the rock as indicators. Lichens (such as the widely employed *Rhizocarpon geographicum*) are long-lived, are the primary colonizers of rock surfaces, and within an unchanging environment grow slowly but at a constant rate. Thus the size of a lichen, and in particular its diameter, is a good guide to its age. Lichenometry is based on the construction of a *growth curve*, involving measurements of lichens on surfaces of known age (such as buildings and tombstones); on the GRAPH, *diameter* is plotted against *time*. At sites of unknown age (for example, an abraded rock surface exposed by glacial retreat, or boulders on a glacial moraine), a number – usually five – of the largest, and therefore the oldest, lichens present are measured, averaged, and age is read off from the growth curve. Lichenometry has been applied widely to the problem of dating retreat stages of ICE SHEETS and glaciers during the POSTGLACIAL period, but can also be used to determine the chronology of changes in other landforms, such as river channels.

life cycle Possibly more appropriately referred to as the *family life cycle*, since it is a concept based on the idea that most families or households go through a sequence of changes in their lifetime, which are particularly significant in terms of housing needs and housing moves. Most researchers recognize five significant stages in the history or life cycle of a family. (i) *Marriage and the pre-child stage* – the fact of marriage creates a new household and therefore a demand for a dwelling unit. Space demands are small, because both partners are probably working and the combined income is modest (thereby restricting access to housing only in the lower price or rental range). Taking a residential location close to the city centre is one way of saving on COMMUTING costs. For these reasons, young households are encouraged to take up cheap rented accommodation in a central location. (ii) *Child-bearing stage* – demand for more space and increased awareness of environmental quality probably encourage a move to home ownership in the suburbs; this is made possible by accumulated savings, higher income and better credit rating as regards raising a mortgage. (iii) *Child-rearing and child-launching stage* – there may be another move at this stage to more spacious accommodation, precipitated by increased family space requirements and made possible by career progress yielding high levels of income. (iv) *Post-child stage* – as children leave home, so space needs fall again, and this may encourage a move to a smaller property, but possibly one in an even more expensive area. (v) *Widowhood stage* – after the death of one partner, there may be a move into either some form of SHELTERED HOUSING or into the house of one of the children.

Some researchers have recognized other types of life cycle not so closely tied to changing family circumstances and housing. These include the life cycle of *consumerism*, in which people opt for the good life, and the life cycle of *careerism*, in which the main objective is improvement of SOCIO-ECONOMIC STATUS (sometimes participants are referred to as *spiralists*, because they move up and out).

life expectancy The average number of years a person might be expected to live; a prediction generally made on the basis of *life table* calculations. Improved diet and better medicine have contributed to a significant rise in life expectancy in most parts of the world, the expectancy almost always being greater for women. In Britain the life expectancy for men is 70 years and for women 76 years; for Bolivia the values are 42 and 48 years respectively. Because of the impact of INFANT MORTALITY, life expectancy is normally greater a year after birth than at birth; thereafter expectancy decreases with age.

lifestyle The total pattern of day-to-day activities that make up an individual's way of life. The pattern is reflected in such things as age, dress, place of residence, type of work, socio-economic status, leisure interests, and choice of expenditure on goods and services.

light industry The manufacture of articles of relatively small bulk, using small amounts of RAW MATERIALS, e.g. the making of tools and televisions, the processing of food, etc. Ct HEAVY INDUSTRY.

limestone An important type of SEDIMENTARY ROCK, frequently but not always of marine origin (hence *marine* and *freshwater limestones*). Limestones, which are dominantly composed of calcium carbonate and are also referred to as calcareous rocks, are of

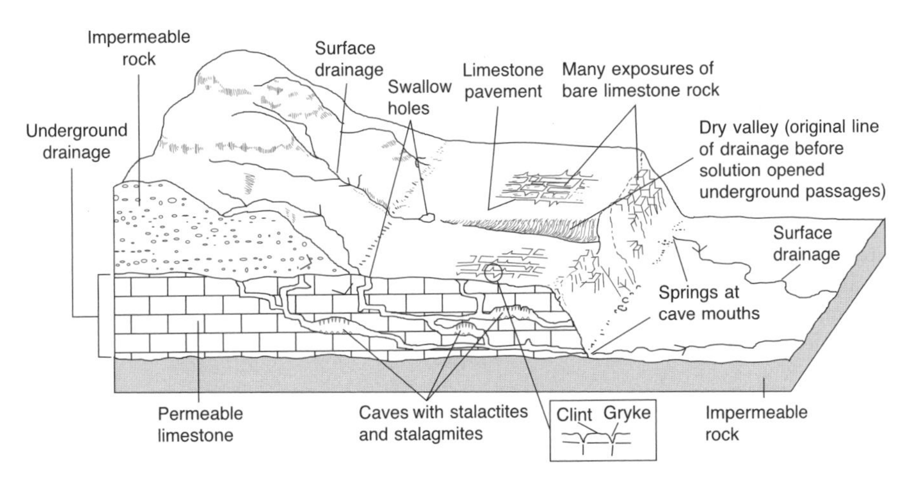

Limestone features

many types, including: *shelly limestone* (comprising masses of whole or broken shells); *coral limestone* (the remains of reef-building corals); *oolitic limestone* (consisting of 'ooliths', or grains of calcareous material deposited in concentric fashion around a nucleus such as a SAND particle); *magnesian limestone* or *dolomite* (resulting from PRECIPITATION of the double carbonate of calcium and magnesium); *chalk* (a pure, soft limestone formed from a calcareous mud deposited on the sea-floor, see CHALK).

Limestones vary considerably in age and thus compaction, and this has a bearing on landforms. Limestone and chalk in particular are associated with characteristic landscape features (see KARST). [*f*]

limestone pavement A horizontal or gently inclined bare limestone surface, broken by numerous small-scale solutional features. LIMESTONE pavements appear to coincide with major BEDDING PLANES that have been exposed by glacial erosion (*glacial stripping*). The process has been most effective on the lee slopes of limestone hills overridden by ICE SHEETS. Solutional processes exploit the rectilinear jointing pattern to form blocks called CLINTS, separated by enlarged joints called GRIKES. Eventually, destruction of the clints and enlargement of the grikes by continuing solution will produce a rubble layer, on which vegetation can flourish. In this sense, limestone pavements are comparatively short-lived landforms, and many will disappear in the not too distant future. [*f* LIMESTONE]

limiting angle An angle of slope that defines the limits above or below which certain processes operate. For example, SCREE slopes are often restricted to a maximum of 36–38° because, above this angle, instability will produce rapid debris sliding. Active SOIL slippage, by contrast, is rare on slopes below 20° and SOLIFLUCTION under PERIGLACIAL conditions may require slopes of 6–8° in present-day Arctic environments.

limits to growth The title given to the first report of the CLUB OF ROME, published in 1972, which examined the five basic factors that determine, and therefore ultimately limit, demographic growth: population, food production, natural resources, industrial production and pollution. Most of these factors have been observed to grow at an EXPONENTIAL GROWTH RATE. The principal conclusions reached in the report were as follows.

(i) If the present growth trends in world population, industrialization, pollution, food production and resource depletion continue unchanged, the limits to growth on this planet will be reached sometime within the next 100 years. The most probable result will be a sudden and uncontrollable decline in both population and industrial capacity.
(ii) It is possible to alter these growth trends and to establish a condition of ecological and economic stability that is sustainable far into the future. The state of global equilibrium could be designed so that the basic material needs of each person on Earth are satisfied and each person has an equal opportunity to realize his individual human potential.
(iii) If the world's people decide to strive for the second outcome rather than the first, the sooner they begin working to attain it, the greater will be their chances of success.

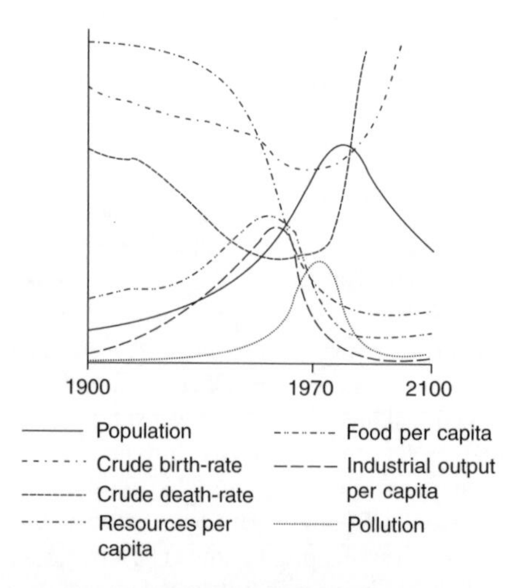

Limits to growth: the world model. [This 'standard' model assumes no major change in the physical, economic or social relationships that have historically governed the development of the world system.]

The general drift of the conclusions reached in the report are strongly reminiscent of Malthus' view put forward nearly 200 years earlier (see MALTHUS' THEORY OF POPULATION GROWTH). [*f*]

limon A fine-grained deposit, occurring on PLATEAUS and INTERFLUVES in Belgium and N France, and giving rise to easily worked, fertile LOAMS. Limon is largely of AEOLIAN origin, and was transported to its present locations by the wind under cold and dry conditions from extensive GLACIO-FLUVIAL silts (the product of glacial ABRASION) laid down at the margins of the Pleistocene ICE SHEETS (see LOESS).

linear eruption See FISSURE ERUPTION.

linear pattern Where the points of a distribution pattern are organized in lines, e.g. the distribution of shops along a main street (see RETAILING RIBBON) or of SETTLEMENTS along a river bank.

linear regression See REGRESSION ANALYSIS.

line-squall A period of gusty winds and heavy showers (commonly of SLEET or HAIL) marking the passage of a particularly well-defined COLD FRONT. See SQUALL.

linkage The connection between FIRMS involved in the same line of production or service. No enterprise is entirely self-contained. Inevitably, therefore, every enterprise has links with other firms, by virtue of either receiving goods and services and/or supplying goods and services to other firms. Thus the linkages of a firm are of two kinds, namely *backward* or *input linkages* (where the firm 'receives') and *forward* or *output linkages* (where the firm 'supplies'). Functional linkages will vary considerably in their scale and complexity, but no matter what the realm of economic activity, these linkages will bind firms into PRODUCTION CHAINS of varying proportions. The advantage of linkage chains is that they encourage specialization which, in turn, results in the reduction of costs and in greater efficiency. The disadvantage of such chains is the proverbial one, namely that the chain is only as strong as its weakest link; default by one firm repercusses through the chain and to the detriment of other linked firms. [*f*]

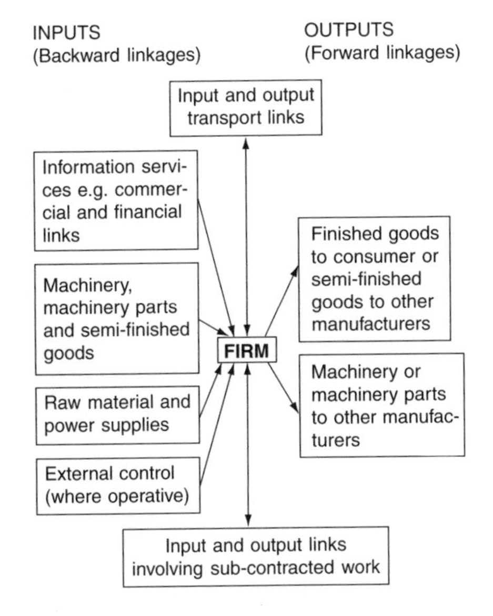

Functional linkages

listed building A building of special architectural or historical interest that, as a result of being 'listed' by the Department of the Environment, enjoys special protection under the British planning system. In effect, 'listing' leads to PRESERVATION or CONSERVATION. See URBAN CONSERVATION.

literacy The ability of a person to read and write. The *adult literacy rate* in a country is a widely used measure of its level of DEVELOPMENT. The rate reflects the quality of education, which in turn reflects the country's wealth and social values. Adult literacy rates currently range from close to 100% in most MEDCS to less than 30% in a number of African countries. Nearly two-thirds of non-literate adults in the THIRD WORLD are women, a reflection of overt GENDER discrimination.

lithification The process whereby loose sediment turns into SEDIMENTARY ROCK following compaction and cementation.

lithosphere The outer layer of the Earth, some 100 km thick, comprising the crust and the upper mantle. The lithosphere forms the plates associated with the theory of PLATE

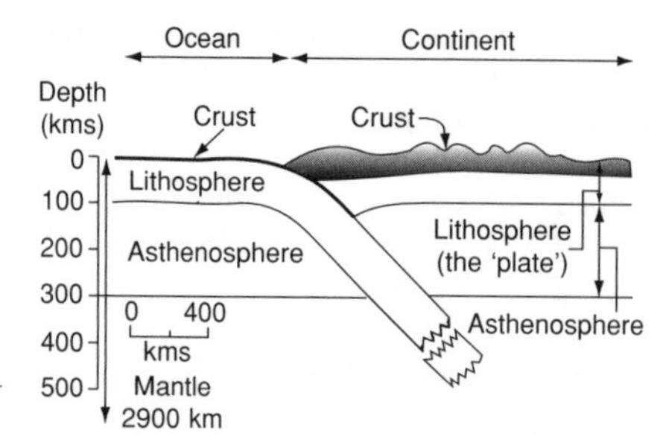

The structure of the Earth from the mantle to the crust

TECTONICS. Beneath the lithosphere is believed to be a less dense layer (the upper ASTHENOSPHERE), which facilitates the movement of the overlying rigid plates. [*f*]

litter The surface deposit or layer of vegetation (e.g. dead grass, leaves and twigs) that, over time, is broken down by chemical and biological processes, and gradually incorporated within the SOIL. See DECOMPOSERS, HUMUS.

littoral (sediment) cell system An system of interlinking sediment cells found adjacent to the coastline. Typically, a sediment cell involves INPUTS of sediment from CLIFF erosion or river DISCHARGE, sediment transfer (e.g. LONGSHORE DRIFT) and sediment DEPOSITION in zones termed *sinks* (e.g. a BEACH or BAR). The concept of large interconnected cells is important, and for this reason it is now incorporated into a more holistic approach to coastal management and planning. [*f*]

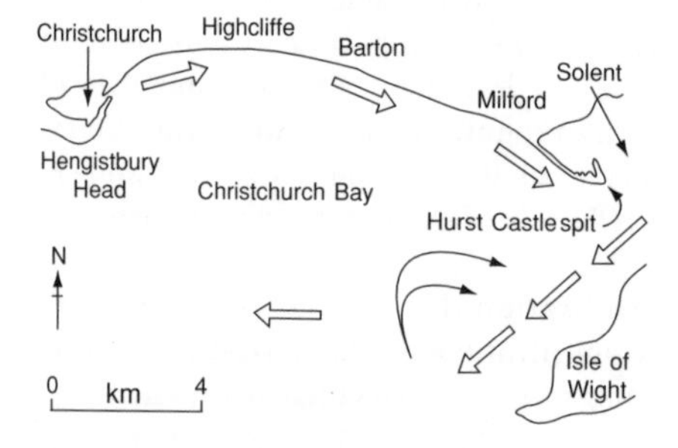

Sediment circulation in Christchurch Bay

LLDC (least developed country) See FOURTH WORLD; ct LEDC, MEDC, THIRD WORLD.

LNG (liquefied natural gas) Natural gas derived from either gas-rich oilfields or separate gas fields that is frozen into a liquid form. Doing this reduces the gas to one six-hundredth of its original volume. Thus, in this liquefied form, natural gas becomes much easier and cost-effective to transport from producing to consumer areas either by pipeline or special tanker ships.

load The material (dissolved or solid) being transported by a river. Load is carried: in SOLUTION; as suspended sediment (see SUSPENDED SEDIMENT LOAD); as traction load dragged, rolled or bounced along the river bed (see BED LOAD). See also COMPETENCE, CAPACITY.

loam A 'medium-textured' SOIL, comprising mixtures of SAND, SILT and CLAY, and combining the most favourable characteristics of both sandy and 'clayey' soils. In other words, loams are well aerated, possess inherent fertility, do not drain too freely, warm up quite rapidly in spring, and can easily be ploughed. They are thus agriculturally useful soils. Different types of loam can be identified, according to the precise content of sand, silt and clay (hence *sandy loam*, *silty loam* and *clay loam*).

local authority The body responsible for the government and administration of public affairs at a local level. In most countries, local authorities are subject to control by the central or national government, but are allowed some freedom of action. In the UK, local authorities have particular responsibility for providing a range of services: amenity (e.g. leisure facilities), environmental (e.g. waste disposal), protective (e.g. enforcing safety regulations) and trading (e.g. day nurseries).

local climate The climate of a small area, which shows some significant contrasts (in terms of temperature, rainfall, wind speed and direction, susceptibility to FOG and frost, etc.) with adjacent areas. These contrasts are the product of RELIEF, slope, ASPECT, ALBEDO (determined by snow cover, soil colour and vegetation), INDUSTRIALIZATION and URBANIZATION (see HEAT ISLAND). Ct MACROCLIMATE, MICROCLIMATE.

localization economies A type of AGGLOMERATION ECONOMY involving potential savings gained by FIRMS in a single INDUSTRY or a set of closely linked industries at a single location. These economies accrue to the individual firm through the enlarged output of the industry as a whole at that location. Cf EXTERNAL ECONOMIES, REGIONAL SPECIALIZATION.

location The geographical situation of a particular phenomenon; its point or position in SPACE.

location coefficient See LOCATION QUOTIENT.

location quotient A simple statistical measure of the degree to which a particular phenomenon (e.g. an economic activity) is concentrated in a given area. The statistic compares the percentage of the overall 'population' (P) found in a sub-area with the percentage of the total area (A) occurring in that sub-area, so that location quotient is

$$\frac{\% \mathrm{P}}{\% \mathrm{A}}$$

The more the location quotient exceeds a value of 1, the more concentrated is the phenomenon in the given sub-area than in the area as a whole. If the value is less than 1, then the phenomenon is more dispersed in the sub-area than in the area as a whole. The location quotient is based on the LORENZ CURVE. Cf INDICES OF SEGREGATION.

location theory See INDUSTRIAL LOCATION THEORY.

locational inertia See INDUSTRIAL INERTIA.

locational interdependence The notion that, when it comes to locational DECISION-MAKING, few FIRMS can act totally independently; rather, locational choice will be influenced, to varying degrees, by the location of other firms with which there are LINKAGES.

locational polygon See VARIGNON FRAME.

locational rent See ECONOMIC RENT.

locational triangle See WEBER'S THEORY OF INDUSTRIAL LOCATION.

lodgement till See GROUND MORAINE.

loess A well-sorted and well-rounded brownish-yellow sandy LOAM, often rich in lime, and forming fertile SOILS. Loess covers wide areas of Europe and China. In the latter location, loess is a massive formation, attaining a thickness of 150–300 m and completely blanketing the pre-existing landscape of hills and valleys. Loess is regarded as a PLEISTOCENE wind-blown SEDIMENT, derived from the SANDUR plains south of the ICE SHEET.

logarithmic scale A logarithmic scale differs from a simple *arithmetic scale* in that, whereas a given numerical difference is shown by a constant interval on an arithmetic scale, on a logarithmic scale a constant interval is used to depict a given proportional difference. A logarithmic scale should be adopted in a GRAPH if the aim is to show, for example, proportional change between points in a time series. In such a case, *semi-logarithmic* graph paper should be used, with the phenomenon under investigation plotted against a vertical logarithmic scale and time plotted on an arithmetic horizontal scale. Figures that are increasing at a constant proportional rate per unit of time (doubling every 10 years, say) are shown as a progressively steepening curve when graphed on arithmetic scales, and as a straight line when graphed semi-logarithmically. See also LOGARITHMIC TRANSFORMATION. [*f*]

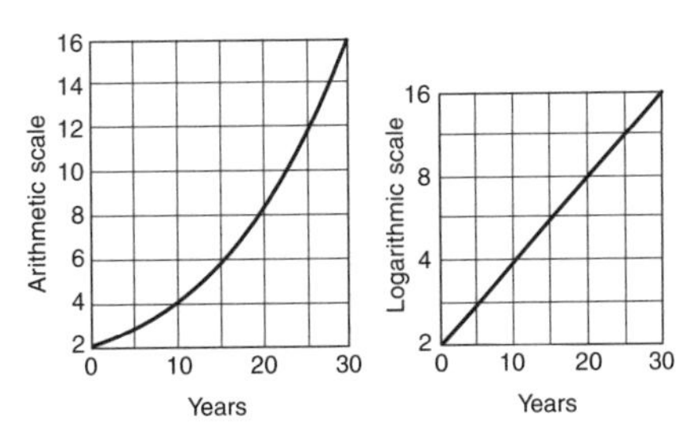

Arithmetic and semi-logarithmic scales

logarithmic transformation This takes place when data are plotted on a GRAPH or are analysed, not at their face or arithmetic value, but in terms of their logarithms (see LOGARITHMIC SCALE). Such a procedure is frequently used where data are not normally distributed, being either positively or negatively skewed (see FREQUENCY DISTRIBUTION). In these instances, the logarithmic transformation is intended to compensate the 'skewness' and to make the frequency distribution appear more normal. As a consequence, such data become more amenable to the STUDENT'S T TEST and other PARAMETRIC TESTS.

lognormal distribution A CUMULATIVE FREQUENCY distribution that, when plotted on either *logarithmic* or *logarithmic probability* GRAPH paper, appears as a straight line. For example, a lognormal CITY-SIZE DISTRIBUTION is produced when frequency decreases at a constant proportional rate with increasing city size. Such *lognormality* was thought to be characteristic of large, developed and highly urbanized countries (e.g. the USA). Research has, however, clearly demonstrated the need to be circumspect about making such a correlation, for countries as diverse as India and Italy, Belgium and Brazil, Switzerland and S Africa all show such city-size distributions. See BINARY PATTERN, RANK-SIZE RULE; ct PRIMATE CITY. [*f* RANK-SIZE RULE]

longitude The angular distance from the *prime meridian* (0°, or Greenwich Meridian), measured up to 180° E or W. A circle drawn around the Earth, passing through each pole, is a *meridian of longitude*; all meridians intersect parallels of LATITUDE at right angles. Meridians (unlike lines of latitude) are *not* parallel to each other, but converge at the poles. Thus, while degrees of latitude have a virtually constant dimension (1° = 110.551 km at the Equator, and 111.698 km close to the poles), degrees of longitude decrease polewards from 111.320 km at the Equator, to 78.848 km at 45°, and 38.187 km at 70°.

long-profile The longitudinal section through a river's course, from its source to its mouth. River long-profiles may be of irregular form, with numerous breaks of slope (see KNICKPOINT), but are believed to tend towards a smooth curve representing a state of equilibrium (see GRADE). The concave form of many river profiles is explained as the outcome of several factors, including the following. (i) As the SEDIMENT load is comminuted downstream, it can be transported over a gentler slope. (ii) As DISCHARGE increases downstream with the entry of tributaries, energy also increases, facilitating the transport of LOAD. (iii) As the river increases in size downstream, its CHANNEL becomes more efficient (mean velocity increases with distance downstream), as the effects of increased *hydraulic radius* are felt, and the impedance to flow by channel roughness is reduced.

longshore drift (also **littoral drift)** The movement of beach SEDIMENTS along the shore by the action of breaking waves. Where waves approach the coast obliquely, the SWASH of the breaking wave is directed diagonally up the BEACH. However, the BACKWASH of the returning water – under the influence of gravity – tends to run more directly down the beach. There is thus, over a period of time, the net transference of large quantities of beach sediment in a down-drift direction. The reality of longshore drift is shown by the piling up of shingle on the windward sides of obstacles such as beach groynes, and the development of coastal landforms such as SPITS. On many coasts the longshore drift is predominantly in one direction (for example, from W to E along the S coast of England, in response to the dominant southwesterly winds and associated waves) but for brief periods a return drift may operate (in S England when winds blow from the E for several days). [*f*]

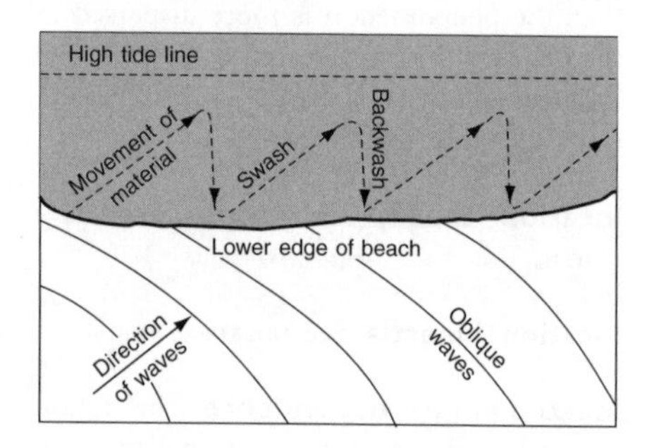

Longshore drift

lopolith A saucer-shaped intrusion, usually developed along BEDDING PLANES and resembling a large down-warped SILL. Where IGNEOUS ROCKS are harder than the country rock, subsequent DENUDATION will give rise to out-facing ESCARPMENTS.

Lorenz curve A simple graphical method used for comparing a given DISTRIBUTION with a perfectly even one, with a view to establishing the degree of concentration or segregation shown by the distribution. A Lorenz curve is drawn on GRAPH paper and with the *x* and *y* axes employing comparable scales. In the accompanying figure, the *x* axis is the percentage of population and the *y* axis the percentage of area. If the distribution of population is perfectly even (i.e. totally dispersed or segregated), then the curve will be a straight line sloping at 45° to the horizontal. If the population were concentrated in one sub-area, then the curve would assume the line of the horizontal or *y* axis. The Lorenz curve will always lie within the shaded portion of the graph, i.e. between the two extremes of concentration and segregation. The nearer the curve is to the diagonal, the less concentrated is the distribution. In the particular case shown, it will be seen that the distribution is quite concentrated (i.e. that 80% of the population is confined to 20% of the area). The Lorenz curve may also be used to investigate non-spatial distribution – as, for example, the distribution of income in a working population. See also INDICES OF SEGREGATION. [*f*]

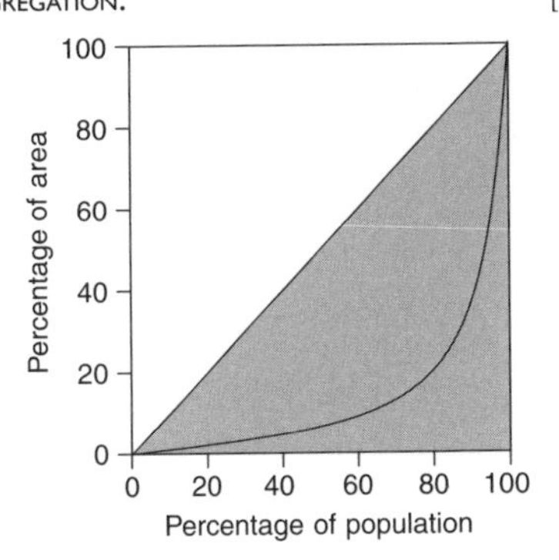

A Lorenz curve

Lösch See CENTRAL PLACE THEORY, K-VALUE.

loss leader A good sold by a retailer below the recommended retail price, not for the purpose of making a profit on the sale, but for the purpose of either attracting customers likely to purchase other goods at the shop or extending the MARKET AREA.

low A term used widely for an area of low atmospheric pressure, such as a mid-latitude FRONTAL DEPRESSION or tropical CYCLONE.

low-income country See LEDC.

low-order goods and services Goods and services with low THRESHOLD and RANGE values, most of them basic to subsistence and being sought at frequent intervals. Outlets of such goods and services will be found concentrated in *low-order* CENTRAL PLACES. See also CENTRAL PLACE THEORY; ct HIGH-ORDER CENTRAL PLACE.

low technology Simple methods and equipment used by traditional societies in the exploitation of local, easily available resources. Ct HIGH-TECHNOLOGY INDUSTRY.

lynchet A man-made TERRACE on a hillside, usually parallel to the contours. It is ascribed to ancient cultivation practice (from the IRON AGE or earlier) and is thought to have been constructed to provide a level, well-drained strip of land with a southward aspect and to check SOIL EROSION.

lysimeter A device for measuring PERCOLATION through a particular SOIL. A receptacle is placed in the ground, and filled with the soil in question, in a state of natural compaction. Water is then fed to the surface of the lysimeter, and the amounts that percolate into a measuring container at the base are recorded. Lysimeters are used in the calculation of EVAPOTRANSPIRATION rates.

M

macroclimate A large-scale climatic region in which controlling factors are latitude, the distribution of large land-masses and oceans, and the general atmospheric circulation. Ct LOCAL CLIMATE, MICROCLIMATE.

magma Molten rock material, highly charged with gases, derived from considerable depths within the Earth. The magma, which is under great pressure from overlying rocks and initially very stiff, becomes more liquid when it is able to penetrate lines of weakness in the crust. It may solidify beneath the surface to form an INTRUSIVE ROCK (e.g. GRANITE) or escape to the surface as LAVA (see

EXTRUSIVE ROCK). The chemistry of magma and its water content are important controls on the explosiveness of volcanic ERUPTIONS. Magmas rich in silica and with a high water content tend to be associated with the most violent eruptions.

magnetic stripes Parallel bands within basaltic IGNEOUS ROCK, occurring on the ocean floor, which are characterized by differing magnetic polarities. These bands are formed at a constructive plate margin, such as the Mid-Atlantic Ridge, and record periodic reversals in the Earth's magnetic field; they also provide compelling evidence of SEA-FLOOR SPREADING (see MID-OCEAN RIDGE, PALAEOMAGNETISM).

malaria An infectious disease caused by the presence of parasitic protozoa in the red blood cells. The disease is transmitted via blood-sucking female mosquitoes and is now confined to tropical and sub-tropical areas. It is the most prevalent and devastating parasitic disease to afflict the human race. Up to 2 million people die each year from the disease and it is estimated that half the world's population is at risk of contracting it.

malnutrition A condition resulting from a DIET that is inadequate either in terms of quantity or vital ingredients (minerals, proteins and vitamins) (see DEFICIENCY DISEASE). Although in itself it is rarely fatal, by weakening the human body it makes people far more vulnerable to killer diseases and illnesses. Cf FAMINE, STARVATION.

Malthus' theory of population growth Malthus (1798) based his theory on two principles: that in the absence of any checks, human populations can potentially grow at a geometric rate – in other words, a population can double every 25 years; and that, even in the most favourable circumstances, agricultural production can at best be expected to increase only at an arithmetic rate. Thus population growth may be expected to outstrip any increase in food supply. Since the rule of diminishing returns applies, any country may be regarded as having a finite food-producing potential and this, in its turn, creates a sort of 'ceiling' to the growth of population within it. Malthus suggested that, once population had reached this ceiling, two types of check would come into effect. (i) *Preventive checks* – these included abstinence from, or delay in the age of, marriage, which would reduce the fertility rate. Malthus also noted that as food became more scarce and therefore more expensive this would tend to delay the timing of marriage. (ii) *Positive checks* – these included FAMINE, disease, war and infanticide, which would help boost the DEATH RATE. A somewhat contrary view of the relationship between population growth and food production has been suggested by Boserup (see BOSERUP'S THEORY), but the later LIMITS TO GROWTH report (1972) rehabilitated Malthus' theory (see NEO-MALTHUSIANS). [*f* BOSERUP'S THEORY]

mangrove A type of tree that has the ability to colonize areas of mud, on the coast and within ESTUARIES, that are exposed at low tide but are otherwise normally inundated by salt or brackish water. The trees are supported by a mass of 'prop' roots, extending a metre or so above the mud surface before uniting to carry the stumpy trunk. The root network of a *mangrove swamp* is close and confused, and is very effective at trapping mud washed in at high tide or from fluvial sources. In some parts of the world, mangroves have been cleared to make way for coastal developments and this has increased the vulnerability of coastal regions to seawater inundation during CYCLONES. In October 1999 a cyclone struck the E coast of India killing thousands, and one reason for the scale of the disaster was believed to be the clearance of mangroves in order to expand the local shrimp fishing industry.

Manning equation A flow equation, designed to relate the velocity of a stream to its controlling factors of hydraulic radius, channel slope and channel roughness (see MANNING ROUGHNESS COEFFICIENT):

$$V = \frac{1.45}{n} R^{\frac{2}{3}} S^{\frac{1}{2}}$$

in which *V* is mean velocity, *R* is hydraulic radius, *S* is channel slope and *n* is channel roughness (the *Manning 'n'*).

Manning roughness coefficient Also referred to as the *Manning 'n'*, this is a function of the roughness of a stream channel, i.e. the degree to which, by way of friction, the channel configuration resists the flow of water in the stream. It is determined empirically, some values for 'n' being:

- a straight channel with no bars – 0.03
- a curved channel with bars – 0.04
- a mountain stream with steep banks and a boulder-strewn bed – 0.05.

Mann's model A model of the British city, proposed by Mann, which combines elements of the CONCENTRIC ZONE MODEL and the SECTOR MODEL. In the model, the concentric zones relate to land use and to age of development, while the sectors relate for the most part to SOCIAL CLASS. The model assumes a prevailing wind from the W, so that INDUSTRY, and its associated poor working-class housing are located on the leeward side of the city. For these reasons, high-class residential development becomes concentrated on the windward, and therefore the relatively pollution-free, side. [*f*]

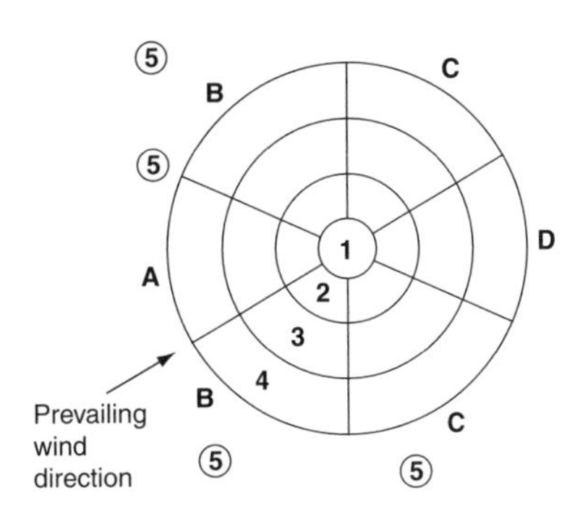

A middle-class sector
B lower middle-class sector
C working-class sector (including council estates)
D industry and lowest working-class sector
1 CBD
2 transition zone
3 zone of small terraced houses in sectors **C** and **D**; larger by-law housing in sector **B**; large old houses in sector **A**.
4 post-1918 residential areas; with post-1945 housing on the periphery
5 dormitory towns within commuting distance

Mann's model of a British city

Mann–Whitney U test A NON-PARAMETRIC TEST used to determine whether the difference between the means of two independent random samples is statistically significant (see SIGNIFICANCE); i.e. whether the two samples represent different POPULATIONS. Its use is recommended when the assumptions of the STUDENT'S T TEST cannot be satisfied, when there is a wish to avoid the rather lengthy calculations of that test, and when only simple measurement by ranking is possible.

mantle A layer of ultrabasic rocks, of high density and considerable thickness (over 2 500 km), lying between the *crust* and *core* of the Earth. The upper surface of the mantle is the MOHOROVICIC DISCONTINUITY and its lower limit the GUTENBERG CHANNEL.

manual worker A person who works with their hands, often in a factory. Three types of manual worker are recognized: (i) *skilled* – electricians, plumbers, welders; (ii) *semi-skilled* – production line workers, drivers; (iii) *unskilled* – cleaners, sweepers.

manufacturing industry The conversion of RAW MATERIALS into fabricated products (e.g. the making of steel or paper, oil refining); the making and assembly of parts and components (e.g. engineering and motor vehicle industries). The essential component of the SECONDARY SECTOR. Ct EXTRACTIVE INDUSTRY, SERVICE INDUSTRY.

map A representation on a flat surface (usually of paper) of the features of part of the Earth's surface, drawn at a specific scale. It involves certain degrees of generalization and exaggeration, of selective emphasis and stylized representation, according to the scale and detail involved. Maps may be prepared in many different forms and for a wide variety of purposes. For example, they can range from general topographic maps (such as the Ordnance Survey 1:50,000 and 1:25,000 sheets) to selective representation of things such as Atlantic weather systems, the distribution of population density at a national level or a city's public transport network (see TOPOLOGICAL MAP). See also CARTOGRAPHY, GRAPHICACY, MAP PROJECTION.

map projection A method that allows the whole of the spherical Earth to be represented on a flat MAP. Parallels and meridians are transformed into a net (*graticule*) on a plane surface. In fact, it is a geometric impossibility to produce a map of the world that is true in terms of shape, bearing and distance. Many different map projections have been devised over the centuries to minimize these difficulties, but inevitably none is perfect. For example, Mercator's projection exaggerates the high latitudes, but bearings and shapes are correct (*orthomorphic projections*), while Mollweide's and Peters' distort direction and shape, but are correct in terms of area (*equal area projections*).

margin of cultivation See MARGIN OF PRODUCTION.

margin of production This is said to occur where the revenue from a particular economic activity is equal to the costs. In such a situation, there will be no incentive to participate in that activity; i.e. the ECONOMIC RENT will be zero. Where AGRICULTURE is concerned, the margin of production is frequently referred to as the *margin of cultivation.*

margin of transference As used in VON THUNEN'S MODEL, this represents the point at which, moving away from the market, the ECONOMIC RENT derived from one type of agricultural production is surpassed by that derived from an alternative. In short, the margin of transference represents the boundary between adjacent agricultural practices.

marginal channel A channel formed by meltwater flowing along the edge of a glacier or ICE SHEET. It may be formed wholly within the ice. Former marginal channels have been used to reconstruct stages in the DEGLACIATION of an area, since they record the ice margins at successive stages of retreat.

marginal costing The cost or expenditure incurred in producing an additional unit of a product or service.

marginal land Land that is barely worth cultivating, or that may or may not be cultivated according to changes in economic conditions, government subsidies or according to the length and nature of wet and dry seasons. Some land (e.g. bordering a desert) may be *fluctuatingly marginal*; other land (e.g. upland areas) may be *permanently marginal.*

marina A purpose-built HARBOUR providing moorings for yachts and other leisure craft, together with shore-based facilities such as parking, chandlery and sometimes housing.

maritime The conditions associated with proximity to the coast. A maritime CLIMATE tends to be moderated by the influence of the sea and does not have the extremes of temperature associated with continental climates. PRECIPITATION tends to be higher but more evenly distributed than in continental interiors. A maritime AIR MASS is generally associated with moist conditions. The polar maritime air mass brings cold conditions from its source area over Greenland, whereas the tropical maritime brings much warmer conditions as it moves from the SW towards the UK.

market area The area in which there is a demand for a given product or service; the area over which it will be supplied (see CENTRAL PLACE THEORY). See also HINTERLAND.

market area analysis A technique employed in predicting the MARKET AREA of a particular product or service, taking into account the delivered price, the maximum price the consumer is prepared to pay for it, and the location of competing firms.

market economy See FREE MARKET ECONOMY, CAPITALISM.

market forces The forces of supply and demand, which together determine the price at which a product is sold and the quantity that will be sold.

market gardening A form of INTENSIVE AGRICULTURE involving the production of vegetables, fruit and flowers (sometimes resorting to cultivation in glasshouses), and traditionally undertaken close to urban markets. In N America market gardening is referred to as *truck farming*. See HORTICULTURE.

market-oriented industry An economic activity characteristically locating close to the market. The pull of the market is particularly strong for those enterprises where: there is little *weight loss* or wastage during production (see MATERIAL INDEX, WEBER'S THEORY OF INDUSTRIAL LOCATION); the industry is *weight gaining* (e.g. the assembly of a product involving components from dispersed locations); the product is costly to transport; there is a need to be in close touch with the market so as to be accessible to clients and to stay abreast of changing fashions, new trends, etc. Ct MATERIAL- ORIENTED INDUSTRY.

market principle One of the three principles underlying Christaller's CENTRAL PLACE THEORY and governing the spatial arrangement of CENTRAL PLACES relative to their MARKET AREAS. This particular arrangement is claimed to maximize competition within the central place system and to be the most efficient from a marketing viewpoint. The arrangement, having a K-VALUE of 3, ensures that three central places compete for the trade forthcoming from lower-order DEPENDENT PLACES. Thus, from a consumer's point of view, competition is maximized and distance minimized. Ct ADMINISTRATIVE PRINCIPLE, TRAFFIC PRINCIPLE. [*f* ADMINISTRATIVE PRINCIPLE]

market processes A term usually used where the ability to pay the asking price takes precedence over local and/or national interests. This sort of situation most often occurs in a PLANNING context where objectors to a planning proposal are unable to outbid the developer.

marl A rock intermediate in composition between CLAY and LIMESTONE; a limey clay.

marriage rate The number of marriages per thousand population in a year. The frequency of marriage (sometimes referred to as *nuptiality*) is a major factor in determining the fertility of a population (see FERTILITY RATIO). However, in Britain and other parts of the world it is becoming less so as an increasing number of children are born outside marriage.

Marxism A perspective on society, propounded by Marx in the mid-19th century, which views the economic base, or MODE OF PRODUCTION, as providing the key to understanding society, its institutions, CLASS structures, patterns of behaviour, beliefs and, indeed, the course of human history. In his analysis of the economy, Marx placed particular emphasis on the importance of labour as a FACTOR OF PRODUCTION. He regarded labour simply as a commodity, for which capitalists paid wages, the *exchange value* of labour being determined by the costs of 'producing' it (i.e. the costs of raising, feeding, clothing, educating and housing the worker). In return for meeting these costs, the capitalist benefits from the labourer's *use value*. Marx went on to claim that the use value of labour to a capitalist exceeds the exchange value, so that labour eventually yields a *surplus value*. In this respect, labour is the only factor of production to command a surplus value.

Marx interpreted history in terms of a series of class struggles, arguing that in each period there is a dominant economic class. In time, open conflict breaks out between the dominant class and a 'rising' class, resulting in the overturn of the old ruling class and the establishment of a new dominant class. In this manner, the capitalist class replaced the feudal aristocracy as the dominant class in the West. However, Marx did not regard the class struggle as an unending process. He maintained that industrialized, capitalist societies were, in the 19th century, becoming increasingly polarized into two classes, namely the *bourgeoisie* (the dominant capitalist class) and the PROLETARIAT (the working masses). He predicted that, eventually, the proletariat would overthrow the bourgeoisie and establish a classless society.

mass balance (glacier budget) The relationship between annual accumulation of snow and ice on a glacier or ICE SHEET and the losses – mainly in the form of meltwater – resulting from ABLATION. A *positive mass balance* (where accumulation exceeds ablation) will tend to produce glacier advance, though there is usually a considerable delay in the response at the glacier snout. By contrast, a *negative mass balance* will favour glacier recession. The *total budget* (annual accumulation plus ablation) helps to determine the 'activity' of a glacier. Where the budget is very large, rapid flow velocities will be required to transfer mass from the ACCUMULATION ZONE to the ABLATION ZONE. See PASSIVE GLACIER. [*f*]

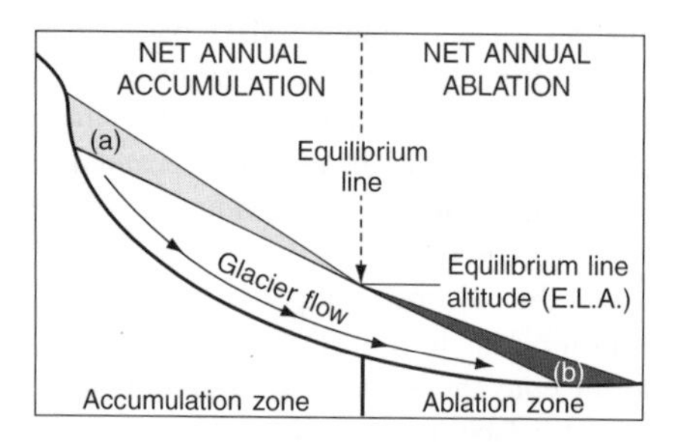

Mass balance of a glacier. If the net annual accumulation (a), adjusted for glacier surface area, exceeds net annual ablation (b), also adjusted, there is a positive mass balance

mass movement An inclusive term, covering several types of process responsible for the transport of weathered materials on slopes, without the direct action of running water. Mass movement occurs when the gravitational force acting on particles on the slope exceeds the resistance of those particles to displacement. The latter is reduced by the presence of moisture (high PORE WATER PRESSURE effectively separates particles and reduces friction between them), by disturbance (for example, heaving of particles owing to expansion and contraction of the SOIL, the formation of GROUND ICE, wetting and drying, and the impact of falling rocks) and by steepening of the slope. Mass movement occurs on a variety of scales and at widely differing rates (see ROCK FALL, LANDSLIDE, SOIL CREEP, SOLIFLUCTION). Slopes that are affected mainly by WEATHERING and mass movement are said to experience *mass wasting*. This produces the smoothly rounded profiles characteristic of many humid temperate landscapes. [*f*]

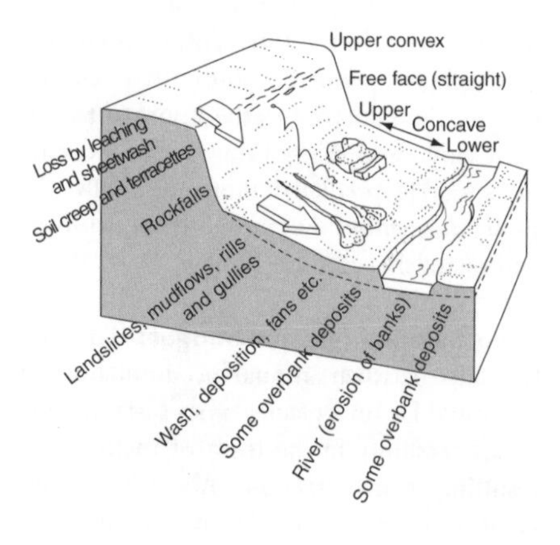

Slope processes and form

mass production The large-scale manufacture of a standardized product by the use of specialized LABOUR and CAPITAL equipment. Mass production has been greatly helped by ASSEMBLY-LINE PRODUCTION and AUTOMATION.

mass wasting See MASS MOVEMENT.

massive A term used to describe a rock in which there are few BEDDING PLANES, JOINTS or other fissures. A massive rock is often IMPERVIOUS, and is highly resistant to processes such as internal SOLUTION and BLOCK DISINTEGRATION.

material index A crucial measure adopted in WEBER'S THEORY OF INDUSTRIAL LOCATION to differentiate between industries where there is much and little *weight loss* (i.e. waste) in the processing of materials into a product. The material index is derived by dividing the total weight of localized materials used per product by the weight of the product. For industries using pure materials (i.e. with no waste), the index equals 1. Where there is substantial weight lost during manufacture, the index will be much higher than 1. For the latter industries, the cost of transporting materials will be much higher than the cost of moving the product; therefore the LEAST-COST LOCATION will tend to lie towards material sources rather than the market (see MATERIAL-ORIENTED INDUSTRY). On the other hand, industries with a material index of 1 or close to it will tend to locate close to the market, since for them the cost of transporting the product (deemed to be *weight gaining*) will be much greater than the cost of transporting any one of the pure materials from which the product is derived (see MARKET-ORIENTED INDUSTRY).

material-oriented industry An industry using low-value RAW MATERIALS, which lose weight during processing; i.e. an industry characterized by a high MATERIAL INDEX (see also WEBER'S THEORY OF INDUSTRIAL LOCATION). In order to keep transport costs to an acceptable level, and because these costs account for a significant proportion of total costs, such an industry will tend to locate close to the sources of raw material supply. In those cases where some of the materials are imported, there will be a tendency for firms to occupy PORT locations (see BREAK-OF-BULK POINT). Ct MARKET-ORIENTED INDUSTRY.

matrix The word has various meanings, but in geography is increasingly used to describe an orderly array or tabulation of symbols or numbers by rows and columns (e.g. BEHAVIOURAL MATRIX). [*f*]

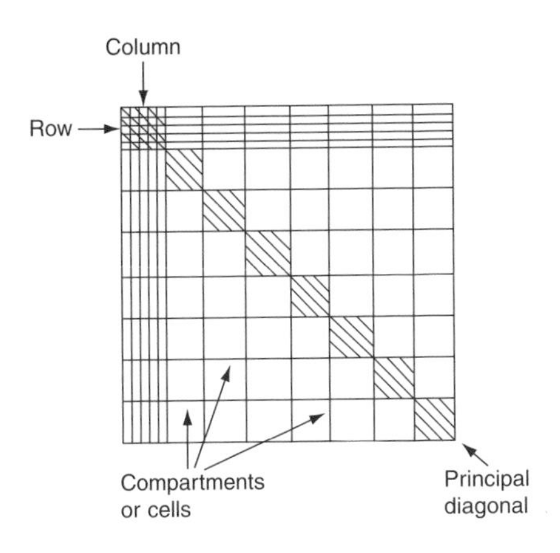

A matrix

maximum sustainable yield A term used in connection with the exploitation of renewable RESOURCES to indicate the maximum yield that may be derived from that resource over a given period of time if it is to maintain the same level of productivity in the future. The concept is important, for example, in the context of the CONSERVATION of fish stocks and RAINFORESTS, for if the maximum sustainable yield is exceeded, then there is every prospect of resource depletion and dwindling stocks. Cf SUSTAINABLE DEVELOPMENT.

mean See ARITHMETIC MEAN, GEOMETRIC MEAN.

mean centre This is a measure of CENTRAL TENDENCY within a SPATIAL DISTRIBUTION, and is calculated in a similar way to the ARITHMETIC MEAN of a numerical distribution. The location of a particular point in a given spatial distribution can be defined accurately by means of two coordinates (x, y), representing the distance of that point both horizontally and vertically from a fixed reference point (i.e. in the manner of a grid reference). The mean centre of a point pattern is defined as a point that has as its coordinates (x, y), the respective means of all the x and y coordinates of all the points in the distribution. Thus by combining the mean values of two separate numerical distributions, scaled along different axes, the mean centre of a spatial distribution is located.

meander A sinuous river CHANNEL, resulting from long-continued bank EROSION at alternating points along the channel margins. The erosion is concentrated where the threads of water flowing at a high velocity come into contact with the bank. Opposite such points, DEPOSITION often occurs, leading to the formation of a *point bar*. Over time, meanders tend to become more pronounced and they also tend to migrate slowly down-valley. The degree of meandering can be expressed by the *sinuosity ratio* (the relationship between stream length and valley length).

It was once thought that meanders were 'chance' features, but it is now known that the main parameters of meanders, such as *meander wave length*, *meander amplitude* and *radius of meander curvature* are related to channel size – itself a function of stream DISCHARGE. Thus it has been shown that meander wave length is usually within the range 7 to 11 times channel width. Moreover, individual meanders resemble mathematical curves (such as 'sine-generated' curves). There may be a relationship between meander development and POOL-AND-RIFFLE sequences in streams. Pools commonly occur at the outside of meander bends, whereas riffles are sited at the *crossover points* between meanders. Other factors favouring meander formation are: comparatively regular discharge; stable banks (comprising SILT and CLAY);

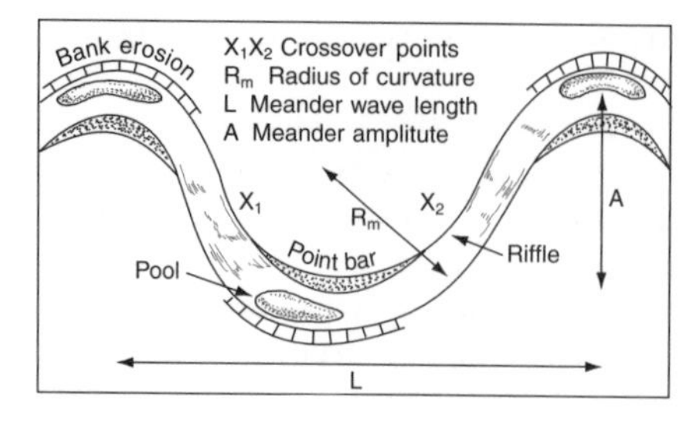

A river meander, with a pool-and-riffle sequence and point bars

gentle stream gradients. See also CUT-OFF, HELICOIDAL FLOW). [*f*]

mechanical (physical) weathering The disintegration of rocks into fragments by entirely mechanical means (i.e. without any chemical change taking place) such as expansion and contraction (ct CHEMICAL WEATHERING). *Biotic weathering* is sometimes recognized as a separate type, though it can involve a physical action (such as the splitting of a rock by roots penetrating a JOINT). The principal types of mechanical weathering include: FREEZE-THAW WEATHERING; *insolation weathering*, resulting from the daytime heating of rock surfaces by the sun, followed by nocturnal chilling (though there are now doubts as to the effectiveness of this process); SALT WEATHERING; DILATATION (not strictly speaking a process of weathering, but having the same effect).

Mechanical weathering, which tends to produce relatively coarse debris, is usually regarded as most active in hot DESERTS, high mountain ranges and high-latitude regions such as the Arctic. However, even here it is often aided by chemical processes (as in freeze-thaw weathering, where widening of joints by SOLUTION and other reactions allows the ingress of water, prior to freezing).

MEDC (more economically developed country) Countries, sometimes also referred to collectively as ADVANCED COUNTRIES, the FIRST WORLD or the NORTH, enjoying considerable wealth and high standards of living derived from well-developed SECONDARY and TERTIARY SECTORS, and increasingly from QUATERNARY SECTOR activity. Characteristics also include high levels of energy consumption, high rates of LITERACY, and considerable economic and political influence at the global scale. The description MEDC would include most European countries, the USA, Canada, Japan, Australia and New Zealand. Ct LEDC; see also BRANDT COMMISSION, CLUB OF ROME.

medial moraine A morainic ridge, developed on the surface of a valley glacier formed by the joining of two separate ice streams (the LATERAL MORAINES of which unite to form the medial moraine). The debris covering the ridge may be derived from WEATHERING and collapse of the slopes above the two ice streams. However, in many instances, there is also a zone of concentrated englacial debris along the line of the medial moraine; this is released at the glacier surface by ABLATION and contributes to the development of the medial moraine. Medial moraines are invariably ice-cored ridges, with a veneer of debris up to 0.5 m in thickness. The ridges rise to maximum heights of 20–30 m above adjacent bare ice.

median The value that is central in an ordered series of values, having an equal number of values above and below it. In a CUMULATIVE FREQUENCY distribution, the median is the value at 50%. [*f*]

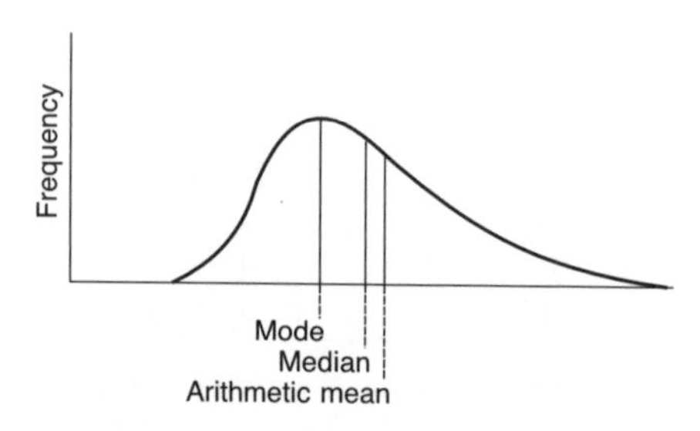

Arithmetic mean, median and mode

medical geography The study of: the spatial incidence of disease, MORBIDITY and mortality; the environment as it affects human health; the spatial organization of health care – the provision of medical centres, clinics, hospitals and so on – as related to the distribution of population, access, etc.

megalopolis A term originally applied to the almost continuous extent of densely populated, URBAN and SUBURBAN area stretching from Boston to Washington DC, USA (sometimes referred to as *Bowash Megalopolis*). The term is used particularly to denote the growing together and integration of large urban AGGLOMERATIONS into some higher order of urban structure or complex. Megalopolitan areas are now discernible in many highly urbanized countries, as for example in the AXIAL BELT of Britain, along the Pacific coastlands of S Honshu, Japan, along the Rhône valley between Lyons and Marseilles, France, and between Los Angeles and San Diego, USA.

meltwater erosion Erosion effected by running water from the melting of glacier ice, snow patches and more extensive snow accumulations. While meltwater issuing from glacier snouts may cause some erosion (though this may be limited by the large SEDIMENT load), the most spectacular examples of meltwater erosion (large, deep POT-HOLES and gorge-like channels, often incised into hard rock such as GRANITE and GNEISS) are formed beneath the ice. They have been carved by SUBGLACIAL streams, which flow with great speed and under considerable hydrostatic pressure, causing erosion by CAVITATION. In PERIGLACIAL environments, meltwater erosion is achieved mainly during spring, when the rapid thawing of winter snow leads to the so-called *nival flood*. It is widely believed that many dry valleys in the English CHALK country are the product of past meltwater erosion. See DRY VALLEY.

mental map A person's image of an area or place carried inside the head, derived either from first-hand experience of that location or from information about it received through various media (films, books, newspapers, etc.); sometimes referred to as *cognitive maps*. In most cases, the mental MAP will be substantially different from atlas maps. Distance and direction will be distorted, some parts of the area will be well known and therefore mapped in detail, while others will be little known and the maps distinctly vague. The mental maps that people carry, since they embody a person's individual PERCEPTION of an area, will frequently influence various aspects of locational DECISION-MAKING (i.e. how they 'rate' different areas and how they discriminate between different areas). Thus mental maps may have a considerable bearing on residential preferences and on the selection of an area in which to live. [*f*]

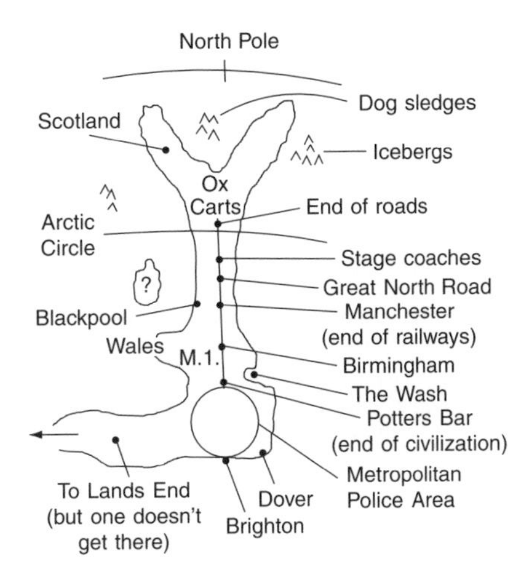

A mental map of Britain

Mercalli (modified) scale A measure of the effects of an EARTHQUAKE as opposed to the RICHTER SCALE, which is a measure of earthquake magnitude and energy. Originally devised in 1902, it was modified to take account of urban lifestyles in 1931, thereafter being known as the 'modified Mercalli scale'. It is a subjective 12-point scale (from I–XII) that describes the damage done by an earthquake, where I represents 'only felt by a few' and XII represents 'total damage'. [*f*]

meridian See LONGITUDE.

mesa A flat-topped hill, larger in extent than a BUTTE, formed in horizontal or near-horizontal structures as a result of stream dissection and slope retreat. Mesas represent early stages in the formation of buttes, and the processes responsible are the same.

Mesolithic period A cultural period that followed on from the PALAEOLITHIC PERIOD and lasted from 10,000BC to 4000BC. It is thus associated with the end of the PLEISTOCENE. It is characterized by the use of small stone implements, which have been found in an area ranging from Mesopotamia to the Baltic and the New World. Other features include the development of fishing and the domesti-

I	Not felt except by a very few under especially favourable circumstances.
II	Felt only by a few people at rest, especially on upper floors of buildings.
III	Felt quite noticeably indoors, especially on upper floors of buildings, but many people do not recognize it as an earthquake. Standing vehicles may rock slightly. Vibration like passing truck. Duration estimated.
IV	During the day, felt indoors by many, outdoors by a few. At night some awakened. Dishes, windows, doors disturbed; walls make cracking sound.
V	Felt by nearly everyone, many awakened. Some dishes, windows, etc. broken; a few instances of cracked plaster; unstable objects overturned. Pendulum clocks may stop.
VI	Felt by all, many frightened and run outdoors. Some heavy furniture moved; a few instances of fallen plaster or damaged chimneys. Damage slight.
VII	Damage negligible in buildings of good design and construction; slight to moderate in well-built ordinary structures; considerable in poorly built or badly designed structures; some chimneys broken.
VIII	Damage slight in specially designed structures; considerable in ordinary substantial buildings, with partial collapse; great in poorly built structures. Fall of chimneys, factory stacks, columns, monuments, walls. Heavy furniture overturned.
IX	Damage considerable in specially designed structures; well-designed frame structures thrown out of plumb; great in substantial buildings, with partial collapse. Buildings shifted off foundations.
X	Some well-built wooden structures destroyed; most masonry and frame structures destroyed with foundations; ground badly cracked. Rails bent.
XI	Few, if any (masonry) structures remain standing. Bridges destroyed. Rails bent greatly.
XII	Damage total. Practically all works of construction are much damaged, or destroyed.

Mercalli (modified) scale

cation of the dog. In Britain, the Mesolithic period lasted from the 8th to the 4th millennium BC. During this time, England became separated by sea from the rest of the continent of Europe (*c.* 5000BC) and the moister Atlantic climatic stage began. Ct IRON AGE, NEOLITHIC PERIOD.

mesophyte A plant requiring a moderate amount of moisture for successful growth (ct HYDROPHYTE, XEROPHYTE). Most trees are *mesophytic.*

metamorphism The process by which existing rocks are changed (metamorphosed), in terms of texture, composition or structure, as a result of intense pressure and/or heat.

meteorology The scientific study of the phenomena and processes of the ATMOSPHERE; in simpler terms, the study of weather.

metropolis A CITY that predominates as a seat of government, of ecclesiastical authority, of commercial activity, or of culture. Strictly the chief city (but not necessarily the capital) of a country, STATE or REGION. The term tends to be used loosely to refer to any large city, as in *metropolitan county* and *metropolitan region*, i.e. the area served or governed by such a city.

microchip revolution Based on revolutionary developments within the microelectronics industry, and starting in 1948 with the invention of the transistor. It was furthered in the early 1950s with the adoption of silicon as the dominant semi-conductor raw material. By the end of the 1950s semiconductors had appeared in computers and in complex defence systems. There then began a steady drift of firms from the E to the W coast of the USA, and in particular concentrating in *Silicon Valley* (the Santa Clara valley) near San Francisco. Climatic conditions there seemed better suited to semiconductor processing; in addition, many key entrepreneurs wanted to live there rather than on the E coast (see SUNBELT). In the late 1950s techniques were developed capable of forming transistors and their interconnecting patterns (i.e. integrated circuits) on wafers of silicon, cut into *chips* approximately 5 mm^2. One such integrated circuit is the microprocessor. The cheapness, size and flexibility of these integrated circuits enables them to be used nowadays in a wide range of equipment (e.g. defence systems, computers, office equipment, telecommunications, domestic appliances, and so on). Their manufacture now involves elaborate PRODUCTION CHAINS spread over large areas of the globe and involving both LEDC and MEDC locations.

The labour repercussions of the microchip revolution are ambivalent. While the application of micro-electronics continues to reduce employment in factories through the AUTOMATION of various jobs, the microchip industry itself continues to create demands for new products (word-processors, fax and e-mail devices, scanners, printers, etc.) as well as for labour-intensive support industries, especially firms involved in the design and development of computer software.

microclimate The climate of a very small area within a few metres of the ground, e.g. the immediate vicinity of a small plant or group of plants, the lee of an obstruction such as a hedge or wall, or within the foliage of a tree. Strictly speaking, microclimatology operates at a much more restricted scale than LOCAL CLIMATE (e.g. an URBAN CLIMATE), however, the terms are now often used synonymously.

mid-ocean ridge A prominent ridge (in effect a submarine mountain range), formed largely of basaltic rocks and associated with shallow EARTHQUAKES. The mid-ocean ridge systems comprise several elements, including the mid-Atlantic ridge, the mid-Indian Ocean ridge (continuing S of Australia as the Indian–Antarctic ridge), and the Pacific–Antarctic ridge (ultimately trending northwards towards the coast of the W USA). The ridges are mainly covered by water (up to 1000–2000 m in depth for the most part), but locally rise above the ocean surface (e.g. the Azores, Ascension Island and Tristan da Cunha, all of volcanic origin, along the mid-Atlantic ridge). Mid-ocean ridges are the product of SEA-FLOOR SPREADING, and the extrusion of basic LAVA in vast quantities along the lines of fissure, thus creating new oceanic crust. On either side of the ridge, parallel MAGNETIC STRIPES of the ocean floor display alternate 'positive and 'negative' magnetization, with the patterns being exactly symmetrical about the ridge. This is the result of 'freezing' of the magnetic polarity (which at certain times has become reversed, and at others remained normal; see PALAEOMAGNETISM). These magnetic patterns have been used to date the growth of the mid-ocean ridges, and to calculate the rate of expansion of the ocean floor (from 1 to 6 mm yr^{-1} over the past 5 million years). See also PLATE TECTONICS. [*f*]

migration The movement of animals and people. In the case of the latter, a common distinction is made between *internal migration* (within a country) and *external* or *international migration* (to and from a country). At a regional and local level, a distinction is frequently made between *in-migration* and *out-migration*. Migration is usually interpreted as a response to two sets of reciprocal forces: *push factors* operating in the place of departure; and *pull factors* at work in the place of destination. Although the decision to migrate is essentially unique to each individual, people frequently move in groups and share in group decisions. When this occurs, patterns of movement may be identified, and these have led geographers to formulate theories that attempt to explain the principles underlying migration (see, e.g., RAVENSTEIN'S LAWS OF MIGRATION, MOBILITY

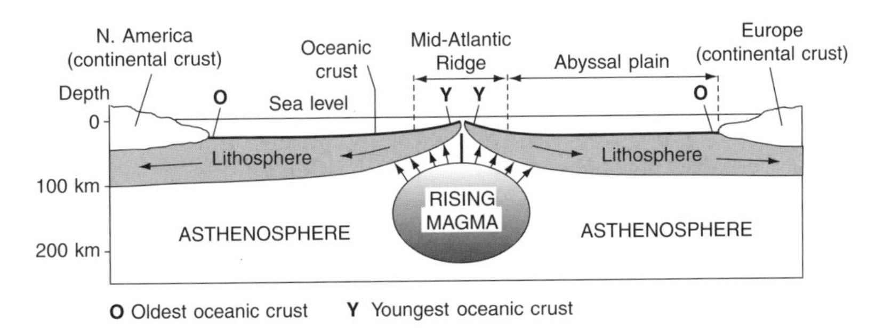

Mid-ocean ridge

TRANSITION). Ct CIRCULATION. See also EMIGRATION, IMMIGRATION.

migration balance The difference between the number of migrants entering an area over a given period and the number leaving. A *positive migration balance* (also referred to as *net in-migration*) occurs when the number of arrivals exceeds departures; a *negative migration balance* (or *net out-migration*) occurs when the balance is reversed. See MIGRATION; ct NATURAL INCREASE.

migration chain The tendency for migrants to follow in the footsteps of those who have previously migrated, so that migrants from a particular area in the country of origin are likely to have a common destination in the country of immigration. The link is forged partly by a feedback of information from those who have already moved, and partly as a result of kinship and friendship ties. The latter often involve money being sent back to the country of origin to pay the fares of those wishing to follow the same migration route. Migration chains have been evident in the movement of New Commonwealth immigrants into Britain; as such they have played a significant part in the emergence of GHETTOS. See FILL-IN MIGRATION, MIGRATION.

Milankovitch's theory A theory that seeks to explain CLIMATIC CHANGE, particularly the ICE AGES, in terms of changes in the Earth's orbit around the Sun and in the 'tilt' and 'wobble' of its axis. These three changes affect the amount of INSOLATION and each produces a cycle of 95,000, 42,000 and 21,000 years respectively. The ice ages are claimed to occur when there is a particular coincidence of the three cycles; that is, when the Earth's orbit is circular rather elliptical, the tilt is small, the wobble makes the Northern Hemisphere

The 95,000-year stretch

The Earth's orbit stretches from being nearly circular to an elliptical shape and back again in a cycle of about 95,000 years. During the *Quarternary*, the major glacial – interglacial cycle was almost 100,000 years. Glacials occur when the orbit is almost circular, and interglacials when it is a more elliptical shape.

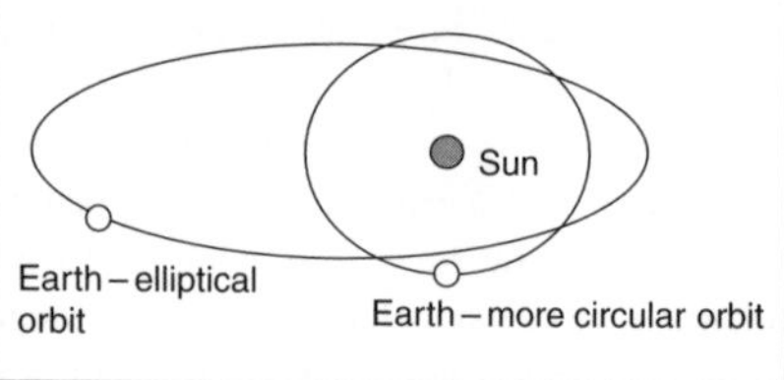

Sun
Earth – elliptical orbit
Earth – more circular orbit

The 42,000-year tilt

Although the tropics are set at 23.5° N and S to equate with the angle of the Earth's tilt, in reality the earth's axis varies from its plane of orbit by between 21.5° and 24.5°. When the tilt increases, summers will become hotter and winters colder, leading to conditions favouring interglacials.

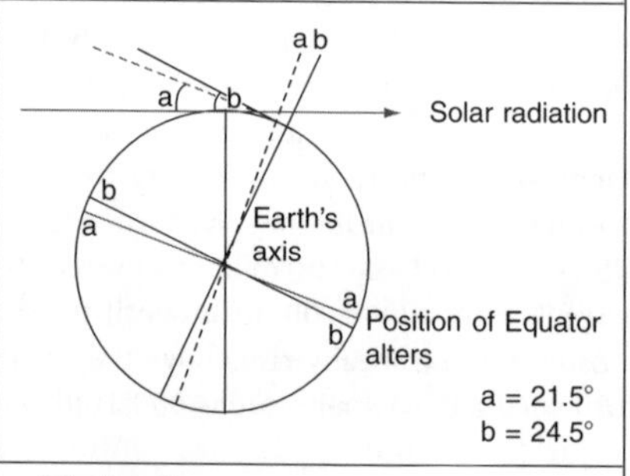

Solar radiation
Earth's axis
Position of Equator alters
a = 21.5°
b = 24.5°

The 21,000-year wobble

As the Earth slowly wobbles in space its axis describes a circle once in every 21,000 years. At present the orbit places the Earth closest to the Sun in the northern hemisphere's winter and furthest away in summer. This tends to make winters mild and summers cool. These are ideal conditions for glacials to develop.
12,000 years ago the position was in reverse and this has contributed to our present warm 'interglacial'.

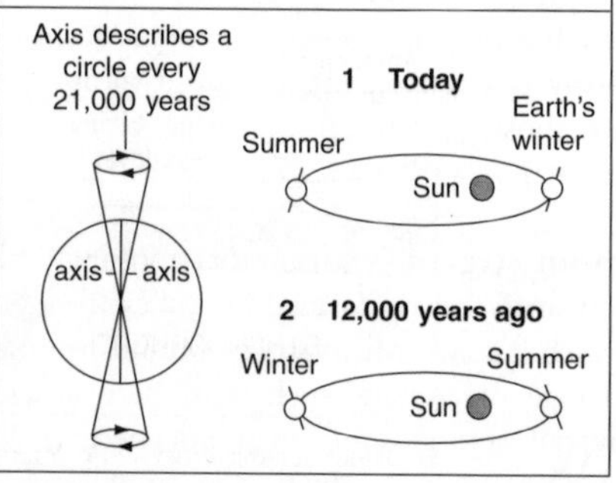

Axis describes a circle every 21,000 years
axis
axis
1 Today
Summer
Earth's winter
Sun
2 12,000 years ago
Winter
Summer
Sun

Milankovitch's theory

furthest from the Sun in summer. There is now a considerable amount of evidence to support the existence of the three cycles. [*f*]

million (or **millionaire) city** A city with a population of a million or more. In 1800 there was only one million city and that was London; at the time of writing (2000) there are estimated to be around 300 such cities. These are by no means confined to the highly urbanized MEDCs. In fact, roughly half are to be found in LEDCs, particularly in S and E Asia (e.g. Tehran, Karachi, Bombay, Calcutta, Manila, Djakarta, Canton, Seoul).

minamata disease A form of mercury poisoning named after a port in S Japan. Between 1953 and 1956, people died and many more became brain-damaged because they ate fish contaminated by the waste discharged into the sea by a FERTILIZER factory.

mineral A natural inorganic substance, either element or compound, that takes the form of crystals. Minerals are regarded as being the 'building blocks' of rocks.

minimum effort See LEAST EFFORT.

mining The extraction of mineral RESOURCES (coal, tin, copper, etc.), but not including the working of building stone (usually referred to as *quarrying*).

minority A group of people living in a country different in any aspect of RACE, religion, language, social customs and national sympathies from the majority of the people (see ETHNIC GROUP). The USA is renowned for its many minority groups, a simple reflection of the fact that during the last 100 years it has drawn population from many different countries; those minorities are most evident in the large cities (see GHETTO).

misfit stream A stream that appears to be too small to have formed the valley through which it flows. Misfit streams result from: a change of climate, such that there is less overall RUN-OFF (this may not necessarily involve simply a reduction in rainfall, since vegetation cover and EVAPOTRANSPIRATION are also important factors); RIVER CAPTURE (in which there is a loss of CATCHMENT to a neighbouring expanding stream network); underground ABSTRACTION of water (e.g. in LIMESTONE and CHALK regions); past episodes of glacial EROSION in which mountain valleys were greatly widened and deepened, so that the POSTGLACIAL streams appear as 'underfit'.

mist See FOG.

mistral A cold, dry and often very powerful wind blowing from the NW or N along the 'funnel' of the lower Rhône valley in France, and affecting the Rhône DELTA and adjacent coastal lowlands. The mistral typically develops when cold continental air over Europe is drawn in by a depression passing eastwards through the Mediterranean Basin. It causes considerable discomfort, and also much damage to vines and fruit trees; hence the construction of numerous wind-breaks (lines of poplars and cypress hedges) in the lower Rhône valley.

mixed economy An economy in which resources are allocated partly through the decisions of private individuals and privately owned enterprises (see PRIVATE SECTOR) and partly through the decisions of government and state-owned enterprises (see PUBLIC SECTOR). A capitalist economy that has been subjected to a degree of GOVERNMENT INTERVENTION. The British economy is a good example, with the actual balance between the two sectors fluctuating according to the changing political inclinations of successive governments. See FREE MARKET ECONOMY, PLANNED ECONOMY.

mixed farming AGRICULTURE involving both crops and livestock. This is not to be confused with *mixed cultivation*, implying merely the growing of a series of different crops.

mixing ratio See RELATIVE HUMIDITY.

mobility The quality of being able to move about. In geography, the term can be used in a number of different contexts: *physical mobility* – in the sense of a person either having access to personal transport (i.e.

through car ownership) or being able to afford full use of public transport services; *residential mobility* – in the sense of moving house; *social mobility* – in the sense of commanding opportunities for social betterment; *mobility of capital* – where the movement of capital is not hindered by institutional barriers; *mobility of labour* – the willingness of labour either to shift to a new location or to change to new types of work (i.e. learn new skills); *mobility of technical knowledge* – the efficiency with which communications networks can transmit new ideas and techniques.

mobility transition A model, put forward by Zelinsky, based on the idea that national or regional progress is marked by a change in the type and amount of migration. The model recognizes five phases that can be related to the stages of DEMOGRAPHIC TRANSITION and originally defined five different types of movement (international, frontierward, rural–urban, inter- and intra-urban, and CIRCULATION). However, given more recent developments, some would argue that a sixth type should be added: urban–rural migration (see COUNTERURBANIZATION).

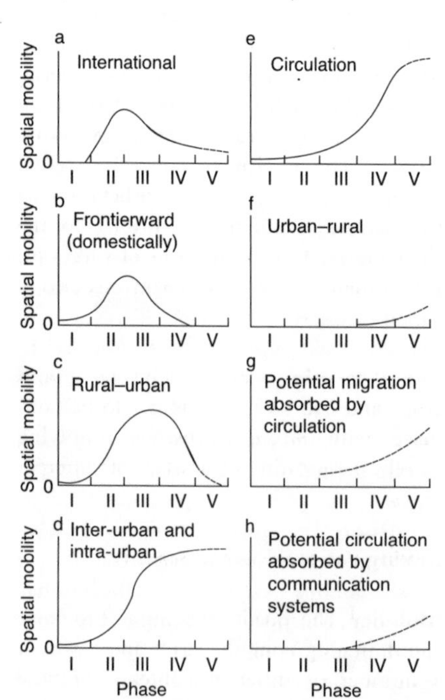

The mobility transition

The general transition is from very limited geographical MOBILITY towards much wider and more elaborate forms of movement. The model also recognizes that, in the later phases: the need to migrate is reduced by developments in transport (i.e. MIGRATION may be substituted by CIRCULATION); the need to circulate is reduced by developments in COMMUNICATIONS (i.e. media such as the telephone, fax, e-mail, etc. increasingly provide a substitute for person-to-person contact).

[*f*]

mode (i) Way, manner, means, as in *mode of transport*, e.g. car, train, bus, etc. See MODE OF PRODUCTION. (ii) The value that occurs most frequently in a data set. Its main importance lies in indicating the value of a substantial part of a data set, and it is most useful for interpretative purposes when most of the values tend to cluster around the modal value. It is the peak of any FREQUENCY DISTRIBUTION curve. Cf ARITHMETIC MEAN, MEDIAN.

[*f* MEDIAN]

mode of production The way in which a society organizes its productive activities (i.e. ECONOMY) and its social life. Four modes of production are generally recognized: CAPITALISM, COMMUNISM, *feudalism* and *slavery*. See MARXISM.

model A representation of the real world, necessarily simplified in detail and by definition usually reduced in scale. Models of various kinds are widely used in modern geography, in an attempt to depict *general* rather than *unique* situations. For example, as a substitute for studying all the individual features of actual cities, Burgess introduced the CONCENTRIC ZONE MODEL, which combines the major features of many cities in a relatively simple model. Thus 'model building' has helped geographers to make generalizations, to formulate laws relating to physical and human geography, and to increase the possibility of accurate prediction.

Among the models used by geographers are: *simple linear models*, involving the fitting of regression lines to the scatter of points on a GRAPH showing relationships between variables (e.g. stream DISCHARGE and CHANNEL width); *simulation models*, which involve attempts to represent the operation of the real world, using 'counterfeits' (as in the simulation of insequent drainage patterns by the random walk method); *working scale models* can be constructed in laboratories (e.g. stream channels of variable shape and size can be 'fed' with different discharges and SEDIMENT loads, and the resultant changes monitored); *conceptual models*, which are derived from observations of real situations (as in models of city structure) or from deductive reasoning (e.g. VON THUNEN'S MODEL of land use zonation around a city). The latter type of model provides hypotheses that can be tested by comparison with the real world.

modernization Literally, the adaptation or updating of something to meet present needs and conditions. In geography, the term tends to be used in the more restricted sense of a process of change in society, in which diffusion and adoption lead to a society generally progressing in terms of DEVELOPMENT. Specifically, modernization might involve greater social MOBILITY, more efficient social organization (often in the cause of raising economic production) and changing social values. It is often implied that modernization is undertaken in order to emulate what are deemed to be the more advanced societies.

modified Mercalli scale See MERCALLI SCALE.

mogote A prominent, usually forested, hill of LIMESTONE, with marginal slopes at 60–90°. In TOWER KARST groups of mogotes are separated by a more or less flat, alluviated plain resulting from long-continued SOLUTION of limestone at the level of the WATER TABLE.

[*f* COCKPIT KARST]

Mohorovicic discontinuity The line of junction between the Earth's crust and the MANTLE lying at a depth of approximately 40 km beneath the continents and 6–10 km beneath the ocean floors The discontinuity affects the speeds at which EARTHQUAKE waves travel, and was discovered from study of a particular Balkan earthquake by Mohorovicic in 1909. In common usage, the term Mohorovicic is frequently abbreviated to *Moho.*

Mohs' scale See HARDNESS SCALE.

monetarism A school of economic thought built around the belief that economies left to their own devices are stable and therefore require the minimum amount of GOVERNMENT INTERVENTION to ensure their smooth running. Faith is placed in the market as the best way of achieving required levels of supply. Distortion of the market, by government, MONOPOLIES or trades unions, is regarded as being highly undesirable. In the monetarist view, the role of government is simply to provide a stable business environment through the control of INFLATION and the encouragement of free trade.

monocline Literally a 'one-limbed' fold. A monocline is thus strongly asymmetrical, with a very steep angle of DIP on one side and virtually horizontal strata on the other.

monoculture An emphasis on one dominant crop, either large-scale and extensive (see EXTENSIVE AGRICULTURE) as in parts of the Canadian PRAIRIES, or small-scale and intensive (see INTENSIVE AGRICULTURE) as with rice cultivation in Monsoon Asia and VITICULTURE in parts of France. Monoculture has potential pitfalls as a farming practice, partly because of dangerous dependence on one crop (i.e. on its yield and the market price in any one year) and partly because the SOIL can readily become impoverished.

monopoly A monopoly exists where an organization has control over a sufficiently large proportion of the total OUTPUT of a commodity or service to enable it to raise the price of that commodity or service by restricting output or provision. This implies that there are no close substitutes for the commodity and that the FIRM in question is not threatened by competitors.

monsoon The seasonal reversal of winds over land-masses and adjacent oceans. The most famous example is the Indian Monsoon, which is related to the alternating development of high- and low-pressure systems, the movement of the inter-tropical convergence zone (see ITCZ), and the rearrangement of the westerly JET STREAMS. In

Asia there is in winter a layer of cold, high-pressure air giving dry 'outblowing' winds over much of the Indian subcontinent. At this time the upper air westerlies form two currents, N and S of the Tibetan Plateau; the southern jet steers depressions across the Middle East to give some winter rainfall in N India and Pakistan. In spring, intense heating of the land begins, setting up a 'thermal low' over N India, which begins to cause a reversal in the wind direction, drawing in moist air from the SW. In early summer, the southern jet begins to break down, allowing the movement northwards of the Equatorial Low Pressure (associated with the ITCZ) and the influx of rain-bearing winds from the Indian Ocean. The southwesterly monsoon typically arrives over the Indian subcontinent in June.

monsoon forest A tropical FOREST adapted to the seasonal drought characteristic of MONSOON climates. There is luxuriant growth during the heavy rains of the wet season, but leaf-shedding occurs in the dry season (hence use of the term *tropical deciduous forest*). Forest growth is more open than in the tropical RAINFOREST and there is more under-storey vegetation in the form of dense shrub thicket. Tree height is also reduced (up to 35 m at maximum). A typical monsoon forest tree is teak.

moor (moorland) An area of largely open country, differing from HEATHLAND in its occurrence at greater altitudes (in Britain usually above 200 m, though the term has been applied to lowland wet areas, as in Sedgemoor, Somerset) and in its greater degree of wetness due to heavier PRECIPITATION (heath is essentially a 'dry' environment). Moorland is a 'degenerate' type of ECOSYSTEM, resulting in many areas from FOREST clearance or the decline of birch-pine woodland (see PLAGIOSERE). There has in the past been considerable accumulation of peat (up to 10 m in thickness, contrasting with the few cm of peat in heathland), though this is often undergoing EROSION today. Three main types of moor are recognized: (i) *heather moor*, relatively dry, and dominated by *Calluna vulgaris*, bell heather and bilberry; (ii) *bilberry moor*, usually at higher elevations; (iii) *cotton-grass moor*, formed on saturated peat and associated with 'blanket bogs' comprising sphagnum moss. Heather moorland is a good example of a PLAGIOCLIMAX.

mor humus See HUMUS.

moraine Mixed debris comprising rounded, sub-angular and angular BOULDERS, GRAVELS, SAND and SILT, being transported on, within and beneath a glacier or ICE SHEET (ct TILL, which is morainic material actually deposited by the ice). Various types of moraine can be identified: *supraglacial moraine* is composed largely of material from the slopes above the glacier, and is predominantly angular and coarse; *englacial moraine*, carried within the ice, comprises SUPRAGLACIAL debris that has been incorporated into the ice in the ACCUMULATION ZONE; *subglacial moraine* is carried beneath the ice and is mainly the product of plucking and ABRASION. The term moraine is also used to describe the landforms associated with the various types of morainic debris (see LATERAL MORAINE, MEDIAL MORAINE, TERMINAL MORAINE).

morbidity The incidence of sickness in a population.

morphological map A MAP designed to depict objectively detailed surface forms, usually of a small area. Breaks and changes of slope, slope gradients, etc. are determined by field survey, and shown in symbolic fashion. A distinction is often drawn between morphological maps, which simply record form, without attempting to show the age and origin of the landforms, and *geomorphological maps*, which are 'genetic' in character, depicting features such as EROSION SURFACES and containing information as to the age and origin of these. Morphological maps are useful in the interpretation of relationships between landforms, and SOIL and vegetation distributions.

morphology In the context of GEOMORPHOLOGY, the 'study of form' – a principal aim of which is to study the shape, size and origin of landforms. This is often attempted in a regional context – for example, the landform assemblages (or landscapes) associated with particular climatic regions, such as humid tropical, SAVANNA, humid temperate or PERIGLACIAL. These are referred to as *morphological regions* or *morphogenetic regions*. In URBAN GEOGRAPHY, morphology refers to the three-dimensional form of the BUILT-UP AREA.

mortality rate See DEATH RATE.

moulin A SINK HOLE in a glacier, caused by the melting out of a CREVASSE or ice fracture, down which a SUPRAGLACIAL stream may flow to join the ENGLACIAL drainage system. Moulins are often near-vertical shafts, circular in cross-section, and sometimes with a distinctive 'spiral' form produced by the motion of the rapidly descending water. Large pot-holes, formed in solid rock beneath glaciers, are also termed moulins. One view is that these form where a descending englacial stream, flowing at a high velocity, strikes the BEDROCK.

mountain building See OROGENY.

mountain front The steep edge of an upland, standing above a PEDIMENT or BAJADA, in a desert region (particularly in the arid SW of the USA). The mountain front appears to undergo parallel retreat, either as a result of uniform WEATHERING over the slope, or active basal undercutting by occasional RUN-OFF on the pediment. See PEDIMENT, PIEDMONT.

moving average A method of calculating CENTRAL TENDENCY over time. For example, mean annual precipitation is calculated for the period 1990–95. The moving average for subsequent years is found by calculating the ARITHMETIC MEAN for 1991–96, 1992–97, 1993–99 and so on. The benefit of the moving average is that, by smoothing out short-term fluctuations (in this case annual variations), longer-term trends can more readily be identified. Sometimes referred to as a *running mean*.

mudflow A rapid form of MASS MOVEMENT involving the collapse and downhill 'flow' of often saturated, fine-grained material (e.g. mud, CLAY or volcanic ASH). Mudflows are usually triggered by sudden events such as an EARTHQUAKE or a torrential storm. They are capable of inflicting considerable destruction, particularly where houses occupy hillsides. In December 1999 heavy rains triggered catastrophic *mudslides* (involving movement of materials as a single unit across a slide plane) and mudflows in N Venezuela killing over 30,000 people and destroying 26,000 homes. A mudflow frequently shows a lobate form at its furthest extent (see also LAHAR).

mudslide See MUDFLOW.

mud volcano A small cone formed at a volcanic vent where hot water and mud are emitted. Mud-pots are pools of boiling mud, which bubble away in areas of minor volcanic activity (e.g. Yellowstone National Park, USA).

mull humus See HUMUS.

multicultural society See MULTI-ETHNIC SOCIETY.

multi-ethnic (multicultural) society A society containing a mix of people of different racial origins, language, religion or national extraction. For example, during the postwar period, Britain has become a multiracial society as a result of the immigration of people from the W Indies, from the Indian subcontinent and from SE Asia.

multi-family occupation Where a dwelling unit is occupied by more than one family or household; a type of occupancy more frequently encountered in the INNER CITY, where properties formerly occupied by single and more affluent families are subdivided into flats and bedsits. Multi-family occupation is part of the explanation of the paradoxical situation in most cities of the poorer families living in those areas where BID-RENTS are high (see ALONSO MODEL).

multilateral Used to describe international negotiations and agreements between groups of countries (e.g. as between the EU and NAFTA) rather than just between two countries (*bilateral*). See AID.

multinational corporation See TNC.

multinational state A country in which the population contains one or more significant MINORITY groups. While racial characteristics often readily identify such groups, in many multinational states it is language and national extraction that sustain internal cohesion and a sense of separate identity (e.g. French-speaking Canadians, the Flemish in Belgium, the Basques in Spain).

multiple correlation See CORRELATION, FACTOR ANALYSIS.

multiple deprivation See DEPRIVATION.

multiple-nuclei model A model of CITY structure put forward by Harris and Ullman, and acknowledging that city structure is far more complex than countenanced by either the CONCENTRIC ZONE MODEL or the SECTOR MODEL. It stresses the cellular structure of cities and the tendency for like activities and like people to agglomerate, thereby defining specialized LAND USE and social REGIONS within the BUILT-UP AREA. The model makes no prescription about the spatial arrangement of these individual cells; the pattern can assume innumerable variations. [*f*]

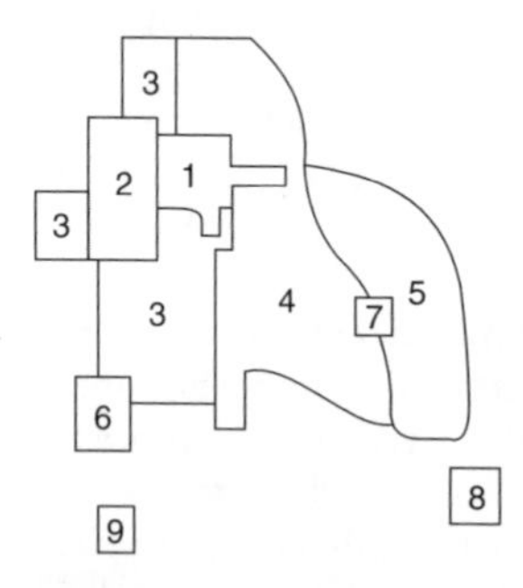

District

1 Central business district
2 Wholesale, light manufacturing
3 Low-class residential
4 Medium-class residential
5 High-class residential
6 Heavy manufacturing
7 Outlying business district
8 Residential suburb
9 Industrial suburb

Harris and Ullman multiple-nuclei model

multiple occupation See MULTI-FAMILY OCCUPATION.

multiple regression The statistical investigation of interrelationships between more than two VARIABLES. See REGRESSION ANALYSIS.

multiplier effect Originally, a term used by economists, but now quite loosely used in geography to denote the direct and indirect, the intended and unintended, consequences or repercussions of an action or decision, particularly where the ramifications are cumulative or take the form of an *iterative loop*. Not readily distinguishable from KNOCK-ON EFFECT.

multivariate analysis The use of statistical methods for examining and evaluating the VARIABLES in any problem, especially where more than two are involved. Techniques of multivariate analysis include FACTOR ANALYSIS, MULTIPLE REGRESSION and principal components analysis.

Myrdal See CUMULATIVE CAUSATION.

N

NAFTA (North America Free Trade Agreement) Membership comprises Canada, the USA and Mexico. It was established in 1988 with the intention of creating a single market in North America that would rival the EU in Europe. Along with the EU, NAFTA is one of three pivotal points of the so-called *global triad* (the third is E and SE Asia).

nappe A gigantic RECUMBENT FOLD, broken by a low-angled THRUST FAULT, where the rocks on the upper limb (above the FAULT) are displaced 'forwards' by many kilometres.

nation A group of people associated with one another by ties of history, sentiment, descent and sometimes language, frequently recognized as a separate political unit. Nations have also been created by the colonial activities of imperial powers (e.g. in Africa). It is common practice to equate a nation with a separate independent STATE (as in the UNO), but a state may in fact comprise two or more national groups (e.g. the Walloons and Flemings in Belgium; the

Czechs and Slovaks in what was Czechoslovakia).

national grid (i) A GRID, based on the Transverse Mercator MAP PROJECTION, used on Ordnance Survey maps. It is a convenient and precise reference system, allowing the exact definition of any place in terms of a code of letters and numbers (a combination of *eastings* and *northings* read from the margins of the map). (ii) The network for supplying electricity to all parts of Britain as originally set up by the Central Electricity Generating Board (now privatized).

national park An area set aside for the CONSERVATION of scenery, vegetation, wildlife and historic objects in such a way and by such means (i.e. PLANNING controls and other regulatory devices) as will leave them unimpaired for future generations, both for scientific purposes and public enjoyment. The status and specific aims of national parks vary from country to country. In the USA the concept began with the designation of Yellowstone in 1872. There are now some 35 parks, mainly WILDERNESS areas, embracing about 60,000 km^2. In England and Wales, national parks were first established following the National Parks and Access to the Countryside Act (1949). Their main aim is to minimize the impact of permitted LAND USES (farming, forestry, mineral extraction, TOURISM) on the intrinsic character of scenic landscapes. There are now 10 designated parks (including the Lake District, Snowdonia and the Pembrokeshire coast), covering an area of over 13,000 km^2, but two more areas (the Norfolk Broads and the New Forest) have recently been given a status closely akin to that of a national park. National parks have also been established in THIRD WORLD countries, particularly for the protection of animal life, as in Kruger (S Africa), Serengeti (Tanzania) and Tsavo (Kenya).

nationalization Transferring the production of goods and the provision of basic services from the PRIVATE SECTOR to the PUBLIC SECTOR. Extending state ownership of the ECONOMY. See COMMUNISM; ct PRIVATIZATION.

NATO (North Atlantic Treaty Organization) Created by a defence treaty originally signed in 1949 by Belgium, Canada, Denmark, France, Iceland, Italy, Luxembourg, The Netherlands, Norway, Portugal, the UK and the USA. Its principal aim was to counteract the military strength of the Soviet bloc and the perceived threat that it posed in the strategic arena of W Europe and the N Atlantic. In the 1990s, with the break-up of the Soviet Union and the decline of COMMUNISM, there was some scaling down of its armaments, but a broadening of its membership to include some former members of the Soviet bloc.

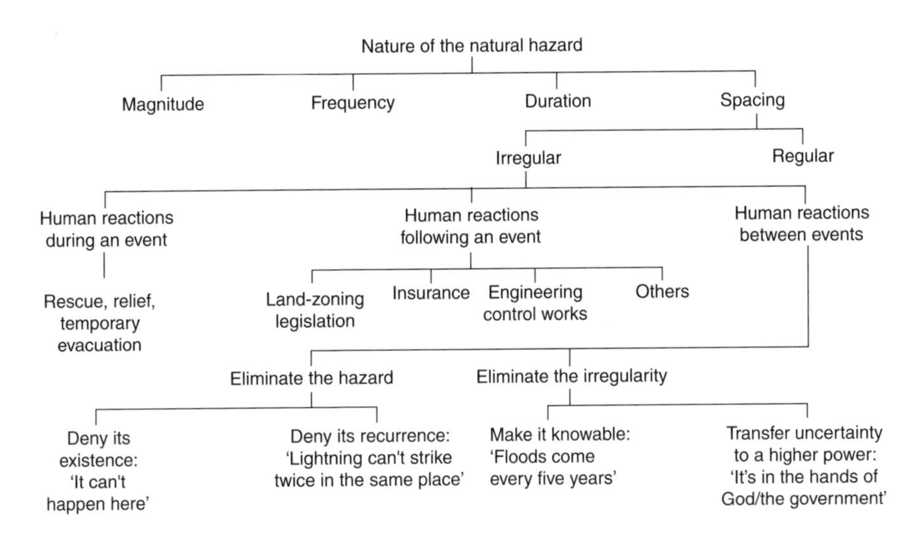

Human responses to natural hazards

natural (environmental) hazard A natural occurrence (event) that threatens, or actually causes damage and destruction to, human SETTLEMENTS, man-made structures (such as roads and dams), economic activities (such as AGRICULTURE and MINING), etc. Among the most serious environmental hazards are river floods, EARTHQUAKES, TSUNAMIS, volcanic ERUPTIONS, LANDSLIDES and GLACIAL OUTBURSTS. The ways in which people react to such hazards is an important focus in modern geographical study. It is also worth remembering that there are hazards (sometimes referred to as *technological hazards*) entirely due to human activities. [*f*]

natural increase The growth in the population of an area resulting from an excess of births over deaths, as distinct from growth attributable to a positive MIGRATION BALANCE. See STABLE POPULATION.

natural resource Any commodity that exists in the natural state and that is useful to people, e.g. minerals, rocks, SOIL, water, plants, animals and air. Whether a commodity becomes an exploitable RESOURCE depends on the ability of people both to discover its whereabouts and to provide the necessary technology to exploit it. Viewed collectively, natural resources provide the link between people and the physical ENVIRONMENT. An important distinction may be made between *renewable* and *non-renewable resources.* The former (e.g. water and air), although inexhaustible, are subject to abuse (e.g. pollution), while the latter (e.g. minerals and soils) are finite, exhaustible and being depleted at increasing rates by rising levels of population and technology. See also RESOURCES. [*f*]

natural vegetation Vegetation that has been unaffected by humans and their activities. It seems likely that *true* natural vegetation, in perfect equilibrium with the existing climatic (CLIMATIC CLIMAX VEGETATION) and edaphic conditions, is rare or non-existent, certainly in areas such as W Europe that have been occupied and utilized by humans for thousands of years. A particularly good example of what appears to be natural vegetation, but is in fact much altered, is SAVANNA.

nature reserve An area preserved so that botanical and zoological communities may survive. This does not necessarily mean an untouched WILDERNESS but may involve careful control to maintain a particular HABITAT or the organized protection of rare or ENDANGERED SPECIES. Regulations are normally strictly imposed to limit public access to nature reserves and so to minimize disturbance. In Britain, English Nature, Scottish Natural Heritage and the Countryside Council in Wales are responsible for the nation's SSSIS and for a network of *National Nature Reserves,* while the Royal Society for the Protection of Birds (RSPB), the largest voluntary conservation organization in Europe, holds around 150 reserves.

neap tide When the Earth, the Sun and the Moon are in quadrature (i.e. at right angles,

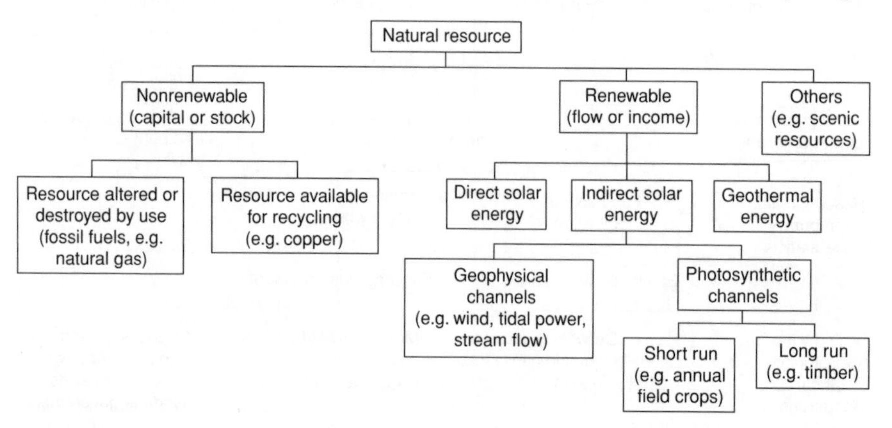

Types of natural resource

with the Earth at the apex) and tide-generating forces are at a minimum. High tides will be relatively 'low', and low tides relatively 'high' – in other words, the tidal range will be much reduced. Neap tides occur every 14.75 days, and coincide with the first and last quarters of the moon (i.e. at times of half-moon). Ct SPRING TIDE. [*f* TIDE]

nearest-centre hypothesis The essence of this hypothesis is that consumers of goods and services seek to minimize the distance travelled in order to secure a particular good or service. In other words, in order to keep travel costs and time to a minimum, goods and services will be obtained from the nearest outlet. This is one of the critical assumptions made in CENTRAL PLACE THEORY. The assumption has, however, been seriously challenged by the results of studies which show that consumers do not always behave in this economically rational way. Rather, the patterns of consumer behaviour reflect the consumers' perceptions and preferences, the degree of personal MOBILITY and affluence, etc. None the less, what does emerge from these studies is that awareness of the desirability of reducing distance travelled is strongest where LOW-ORDER GOODS AND SERVICES are being sought. See LEAST EFFORT.

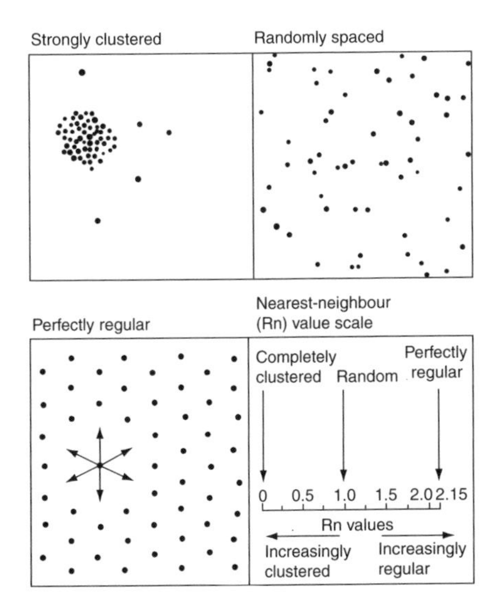

Point pattern examples and the nearest-neighbour value scale

nearest-neighbour analysis A statistical test used for establishing the character of a spatial POINT PATTERN (e.g. of SETTLEMENTS, FIRMS, DRUMLINS, etc.). The *nearest-neighbour statistic* (*Rn*) is found by the formula

$$R_n = 2D\sqrt{\frac{n}{a}}$$

where *D* is the mean distance between neighbouring points, *n* the number of points and *a* the size of the area concerned. The value of *Rn* can range from 0 (totally clustered) to 2.15 (regularly spaced), with values in the order of 1 interpreted as indicating a RANDOM distribution. [*f*]

neck (volcanic neck) A 'plug' of solidified LAVA occupying the pipe of a former active VOLCANO and frequently exposed as a striking landform owing to removal of the surrounding material by denudational processes.

need-satisfaction curve This GRAPH shows the relationship between LEVELS OF LIVING and WELL-BEING, with the former being seen as an input to the latter. Along the *y* axis, four different states of well-being are defined. The curve shows the inputs, or level of living, required to sustain a given output or state of well-being. The form of the curve indicates that, as the level of living increases, the impact on well-being is less than proportional. The equal spacing between *A* and *B*, *B* and *C* along the *y* axis is not repeated along the *x* axis. The essential message conveyed by

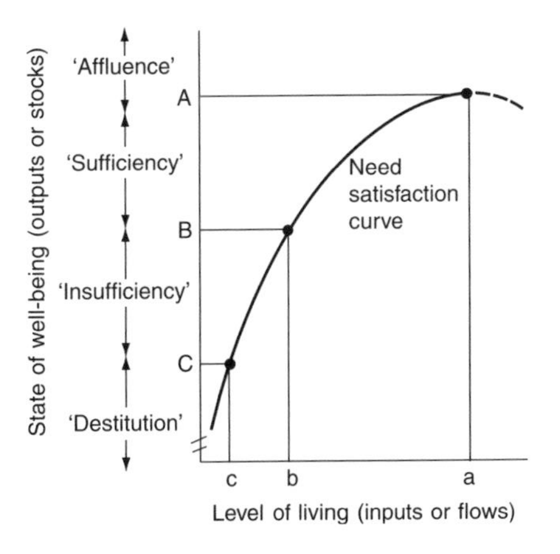

Need-satisfaction curve

the curve may be illustrated by the fact that once hunger is satisfied, further increases in food supply are unlikely to do much to improve well-being. Indeed, there comes a point when excessive feeding has a detrimental impact on health. Hence the eventual downward turn of the curve. [*f*]

needle ice (also known as **pipkrake)** A small ice crystal up to several cm in length, formed when freezing affects the topmost layer of the SOIL. Needle ice is often seen raising individual stones or soil particles above the surrounding soil surface. The development and melting of needle ice in successive freeze-thaw cycles is an important factor in SOLIFLUCTION creep in cold climates.

negative correlation See CORRELATION.

negative externality See EXTERNALITIES.

negative feedback The mechanism by which open systems undergo *self-regulation* (see GENERAL SYSTEMS THEORY). In simple terms, negative feedback can been defined as a 'feedback loop' that counters or works against the processes operating, such that their effectiveness is reduced. For example, on a beach, DESTRUCTIVE WAVES tend to be associated with steep beach gradients. However, their action leads to a reduction in the gradient as material is removed and carried away. In this way, they become less effective. The concept of negative feedback is important in that it demonstrates the ability of natural systems to move towards an equilibrium state. However, intervention by people often disrupts this natural form of self-regulation. Ct POSITIVE FEEDBACK. [*f*]

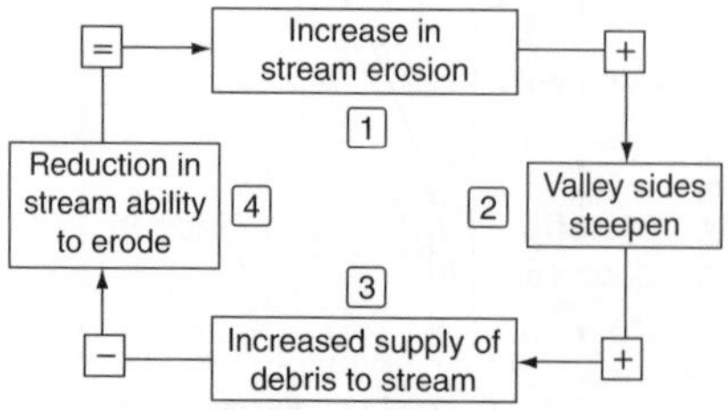

Negative feedback

negative skew A bias within a DISTRIBUTION towards high values, as opposed to *positive skew*, which is a bias towards low values.

neighbourhood effect A key aspect of the SPATIAL DIFFUSION process, whereby the adoption of some innovation is likely to occur most readily around those who have already adopted it. The adopters create a sort of ripple or KNOCK-ON EFFECT. Ct CONTAGIOUS DIFFUSION.

neighbourhood unit An idea that has considerably influenced the design of postwar residential areas and NEW TOWNS, particularly in Britain and N America. The idea had its roots in the GARDEN CITY concept and was experimented with as part of the RADBURN LAYOUT. The intention is to produce a cellular structure of small-scale communities within TOWNS and CITIES ('cells within cells'). The sense of local community and focus is encouraged by the provision of centralized, low-order services and by clearly defining the physical limits of the community (by arterial roads, site features, etc.). Concern for environmental quality is reflected in the setting aside of parks and play areas, the provision of community facilities (meeting hall, library, medical centre, etc.) and designing traffic flows in such a way that there is no penetration of residential areas by through traffic. [*f*]

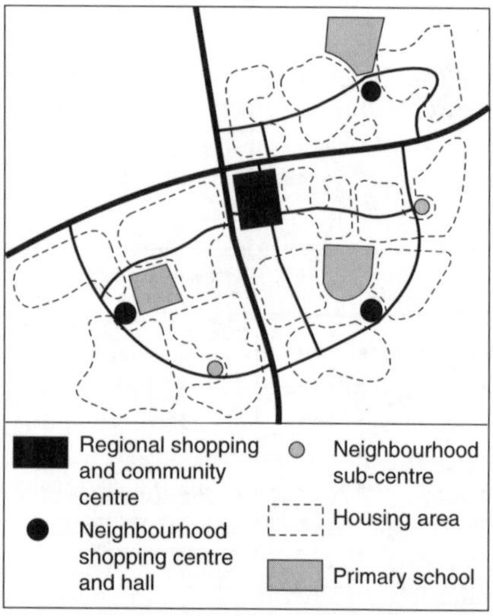

Diagrammatic plan of a neighbourhood unit

neocolonialism Economic and political control exerted mainly by the so-called 'superpowers' (the USA and the former USSR) over apparently 'independent' LEDCS. The control may take a variety of forms, both direct and indirect, from technical to financial AID, from trade to military defence. The ultimate objective is to secure spheres of influence in supposedly uncommitted areas of the world, for example as part of the *Cold War* conducted between East and West during the second half of the 20th century. It is claimed by some that the activities of TNCS represent a form of neocolonialism.

Neoglacial A period of relatively cold climate, associated with several minor glacial advances, which began approximately 5000–4000BP. These advances have also been referred to as the 'little ICE AGE' (particularly in N America), though in Europe the term has been reserved for the advances that affected many Alpine glaciers between 1550 and 1850. Prior to the Neoglacial period there was a relatively warm period known as the *Climatic Optimum*.

neo-imperialism See NEOCOLONIALISM.

Neolithic period A cultural period, following the MESOLITHIC PERIOD, from the latter part of the 4th millennium BC until the onset of the BRONZE AGE. Characterized by the addition of polishing and grinding of stone tools (notably FLINT) to the earlier percussion and pressure-flaking methods. Its distinguishing feature was the beginning of the domestication of animals, the cultivation of crops and the making of pottery, and in Britain the construction of long barrows, megalithic tombs and great religious sanctuaries such as Woodhenge and the first part of Stonehenge.

neo-Malthusians Although Malthus was writing towards the end of the 18th century, there are many today who believe that unsustainable rates of population growth are increasingly the cause of much human misery. One observer has written: 'Each year food production in the underdeveloped countries falls a bit further behind burgeoning population growth, and people go to bed a little bit hungrier.' In short, food supply chains in many LEDCS are unable to keep pace with the subsistence demands of an increasing population. See MALTHUS'S THEORY OF POPULATION GROWTH.

neotechnic era The second phase of the Industrial Revolution (ct PALAEOTECHNIC ERA), which in Britain started *c.* 1870 and involved: technological advances, such as the increasing use of ELECTRICITY and the development of the motor vehicle; the proliferation of light, MARKET-ORIENTED INDUSTRIES; the expansion of markets; the rapid outward spread of the BUILT-UP AREA. See also KONDRATIEFF CYCLES.

nested sampling A SAMPLING method that involves dividing a study area into a HIERARCHY of sampling units that nest within one another. RANDOM processes select the large, 1st-order units, the middle or 2nd-order units within these, and so on. The final location of points within each small, or 3rd-order, unit can be randomly determined. Nested or *hierarchic sampling* offers the advantage of reduced field costs (only a small part of the study area needs to be covered) and of being able to investigate scale variations (as between different orders of spatial unit), but it suffers from high sampling error. Ct STRATIFIED SAMPLING, SYSTEMATIC SAMPLING. [*f*]

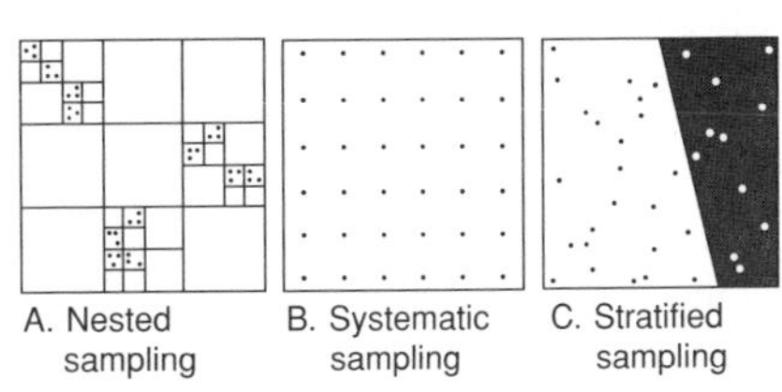

Three sampling strategies

nesting An arrangement of MARKET AREAS or HINTERLANDS in which those of a low order fit, in cellular manner, within higher-order units. For example, in CENTRAL PLACE THEORY, the hinterland of, say, a CITY will comprise the nesting hinterlands of its DEPENDENT PLACES.

net benefit See COST-BENEFIT ANALYSIS.

net migration See MIGRATION BALANCE.

net reproduction rate In population studies, the net reproduction rate affords a more exact indication of whether a generation can reproduce itself than is given by the GROSS REPRODUCTION RATE. Allowance is made for the deaths of females before attaining child-bearing age, later deaths, non-marriage and infertility, and applied as a correction to the gross reproduction rate. A net reproduction rate of 1 indicates an ability to maintain a population, and a value in excess of 1 indicates the likelihood of population increase.

network A SYSTEM of interconnected places, people or objects (e.g. rivers, roads, telephones, radio stations, etc.). The character of any network is dependent not only on the pattern generated by the spatial arrangement of nodes, but also on its *density* and CONNECTIVITY. The density of a network is simply the total length of the network expressed in terms of length per unit area; in the case of a transport network it gives an indication of the degree of ACCESSIBILITY afforded in the area served by the network The connectivity gives some measure of the directness of movement that is possible within the network. See also NETWORK ANALYSIS.

network analysis Establishing the fundamental character of a NETWORK using a range of possible measures to assess such properties as its CONNECTIVITY (e.g. *alpha index, beta index, cyclomatic number*) and its ACCESSIBILITY (e.g. DETOUR INDEX, *König number, Shimbel index*). Such measures are useful in comparing networks over time and space, while TOPOLOGICAL MAPS are useful in depicting networks.

nevé See FIRN.

New Commonwealth countries A distinction made in the British census under the heading of 'country of birth'. The title embraces all Commonwealth countries except the *Old Commonwealth* countries of Australia, Canada and New Zealand – i.e. principally Commonwealth countries in Africa, the Caribbean, the Pacific, and S and SE Asia.

new town A comprehensively planned URBAN community, built from scratch in a relatively short time, and created for some specific purpose – e.g. relocating a national seat of government (Brasilia), providing a residential dormitory for a major city (Tama on the outskirts of Tokyo), exploiting the resources of a peripheral region (Bratsk in Siberia).

The modern new town movement began in Britain in the early 1900s with the building of GARDEN CITIES. During the postwar period, and following the New Towns Act (1946), some 28 new towns were officially designated and built in Britain, for three principal reasons: (i) to take OVERSPILL from London and other CONURBATIONS, e.g. the ring of eight self-contained new towns located around London (Crawley, Harlow, Stevenage, etc.); (ii) to act as a GROWTH POLE or catalyst in the planned revival of DEPRESSED REGIONS, e.g. Aycliffe, Peterlee and Washington in NE England; (iii) to accompany a large-scale expansion in basic INDUSTRY, e.g. Corby and the iron and steel industry in Northamptonshire.

Two important trends in the development of British new towns emerged during the 1960s, namely a move to much larger schemes (population targets set in the 150,000–250,000 range rather than around 80,000, as was the case with the *first-generation* new towns designated during the late 1940s) and the expansion of sizeable existing towns as 'new town' schemes (e.g. as in the cases of Northampton and Peterborough). Milton Keynes, with a population of nearly 200,000 is probably one of the most successful of Britain's later new towns. However, since the 1970s no new designations have been made.

NGO (non-government organization) An organization with significant influence and responsibilities created by volunteers rather than government legislation. Some NGOs are leading charities operating at an international level and helping, for example, in major disaster situations (e.g. Oxfam, the Red Cross) or protecting the environment (e.g. Greenpeace, WWF). Others, such as the National Trust and the Royal Society for the Protection of Birds (RSPB) operate at a national level.

NIC (newly industrializing country) A term used when referring to a select band of developing countries (see THIRD WORLD), which over the last three decades have sustained high rates of economic growth. They have out-performed the MEDCS, principally as a result of their competitive edge in manufacturing. Particularly noteworthy is the cluster of NICs in E and SE Asia – Hong Kong, Malaysia, Singapore, S Korea and Taiwan – sometimes referred to as the *Asian Tigers*. Cf RIC.

nimbus A type of CLOUD from which rain is falling. The term is usually used in conjunction with other cloud types, as in CUMULONIMBUS and *nimbostratus* – a uniform, low, dark-grey cloud, giving continuous rain or snow, usually in association with a WARM FRONT.
[*f* CLOUD]

NIMBY Short for 'not in my back yard'. A term used where individuals object to a particular development on the grounds that it is too near to where they live and will disturb them. The classic example is the bypass proposal where there is general agreement about the need to build one, but a precise routing cannot be agreed because everyone objects to it running past their property.

nitrates Chemicals derived from the use of nitrogen as a FERTILIZER in farming, which since the advent of INTENSIVE FARMING have become increasingly concentrated in rivers and AQUIFERS. Nitrates are thought to be responsible for the process of EUTROPHICATION of rivers and lakes. Some people believe that there may be harmful effects on human health, with links being suggested between nitrate concentration in water and the incidence of stomach cancer and the 'blue-baby syndrome'. It is now generally agreed that levels of nitrates should be reduced by cutting down on the use of fertilizers and reducing the intensity of farming. However, as percolation of water containing chemicals, such as nitrates, through the ground to GROUNDWATER takes up to 40 years, it will be some time before the effects of any change will register in underground water supplies.

nitrogen cycle A circulation of nitrogen that starts with NITRATES in the soil being taken up by the roots of plants. These plants are then consumed by animals and the nitrogen released as ammonia in their excretions. The excreta decays on the soil, and bacteria convert the ammonia back into nitrates, which are then leached back into the soil. The cycle then starts again. See NUTRIENT CYCLE.

nivation Sometimes referred to as 'snow-patch EROSION', nivation is a complex group of processes involving freeze-thaw action, SOLIFLUCTION, TRANSPORT by running water and (possibly) CHEMICAL WEATHERING beneath and at the edges of a snowpatch, formed in a hillside hollow. With seasonal enlargement and shrinkage of the snowpatch, surrounding areas will be exposed to frost action and the resultant debris will be removed (mainly in the thaw season) by solifluction and rivulets of meltwater escaping from the toe of the snow. Over a long period, the snowpatch will 'eat back' into the hillside, forming a rounded *nivation hollow* with a comparatively steep head. Nivation hollows are regarded as early stages in the formation of NIVATION CIRQUES and true glacial CIRQUES.

nivation cirque A transitional form between a NIVATION hollow and a glacial CIRQUE. Nivation cirques possess steep, frost-shattered headwalls, sometimes more than 100 m in height, and in plan are semi-circular. Most show a preferred orientation, facing northeastwards in the Northern Hemisphere. In contrast to fully developed

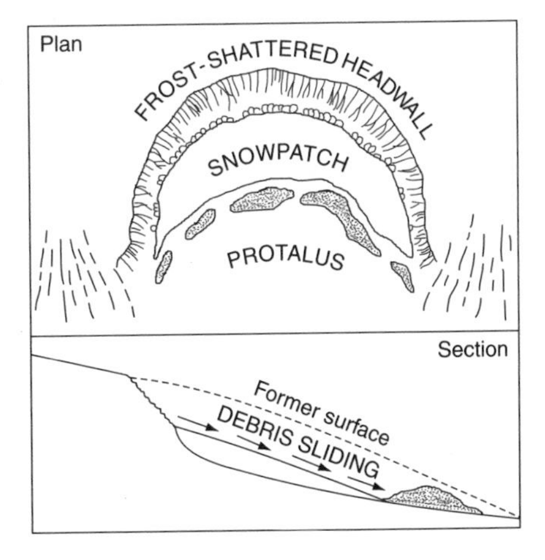

Nivation cirque with protalus formation

glacial cirques, nivation hollows do not possess rock basins and lips. They are, however, characterized by the presence of MORAINE-like mounds and ridges, composed of angular debris (the product of FREEZE-THAW WEATHERING of the headwall), which has slid across the surface of the snowpatch to accumulate along a line roughly parallel to the headwall (see PROTALUS). [*f*]

niveo-fluvial A term used to describe a group of PERIGLACIAL processes that operated in S England at the close of the final glacial period. A relatively mild climate (by comparison with full glacial conditions), with heavy winter snowfall on high ground but a rapid early summer thaw, resulted in many fluctuations of temperature about 0°C, intense frost weathering, rapid solifluction and periods of powerful surface run-off. This led to heavy and selective EROSION at some points.

nodality The degree or extent to which lines, roads or any set of things having a linear character approach each other or converge at a point. Cf CENTRALITY.

nomadism A way of life followed both by people (e.g. the Bedouin Arabs) and animals (e.g. the migrating herds of caribou in Canada), and involving more or less perpetual movement in search of food and in response to various pressures (climatic, ecological and political).

nominal data One of three major categories of data. It includes data referring to classifications that do not imply quality or relative order of magnitude. The distinction between forest, grassland and heathland ECOSYSTEMS or between SETTLEMENTS of different shapes would constitute nominal data. The most common statistical measure of nominal data is frequency – i.e. the number of times a particular event occurs. Ct INTERVAL DATA, ORDINAL DATA.

non-parametric tests These are statistical methods designed to take the place of standard PARAMETRIC TESTS when the assumptions required by these tests cannot be satisfied. They are called non-parametric tests because they do not involve any assumptions being made about PARAMETERS of the POPULATION that is under investigation. Non-parametric tests include the CHI-SQUARED TEST and the MANN–WHITNEY U TEST.

non-renewable resources See NATURAL RESOURCES.

normal distribution A FREQUENCY DISTRIBUTION that is symmetrically bell-shaped and in which most values lie near the MEAN of the values at the apex of the curve. It is a continuous frequency distribution, unlike a BINOMIAL DISTRIBUTION or POISSON DISTRIBUTION. One of the assumptions made in most SIGNIFICANCE TESTS is that the values under investigation are normally distributed (see PARAMETRIC TESTS). Deviation from the symmetry of the normal distribution is referred to as 'skewness' (see SKEWED DISTRIBUTION).

[*f* FREQUENCY DISTRIBUTION]

normal fault See FAULT.

North Countries – sometimes also referred to collectively as ADVANCED COUNTRIES, the DEVELOPED WORLD, the FIRST WORLD or MEDCS – enjoying considerable wealth and high standards of living derived from well-developed SECONDARY and TERTIARY SECTORS and increasingly from QUATERNARY-SECTOR activity. Characteristics also include high levels of energy consumption, high rates of literacy and considerable political influence at the global scale. The North would be defined as embracing most European countries, the USA, Canada, Japan, Australia and New Zealand. Ct DEVELOPING WORLD, SOUTH; see also BRANDT COMMISSION, CLUB OF ROME.

[*f* BRANDT COMMISSION]

North American Free Trade Area See NAFTA.

North Atlantic Treaty Organization See NATO.

nuclear power (i) The use of radioactive energy created in the process of nuclear fission to generate electric power or to provide propulsion for ships. In both instances, the heat released by such energy is harnessed to produce steam which, in turn, is used to drive turbines. The use of nuclear power is a highly controversial issue, there being public concern about possible radiation hazards and about the safe disposal of radioactive waste. (ii) Any country that has the capacity to manufacture nuclear weapons (e.g. the UK, USA, the Russian Federation, China, France and India).

nucleated settlement A SETTLEMENT in which there is a close juxtaposition of dwellings and other buildings, and where there is a well-defined break between the settlement itself and the surrounding countryside. Historically, this *clustered* form may have been a response to a variety of circumstances, ranging from association with the open field system (e.g. the English village prior to the enclosure acts) to the exploitation of highly localized resources (e.g. the mining town); from the necessity for defence (e.g. the walled towns of the Middle Ages) to an involvement in trade, encouraging settlement to become clustered around nodal points in the developing transport network (e.g. the market town). The layout of a nucleated settlement can show a variety of different forms, often determined by the pattern of routeways around, or along, which it grows. For example: *linear*, as along one major routeway; *stellate*, as around a multiple intersection; or *T-shaped*, as at one intersection. Ct DISPERSED SETTLEMENT; see also SETTLEMENT FORM.

nuée ardente A catastrophic blast of hot gas, steam and burning dust, released by a violent ERUPTION and descending the flanks of a VOLCANO as a high-velocity 'incandescent cloud'. During the eruption of Mont Pelée in Martinique, Caribbean, in 1992, the town of St Pierre was overwhelmed without warning by a nuée ardente. Of its 25,000 inhabitants only one (occupying the local jail) survived.

null hypothesis This is used to determine the SIGNIFICANCE of the differences observed between two or more samples of a POPULATION. The null hypothesis assumes that there is no significant difference between the samples and that the observed differences are no more than chance variations. In other words, the null hypothesis is a negative assumption that an effect is not present in the samples. Appropriate SIGNIFICANCE TESTS are applied to all the samples, and as a result the hypothesis may or may not be rejected. If the null hypothesis is rejected, then it can no longer be assumed that there is no significant difference between the samples. If, on the other hand, the null hypothesis is not rejected, then the case is 'not proven'. In short, a null hypothesis cannot be proved to be correct, it can only be shown to be false.

nunatak A rocky peak projecting above the surface of an ICE SHEET as in Greenland and along the margins of the Antarctic ice sheet. Nunataks are subjected to FREEZE-THAW WEATHERING, but remain largely free of snow and ice owing to their precipitous slopes, and the presence of dark rocks that absorb solar radiation and, by re-radiating heat, promote rapid snow ABLATION.

nuptiality See MARRIAGE RATE.

nutrient cycle The process whereby mineral elements necessary for plant growth (such as carbon, hydrogen, oxygen, nitrogen, phosphorus and potassium) are taken up

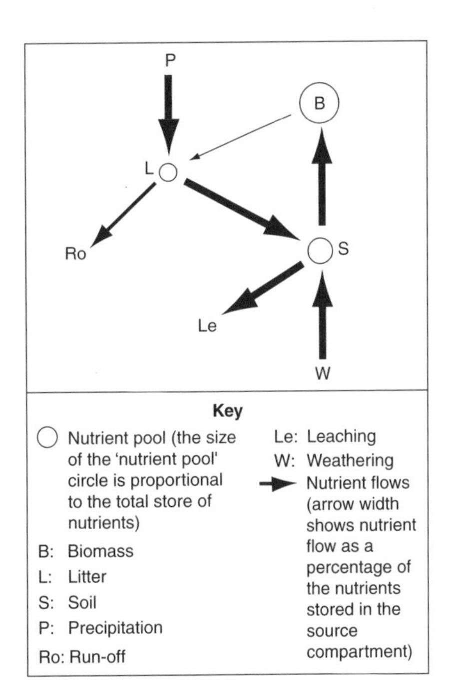

Tropical rainforest nutrient cycle

from the SOIL, and then returned when the plants shed vegetal matter (such as leaves) or die. The decomposition of the resultant plant litter, by the action of micro-organisms and fungi, releases the elements to the soil, forming a source of nutrients for future plant growth. The nutrient cycle is sometimes virtually closed (as in the RAINFORESTS of tropical regions), in which case interruption of the cycle by DEFORESTATION and agriculture (which will lead to removal of part of the vital stock of nutrients) will have serious adverse effects on soil fertility. However, more usually, as the nutrient cycle operates, elements are continually added to the system (e.g. by rock WEATHERING) or lost from it (by escape of nitrogen to the ATMOSPHERE and the process of LEACHING) so that, over a period of time, the balance in the supply and consumption of nutrients is maintained. See DECOMPOSER, NITROGEN CYCLE. [*f*]

O

oasis A place in a desert where water is available by way of SPRINGS or wells. Oases occur when the WATER TABLE approaches the surface (for example, on the floor of a hollow resulting from DEFLATION).

OAU (Organization of African Unity) This alliance was set up in 1963 with the objectives of: furthering African unity through the coordination of a range of policies relating to politics, economics, culture, health, science, defence, etc.; eliminating COLONIALISM from the continent. Nearly all African nations belong to the organization, with the notable exception of the Republic of S Africa.

obsequent stream A stream whose course runs opposite to the 'original' slope of the land surface (ct CONSEQUENT STREAM). Obsequent streams are particularly associated with TRELLISED DRAINAGE, in areas of scarp-and-vale scenery. The initial pattern of consequent streams is disrupted by the extension of subsequents along outcrops of weak rock, leading to the process of CAPTURE. Obsequents then work back headwards, from the subsequents, frequently along the lines of captured consequents. They may take advantage of COLS, eroding into the scarp face as it undergoes recession.

occluded front A type of FRONT developed in a mid-latitude FRONTAL DEPRESSION, resulting from a more rapidly advancing COLD FRONT overtaking a more slowly moving WARM FRONT. The effect is to raise the tropical maritime air of the WARM SECTOR well above the Earth's surface. The formation of an occluded front thus marks the beginning of the decay phase of a depression. There are two types of occlusion: (i) *warm occlusion* – where the advancing cold air is warmer than the cold air ahead of the warm front – in this case, it will ride above the leading cold air; (ii) *cold occlusion* – where the advancing air is colder than the cold air ahead of the warm front – in this case it will undercut the leading cold air. [*f*]

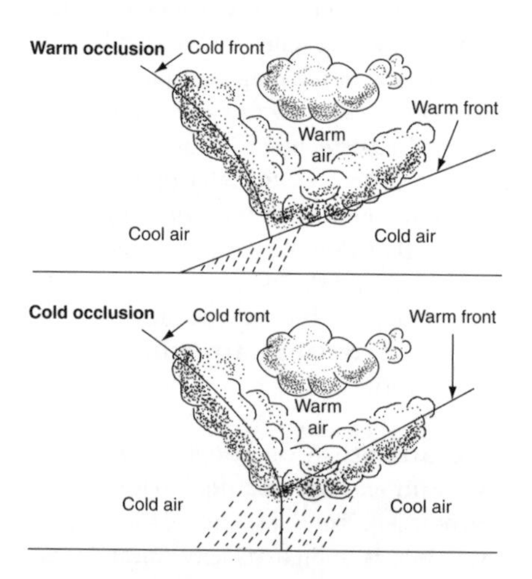

Occluded fronts

ocean currents Relatively fast flows of water in the oceans responsible for distributing heat between the latitudes, thereby assisting to maintain the atmospheric HEAT BALANCE. Warm ocean currents flow from the Equator towards the poles (for example, the Gulf Stream/North Atlantic Drift extends from the Caribbean to the W of the UK and on to W Norway) and cold ocean currents (for example, the Peruvian current to the west of S America) from the poles towards the Equator. In transferring temperature

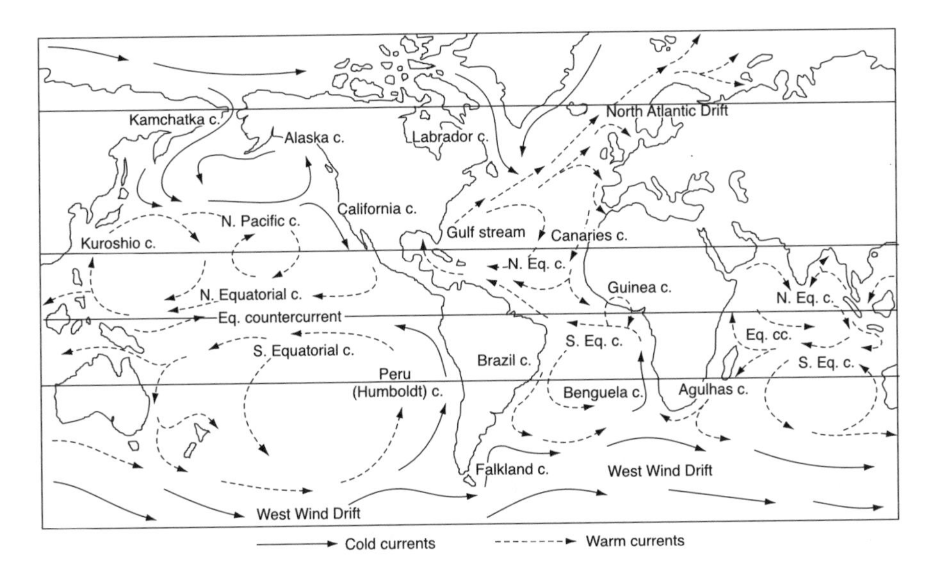

Ocean currents

characteristics, ocean currents have a considerable effect on the climates of coastal regions – for example, the North Atlantic Drift keeps western Scandinavia warmer than other regions at the same latitude. The propagation of ocean currents is still not fully understood but the effect of prevailing winds and the movement of water of different densities are believed to be important factors. Periodically, changes may occur in the patterns of ocean currents; for example, the EL NIÑO effect results in the replacement of the cold Peruvian current by much warmer waters. [*f*]

ocean-floor spreading See SEA-FLOOR SPREADING.

ocean-floor trench See ISLAND ARC, PLATE TECTONICS.

oceanography The scientific study of the seas and oceans. *Physical oceanography* is concerned with the form of ocean basins, the structure, RELIEF and deposits of their floors, movements of seawater in the form of ocean currents, ocean temperatures and salinity. *Biological oceanography* is the study of life in the oceans.

occupancy rate A measure of housing conditions and overcrowding derived by dividing the number of people in a dwelling by the number of 'habitable' rooms (i.e. normally taken to be bedrooms and living rooms).

OD (ordnance datum) The mean sea-level, calculated from hourly tidal observations at Newlyn, Cornwall (hence occasional use of the term *Newlyn Datum*). Heights on Ordnance Survey maps are expressed in feet (formerly) or metres (on current maps) 'above OD'.

OECD (Organization for Economic Cooperation and Development) Set up in 1961 to promote economic and social WELFARE throughout the area of member countries by the international coordination of appropriate policies. Its membership comprises nearly all the countries of W Europe, together with Canada, the USA, Japan, Australia and New Zealand.

office activity A rather loosely defined part of the TERTIARY SECTOR broadly concerned with the collection, processing and exchange of information. Within it, it is possible to distinguish a number of different categories, such as: company administration; finance (banking, insurance, accountancy, stock-broking, etc.); PROFESSIONAL SERVICES (legal, real estate, advertising, market research, etc.);

PUBLIC-SECTOR activity (local and central government, administration of public utilities, QUANGOS, etc.). An office, as such, may range from the prestigious headquarters of a TNC to the one-roomed premises of a private detective.

offshore assembly A form of manufacturing largely encountered in the THIRD WORLD involving the assembly of a finished product made from parts and components produced in and imported from MEDCs. It is increasingly undertaken by TNCs. By locating the labour-intensive work of assembly in countries offering cheap labour, these businesses are able to maintain a competitive pricing of their finished products, as well as improve their penetration of the domestic markets. Today, many Japanese companies are involved in offshore assembly in SE Asia, Africa and Latin America.

offshore bar A bank of SAND and/or SHINGLE, developed some distance offshore on a very gently shelving coastline. One explanation of its formation is that larger waves will tend to break before reaching the BEACH, thereby eroding the sea bed and throwing up the resultant debris to form the bar. The bar may be lengthened by LONGSHORE DRIFT, and stabilized by the formation of SAND DUNES. Inland from the bar, a shallow lagoon will be formed, which may eventually be infilled by mud and SILT, and a SALT MARSH may develop. See also BREAK-POINT BAR.

offshore economy The outcome of TNCS and other businesses in one country investing overseas, be this by setting up BRANCH PLANTS, entering into joint ventures or acquiring equity (stock and shares) in foreign businesses. During the 1980s and 1990s, much of the Japanese economy was moved offshore by these means.

O-horizon The uppermost layer of a SOIL PROFILE comprising organic matter (hence the 'O') in the form of litter (*L-horizon*), fermenting matter (*F-horizon*) and humus (*H-horizon*). The O-horizon is generally dark brown or black in colour and is slowly incorporated into the A-HORIZON by earthworms, providing much-needed nutrients and the ability to retain water. [*f* SOIL PROFILE]

oil crises See ENERGY CRISES.

old-age index A measure of the 'greying', 'silvering', 'wrinkling' or ageing of a population, calculated by dividing the number of persons over retirement age by the number of people of working age.

old fold mountains Fold mountains created by earth movements prior to the Alpine OROGENY of middle Tertiary times (see YOUNG FOLD MOUNTAINS). The old fold mountains of the British Isles, W Europe and elsewhere were created mainly during the Hercynian (late Carboniferous–early Permian times) and Caledonian (late Silurian–early Devonian times) orogenies. The structural features (such as folds, NAPPES, FAULTS, etc.) of old fold mountains are similar to those of young fold mountains. However, owing to their great age, the former have been subjected to immense amounts of EROSION, in the course of which they have been reduced to PENEPLAINS. They have also experience episodes of uplift and rejuvenation, and are now often of PLATEAU-like form, with broad INTERFLUVES separated by deeply incised river valleys. In some instances, old fold mountains were inundated by marine transgressions and became covered by younger SEDIMENTARY ROCKS. Following uplift, these have been stripped away (except around the margins), revealing *exhumed surfaces of erosion*. The drainage systems developed on the younger rocks have been superimposed on to the older structures to give strikingly DISCORDANT patterns. However, the landforms of old and young fold mountains are not wholly contrasting. In many cases (for example, the European Alps and the Scandinavian Highlands), similar processes have operated; thus PLEISTOCENE glaciation has produced CIRQUES, ARÊTES and glacial troughs in both regions.

oligopoly An economic activity in which a small number of FIRMS account for a large proportion of OUTPUT and employment; e.g. the motor vehicle industry. Cf MONOPOLY.

oligotrophic lake A lake that is characterized by a relative lack of dissolved minerals and a generally low nutrient status. Many deep lakes in upland granite regions, such as parts of Scotland and Norway, are oligotrophic. The water, lacking in microscopic algae, is clear, and sunlight penetrates to considerable depths, allowing the growth of aquatic plants. However, because of the lack of nutrients the flora and insect life are impoverished (with certain exceptions, such as mayflies), and salmon and trout are the only large fish species present. See EUTROPHICATION.

onion-skin weathering The detachment of curved sheets of rock from a large BOULDER or exposed face as a result of 'unloading' (see DILATATION), surface expansion and contraction related to temperature changes, or pressure from within the rock exerted by hydration of minerals or salt crystal growth. See EXFOLIATION.

OPEC (Organization of Petroleum Exporting Countries) Formed in 1960 with the aim of unifying the oil policies of member countries and of determining the best means of safeguarding their interests. Its membership comprises Algeria, Ecuador, Gabon, Indonesia, Iran, Iraq, Kuwait, Libya, Nigeria, Qatar, Saudi Arabia, the United Arab Emirates and Venezuela.

open system See CASCADING SYSTEM, GENERAL SYSTEMS THEORY.

opencast mining A form of extensive excavation, only practicable where the mineral deposits lie at or near the surface. The overlying material (*overburden*) is removed by large-scale machinery and the seams or deposits thus revealed are then quarried. The equivalent American terms are *strip-mining* or *open-cut*. It is a technique used particularly for working coal, brown coal and iron ore.

optimal city size The ideal CITY size that may: maximize benefits; minimize costs; generate the greatest net benefit. The notion is based on the claim that there are certain trends in both costs and benefits accompanying increasing city size, the cost curve being depicted as U-shaped and the benefit curve as S-shaped. Thus on a GRAPH, certain key points may be identified along the city-size axis. *P1* is the threshold at which benefits exceed costs (i.e. the onset of net benefit); *P2* is the city size that incurs least costs; *P3* is the city size that attains the greatest net benefit (i.e. largest margin of difference between costs and benefits); *P4* is the city of greatest benefit; while *P5* is the point at which costs once again begin to outweigh benefits. So *P1* and *P5* might be specified as the desired lower and upper limits to city size, while *P2*, *P3* and *P4* might be claimed as indicating optimal city sizes. [*f*]

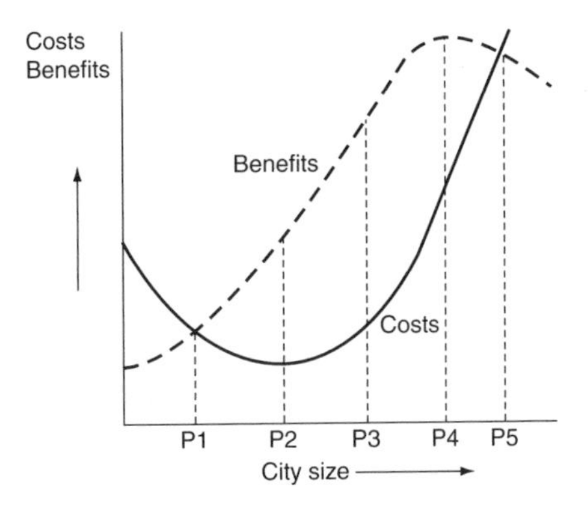

Optimal city size

optimal location A term much used in ECONOMIC GEOGRAPHY (particularly during its normative phase) variously interpreted to mean the best location from an economic viewpoint, be it the LEAST-COST LOCATION, the location yielding maximum benefit or the greatest profit, or the location that best satisfies the particular needs of a given activity.

optimizer concept A concept which holds that the ENTREPRENEUR or decision-maker always strives for maximum productivity and profit in the conduct of business. It dominated ECONOMIC GEOGRAPHY during the mid-20th century and is an integral part of neoclassical economics. Ct SATISFICER CONCEPT.

optimum population A concept that is open to differing interpretations. It might be defined as the number of people who can live most effectively in any area in relation to the possibilities and RESOURCES of the ENVIRONMENT and enjoy a reasonable standard of living. That number of people will be influenced by many factors, including the nature and productivity of the MODE OF PRODUCTION, the level of technology, etc. An alternative and more exacting definition would be that size of population in a given area which allows the maximum utilization of resources, and which achieves the greatest output per head and the highest living standards. Both definitions given here are highly subjective. See SUSTAINABLE DEVELOPMENT. [*f*]

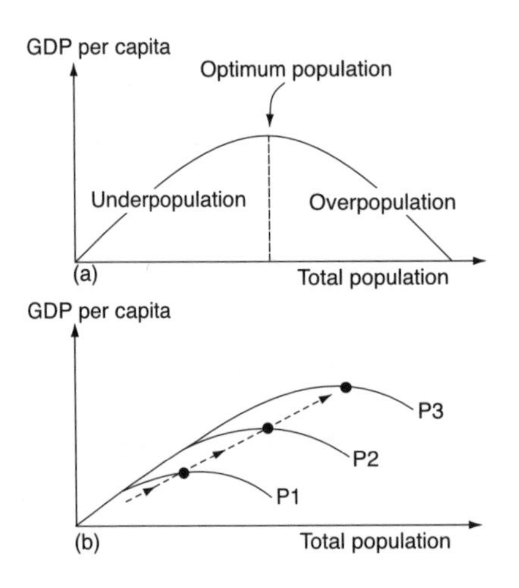

(a) Optimum population and (b) the effect of improvements in technology on optimum population

order In CENTRAL PLACE THEORY this relates to the hierarchic status of individual goods, services and central places. In the case of goods and services, order is largely determined by their THRESHOLD and RANGE VALUES, while with central places it is allied to position in the CENTRAL-PLACE HIERARCHY. See also HIGH-ORDER CENTRAL PLACE. LOW-ORDER GOODS AND SERVICES, STREAM ORDER.

ordinal data Data that are presented in terms of their relative importance (or order of magnitude or rank) rather than their absolute value. For example, the classification of BED LOAD material into fine, medium and coarse categories, or the grouping of SETTLEMENTS into size classes would constitute ordinal data. This is one of three major categories of data. Ct INTERVAL DATA, NOMINAL DATA.

ordnance datum See OD.

organic farming A type of FARMING that does not use chemical FERTILIZERS, pesticides, herbicides and fungicides. Instead, use is made of animal and green manures, and fertilizers made from fish and bone meal. With growing concern about chemicals entering the human food chain and adversely affecting health, organic farming is currently experiencing something of a boom. It is also thought to be much less harmful to wildlife and the environment. Because yields are generally lower than with modern intensive farming, products are more expensive. But it seems that more and more people are willing to pay that little bit more for what is seen as 'safe' food.

organic weathering See BIOLOGICAL WEATHERING.

Organization for Economic Cooperation and Development See OECD.

Organization of African Unity See OAU.

orogeny A major episode of mountain building, involving powerful compressive folding and faulting of previously formed sediments, followed by uplift and EROSION. The main orogenies to affect Europe since the Pre-Cambrian have been the *Caledonian*, *Hercynian* (or *Variscan*) and *Alpine* (see YOUNG FOLD MOUNTAINS, OLD FOLD MOUNTAINS).

outlier An outcrop of younger rock completely surrounded by older rock. Outliers are commonly formed as a result of differential erosion of ESCARPMENT faces. As systems of stream valleys are eroded back into the SCARP, parts of the CAP-ROCK of the latter may become isolated (as in the case of Bredon Hill, near Tewkesbury, which is capped by an outlier of inferior oolite LIMESTONE and has become detached from the main escarpment of the Cotswolds). [*f*]

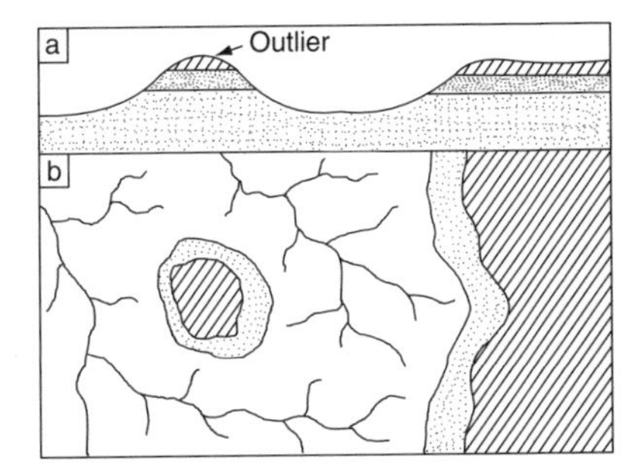

Outlier in (a) geographical cross section and (b) plan

out-of-town shopping centre A large retailing complex located at the edge of the BUILT-UP AREA or just beyond it. Factors encouraging such a development include the relatively low costs of land (as compared within the CBD), room for expansion and ACCESSIBILITY from a much wider area than just the nearby town or city. Some of the largest centres in the UK (e.g. Gateshead's Metro Centre and Kent's Bluewater) contain not just shops but a whole range of leisure amenities in order to encourage a day out for the family. See HYPERMARKET, RETAIL PARK.

outport A subsidiary PORT developed, often at the mouth of an ESTUARY, to service (or even replace) an established port that is proving in some way inadequate. Outports have been developed: as shipping has grown in size and the original port has become increasingly difficult to access for modern shipping (e.g. Avonmouth in Bristol); as estuaries have become progressively silted up (e.g. Zeebrugge superseding Bruges); as passenger traffic has required speedy transport by rail or road to its destination, instead of proceeding slowly by ship further up the estuary (e.g. Cuxhaven for Hamburg); where large areas of land have been required for the development of new port facilities (such as bulk handling) and of large-scale PORT-RELATED INDUSTRIES (e.g. Europoort for Rotterdam).

output (i) The quantity of goods produced or services provided during a given period by a FIRM, an INDUSTRY, a REGION or country. (ii) See GENERAL SYSTEMS THEORY.

outwash plain A gently sloping area (sometimes fan-shaped) comprising SANDS and GRAVELS deposited by meltwater streams flowing from the stationary margins of a glacier or ICE SHEET. Large outwash plains, related to contemporary glacial conditions, are found in S Iceland and are called SANDUR. These are being built up by streams that migrate slowly across the plain from one side to another; indeed the outwash plain as a whole will be inundated only by a major GLACIAL OUTBURST.

overbank flow See BANKFULL.

overburden See OPENCAST MINING.

overcapacity An excess of productive capacity that is likely to arise in an INDUSTRY subject to constant fluctuations in the demand for its products, when the level of demand falls below the level of OUTPUT, i.e. surplus production.

overflow channel (also **spillway)** A CHANNEL cut by water draining from a GLACIAL LAKE. Some features once interpreted as overflow channels are now believed to have resulted from SUBGLACIAL stream EROSION.

overgrazing In farming, putting too many animals on land to graze, so that the vegetation cover is gradually destroyed. In an area of uncertain or short seasonal rains, this removal of the protective mat of vegetation can lead to serious SOIL EROSION.

overhead costs See FIXED COSTS.

overland flow The proportion of total rainfall that is not intercepted by vegetation and does not infiltrate the SOIL, but runs over the ground surface as sheet or rill wash. The volume of overland flow increases with RAINFALL INTENSITY, and is usually at a maximum during convectional storms in semi-arid and tropical regions. Overland flow occurs either when rainfall exceeds the INFILTRATION capacity of the soil (*infiltration-excess overland flow*) or, more commonly, when the pore spaces of the soil are saturated and unable to absorb additional water (*saturation-excess overland flow*). Overland flow represents very rapid transfer of water, and is associated with the rising limb and peak of a storm hydrograph.

It is most likely to occur in areas of steep slopes, thin soils, impermeable bedrock, lack of vegetation, frozen soil, intense rainstorms, etc.

overpopulation A condition in which the population of an area exceeds the OPTIMUM POPULATION, and thus cannot adequately be supported by the available RESOURCES, technology and MODE OF PRODUCTION. As a consequence, the standard of living is low and tends to decline still further. Overpopulation is not necessarily equated with dense population; for example, high-density cities are not necessarily overpopulated, while, equally, some mountain and DESERT areas may be overpopulated with a very low POPULATION DENSITY. Ct UNDERPOPULATION. [*f* OPTIMUM POPULATION]

overspill Surplus population and employment from a densely crowded URBAN area that can no longer be effectively accommodated and maintained there, and which removes either voluntarily or in response to statutory plans. Such decentralizing people and FIRMS either relocate in new SUBURBS and in nearby, less congested TOWNS, or remove to planned overspill reception schemes (e.g. NEW TOWNS, EXPANDED TOWNS, DEVELOPMENT AREAS) located further afield.

overthrust See NAPPE.

overurbanization A state of affairs thought to exist mainly in some THIRD WORLD countries, where it is claimed that the degree of URBANIZATION is too high relative to the level of economic DEVELOPMENT. The particular problem resulting from such a situation is that substantial rural-to-urban MIGRATION creates a large URBAN labour force that cannot be fully employed by the SECONDARY and TERTIARY SECTORS at their current levels of development. As a consequence, there are unacceptable levels of unemployment, from which spring such things as poverty, SLUM housing (see SHANTY TOWN), crime and social unrest.

oxbow lake (also **oxbow)** A crescent-shaped lake or marshy area, occupying a former river MEANDER that has become cut off by the breaching of the meander neck. Oxbows are characteristic of most large FLOOD PLAINS (as in the Severn and Trent valleys in Britain, and the Mississippi valley in the USA).

oxidation A common process of CHEMICAL WEATHERING, whereby rock minerals (particularly those containing iron) combine with oxygen, usually dissolved in infiltrating SOIL water. The reverse process to oxidation is *reduction.*

ozone layer Ozone is a form of oxygen (O_3), very faintly blue in colour, found in greatest quantities at a height of 20–25 km in the Earth's atmosphere. Ozone is believed to result from photochemical changes through the absorption of ultraviolet radiation by oxygen. Indeed, the ozone layer acts as a filter, protecting the lower ATMOSPHERE and ground surface from potentially harmful ultraviolet rays. Recent research has shown that concentrations of ozone in the ozone layer are being seriously depleted, largely as a result of the build-up in the atmosphere of chlorofluorocarbons (CFCS) from sources such as domestic refrigerators and aerosol spray cans. So-called 'holes' in the layer have been discovered, first over Antarctica and, more recently, over the North Pole.

Recently, depletion of the ozone layer has begun to affect even parts of Europe and N America, giving rise to fears concerning possible adverse effects on human health (e.g. an increased incidence of skin cancers). Accordingly attempts are being made internationally to restrict the use of CFCs. Under the Montreal Protocol (1987) CFC consumption was to be phased out by an agreed timescale. For MEDCS, the agreed date was 1996 and, for LEDCS, 2010. (Funding has been made available to help the poorer countries meet this target.) According to the Scientific Assessment Panel of the Montreal Protocol, the ozone layer is projected to recover to pre-1980 levels in 2050 (assuming that all countries abide by the agreed dates for phasing out the damaging gases). Since 1986, total consumption of CFCs has fallen by 84% worldwide and by 97% in the industrialized countries.

At lower levels, ozone is a harmful pollutant, particularly in urban areas in the summer (it results from the reaction of certain chemicals to sunlight) and weather forecasts now frequently include references to 'ozone pollution'.

P

Pacific rim The countries bordering the Pacific Ocean. These include Japan, China, the ASIAN TIGERS and other countries of SE Asia, together with the USA, Canada, Mexico and the countries of Central America. In the Southern Hemisphere, there are Australia, New Zealand and the S American countries of Colombia, Ecuador, Peru and Chile. Because of economic growth during the last quarter of the 20th century, that part of the Pacific Rim lying in the Northern Hemisphere now challenges as the leading global *sunrise region* (see SUNBELT).

Pacific ring of fire The concentration of VOLCANOES and other features of tectonic and seismic activity around the margins of the Pacific Ocean corresponding with plate margins (see PLATE TECTONICS).

Pacific-type coast A type of coastline that is formed where the trend of mountain ridges and valleys is parallel to the coastline as a whole, e.g. the coast of British Columbia. Such a coast tends to be straight and regular, or (when a rise of sea-level has occurred) is characterized by lines of islands and straits. Ct ATLANTIC-TYPE COAST.

package tourism The sector of the tourist industry (see TOURISM) in which holiday travel and accommodation are put together as a relatively cheap package. It is aimed at the mass market and the most popular tourist destinations (as in the Mediterranean), and benefits from ECONOMIES OF SCALE.

palaeoecology The science of the reconstruction of past ecological conditions, particularly during the PLEISTOCENE and postglacial periods, from floral and faunal remains preserved in deposits such as PEAT. The reconstruction of former vegetation conditions can provide the basis for inferences about former climates and climatic change. See POSTGLACIAL, DENDROCHRONOLOGY.

Palaeolithic period The Old Stone Age, the earliest period of human prehistory, coinciding with the greater part of the PLEISTOCENE Ice Age. A series of culture periods has been distinguished, related to the various glacial and INTERGLACIAL phases. The earliest implements were roughly fashioned FLINTS, but gradually these became more highly fabricated, and at some stages bone was used. Cave paintings are a significant feature, especially in SW France. See also MESOLITHIC PERIOD, NEOLITHIC PERIOD, BRONZE AGE, IRON AGE.

palaeomagnetism The phenomenon of 'fossil magnetism' in rocks. The needle of a normal magnetic compass is influenced by two related forces. It will point in the direction of the magnetic North Pole (*declination*) and, if freely pivoted, will dip at an angle to the horizontal (*inclination*), the precise amount of dip depending on latitude. Thus, at the Equator, dip is zero, and will increase progressively to become vertical at the pole. When IGNEOUS ROCKS cool, magnetic minerals (comprising iron) will, as they crystallize out, behave like freely suspended magnetic needles, thus recording the direction and distance from the magnetic pole at the time of cooling; the magnetism forms a 'fossil' record for that time. Similarly, magnetic grains in SEDIMENTARY ROCKS become oriented at the time of DEPOSITION; as the SEDIMENTS become lithified to form SANDSTONE or SHALE, the magnetism will again be preserved, though in a weaker form. Studies of the *residual magnetism* of rocks provide information as to changes in the Earth's magnetic field over geological time. This field has been reversed many times; it is as if the North and South Poles exchanged places. Palaeomagnetism has been used to provide evidence supporting the concept of CONTINENTAL DRIFT and the current theory of PLATE TECTONICS (see POLAR WANDERING CURVE).

palaeotechnic era The first phase of the Industrial Revolution, during which: technological developments encouraged the growth of HEAVY INDUSTRY (located for the most part on coalfields); transport NETWORKS (canal and rail) were still relatively primitive; markets were highly localized; URBAN settlements were comparatively compact. In Britain, the palaeotechnic era lasted from *c.* 1750 to *c.* 1870. Ct NEOTECHNIC ERA; see also KONDRATIEFF CYCLE.

pallid zone A widely identified WEATHERING horizon, lying between the overlying mottled zone (a reddy-yellow horizon typical of tropical weathering environment), and the underlying partially decomposed BEDROCK, in tropical REGOLITHS. The pallid zone may reach a thickness of 60 m or more, but is usually less than 25 m. The horizon is referred to as the pallid zone because of its very pale appearance when exposed in section.

palsa A conical or elongated mound, containing a lens of ice and occurring in peat bogs in PERIGLACIAL regions. Palsas are several metres in height, and up to 100 m in diameter. The palsa core comprises SILT, underlying the PEAT, which freezes as the surface cold penetrates the bog; as a result, the peat layer is uplifted and domed.

pampa An extensive grassy plain in Argentina and parts of Uruguay, S America. The term is sometimes regarded as synonymous with temperate grassland.

pandemic A disease, such as AIDS, affecting large numbers people over much of the world.

Pangaea See CONTINENTAL DRIFT.

parabolic dune A type of DUNE, characteristic of sandy coastlines, which is curved in plan and with 'tails' that are pointing upwind (ct BARCHAN). Parabolic dunes have been explained in terms of stabilization at each end by marram grass, while the central higher part of the dune has continued to migrate inland under the influence of AEOLIAN transport of SAND from the windward to the lee slope.

parallel drainage A pattern of drainage, developed on a uniformly sloping surface, in which the individual streams run virtually parallel to each other. Parallel drainage is found on DIP-slopes (see CUESTA) and gently sloping LAVA flows.

parallel sequence A series of former glacial OVERFLOW CHANNELS, running approximately parallel to each other and cutting through an upland ridge between valleys once occupied by GLACIAL LAKES.

parameter (i) Any numerical characteristic (MEAN, MODE, STANDARD DEVIATION, etc.) of an entire POPULATION (i.e. a complete data set). Where the statistical measures refer only to a sample or samples of that population, then the term *sample statistic* or *sample estimate* should be used. (ii) A constant quantity in the equation of a curve.

parametric tests One of two major families of SIGNIFICANCE TEST made up of those that can only be applied on an interval scale (see INTERVAL DATA). They make certain assumptions about PARAMETERS of the POPULATION from which samples have been drawn for testing. Most commonly, these assumptions include that the population data are normally distributed, that the observations are independent of each other and, where populations are being compared, that they have the same variance. If parameter assumptions such as these cannot be made, then use should be made instead of NON-PARAMETRIC TESTS. Possibly the most widely used of all the parametric tests is the STUDENT'S T TEST.

parent material The mineral matter, either solid bedrock or superficial deposits (e.g. SANDS and GRAVELS, ALLUVIUM, TILL) from which a soil is formed. Minerals derived from the weathering of parent material may account for up to 50% or more of the total soil volume. Parent material is an important determinant of SOIL TEXTURE.

partial correlation See CORRELATION.

participation rate The proportion of a group of people or whole population engaging in a particular activity. While ACTIVITY RATES apply specifically to employment, participation rates are used to measure the degree to which people are involved in other pursuits, such as outdoor recreation, voluntary service, adult education or political protest.

particle-size analysis A geomorphological technique for analysing the sizes of particles contained within a deposit (such as river ALLUVIUM, a coastal BEACH, or a glacial TILL). A sample of the deposit is taken in the field, the sediment sorted in the laboratory (for example, by the use of a series of sieves with different mesh sizes), and the weight of the sediment in each particle-size class determined. The *particle-size distribution* is then plotted on a frequency HISTOGRAM (showing percentage by weight in each category). The main categories used are as follows: *clay* (0.00024–0.004 mm particle diameter); *silt* (0.004–0.062 mm); *sand* (0.062–2 mm); *gravel* (2–64 mm); *cobbles* (64256 mm); *boulders* (greater than 256 mm).

passive glacier A glacier with a very low rate of flow, owing to a combination of gentle gradient (e.g. on a PLATEAU surface) and small inputs of snow and ice, and limited melting. See also ALPINE GLACIER, MASS BALANCE.

pastoralism A type of farming concerned mainly with the rearing of livestock, whether for meat, milk, wool or hides. It may be *extensive* (e.g. the sheep farms of Australia, the ranches of USA) or *intensive* (e.g. the dairy herds of W Europe). It may be nomadic and of low technology (as practised in parts of N Africa and the Middle East) or highly scientific (as in W Europe) with careful breeding programmes, close monitoring of feeding and livestock performance, as well as involving the use of sophisticated machinery. See also TRANSHUMANCE.

pasture An area of land covered with grass used for the grazing of domesticated animals, as distinct from that which is mown for hay (*meadow*), although the same field may be used for both purposes at different times of the year. Some pasture may be 'natural' (e.g. an ALP or mountain pasture), but most areas of grazing are usually improved in some way (by liming, fertilizing, draining and periodic reseeding).

paternoster lakes A series of lakes in a formerly glaciated valley, separated from each other by morainic deposits or rock bars, but linked together by streams. The term is derived from the resemblance of such lakes, when viewed in plan, to a string of beads (paternoster = a bead in a rosary).

pattern See SPATIAL PATTERN, DRAINAGE PATTERN.

patterned ground A surface and subsurface pattern of coarse and fine debris, most evident in areas with little or patchy vegetation cover, which is produced by *lateral sorting processes*, most frequently in a PERIGLACIAL environment. Patterned ground is highly variable in character, and includes such features as STONE POLYGONS, fissure and ICE WEDGE polygons, and STONE STRIPES. [*f*]

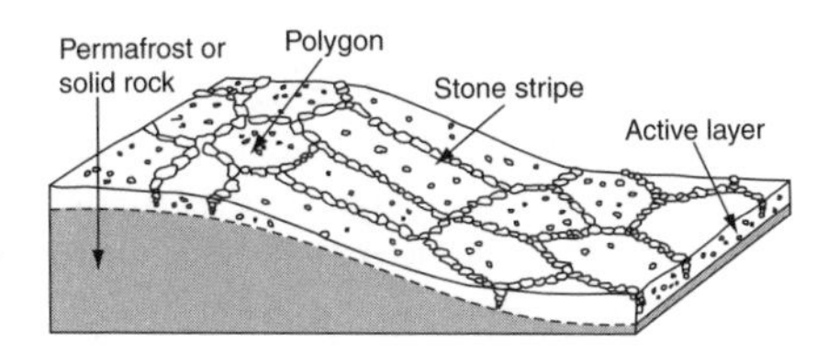

Sorted stone polygons and stripes

peak land-value intersection (point) The highest point on the URBAN LAND-VALUE SURFACE; i.e. the point within the TOWN or CITY where land values are greatest. It usually occurs at a major traffic intersection within the CBD, and its location is marked by the most prestigious shops.

[*f* URBAN LAND-VALUE SURFACE]

peat An organic deposit comprising partially decomposed plant materials that have accumulated in a waterlogged (ANAEROBIC) environment under conditions of impeded drainage (e.g. on flat areas or valley floors) and/or during prolonged wet periods over the landscape generally (see BLANKET BOG). Peat is found in cold or temperate climates. In the latter it is often a relict formation, developed under wetter conditions in the recent past. Peat is widely used for fuel (domestically, or in peat-fired power stations) and as a conditioner for garden soils. This has led to the widespread destruction of peat bogs (a distinctive and irreplaceable habitat for certain types of flora and fauna).

Owing to pressure from conservationists, alternatives to peat (such as coir, derived from the fibre waste of the coconut industry) are increasingly being sought and used. See PALAEOECOLOGY.

pedalfer A broad category of SOILS, associated with humid climates, which favour LEACHING and ELUVIATION, and the accumulation of aluminium and iron oxides in the soil (hence the *al* and *fer* of the term). Pedalfers are found mainly where the annual rainfall exceeds 600 mm (ct PEDOCAL).

pedestal rock See PERCHED BLOCK.

pediment A gently sloping surface, with angles of usually less than 7°, developed at the base of steep slopes particularly in arid and semi-arid regions. Pediments are essentially 'rock-cut' surfaces. In cross-profile they are gently concave; this together with the presence of alluvium suggests that running water has played some part in their formation. However, they are controversial landforms, and of the theories to account for them the following two are noteworthy. (i) The pediment is regarded as an active basal slope (or *slope of transport*), left by the recession of the mountain front; its angle is adjusted precisely to allow SHEET FLOODS to transport SEDIMENT from the foot of the mountain front to the peripediment. (ii) The pediment may also result from lateral PLANATION by running water (in the form of stream floods), which undercuts the mountain front, causing it to recede. It has also been noted that pediments are associated with sheet floods, though whether these help to shape the pediment or result from its existence is not clear. [*f* PIEDMONT]

pediplain A widespread surface of EROSION, surmounted by the occasional rocky hill (INSELBERG) or pile of weathered boulders (KOPPIE), associated particularly with semi-arid and SAVANNA landscapes. Pediplains are regarded by King (1949) as the product of major *cycles of pediplanation*, involving the twin processes of *scarp retreat* and *sedimentation.*

pedocal A broad category of soils, associated with relatively dry climates (where evaporation exceeds PRECIPITATION) where the processes of LEACHING and ELUVIATION are absent or weak, and CALCIFICATION leads to the accumulation of calcium carbonate, particularly in the B-HORIZON. Ct PEDALFER.

pedology The scientific study of SOILS and the processes of their formation.

pelagic A term applied to the water of the sea, and the organisms living there, independent of the shore or sea bottom. A *pelagic deposit* comprises very fine materials deposited as an *ooze* on the abyssal floors of oceans, and derived from floating organisms that die and sink, e.g. *globigerina ooze, radiolarian ooze* and *diatom ooze.*

pelean eruption A volcanic eruption accompanied by clouds of incandescent ASH, from the type-example of Mont Pelée (Martinique, Caribbean), which erupted catastrophically in 1902. See NUÉE ARDENTE.

Peltier's diagram A diagram devised by Peltier to show the relationship between WEATHERING processes and CLIMATE (temperature and precipitation). While the diagram is useful in suggesting the existence of broad climatic zones of weathering, it does not take into account fluctuations in temperature (important in mechanical weathering processes such as freeze-thaw) or variations in PRECIPITATION throughout the year. [*f*]

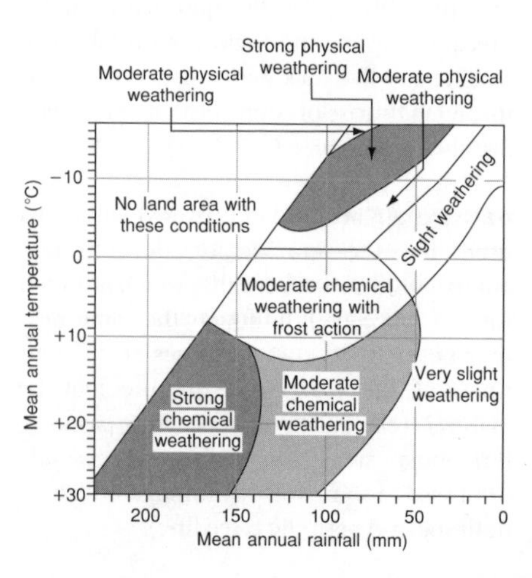

Peltier's diagram

peneplain An extensive, gently undulating surface of EROSION, resulting from a very long period of mass wasting and fluvial erosion under conditions of crustal stability. The formation of peneplains is referred to as the process of *peneplanation.* Although when formed, peneplains lie close to the existing BASE-LEVEL OF EROSION, most are of considerable age and have been affected by uplift, tilting and dissection by streams.

per capita This means 'per head of population', and is often used when comparing countries and regions that differ in extent and overall population. The sorts of measure expressed in per capita terms include GDP, income, diet, and expenditure on health and education.

perceived environment See ENVIRONMENTAL PERCEPTION.

perception See ENVIRONMENTAL PERCEPTION.

perched block A large BOULDER, perched in a delicate state of balance, following its DEPOSITION, usually by a glacier or ICE SHEET. Alternatively, a block can be left upstanding above the adjacent ground following selective WEATHERING and the washing away of surrounding SEDIMENTS. See also EARTH PILLARS, GLACIER TABLE.

perched water table An isolated mass of GROUNDWATER, situated well above the level of the main WATER TABLE and separated from the latter by an unsaturated zone. Perched water tables are formed as a result of lithological and structural variations in the rock. For example, in a PERMEABLE rock such as LIMESTONE, discontinuous MARL bands may locally impede PERCOLATION and give rise to small perched water tables. Alternatively, an area of weakly jointed rock may hold up groundwater, while surrounding well-jointed rock allows free percolation.

percolation The downward movement of water, under the influence of gravity, through the pores, JOINTS and BEDDING PLANES of a PERMEABLE rock.

percoline A line of concentrated water seepage through the SOIL, usually oriented downslope, approximately at right angles to the contours, and constituting an important channel for THROUGHFLOW.

pericline A small ANTICLINE that pitches along its AXIS in opposite directions from a central high point, rather in the manner of an elongated dome. When subjected to the process of breaching (see BREACHED ANTICLINE), periclines form oval-shaped vales, surrounded by a near-continuous in-facing ESCARPMENT.

periglacial A term meaning literally 'around the ice'. The concept of a *periglacial zone,* to include the climatic and geomorphological conditions of areas peripheral to the Pleistocene ICE SHEETS, was introduced by van Lozinski, a Polish geologist, in 1910. However, this definition is now seen as both misleading and too restrictive. For one thing, the zone immediately next to the ice (where GLACIO-FLUVIAL processes are active) is normally referred to as PROGLACIAL. For another, many parts of Siberia, N Canada and Alaska are accepted as typically periglacial, not as a result of proximity to an ice sheet but because low annual temperatures favour the development and/or maintenance of PERMAFROST.

In modern geomorphology, the term periglacial is used to denote an environment in which frost processes are dominant and permafrost occurs in a continuous or discontinuous form. However, it is important to emphasize that other processes (notably water and wind action, and possibly CHEMICAL WEATHERING) are also active in a periglacial regime. The extent of the periglacial zone, thus defined, is very large, currently covering some 20% of the Earth's surface. Moreover, in the past, as the PLEISTOCENE ice sheets advanced, periglacial conditions were established in many present-day temperate regions (amounting to an additional 20%). There is, of course, considerable variability in the importance of geomorphological processes from place to place in the periglacial zone and it is therefore an oversimplification to think of it as a uniform morphogenetic region (see NIVEO-FLUVIAL).

periphery In Friedmann's CORE-PERIPHERY

MODEL, the term periphery is applied to those areas of a country located outside the CORE and that compare unfavourably with it in such terms as level of economic DEVELOPMENT, prosperity, standard of living, etc. In the model, three types of peripheral area are designated: (i) *Upward transitional areas* favoured parts of the periphery benefiting from close proximity to the core and showing intensified use of RESOURCES, positive migration balances and increasing investment; essentially areas that are 'on the way up'. (ii) *Downward transitional areas* – areas unfavourably located relative to the core and the rest of the country, suffering from a poor and often dwindling resource base, low productivity and selective out-migration. (iii) *Resource frontiers* – areas where new resources have been discovered and where exploitation is being instigated, thereby gradually generating economic development and leading eventually to the emergence of *secondary cores.* [*f* CORE-PERIPHERY MODEL]

periphery firm See DUAL ECONOMY.

permafrost Perennially frozen ground in a PERIGLACIAL environment. In summer the upper layers of the ground are thawed, usually to a depth of a metre or more, giving rise to the ACTIVE LAYER. The upper surface of the permafrost is known as the *permafrost table.* It is estimated that between 20% and 25% of the Earth's surface is underlain by permafrost at the present time (mainly in the former USSR, Canada and Alaska). Permafrost reaches considerable depths (in excess of 500 m over much of the Canadian Arctic). Permafrost is a factor of great geomorphological importance, largely because it impedes PERCOLATION and disrupts the hydrological regime, resulting in abundant surface water, particularly during the summer melt season. See also GROUND ICE, THERMOKARST. [*f*]

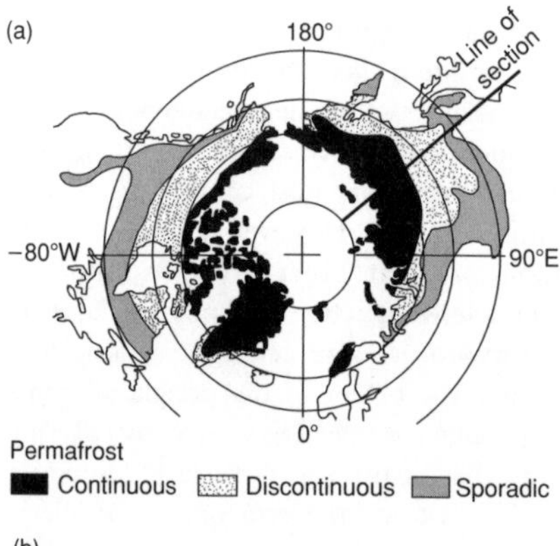

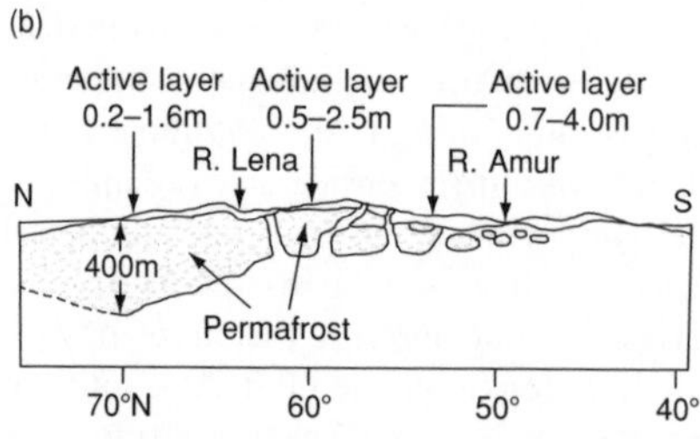

Permafrost: (a) distribution and (b) cross-section along 130°E

permeable A term describing rocks or superficial deposits that readily allow the passage of water, by way of pores, JOINTS and BEDDING PLANES. Permeable rocks include LIMESTONE, CHALK and SANDSTONE, and permeable sediments include SAND and GRAVEL. Rocks that are permeable have few if any surface streams and are therefore eroded at very slow rates. Water can, however, flow on the surface if the WATER TABLE is high or if PERCOLATION is restricted, say during a glacial or PERIGLACIAL period when pores and joints become filled with ice. Ct IMPERMEABLE.

personal enquiry (investigation) A piece of research based on first-hand (i.e. field work) and/or secondary data, often required by GCSE and A-level Geography syllabuses. The report of the enquiry normally follows the following sequence: (i) introduction – defining the topic and its aims, justifying its choice, putting it in context; (ii) data – sources, methods of collection, reliability; (iii) analysis and interpretation – methods used, results obtained; (iv) conclusion – summary of findings, evaluation, link back to initial aims, broader implications. See SOURCES.

personal sector See PRIVATE SECTOR.

personal services A category of TERTIARY SECTOR activity comprising services rendered to individuals on a personal basis, as for example by a beautician, chiropodist, hairdresser or prostitute. Not clearly distinguishable from PROFESSIONAL SERVICES.

pervious The capacity to allow the passage of water, via the JOINTS, cracks, fissures and BEDDING PLANES within a rock, as in CHALK and many types of LIMESTONE. Ct IMPERVIOUS.

photochemical smog A type of SMOG associated not with the burning of fuels such as coal, and the resultant concentration of soot particles in the atmosphere, but with the large-scale consumption of petroleum products, in areas where cars are concentrated. Photochemical smog appears to be cleaner than ordinary smog, but can cause intense eye irritation and damage to plants. It is most common in the tropics and in climates where, at certain times of the year, sunlight is strong and of long duration. Here photochemical decomposition of nitrogen oxide into nitric oxide and atomic oxygen occurs; the latter can in turn react with molecular oxygen to form ozone. Photochemical smog is characteristic of cities such as Los Angeles, where pollution is caused by the very high number of cars, and is trapped by subsiding air associated with the frequent anticyclonic conditions.

photogrammetry The technique of transforming an AERIAL PHOTOGRAPH into a topographical MAP.

photosynthesis The process whereby INSOLATION is converted into energy for plant growth. See CARBON CYCLE.

phreatic water Water lying in the zone of saturation within PERMEABLE rocks, which is sometimes referred to as the *phreatic zone* (that is, beneath the WATER TABLE).
[*f* WATER TABLE]

pH value A measure of acidity, widely applied in the context of SOIL study. Thus a soil with a pH value of less than 4 is *highly acidic*, with a pH of 4–7 is *mildly acidic* or *neutral*, and with a pH value in excess of 7 is *alkaline*. The development of soil acidity (*acidification*) is related to increasing hydrogen ion concentration, which occurs as exchangeable base ions (calcium, magnesium, potassium) in the CLAY-HUMUS COMPLEX are replaced by hydrogen ions from weak organic acids. Acidification is therefore retarded where there is a continual supply of weathered calcium and magnesium ions, but is accelerated by high permeability and rapid soil drainage (as in open-textured SANDS) and retarded humus incorporation. The latter is typical of cool, moist climatic conditions, where organic material accumulates at the soil surface as a raw HUMUS layer. In coniferous forests and heathlands, the formation of mor humus (itself strongly acidic) discourages soil organisms, delays *humifaction* (the process of humus incorporation), and promotes soil acidity (as in PODSOLS).

physical geography The branch of geography concerned with the study of the form of the land surface (GEOMORPHOLOGY); rivers, lakes and underground water (HYDROLOGY); the seas and oceans (OCEANOGRAPHY); atmospheric processes and phenomena (CLIMATOLOGY and METEOROLOGY); SOIL types and formation (PEDOLOGY); and the distribution of plants and animals (BIOGEOGRAPHY).

physical mobility See MOBILITY.

physical weathering See MECHANICAL WEATHERING.

pie diagram A descriptive name for a divided circle GRAPH, a diagrammatic device whereby a circle is divided into sectors. Each sector is drawn proportional in size to the percentage it represents of the total being shown by the circle as a whole. Also known as a *pie chart* or a *pie graph*. [*f*]

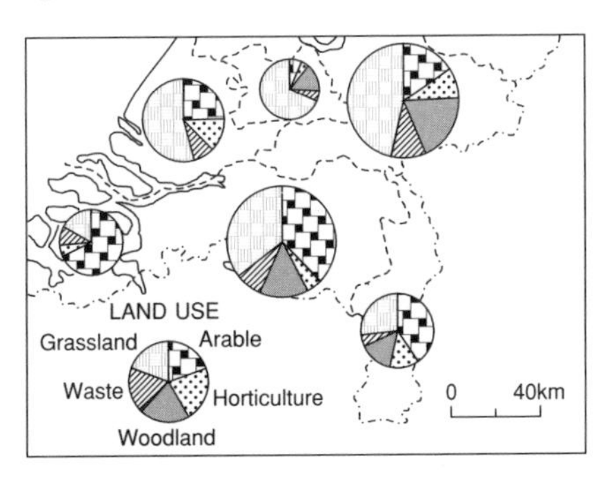

Land use in some provinces of the Netherlands shown by pie diagrams

piedmont A term meaning literally 'mountain foot'. It has been applied to the sequence of landforms along the margins of uplands, notably within the arid SW of the USA (see MOUNTAIN FRONT, BAJADA, PLAYA). The term is also used in glaciology to describe broad glaciers formed where constricted valley glaciers are able to spread laterally when passing into an adjacent lowland. Some piedmont glaciers are characterized by a zone of stagnant marginal ice at the snout (see GLACIER KARST). [*f*]

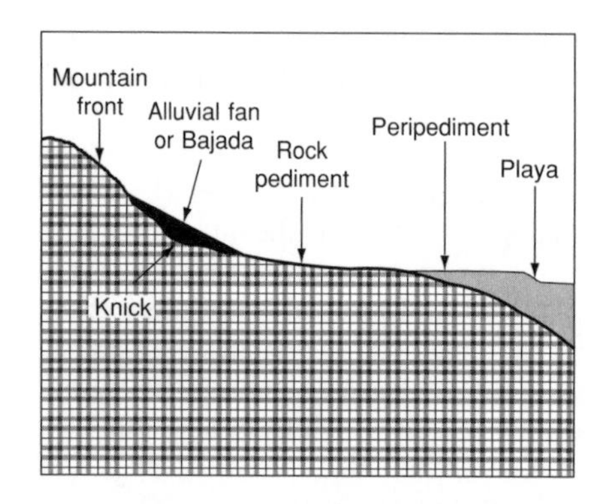

Cross-profile of the desert piedmont

pillow lava LAVA that has solidified very rapidly in contact with water (possibly, but not necessarily, at a considerable depth within the sea), to give a *pillow structure*, resembling a pile of sacks or pillows.

pingo A relatively large, ice-cored hill domed up from beneath either by the intrusion of water (which subsequently freezes) under pressure or by the growth of segregated ice-masses (see GROUND ICE. Pingos are up to 60 m in height and 300 m in diameter, and comprise a SEDIMENT layer (overlaying ice) up to 10 m in thickness, which may be breached at the summit of the pingo to expose the ice beneath.

pioneer community The very first plants to become established on a bare surface, of rock, debris or finer deposited materials. Such an environment is relatively hostile (hot, dry, without soil or humus, salty, and hence infertile), and the colonizing plants, such as lichens, mosses and small grasses, are small and form a discontinuous cover. Over a relatively long period, decay of these plants as they die produces HUMUS, and a primitive soil begins to accumulate, allowing larger and deeper-rooted plants to replace the initial colonists. See VEGETATION SUCCESSION.

pioneer settlement The early colonization and SETTLEMENT of a 'new' country, playing a major part in its opening-up, such as took place westwards across N America during the 18th and 19th centuries. Hence the term *pioneer fringe*, the zone beyond the present settled area and lying in the immediate wake of the *pioneer frontier*.

pipkrake See NEEDLE ICE.

pitch The direction in which the AXIS of a fold declines in height. See PERICLINE.

place names See TOPONYMY.

place utility A term used in MIGRATION studies and in investigations of residential

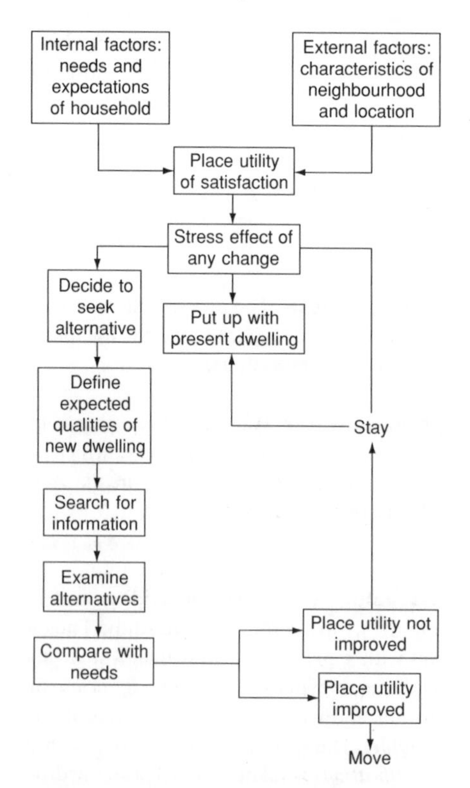

A model of the decision-making process in residential mobility

choice to denote an individual's integration at some point in space; i.e. his or her level of satisfaction or dissatisfaction with respect to a given location. The utility relates to such features as dwelling size and AMENITIES, physical characteristics of the neighbourhood, its socio-economic make-up, etc. Place utility is therefore a composite measure and may at a net level be either positive or negative. If the latter, then this is likely to produce stress which, in its turn, may induce the individual to move to another residential location. [*f*]

plagioclimax The climax stage in a VEGETATION SUCCESSION brought about by the actions of people. A good example of a plagioclimax is heather MOORLAND, which is widespread in upland regions of the UK. Much of this land was once forested but, when cleared to make way for farming, the high rainfall led to severe LEACHING and the soil became acidic and of low fertility. Heather (*caluna vulgaris*) is able to tolerate these harsh conditions and its young shoots provide good grazing for sheep. Through a programme of rotational burning, heather moorland forms an 'artificial' climax vegetation. Were it not for human actions, the moorland would revert to shrubs and eventually trees.

plagiosere A plant community resulting from the deflection of the normal VEGETATION by the direct or indirect activities of people (see PLAGIOCLIMAX) and their animals. For example, in an area where the CLIMATIC CLIMAX VEGETATION is woodland, the pasturing of animals – which eat seeds and saplings – can prevent tree regeneration. Old trees will die and fall, allowing the penetration of sunlight to the woodland floor and promoting the growth of low bushes; if in turn these are browsed and destroyed, grassland may become dominant. Such a vegetation change (from forest to grass) runs counter to the usual succession, and is therefore *retrogressive* in character. It is now known that many of the lowland HEATHS in Britain are plagioseres.

planation The formation by erosional processes, in particular wave EROSION and laterally shifting streams, of a level surface cut independently across the geological structure; hence the term *planation surface.* Planation may operate on a limited scale (as in the formation of WAVE-CUT PLATFORMS or RIVER TERRACES), or may over a very long period of time lead to the reduction of the landscape as a whole to a near-perfect plain. The term planation surface has been used in a general sense to describe all extensive erosional surfaces, whether they be PENEPLAINS, PEDIPLAINS, panplains, etch plains or the product of marine planation.

planned economy An ECONOMY characterized by state control of the FACTORS OF PRODUCTION and centralized state planning. Sometimes referred to as a *command economy.* See COMMUNISM; ct FREE MARKET ECONOMY.

planning Planning is the preparation of decisions for action directed at achieving specified goals by desirable means. It may be seen as having two aims: (i) increasing efficiency in the use and management of RESOURCES; (ii) improving the quality of the material and social ENVIRONMENTS. Such objectives can range from the conservation of countryside and historic TOWNS to the provision of new housing and services, from the construction of new motorways to coastal protection, from the decentralization of employment to the drawing up of regional aid programmes. As to the means by which goals might be realized, planning always has to operate within certain constraints (financial, political, social), balancing costs against benefits, balancing the good that might accrue to the community as a whole against the fact that there may be a group or groups disadvantaged by a certain course of planned action. (See COST–BENEFIT ANALYSIS).

The essential character of planning varies according to the political leanings of the country in which it is practised, ranging from the rigid *central planning* of socialist and communist states, through the more flexible and persuasive planning of the MIXED ECONOMY, to the truly capitalist country where there is little if any GOVERNMENT INTERVENTION and where planning as such is undertaken by firms rather than governments. See also ECONOMIC PLANNING, URBAN PLANNING.

planning blight The adverse impact of a proposed development. For example, the construction of a new motorway may be expected to adversely affect houses located close to the proposed line. Clearly, it will not be easy for homeowners to move other than if they are prepared to sell for a price substantially below that which they would have received before the motorway proposal was made. As a consequence, therefore, the residential prospects of such people have become blighted by the planned proposal. Planning blight is a good example of a NEGATIVE EXTERNALITY.

plant In ECONOMIC GEOGRAPHY, the term refers to a production unit (e.g. a factory, farm or office) in which inputs are converted into output.

plant community An assemblage of plants, of different species, growing together in a particular HABITAT. Plant communities exist because: (i) the physical habitat satisfies their requirements for growth and reproduction; (ii) they are able to live together – in other words, the needs of the species are not exactly similar, so that each makes a slightly different use of the habitat and there is little direct competition. Thus a deciduous woodland will contain not only oak trees, which leaf and fruit in summer, but smaller plants such as primroses, which flower in spring when light conditions at the woodland floor are still favourable. Moreover, the root systems of the trees and the flowers avoid competition by drawing on different levels in the SOIL.

plant succession see VEGETATION SUCCESSION.

plantation (i) An estate on which large-scale production of CASH CROPS (rubber, sugar cane, tobacco, etc.), usually on a monocultural basis, is carried out, generally by scientific, efficient methods, under a manager with an often large force of paid labour, involving a considerable amount of organization and administration. The system of production originated during the colonial period, using European organization, technology and capital. In some cases in the early days slave or forced labour was used (sometimes native, sometimes imported: e.g. the Black slaves imported into USA from Africa to work on the cotton plantations of the South). Some processing of the product is usually required, involving 'factory' buildings – for example, for the drying and maturing of tea or for the expressing of palm oil. The plantation system is found mainly in tropical and subtropical areas, especially where the use of White labour (other than for managerial purposes) was originally thought to be impossible for climatic and health reasons.

(ii) An area of trees, usually quick-maturing conifers, planted (e.g. by the FORESTRY COMMISSION in Britain) to provide supplies of softwood, pulp, etc.

plate tectonics An important geological theory, dating from the 1960s, which proposes that the Earth's crust consists of a number of mobile rigid elements (*plates*). Over a long period of geological time, these plates move relative to each other, causing: the ocean basins to change in size and form; the continents to drift apart (for example, the increasing separation of the African and American plates, to create the Atlantic Ocean); major landforms (such as MID-OCEAN RIDGES and mountain chains) to be formed. Each individual plate comprises a part of the LITHOSPHERE (extending to a depth of 50–125 km). Beneath lies the *asthenosphere*, in which temperatures are raised and rock materials reduced in strength, thus allowing movement laterally of the overlying plates. Studies of major FAULT-lines, EARTHQUAKE zones and VOLCANOES indicate that the Earth's crust comprises six main plates (the Eurasian, African, Indian, Pacific, American and Antarctic), plus 10 to 15 smaller plates.

There are three types of *plate boundary*. (i) *Constructive* or *divergent boundaries* occur where plates are moving away from each other, and new crustal rock is being formed by volcanic activity along the divergence; this is associated with SEA-FLOOR SPREADING and the formation of mid-ocean ridges. (ii) *Destructive* or *convergent boundaries* develop where the plates move towards each other, with one plate being overridden by the other. The overridden plate is 'bent' downwards and thrust into the MANTLE, forming a *subduction zone*. Convergent plate boundaries are

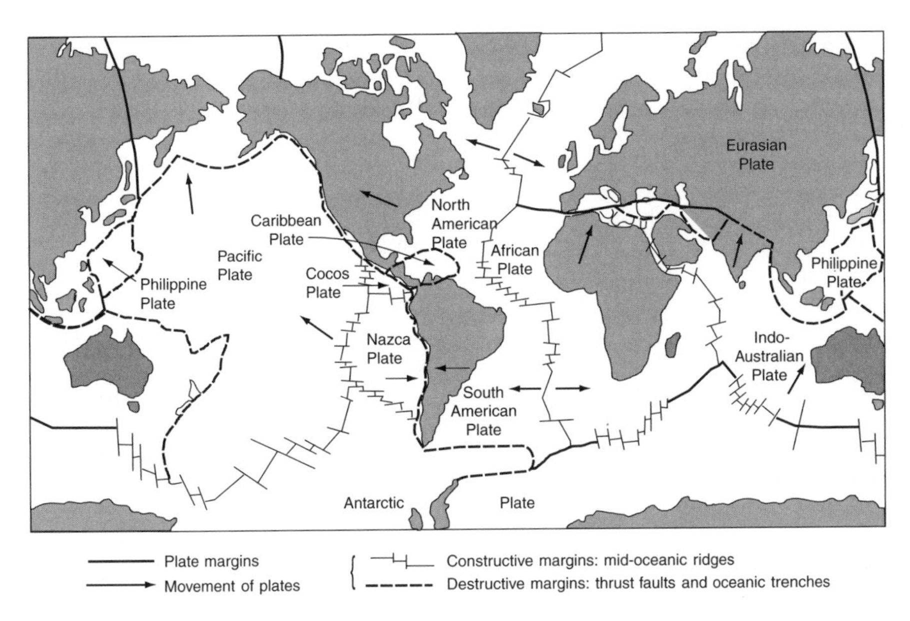

Tectonic plates and their boundaries

associated with intense compressive folding of sedimentary strata, the formation of deep ocean-floor trenches, the emplacement of granite BATHOLITHS, and chains of active volcanoes (as in western S America, where movement of the Nazca Plate beneath the American Plate is responsible for the Andes and the offshore Peru–Chile trench). *Shear boundaries* are formed where plates move parallel to each other, and are *conservative* in the sense that crustal rocks are neither formed nor destroyed (as at *transform faults*, such as the San Andreas Fault in California). The movement of plates is surprisingly rapid; up to 6 cm yr^{-1} of divergence has been recorded along the Mid-Atlantic Ridge. See also CONTINENTAL DRIFT, HOT SPOT. [*f*]

plateau An upland with a near-level summit, which is often bounded by steep margins. Plateaus may result from the EROSION of horizontal structures containing hard SEDIMENTARY ROCKS that act as CAP-ROCKS. Alternatively, they may result from the outpourings of LAVA on a major scale, hence *plateau basalts* (see also FISSURE ERUPTION). When these plateaus are incised by streams, *dissected plateaus* are formed (see BUTTE, MESA).

plateau gravel A spread of GRAVEL, found at a relatively high level on the planed-off summits of INTERFLUVES, by contrast with the gravel deposits (*terrace gravels* or *valley gravels*) found within valleys. Most spreads were laid down by meltwater flows and SOLIFLUCTION during the PLEISTOCENE.

playa A temporary or fluctuating lake, developed in a basin of internal drainage in an arid region (e.g. the arid SW of the USA). During infrequent periods of rain, water drains from surrounding uplands across the PEDIMENT and BAJADA to the central part of the basin (or *bolson*). However, the lake that forms is rapidly depleted by evaporation, and its site may be occupied for most of the time by a saline or alkaline crust or mud-flat. [*f* PIEDMONT]

Pleistocene A geological time period lasting from 2 million years ago until 10,000 years ago and broadly corresponding with the last so-called ICE AGE, which was, in fact, a series of ice advances and INTERGLACIALS.

plucking See JOINT-BLOCK REMOVAL.

plug See NECK.

plunge-pool A large, rounded depression at the base of a WATERFALL, eroded by the hydraulic impact of the descending water.

Powerful eddying and, possibly, CAVITATION may assist in the formation of plunge-pools.

plural economy The economy of a PLURAL SOCIETY within which culturally different groups keep, to a large extent, their own economic systems and tend to be involved in particular occupations (e.g. in Kenya, the association of people of Asian extraction with commerce and retailing).

plural society A society or community made up of culturally different groups. Given substantial immigration from the NEW COMMONWEALTH COUNTRIES during the 1950s and 1960s, Britain arguably has become a plural, or MULTI-ETHNIC, society.

plutonic rock A type of INTRUSIVE ROCK formed by the cooling of MAGMA at considerable depths in the Earth's crust. Plutonic rocks are characterized by a very coarse, large-crystalled texture, as in GRANITE.

pluvial A 'rainy period', lasting some hundreds or even thousands of years, frequently invoked to explain the past development of landforms in areas such as the Sahara and other DESERTS. With the growth of continental ICE SHEETS during the PLEISTOCENE, the climatic zones of the non-glacial areas were displaced towards the Equator. The Sahara was 'invaded' from the N by the winter rains now characteristic of Mediterranean climates, and at the same time the southern margins of the desert encroached on to the SAVANNA. Evidence of former pluvials takes the form of abundant GROUNDWATER in porous SANDSTONES where evaporation now greatly exceeds rainfall and recharge of the AQUIFER is impossible, and botanical remains showing that parts of the Sahara Desert were occupied by open savanna woodland or (at the very least) STEPPE grassland. Many geomorphologists have assumed that desert watercourses and valleys (WADIS) were eroded during the Pleistocene pluvials. However, increased rainfall does not necessarily lead to increased RUN-OFF and EROSION. This has led to the suggestion that desert erosion was very active not during the height of the pluvials, but in the transitional phases at their beginnings and ends, when rainfall was higher than at present but insufficient to nourish a protective vegetation cover.

podsol A type of ZONAL SOIL developed in cool temperate climates where PRECIPITATION is adequate for FOREST cover (coniferous trees such as pines and spruce, and deciduous trees such as birch). In this environment SOIL bacterial activity is limited, clay-humus production weak, and LEACHING and ELUVIATION highly effective, particularly where the parent material comprises coarse, well-drained SAND. There is abundant litter and fermenting organic matter at the soil surface (see HUMUS), providing humic acids that remove bases, colloids and oxides of iron from the A-HORIZON. A typical podsol has clearly demarcated SOIL HORIZONS – perhaps more so than any other type of soil. They are characterized by having a well-developed A2- (or E-) horizon, ashy grey in colour and consisting largely of infertile silica sand. The B-HORIZON is enriched by humus, colloids and oxides from the A2-horizon, and becomes dark in colour (often dark red, reflecting the iron content) and dense in structure, so much so that a clay-pan or even iron-pan may develop. In time, the presence of a 'pan' will impede drainage and counter the process of podsolization, while promoting the process of gleying (see GLEY SOIL). In time, an 'iron-pan podsol with gleying' may form. [*f*]

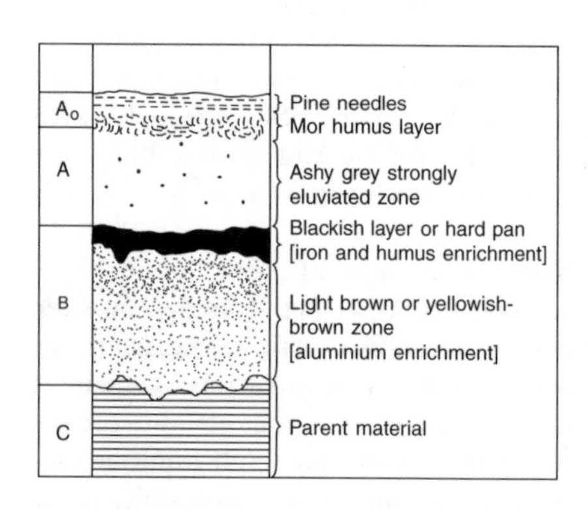

Podsol soil profile

point-bar An accumulation of SILT, SAND and GRAVEL, often BEACH-like in appearance, formed by FLUVIAL deposition on the inside of a river MEANDER. [*f* MEANDER]

point-bound place See CENTRAL PLACE THEORY.

point pattern The spatial arrangement of points. Such points may relate to the locations of any spatially distributed phenomena, be they mountain peaks, SWALLOW HOLES, schools or SETTLEMENTS. In the classification of possible point patterns, a threefold scheme is usually recognized – i.e. *uniform*, *random* and *clustered* patterns. See NEAREST-NEIGHBOUR ANALYSIS, QUADRAT ANALYSIS.

[*f* NEAREST-NEIGHBOUR ANALYSIS]

Poisson distribution A FREQUENCY DISTRIBUTION in which the PROBABILITY of any event occurring is very small compared with the probability that it will not. For example, the probability of a shop being a betting shop is small because there are many more shops than betting shops. The DISTRIBUTION is very asymmetrical (positively skewed); its MEAN is equal to its variance. It is used in the QUADRAT ANALYSIS of distribution MAPS to test whether a set of points deviates from a RANDOM distribution (see NEAREST-NEIGHBOUR ANALYSIS).

polar cell A weak and shallow circulation cell (part of the global atmospheric circulation system) located in each hemisphere between 60° and the pole. At the pole, cold air sinks to the surface and then moves across the surface towards the Equator. At about 60° N and S, it converges with warm tropical air moving polewards and forms an active boundary called the POLAR FRONT. Here it rises and flows back towards the pole.

[*f* HADLEY CELL]

polar front An irregular and discontinuous frontal zone in the N Atlantic and N Pacific oceans, along which polar maritime air masses from the N meet tropical maritime air masses from the S. The polar front migrates seasonally, pushing northwards in winter and southwards in summer, in line with shifts in position of the *polar front jet stream* (see JET STREAM). The formation of small 'waves' along the polar front leads to the generation of large mid-latitude FRONTAL DEPRESSIONS.

polar wandering curve A line plotted on a map of the Earth's surface, linking successive positions of the magnetic poles over geological time, as revealed by the evidence of PALAEOMAGNETISM. The notion of polar wandering, involving a gradual shift of the South Pole from approximately 30°S to its present position since the Carboniferous period, was proposed by Wegener, in the context of his hypothesis of CONTINENTAL DRIFT. More recently, palaeomagnetic studies have provided information on detailed changes in the position of the magnetic North Pole.

polarization The term used by Hirshmann to describe the movement of RESOURCES and wealth from the PERIPHERY and their spatial concentration in the CORE. It is synonymous with the BACKWASH EFFECT. See also CORE-PERIPHERY MODEL; ct TRICKLE DOWN.

polder An area of land near, at or below sea-level reclaimed from the sea, a lake or river FLOOD PLAIN by endyking and draining, often kept clear of water by pumping. The largest scheme ever attempted in Europe is the partial RECLAMATION of the former Zuider Zee in The Netherlands. About 2227 km^2 or 60% of the total area has been reclaimed in five polders, leaving the freshwater area now known as IJsselmeer. Much of Tokyo Bay is currently being reclaimed by a system akin to polders.

political asylum A place of refuge for people forced to leave their home country because of political oppression. Such people are referred to as *asylum seekers*. The rising numbers trying to seek political asylum in the EU and the UK (particularly from eastern Europe, the former Yugoslavia and Africa) give rise to increasing public concern. The problem for governments is to distinguish between those who are the genuine victims of political oppression and those who are simply using this as a pretext for MIGRATION. Cf REFUGEE.

political system The government or administrative set-up in a country. A number of different types may be recognized: democratic – government by the people through elected representatives; totalitarian – rule by an individual or by one party and no opposition; capitalist – see CAPITALISM; socialist – see SECOND WORLD.

polje A large enclosed depression in the Karst region of the former Yugoslavia, normally dry but occasionally partly flooded after heavy rainfall. Poljes are usually elongated in plan, with steep enclosing walls and flat floors. The largest poljes are up to 65 km in length and 10 km in width. One view is that poljes are the result of massive SOLUTION of LIMESTONE, but it seems very unlikely that such major landforms are the outcome of the collapse of underground cavern systems. There is some evidence of structural control, and this has led to the suggestion that poljes result from surface solutional WEATHERING of large downfaulted or downfolded limestone blocks.

pollen analysis A technique used by scientists to discover the characteristics of past environmental conditions using pollen preserved in peat, soils and lake deposits. Pollen analysis has been widely used to reconstruct past climatic conditions, particularly in identifying glacial and interglacial periods during the PLEISTOCENE. It enables relative dating to take place, allowing for the correlation of sediments from different locations. It is not, however, a form of absolute dating (see RADIO-CARBON DATING).

polluter pays principle A view increasingly held by governments that any perpetrator of ENVIRONMENTAL POLLUTION should meet the costs of making good the damage that has been done. Cf PRECAUTIONARY PRINCIPLE.

pollution See ENVIRONMENTAL POLLUTION.

pollution pathway The 'route' regularly taken by pollutants when they are released into the ENVIRONMENT. For example, ACID RAIN represents the final stage of a pathway that starts with the emission of sulphur dioxide from a factory or power station chimney; once in the air, the exact course of the pollution pathway is determined by the prevailing wind direction. [*f* ACID RAIN]

pollution sink Places or parts of a system where pollutants become concentrated and, therefore, where their impacts are most apparent. For example, when artificial FERTILIZERS consisting of soluble nitrates and phosphates are applied to farmland, they are leached from the soil or run off the surface, entering water courses or ditches. They become concentrated there by evaporation during droughts and hot weather. As water percolates down through the soil and PERVIOUS bedrock, the pollutants will also enter GROUNDWATER stores. In this way, water courses and groundwater stores become pollution sinks.

polycyclic A term sometimes used to describe a landscape that has evolved in several cycles (or partial cycles) of EROSION. A polycyclic landscape comprises a series of EROSION SURFACES, standing one above the other.

ponor A deep vertical shaft in the KARST of the former Yugoslavia, leading down from the surface to an underground cavern system in the LIMESTONE. See SINK HOLE.

pool-and-riffle A sequence of GRAVEL bars (riffles), separated by pools of deeper water, formed in stream CHANNELS where BED LOAD of a heterogeneous nature is being transported. Stream velocity is higher over the bars, producing a 'riffled' surface to the water but is reduced through the pools. Field measurements have shown that bars are not spaced randomly, but occur at distances of 5–7 times mean channel width. The positions of successive bars along a channel may be such as to cause a slightly sinuous flow pattern. In time this may become enhanced, and contribute to the formation of MEANDERS, although there is still a good deal of controversy surrounding meander formation. It is noticeable that deep scour pools are associated with the outsides of meander bends, while shallows (equivalent to bars) occur at crossover points between meanders. [*f* MEANDER]

population (i) The people occupying an area. (ii) In STATISTICS, the term is used much more widely to denote any complete collection of individuals, items or measurements defined by some common characteristics (e.g. VOLCANOES, trees, workers, firms, etc.); a complete *data set* from which samples might be drawn.

population density The number of people per unit area (usually per square kilometre). The mean world population density in 2000 was 45 persons km^2, but densities vary enormously from country to country and from region to region. Arguably, Hong Kong, with a population density of 6566 km_2, was the most densely populated state, followed by Singapore (4434). By comparison, countries that are frequently cited as being among the most populated (e.g. The Netherlands, 469; Belgium, 335; and Japan, 334 persons km^2) show relatively modest densities. Indeed, these MEDCs are surpassed by Bangladesh (953) and Puerto Rico (432). At the other end of the scale, there are countries such as Saudi Arabia (9) and Libya (3), Canada (3) and Australia (2). It should be emphasized, however, that many countries with low mean population densities often contain thickly populated areas; e.g. the urbanized eastern coastlands of Australia, ranging from Brisbane to Adelaide.

population development model See DEMOGRAPHIC TRANSITION.

population explosion A sudden and rapid expansion of population in a particular area, usually the result of a marked decrease in the DEATH RATE, but sometimes amplified by a concurrent increase in the BIRTH RATE. Most frequently experienced by those regions or countries passing through Stage 2 of the DEMOGRAPHIC TRANSITION.

population geography The study of populations, particularly of spatial variations in their DISTRIBUTION,VITAL STATISTICS (see POPULATION STRUCTURE), ethnic composition, rates of growth and socio-economic characteristics. It is also concerned with the patterns and flows of MIGRATION between places and with the reasons prompting such movements.

population potential The possible number of people who can live in a specific area with a reasonable standard of living, in relation to the available RESOURCES in that area. Cf CARRYING CAPACITY, OPTIMUM POPULATION.

population pressure The relationship between the population and RESOURCES of an area, particularly its capacity to produce food. Where the pressure of population on resources is great, then a situation of OVER-POPULATION may well exist.

population pyramid See AGE–SEX PYRAMID.

population structure The composition of a population analysed in the strictly demographic terms of age, sex, marital status, family and household size. Cf VITAL STATISTICS.

pore water pressure The pressure exerted on SOIL and rock particles by water contained in pore spaces. Under conditions of saturation, particles are forced apart by pore water pressure, and rapid mass failure may ensue. Disastrous slips, such as that of the coal spoil heap at Aberfan, S Wales, in 1966 are often the result of exceptionally high pore water pressure, related to the subsurface accumulation of water when SPRINGS and seepages are impeded.

porosity The possession by a SOIL or rock of pore spaces between the individual constituents such as pebbles, SAND grains or CLAY particles. Where the pores are relatively large, numerous and interconnected, water will pass readily through them (see PERMEABLE). Porous rocks – for example, CHALK and SANDSTONE – form important sources of water in the UK (see AQUIFER).

port A term usually employed to describe a point on a coast or along a water course where ships can tie up or anchor, and so load and off-load cargoes and passengers. This implies the existence of an adjacent SETTLEMENT, handling facilities, quays and usually docks, systems of transport and communication inland. There is a distinction to be drawn between a port and a HARBOUR: all ports must have a harbour, but the latter can exist without port facilities. Strictly speaking, it would be more appropriate to distinguish between *seaport* and *airport*, since both are *gateways* or *points of entry* to a country or region, and act as points of transfer to and from land transport. Likewise an *inland port*, such as can develop on a major river artery or with reference to a canal network, may be distinguished from a seaport.

port-related industry A broad designation that might be seen as embracing four categories of economic activity: (i) those basic processing INDUSTRIES that are established at PORT locations in order to avoid unnecessary transfer and TRANSHIPMENT costs (e.g. oil and sugar refining, iron and steel manufacturing); (ii) maritime industries that might be regarded as directly servicing the port (e.g. shipbuilding and repairing); (iii) the DISTRIBUTION of imported commodities; (iv) those services that are vital to the day-to-day running of the port, made up essentially of operational services (customs and excise, conservancy, pilotage, cargo-handling, etc.) and ancillary services (shipping company offices, forwarding agents, shipbrokers, welfare institutions for seamen, etc.).

positive checks See MALTHUS' THEORY OF POPULATION GROWTH.

positive correlation See CORRELATION.

positive discrimination A policy designed to help an area or group of people deemed to be in some way deprived. For example, government assistance to a DEPRESSED REGION or giving AID to people made homeless by a severe earthquake or flood.

positive externality See EXTERNALITIES.

positive feedback When an open system (see GENERAL SYSTEMS THEORY) experiences a change in one of the system's VARIABLES, a chain of events may be triggered off that exaggerates the effect of the initial change (ct NEGATIVE FEEDBACK, in which the system adjusts to counter the effect of the initial change). In simple terms, positive feedback is a 'snowball effect'. It has been suggested that positive feedback may be particularly characteristic of glacial systems. For example, an increase of meltwater at the base of a glacier will produce accelerated sliding. This in turn generates heat from friction, thus releasing further meltwater and promoting still faster sliding, and so on. In this situation the glacier will undergo a major SURGE forward, though the surge will cease if the excessive basal meltwater is able to drain away. However, it must be emphasized that positive feedback operates over a relatively short time-span. The accelerating changes cannot normally last indefinitely, and at some point EQUILIBRIUM must be restored to the system. [*f*]

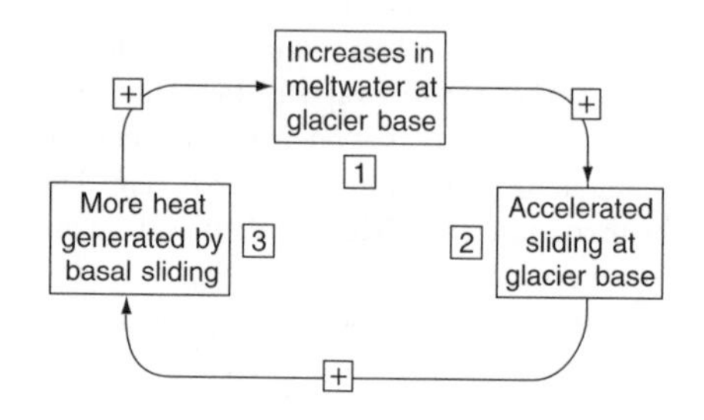

Positive feedback

postglacial The period since the withdrawal of the last Pleistocene ICE SHEET lasting from approximately 10,000BP in N Britain, but for a shorter period in areas such as S Finland, where the postglacial period dates from approximately 8000BP. The postglacial period was associated with a rise of temperature to a maximum (the period of the *Climatic Optimum*, which ended in Europe some 4000–5000 years ago). Subsequently, the climate has become somewhat cooler, leading to minor advances of glaciers (see NEOGLACIAL). Most recently, average temperatures have begun to increase again (see GLOBAL WARMING).

post-Fordist See POST-INDUSTRIAL.

post-industrial A term used to describe a set of changes and processes that have been at work since the 1970s, transforming cities, economies and societies. Implied here is a recognition that the ENERGY CRISES ended one era and began another in which circumstances are substantially different. *Post-modernist* and *post-Fordist* are alternative terms. See DE-INDUSTRIALIZATION, ECONOMIC GLOBALIZATION, GLOBAL ECONOMY, KONDRATIEFF CYCLES.

postmodernist See POST-INDUSTRIAL.

potential energy The energy possessed by a body due to its position, as defined by

$$E_p = m\,g\,h$$

where m = mass, g = gravity and h = height. For example, in streams the potential energy of the water is stated by

$$E_p = W_z$$

where W is the weight of the water and z the height above base-level (sometimes referred to as the *head* of the water). Potential energy is converted into KINETIC ENERGY when the water actually flows downhill, an ocean wave breaks, a glacier slides over its bed, a BOULDER rolls downslope, etc.

potential evapotranspiration See EVAPOTRANSPIRATION.

pot-hole A circular bowl, cut into BEDROCK, in the CHANNEL of a high-velocity stream. Pot-holes are the product of localized eddies, which whirl large stones around at the stream bed, producing concentrated CORRASION or *pot-hole drilling*. Large pot-holes are characteristic of channels incised into solid rock by SUBGLACIAL streams. In some instances they appear to mark the point of impact of meltwater descending vertically through the glacier to the ice-rock face. The term is also used to describe the vertical cave systems in LIMESTONE that are investigated by speleologists (hence 'pot-holers').

poverty trap See CYCLE OF POVERTY.

power (i) The rate of work, as distinct from the capacity for work (see ENERGY). (ii) The ability to control and influence in economic, political and social contexts.

PQLI (physical quality of life index) This is used as an indicator of the level of DEVELOPMENT in a country. It takes into account three different attributes: literacy, LIFE EXPECTANCY and INFANT MORTALITY. The calculation of the PQLI involves a simple indexing system. For each of the three indicators, the performance of individual countries is fitted to a 0 to 100 scale, where 0 equals the 'worst' performance and 100 the 'best'. The composite index is then calculated by averaging for each country the three indicator scores. The resulting PQLI is automatically scaled on an index range of 0 to 100. Cf HDI.

prairie An area of open grassland, from which trees or shrubs are virtually absent except in sheltered moist depressions and along water courses, in the midwest of N America.

precautionary principle The idea that the global community must take action to stave off potential large-scale environmental disasters, such as GLOBAL WARMING, without waiting for scientific proof about their cause and effects. Cf POLLUTER PAYS PRINCIPLE.

precinct An old word used in various ways, as for the enclosed land around a religious foundation. Now used in URBAN GEOGRAPHY and URBAN PLANNING to denote a specialized and defined area within the BUILT-UP AREA; e.g. the *shopping* or *pedestrian precincts* found in many CBDS today. In the USA the term is applied to administrative divisions or districts used for police and electoral purposes.

precipitation The deposition of atmospheric moisture at the surface of the Earth, in the form of DEW, frost, rain, SLEET, HAIL and SNOW. The total amount of precipitation at any place varies enormously, from less than 10 mm a year in the 'hyper-arid' DESERTS (such as the interior of the Sahara) to well over 10,000 mm in some tropical highlands (for example, Cameroon in W Africa and Cherrapungi in Assam).

preference See SPATIAL PREFERENCE.

pre-industrial city A term loosely applied to those early TOWNS and CITIES that developed between the first Urban Revolution (*c.* 5th millennium BC) and the Industrial Revolution (late 18th century). Although crafts, TRADE and AGRICULTURE were conspicuous elements of the city's ECONOMY, it was not principally a seat of economic activity. The pre-industrial city was truly multifunctional, containing religious, administrative, political and cultural activities, as well as being involved in a range of economic pursuits. By present standards, it was small, its size tending to be determined by the agricultural productivity of the local area or by its ability to tap distant food supplies. Key morphological elements in the spatial arrangement of the pre-industrial city were the defensive walls, its subdivision into sectors by further internal walls and moats, its narrow and congested streets, and the general absence of order. Although it would seem that there was residential SEGREGATION on the basis of ethnicity, kinship, occupation and status, the LAND USE pattern showed little evidence of the segregation of different activities.

prejudice See DISCRIMINATION.

preservation This may be defined as the protection of those features of the landscape that have human origins. These are often of an historical character (e.g. old buildings, archaeological remains, canals). In the same vein, CONSERVATION may be regarded as doing the same for the natural features of the landscape.

pressure gradient The difference in atmospheric pressure from one place to another (i.e. between areas of high and low pressure) as revealed by the spacing of the isobars on a synoptic chart. Where the isobars are closely spaced, so that the change in pressure from one point to another is considerable, the barometric gradient is steep (in the same way that closely spaced contours on a map represent a steep slope). Where they are widely spaced, so that the change in pressure from one place to another is slight, the gradient is gentle. A steep barometric gradient commonly associated with mid-latitude FRONTAL DEPRESSIONS or tropical CYCLONES produces gale-force or hurricane-force winds; a gentle gradient (as is usually found under anticyclonic conditions) is associated with gentle breezes.

pressure gradient See BAROMETRIC PRESSURE.

pressure group A group of people, united by a common cause, who actively strive to promote that cause by seeking to influence and generally bring pressure to bear (often by enlisting public support) on the decision-makers and those in authority. Pressure groups are frequently formed in reaction to contentious PLANNING proposals, such as the routing of new motorways, the siting of a new international airport or the construction of nuclear power stations. Ct MARKET PROCESSES.

pressure melting point The temperature at which ice is on the verge of melting, with the result that the exertion of any additional pressure will cause actual melting – for example, when sliding basal ice comes up against a BEDROCK obstacle. Pressure melting point is normally 0°C at the surface of glaciers, but deep within the ice it will be fractionally lowered by the pressure. Most (temperate) ALPINE GLACIERS are at pressure melting point throughout the ice thickness (and are thus WARM-BASED GLACIERS) so facilitating movement by basal slip when meltwater is produced. However, polar glaciers (see COLD GLACIERS), experience temperatures well below 0°C and therefore, in the absence of meltwater, move by internal flow only. See GLACIER FLOW.

pressure release See DILATATION.

prevailing wind The wind that blows most frequently, though not necessarily most strongly, from a specified direction (e.g. southwesterly winds prevail over much of Britain).

preventive checks See MALTHUS' THEORY OF POPULATION GROWTH.

primacy See PRIMATE CITY.

primary energy Energy that is derived from renewable RESOURCES, such as the Sun, wind and wood, and directly from FOSSIL FUELS (natural gas, oil and coal). To be distinguished from *secondary energy* (e.g. petrol, coke, manufactured gas and electricity), which is derived from a processing of primary energy.

primary product Any agricultural product or mineral, in natural or unprocessed form, that enters into international TRADE.

primary sector That part of the ECONOMY made up of activities directly concerned with the collection and utilization of NATURAL RESOURCES, i.e. AGRICULTURE, fishing, forestry, hunting, MINING and quarrying. Ct SECONDARY SECTOR, TERTIARY SECTOR, QUATERNARY SECTOR.

primary sources See SOURCES.

primate city A CITY (often a capital city) that completely dominates the national URBAN SYSTEM of which it is part. In terms of size, an exceptionally large gap exists between the population of the primate city and that of the second-ranking city in a country. As such, the existence of this situation (known either as *primacy* or *urban primacy*) represents a gross exception to the RANK-SIZE RULE. Not only does a primate city attract a large proportion of the national URBAN population, it also accounts for an overriding share of the country's economic functions, its wealth, and its social and cultural activities. This acute CENTRALIZATION, in its turn, provides the primate city with considerable political power.

It is difficult to make generalizations about the conditions that appear to generate urban primacy. Countries that, until relatively recently, have been politically and economically dependent on some foreign power (in a colonial context) tend to have a primate city (e.g. Kenya, Sri Lanka), as do countries that once had control over extensive empires (e.g. Austria, Spain). Certainly it is questionable to make the link (formerly stressed) between urban primacy and UNDERDEVELOPMENT, since highly developed countries (France, Japan and the UK) have primate capital cities. See also CITY-SIZE DISTRIBUTION; ct LOGNORMAL DISTRIBUTION.

primogeniture Inheritance of land and possessions by the first-born child or by the eldest son (*male primogeniture*). Ct *gavelkind* (see LAND REFORM).

principle (i) Something that is fundamental and that frequently provides the basis of a theory; a fundamental truth on which others are founded or from which others spring. (ii) See CENTRAL PLACE THEORY.

principle of least effort See LEAST EFFORT.

prisere A succession of PLANT COMMUNITIES initiated on new, biologically unmodified sites such as MORAINES, SCREE or unweathered rock surfaces. The pioneer community must be able to withstand open, unfavourable conditions, and will include lichens, mosses and small flowers. Eventually, these will be replaced by a more or less continuous ground cover of grasses, herbs and low shrubs, which will accelerate SOIL formation. Ultimately, the improved soil conditions (if the prevailing climatic regime is appropriate) will allow larger shrubs and trees. If the latter form a dense CANOPY, small plants unable to withstand deep shade will disappear. See VEGETATION SUCCESSION.

private sector That part of the ECONOMY made up of activities that are not directly related to local and central government; i.e. the *company sector* (private enterprises and FIRMS) and the *personal sector* (the economic activities of non-profitmaking organizations and private individuals). Ct PUBLIC SECTOR.

privatization The selling-off by government of a nationalized industry or undertaking to private investors and private companies; i.e. *denationalization* (ct NATIONALIZATION). Transferring the production of goods and the provision of basic services from the PUBLIC SECTOR to the PRIVATE SECTOR. During a succession of Conservative Governments in Britain, a range of economic activities became privatized, from steel-making and the railways to water supply and electricity generation. See CAPITALISM.

probability It has been claimed that 'statistics is the mathematical study of probability'. Probability refers to the likelihood of the occurrence of chance or RANDOM events. If the likelihood or probability of the event occurring randomly is low, then the event can be considered significant at the appropriate CONFIDENCE LEVEL. Probability is normally expressed as a number between 0 and 1, or as a percentage; an absolutely certain event would be expressed as 1.0 (i.e. 100% probable), while the likelihood of a tossed coin coming down heads is 0.5 (or 50%

probable). An understanding of probability is absolutely essential in interpreting the results of INFERENTIAL STATISTICS, all of which are expressed in terms of probability, not certainty.

procurement costs The costs incurred by a FIRM or INDUSTRY as a result of buying in materials, specialist services, expertise, etc.; synonymous with ASSEMBLY COSTS. Ct DISTRIBUTION COSTS.

producer goods Goods made for the purpose of producing CONSUMER GOODS, e.g. machinery of all kinds. Sometimes referred to as *capital goods.*

production chain The sequence of activities needed to turn raw materials into a finished product. The chain will include PRIMARY, SECONDARY and TERTIARY SECTOR activities. The activities may be located close together, perhaps in a single plant, or scattered in different locations in order to capitalize on the COMPARATIVE ADVANTAGES of particular locations. The latter are very much a feature of today's GLOBAL ECONOMY. [*f*]

product life cycle The notion that demand varies during the lifetime of a product. Five stages in the cycle are recognized: (i) a long introductory phase during which sales of the product fluctuate and increase only slowly; (ii) a period of rapid growth in sales as market penetration increases; (iii) maturity – this is reached when profit levels peak and sales volumes continue to increase; (iv) market saturation signalled when sales volumes peak; (v) a period of declining sales as the product's market penetration declines. [*f*]

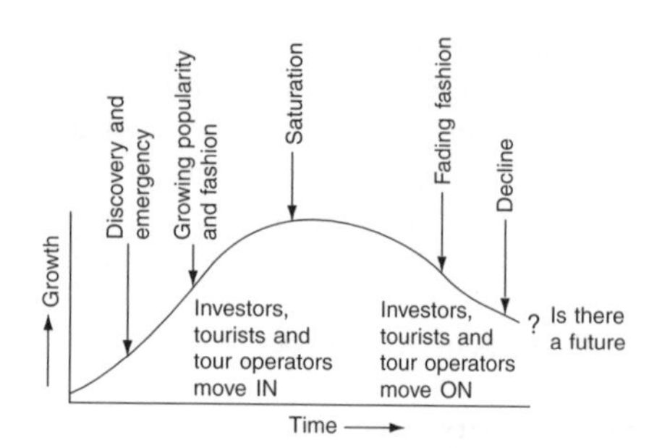

The product life cycle of tourism

production costs Expenditure incurred during the production of a commodity or the provision of a service by way of payments for rent, mortgages, interest on loans, dividends, salaries and wages, buildings, plant and machinery, as well as development and marketing costs. These costs are divided into FIXED COSTS and VARIABLE COSTS. See also FACTORS OF PRODUCTION; ct ASSEMBLY COSTS, PROCUREMENT COSTS.

productivity The efficiency with which the FACTORS OF PRODUCTION are used. The aim of increasing productivity is to produce more of a good at a lower cost per unit of OUTPUT, while retaining quality. It is not easy to measure, but the relationship of output to the number of man-hours expended is often taken nowadays as a rough guide.

professional services A broad category of TERTIARY SECTOR activity comprising expertise that is mainly provided on an individual basis and with face-to-face contact. The professional people providing such services range from the accountant to the stockbroker, from the architect to the dentist, from the solicitor to the estate agent. Most pro-

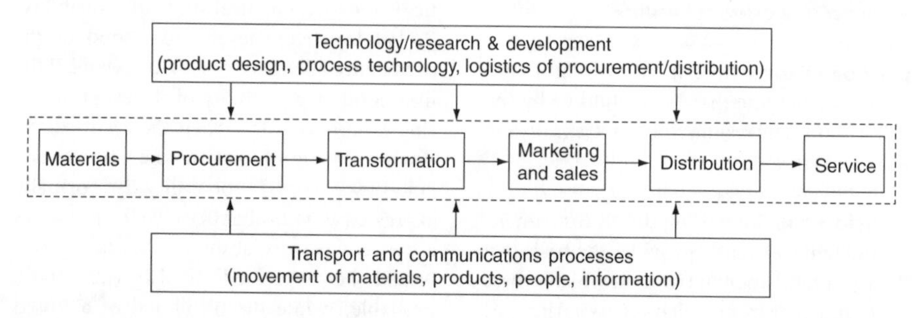

The basic production chain

fessional services may also be classified as OFFICE ACTIVITIES, with the possible exception of medical and dental services. The distinction between professional and PERSONAL SERVICES is a rather clouded one.

proglacial The zone closely adjacent to a glacier snout or ICE SHEET margin, characterized by TERMINAL MORAINES, GLACIO-FLUVIAL activity, OUTWASH PLAINS, GLACIAL LAKES and OVERFLOW CHANNELS. See PERIGLACIAL.

proglacial lake See GLACIAL LAKE.

proletariat The industrial working class. See CLASS, MARXISM.

proportional symbols A representational technique where a symbol (circle, square, sphere, cube, etc.) is drawn proportional in size to the value it represents. Perhaps the most widely used is the proportional circle, which may be used to show such things as the DISTRIBUTIONS of population, towns of different sizes and economic production.
[*f* PIE DIAGRAM]

propulsive industry A term used in REGIONAL DEVELOPMENT theory to denote an economic activity that is thought likely to generate and exert a powerful effect on regional growth, largely through its MULTIPLIER EFFECT. In Perroux's GROWTH POLE theory, a propulsive INDUSTRY or FIRM is defined as one satisfying the following requirements: sufficiently large to generate significant direct and indirect effects; fast-growing; extensive LINKAGES with other industries or firms, in order that its own growth might be widely and speedily transmitted, and hopefully amplified throughout the regional economy; it should be innovative. It is significant to note that the nature of propulsive industry will vary, not just from place to place, but more importantly with changes in the economic climate and with developments in technology. In the 1960s, typical propulsive industries were iron and steel, petrochemicals and motor vehicles; today these have been eclipsed by HIGH-TECHNOLOGY INDUSTRIES. Sometimes referred to as *leader industry*. Ct LAG INDUSTRY.

protalus An accumulation of angular debris (the product of frost WEATHERING of a rockface or the headwall of a CIRQUE or NIVATION CIRQUE) that has slid across the surface of a small glacier or perennial snowpatch. Where the debris has collected beyond the ice or snow to form prominent mounds and ridges, these are known as *protalus ramparts*.
[*f* NIVATION CIRQUE]

protectionism The introduction by government of devices such as import QUOTAS or raised TARIFFS in order to protect its own agriculture or industry from foreign competition. Usually, this happens when the overseas products are cheaper than those from domestic sources. Protectionism, or the threat of it, has been a recurrent feature of TRADE between Japan, the EU and the USA.

public enquiry A procedure used to allow planning applications to be discussed in public, giving both proponents and opponents an opportunity to make their cases before an independent inspector or chairperson appointed by the government.

public goods and services Hospitals, schools, sports centres, meeting halls, etc., and their associated health, educational and recreational services. They are subject to direct public control, usually via local government and its associated PLANNING powers. The availability and ACCESSIBILITY of such facilities and services are becoming increasingly important facets of WELFARE. Ct PUBLIC UTILITIES. See also DEPRIVATION, TERRITORIAL JUSTICE, URBAN MANAGERS.

public policy Any course of action determined by government (at whatever level), frequently pursued in the context of PLANNING and relating to various aspects of the ECONOMY, ENVIRONMENT and society. For example, public policy is likely to be formulated with respect to such diverse issues as taxation, investment in public works, the allocation and use of RESOURCES, regional aid, pollution control, housing improvement, the provision of SOCIAL SERVICES, etc.

public sector That part of the ECONOMY involving central government, local authorities, the nationalized INDUSTRIES and other public corporations. It is a sector of increasing importance even in the so-called capitalist economies (see MIXED ECONOMY). Ct PRIVATE SECTOR.

public utilities Undertakings that provide *essential services* to the community, e.g. water supply, sewage treatment, public transport. They may be described as basically 'physical' services and in this respect possibly distinguished from PUBLIC GOODS AND SERVICES, which are essentially 'social'. Cf INFRASTRUCTURE; ct SOCIAL SERVICES.

purposive sampling See RANDOM SAMPLING.

push-moraine A morainic ridge resulting from the bulldozing action of an advancing glacier snout or ICE SHEET. Push-moraines are usually quite small (up to 10 m in height), simply because larger accumulations cause the ice to ride over the crest of the moraine and flatten it.

push–pull factors See MIGRATION.

pyramidal peak (matterhorn peak or **horn)** A sharply pointed, faceted mountain peak standing high above a glaciated upland, e.g. the Matterhorn (Mt Cervin) in the Pennine Alps on the Swiss–Italian border. Pyramidal peaks form as a result of HEADWARD EROSION by three or four CIRQUE glaciers. Glacial sapping eventually reduces the ridges between the cirques to narrow ARÊTES, which extend steeply upwards towards the crest of the peak. Owing to glacial attack from either side, the arêtes are lowered to the extent that ice may extend over them, leaving the peak as an isolated NUNATAK. Over a long period of glaciation, the peaks are left standing higher and higher, although affected to some extent by frost action.

pyroclastic flow This comprises incandescent gas, ash and rocks and is one of the most devastating products of a volcanic eruption (see VULCANICITY). It can reach temperatures of up to 800°C and is capable of speeds of up to 200 kph. Such flows are most commonly associated with explosive VOLCANOES found at destructive plate margins (see PLATE TECTONICS) and are often the result of a partial collapse of the volcano itself. During the eruption of Mt St Helens in 1980, a pyroclastic flow was responsible for felling fully grown trees up to 25 km away from the volcano. Pyroclastic flows were the main product of the Soufriere Hill volcano, which erupted on the island of Montserrat in 1997.

quadrat analysis A statistical technique used in the analysis of POINT PATTERNS. The area of a point pattern is divided into equal-sized cells (usually grid squares) and the frequency of occurrence of points in each cell is determined. The observed FREQUENCY DISTRIBUTION is then compared with some theoretical or 'expected' pattern, thereby establishing the precise character of the point pattern under investigation (i.e. whether it is uniform, random or clustered) and of the processes producing it. See NEAREST-NEIGHBOUR ANALYSIS.

quadrat sampling A method of SAMPLING usually employed to reduce investigation of a large area to more manageable proportions. The study area is subdivided into a scheme of smaller units of equal size, this subdivision being most easily achieved by superimposing a grid of an appropriate scale. The investigation can then be directed at a sample of those grid-square areas. That sample may be selected either on a RANDOM or on a systematic basis. Cf NESTED SAMPLING.

quality of life A complex concept concerning the general state or condition of a population in a given area. Undoubtedly, it has an important psychological dimension that takes into account such states of mind as satisfaction, happiness, fulfilment and security (sometimes referred to as *social satisfaction*). It also has a physical dimension that embraces such criteria as DIET, housing, access to services, and safety. Other aspects include considerations such as social opportunity, employment prospects, affluence and LEISURE time. Some hold that it is synonymous with WELL-BEING. Others, however, interpret it as a particular expression of well-being, distinguished by an emphasis on the amount and

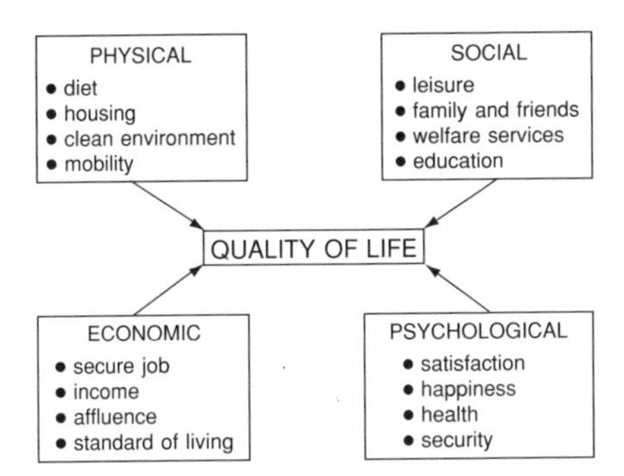

Some components of quality of life

distribution of PUBLIC GOODS AND SERVICES. It is a term commonly and loosely used in the literature of WELFARE GEOGRAPHY. [*f*]

quangos An acronym derived in the 1970s from the abbreviation of the name given to bodies variously referred to as: *qua*si *n*on-*g*overnmental *o*rganizations; *qu*asi *a*utonomous *n*on-*g*overnment *o*rganizations; *qu*asi *a*utonomous *n*ational *g*overnment *o*rganizations. In effect, these three titles refer to the same thing, namely a semi-public administrative body, outside the civil service, but financed by and having members appointed by government. In Britain, examples of quangos would include the BBC, the British Tourist Authority and the Police Complaints Board.

quantitative techniques Statistical methods for abbreviating, classifying and analysing numerical statements of fact in any investigation and making clear the relationships between phenomena. Quantitative techniques fall into two broad classes: DESCRIPTIVE STATISTICS and INFERENTIAL STATISTICS.

quarrying In PHYSICAL GEOGRAPHY, a generic term sometimes applied to the EROSION of a sea- or river-cliff by the combined forces of HYDRAULIC ACTION and CAVITATION. In glaciation, the term has been used to describe the effect of extreme water pressures recorded within the upper part of permeable BEDROCK beneath a glacier, believed to be capable of causing the rock to break apart to produce a jagged and irregular surface. See also MINING.

quartiles The percentiles that divide a distribution into four equal parts (i.e. the 25th, 50th – or MEDIAN – and 75th percentiles). The 25th and 75th percentiles are known respectively as the *lower quartile* and the *upper quartile.* [*f* DISPERSION DIAGRAM]

quartzite A very hard rock, consisting of quartz grains bonded to each other or cemented by secondary silica. Quartzite is particularly resistant to CHEMICAL WEATHERING.

quaternary sector A relatively recently recognized ECONOMIC SECTOR comprising all those personal services that require high levels of skill, expertise and specialization, e.g. education, R&D, administration and financial management. Unlike the TERTIARY SECTOR, this sector is concerned with people and information rather than goods. Ct PRIMARY SECTOR, SECONDARY SECTOR; see also DEVELOPMENT-STAGE MODEL.

questionnaire survey A data collection method involving questions and answers, and used for obtaining information about the way people live and what their views and opinions are on particular issues. Such surveys may be conducted in the street, door-to-door or by post and telephone.

quick clay A clay-rich deposit, comprising fine glacially abraded particles laid down in water, which when subjected to stress change from a solid to a liquefied state, and then undergoes rapid flow. See MUDFLOW.

quickflow The peak on a HYDROGRAPH showing RUN-OFF related to an individual rainstorm.

quota A limit imposed on the quantity of goods produced or purchased. *Import quotas* can be used to restrict the purchase of goods from foreign origins, while *export quotas* have been used to stabilize the export earnings of countries producing primary products by restricting supply and thereby sustaining prices. Quotas may also refer to the minimum level of production required in planned ECONOMIES.

R

R&D (research and development) A vital INPUT to a whole range of INDUSTRIES, particularly HIGH-TECHNOLOGY INDUSTRIES. The *research* is undertaken with a specific commercial aim in mind, while the *development* refers to the work of exploiting that research in either new processes or products, or the improvement of existing processes and products. Expenditure on R&D is crucial if the FIRM is to remain competitive, particularly in the sense of being technically innovative and of playing a leading role in the commercial application of new technology. R&D is regarded by many as the root cause of technical progress, innovation and ECONOMIC GROWTH.

race (i) The rapid flow of seawater through a restricted channel, usually caused by marked tidal differences at either end. (ii) A strongly flowing offshore current swirling round a headland or promontory. (iii) A narrow channel leading water from a river to the wheel of a watermill (*head-race*) and from the mill (*tail-race*). (iv) A large group of people with some basic inherited physical characteristics in common, e.g. skin colour, hair, facial features, head shape, etc. It is a major subdivision of mankind. Anthropologists generally recognize three major racial stocks (*Causasoid*, *Mongoloid* and *Negroid*) and two sub-races (*Australoid* and *Capoid*).

Radburn layout A style of residential layout pioneered at Radburn, New Jersey (USA) between 1928 and 1933, subsequently adopted and adapted in the planning of postwar housing areas in Britain, particularly in NEW TOWNS and EXPANDED TOWNS. Inspiration for the layout came from the *ward* idea incorporated in Howard's GARDEN CITY plans. Its main features include the SEGREGATION of pedestrian and vehicular traffic, 'turned around' housing fronting on to AMENITY space and gardens, and with vehicular access to the rear, loop roads and culs-de-sac. In the British postwar new towns, the Radburn principles were clearly evident in the detailed plans of NEIGHBOURHOOD UNITS. [*f*]

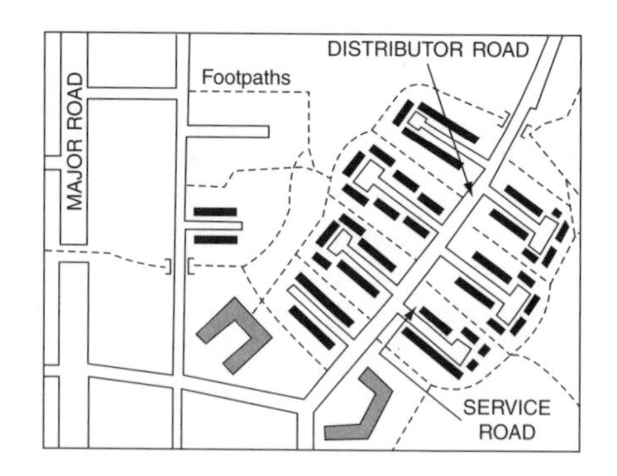

Residential area laid out according to Radburn principles

radial drainage A pattern of drainage in which the streams radiate outwards from the central high point of a structural dome or large volcano (ct CENTRIPETAL DRAINAGE). Good examples of radial drainage are found on large volcanic cones and ancient domes (e.g. Lake District). [*f* DRAINAGE PATTERNS]

radiation fog Fog that is formed at night and early in the morning under conditions favouring free radiation of heat from the Earth's surface. The latter becomes cooled, and the temperature of the overlying air is reduced by conduction of heat to the cold ground surface. If the air is moist, DEW-POINT will be quickly attained (see also DEW), and numerous tiny water droplets, with a very slow settling rate, will be formed by CONDENSATION within the cooled air layer. Radiation fog is associated particularly with anticyclonic conditions (clear skies permitting maximum nocturnal radiation) and light breezes (needed to spread the cooling upwards from the ground). The presence of 'water-attracting' nuclei such as particles of dust, salt and other chemicals may allow condensation to occur before relative humidity reaches 100%. Radiation fogs are often thickest over moist valley bottoms, where the process of formation is intensified by drainage of cold air from adjacent slopes (see KATABATIC WIND). In summer the fog will be dispersed quite rapidly as the sun raises air temperatures above dew-point, but in winter it may persist all day, or even intensify over a period of days.

radio-carbon dating A technique developed for the dating of relatively recent SEDIMENTS containing carbon. Carbon-14, a radioactive isotope of carbon with a half-life of 5570 years, is formed in the upper ATMOSPHERE, subsequently oxidizes to carbon dioxide, and becomes absorbed by living plants. Following the plants' death and the preservation of their remains within sediments under favourable circumstances (buried organic soils, PEAT formations, etc.), the Carbon-14 content diminishes over time at a known rate. From this, the age of the organic remains (and thus the deposit in which they lie) can be calculated within fairly close limits.

raindrop (rainsplash) erosion The displacement of SOIL particles by the impact of large raindrops, particularly under conditions of bare soil affected by intense convectional rainfall, for example in semi-arid environments. The raindrops have the effect of detaching individual particles, which are then moved downslope. As they do so, the finest particles are forced into voids, reducing INFILTRATION, and increasing surface RUN-OFF. Under FOREST conditions, the direct impact of raindrops will be reduced by INTERCEPTION, but water dripping from branches and leaves will still have some effect. In SAVANNA regions, the onset of heavy rains at the end of the dry season will lead to raindrop erosion on baked, unvegetated surfaces.

rainfall intensity The amount of rain that falls in a given time period, usually expressed in millimetres per hour. Rainfall intensity is thus a measure of how 'heavy' the rain is. Maximum values, which may be as high as 100–150 mm per hour, are usually associated with: tropical environments, where prevailing high temperatures and rapid EVAPOTRANSPIRATION result in large values of ABSOLUTE HUMIDITY; intense convective uplift (see THUNDERSTORM), leading to very rapid cooling and condensation of moist air over restricted areas. Rainfall intensity has important implications for landform processes such as rainwash erosion on slopes. See OVERLAND FLOW.

rainfall reliability The likelihood of the mean annual (or monthly) rainfall at a place actually being received in any particular year (or month). Mean values of rainfall are calculated over an extended period of time (30 years or more). If individual annual and/or monthly values for this period are plotted on a DISPERSION DIAGRAM, and median and quartile values calculated, a measure of rainfall reliability can be derived. Thus, the greater the inter-quartile range, and the greater the differences between either QUARTILE values and the median, the less will be the reliability of the rainfall. Indeed, rainfall reliability tends to be particularly low in climates with a low mean annual rainfall (e.g. hot DESERTS), where in one year a single heavy thunderstorm may produce the total annual rainfall, but in other years there may be no rain at all.

rainforest A dense FOREST, comprising tall trees, growing in areas of very high rainfall. Rainforest is found in some temperate regions (e.g. the Pacific Northwest of the USA), but is most characteristic of tropical regions (hence *tropical* and *equatorial rainforest*, and MONSOON FOREST). Tropical rainforest is developed in hot, totally frost-free conditions, where rainfall is both abundant and well distributed throughout the year. Forests are dominated by broad-leafed evergreen trees, which shed old leaves and grow new leaves continuously. There is a great variety of species (it is estimated that there are 2500 tree species in Malaysia alone). Other main features are: a layered CANOPY, consisting of three tiers (with the highest A-layer comprising the crowns of the *dominants*); *lianas* (woody climbing plants) and *epiphytes* (such as ferns and orchids growing on the trees); little undergrowth, owing to reduced light intensity at the forest floor; various minor adaptations (such as buttress roots and 'drip sips' to leaves). The large BIOMASS of tropical forests is not supported by fertile SOILS – in fact the reverse, for the soils are heavily leached and devoid of bases. There is, however, very rapid recycling of plant nutrients, by way of rapid decomposition of leaves at the forest floor with the aid of abundant micro-organisms; the nutrients released in this way are rapidly taken up by the shallow roots of the trees, giving a virtually closed NUTRIENT CYCLE. Tropical rainforests are being seriously depleted and rapidly cleared by human activity (see GLOBAL WARMING). See DECOMPOSER. [*f* NUTRIENT CYCLE]

rainwash (or **sheet flow)** A very thin sheet of water, at least a few mm in depth, running over a slope, and with the capacity to transport fine SOIL particles (hence *sheet flow*). Rainwash is usually not concentrated into CHANNELS, though towards the base of the slope it may be increasingly concentrated into rills and gullies. The role of rainwash in slope development is probably subsidiary to that of MASS MOVEMENT, because it operates intermittently and only after heavy rain, and it occurs rarely on the upper and middle parts of the slope.

raised beach A deposit of SHINGLE, SAND and broken shells, the product of past wave action, standing above the highest level of present-day SPRING TIDES. The raised beach is frequently preserved either by cementation or by an overlying layer of HEAD (formed under cold conditions when sea-level fell subsequent to the formation of the BEACH deposit). Many raised beaches are of PLEISTOCENE age, and are related to the fluctuations of sea-level resulting from the waxing and waning of the CONTINENTAL ICE SHEETS (see GLACIAL EUSTATISM) and from related isostatic crustal movements. [*f*]

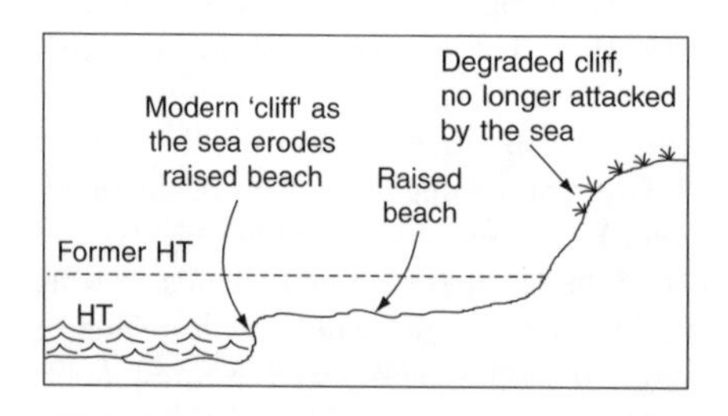

Raised beach

ranching The large-scale rearing of cattle on extensive farms, notably in areas that were originally temperate grasslands, e.g. W USA, W Canada, Argentina and Uruguay. Once the animals roamed on open *ranges*, now they are kept mainly in enclosures, with alfalfa and other fodder crops being grown as feed.

random Literally, without aim or purpose – i.e. haphazard. As used in STATISTICS, the term implies a process of selection applied to a set of objects. The selection is said to be random if it gives to each one of those objects an equal and independent chance of being selected. See also RANDOM SAMPLING.

random noise See FACTOR ANALYSIS.

random sampling A method of overcoming the problem encountered in many investigations of coping with a very large mass of data or of dealing with a very large area (see SAMPLING). In contrast to *purposive sampling* (where typical samples are chosen subjectively), random sampling is such that every item in a data set or every part of the study area has an equal chance of selection, and every item or every part selected is quite independent of all others. In a simple random sample, the objects of study are listed, each is assigned an index number, and a random sample of index numbers is obtained from a table of random numbers. *Randomness* therefore applies more to the process of sample selection than it does to the sample itself. Randomness is essential in statistical sampling because the pattern of PROBABILITY naturally depends on the nature of the data set, and only if a sample is RANDOM may the mathematical laws of probability be applied to the sampling variabilities. If there is a marked clustering of the phenomena under investigation (i.e. within the *data set*), the random sampling method may lead to a biased sample. In such a case, it may become necessary to make a *stratified random sample*, in which the data set is broken down into classes or strata before the sample is taken. In other words, a random sample is taken within each class or stratum (e.g. within both the white and black areas shown in the diagram). [*f* NESTED SAMPLING]

range (i) A line of mountains. (ii) An open area, usually unfenced, used for grazing, as on the High Plains of the USA. (iii) The difference between the maximum and minimum of a series of numerical values, especially climatic elements, such as seasonal and daily temperatures. (iv) The geographical limits of the habitat of a plant or animal. (v) The tidal range between the highest high and lowest low spring tides. (vi) The range of a good (or service) is the distance from which consumers will travel to a CENTRAL PLACE to obtain a good or service; the distance over which a good or service may be marketed. There are two limits to this distance. The inner limit circumscribes the area occupied by the THRESHOLD population, i.e. the minimum number of customers necessary to maintain the profitability of the good or service. The outer limit is the ultimate range of a good or service, in that beyond it people will either go to another central place for that good or service, or will not buy it at all because the gain derived from its purchase will be outweighed by excessive TRANSPORT COSTS. The custom provided by those people living within the range (i.e. between the inner and outer limits) may be seen as constituting the potential profit that might be derived by the ENTREPRENEUR from the marketing of the good or service. See also CENTRAL PLACE THEORY, CENTRAL PLACE HIERARCHY.

rank correlation The mathematical association between paired sets of ranked values, as for example between city size and volume of retail sales. The individual values of each VARIABLE are ranked (i.e. each city is given a rank for its size and for its retail sales) and a rank CORRELATION COEFFICIENT calculated on the basis of those paired rankings (see table below).

City	Rank based on size	Rank based on retail sales
A	1	1
B	2	4
C	3	2
D	4	5
E	5	3

The higher the correlation coefficient, the greater the rank correlation. The statistical tests most widely used for measuring rank correlation are Kendall's Correlation Coefficient and Spearman's Rank Correlation Coefficient.

rank-size rule This rule, formulated by Zipf, states that if all the SETTLEMENTS of a country or region are ranked according to their population size, the population (P_n) of the nth settlement in the RANKING will be $1/n$th that of the largest settlement (P). So $P_n = P/_n$ or in other words, the second largest settlement will have a population half that of the largest. When graphed, this simple rule produces an inverted J-shaped curve between rank and population, which in logarithmic form becomes a straight line (see LOGNORMAL DISTRIBUTION). Empirical studies in different parts of the world have shown that, while relatively few areas faithfully reflect the rule, many approximate to it. [*f*]

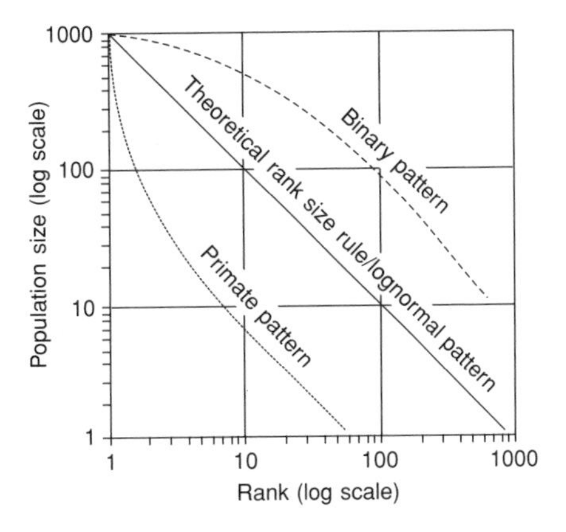

Rank-size relationships

rapids A stretch of rapidly flowing water associated with a steepening of the gradient along a river course. The cause of the rapids is usually geological (for example, the presence of a series of hard rock bands with intervening soft layers, that either DIP quite steeply upstream or dip downstream at an angle somewhat steeper than that of the river profile). Rapids should be contrasted with the more spectacular WATERFALLS, which are usually located at the outcrop of a more massive layer of hard rock or at FAULT-lines.

rapid transit system A form of transport designed to move people speedily around towns and cities and so avoid use of the private car and reduce traffic congestion. Light railways, underground systems and trams are the most commonly used modes of transport. Examples from the UK include the Docklands Light Railway in London, the Metrolink tram system in Manchester and the Tyneside Metro system.

Ravenstein's laws of migration Ravenstein was the first to hypothesize about the character of MIGRATION. On the basis of observations made of population movements in England in the second half of the 19th century, he formulated a set of laws. (i), The volume of migration is inversely proportional to distance; i.e. most migrants travel short distances and numbers of migrants decrease as distance increases. (ii) Migration occurs in stages and in a wave-like motion, i.e. one short movement from an area, leaves a vacuum to be filled by another short migration from another area, and in this way the migrant population gradually progresses towards its eventual destination. (iii) Every migrating movement has a compensating movement, i.e. migration is a two-way process, with net migration being the difference between the two contradictory movements. (iv) The longer the journey, the greater the chance that the movement will end in a large CITY.

While the above are the four principal laws, there are two others worthy of note: town dwellers are less migratory than country dwellers; women are more migratory than men over short distances.

raw material A substance subjected to processing, fabrication or manufacture; it may be natural (animal, vegetable or mineral) or a product of some other activity (e.g. wood pulp, iron and steel, flour).

raw-material orientation See MATERIAL-ORIENTED INDUSTRY.

recessional moraine A morainic ridge, usually comprising coarse 'clumped' SEDIMENTS, formed at the edge of a retreating glacier or ICE SHEET during a brief episode when the ice margin becomes stationary. A series of recessional MORAINES thus records the stages in the overall retreat of glaciers and ice sheets.

reclamation Any process by which land can be substantially improved or made available for some use (e.g. AGRICULTURE) by: the treatment of DERELICT LAND (levelling, landscaping, creating sites for INDUSTRY, or housing, providing AMENITY space); the drainage of temporarily waterlogged land resulting from seasonal flooding; the drainage of marshes; the drainage of lakes or a shallow part of the sea-floor (see POLDER); the improvement of heathlands; the clearance of scrub jungle, RAINFOREST or SAVANNA; the IRRIGATION of arid areas.

recreation Activity voluntarily undertaken during LEISURE time, e.g. stamp collecting, gardening, sport, birdwatching, watching TV, etc. Research into leisure-time activity frequently makes the distinction between recreation and TOURISM, with recreation defined as leisure pursuits involving less than a day's absence from home, whereas tourism involves a longer timescale and often travel over longer distances.

rectangular drainage A pattern of drainage in which sections of individual streams are rectilinear in plan, and stream junctions are at right angles. Rectangular drainage is an example of a pattern that has become adjusted to structure over time, through individual tributaries etching out lines of geological weakness (intersecting major JOINTS and FAULT-lines). Rectangular drainage is characteristic of GRANITE terrain, in which two sets of joints and shatter zones meet at right angles to each other, and is sometimes developed in LIMESTONE and CHALK. [*f* DRAINAGE PATTERNS]

rectilinear slope A slope segment that is straight in profile. The rectilinearity may occur as a small part of a slope profile, or in some circumstances may dominate the profile. Where the underlying geology is complex, there may be several rectilinear segments, each representing a readily weathered outcrop and contrasting with FREE FACES developed on the harder strata.

recumbent fold A type of overfold, resulting from very intense earth movements, in which the limbs of the fold are nearly horizontal.

recurrence interval See FLOOD.

recurve See SPIT.

recycling The re-use of materials (often manufactured goods) once they are discarded by a user. It is an increasingly common activity, encouraged by the depletion of *non-renewable* RESOURCES, by the associated rise in materials prices and by the impact of pressure groups calling for more care in the exploitation of the ENVIRONMENT. The recycling of such things as scrap metal and waste paper has been quite widely practised for many years; less so the re-use of glass and the recycling of garbage (by burning) to generate heat and power for community use. While recycling may appear to be wholly desirable, in practice there are technical problems (as yet unresolved), costs to be borne and much still to be done in the broad field of education and public awareness. See CONSERVATION, SUSTAINABLE DEVELOPMENT.

redevelopment See URBAN RENEWAL.

redlining A practice adopted by some firms concerned with providing financial assistance for would-be home-owners (e.g. building societies, banks, mortgage brokers, estate agents). It involves defining areas of a TOWN or CITY (usually older, inner areas) that are perceived to be in decline and in which lending for house purchase is regarded as a high risk. For these reasons, such areas are starved of finance in the form of mortgage advances. As a result, redlining discriminates against householders in these areas, it exacerbates the decline in housing conditions, as well as possibly helping to inflate the price of housing in those parts of the BUILT-UP AREA where the financial institutions are more willing to provide mortgages. Redlining is also undertaken by insurance companies who turn down applicants on the basis of their postcode. See GATEKEEPERS.

reduction See OXIDATION.

reforestation See AFFORESTATION.

refugee A person who flees their own country for political or economic reasons, or to avoid war or oppression. In 1995 the number of refugees around the world was estimated at over 27 million. About 40% of them were in Africa and 35% in Asia. Many nations now use legislation or screening procedures to limit the in-flow of refugees. See POLITICAL ASYLUM.

refurbishment Improvement of an URBAN area through investment aimed at improving its general appearance. Usually a part of URBAN RENEWAL.

reg A stony pavement in a hot DESERT (see HAMMADA).

regelation A process of refreezing operating beneath glaciers. It occurs either where basal ice at PRESSURE MELTING POINT is forced against a rock obstacle, thus releasing meltwater that freezes again on to the sole of the glacier beyond the obstacle (where pressure is reduced), or where basal meltwater beneath warm ice migrates beneath cold-based ice and freezes on to the glacier base (a phenomenon noted beneath some sub-polar glaciers), giving rise to a *regelation layer*.

regeneration The investment of capital in revitalizing the economic, social and environmental condition of an INNER CITY area or DECLINING REGION. Cf URBAN RENEWAL, REGIONAL DEVELOPMENT.

region An area of the Earth's surface differentiated and given unity by a specific characteristic or a set of criteria. The potential bases for such differentiation are innumerable (e.g. RELIEF, CLIMATE, SOIL, POPULATION DENSITY, LAND USE, standard of living). The study of regions was regarded as the principal aim of geography (see REGIONAL GEOGRAPHY). While this may be a much less widely held view today, there are aspects of regional study that are currently deemed to be important, e.g. REGIONAL DEVELOPMENT.

regional convergence The process of reducing or eliminating REGIONAL IMBALANCES. This might be as a result of direct GOVERNMENT INTERVENTION introducing curbs on growth in CORE regions at the same time as encouraging growth in LAGGING REGIONS (see also DEVELOPMENT AREA). Alternatively, there are theories, such as the CORE-PERIPHERY MODEL, which suggest that, with increasing economic development, the spatial economy of a country eventually reaches a sort of EQUILIBRIUM. At this point, differences in regional performance become minimal, if not non-existent.

regional development ECONOMIC GROWTH, social and cultural change within a specific REGION or a national system of regions. Research into this topic has given rise to the formulation of various theories and models, e.g. CORE-PERIPHERY MODEL, CUMULATIVE CAUSATION and GROWTH POLE theories. Regional development is, to an increasing extent in the western world, stimulated, directed and even controlled by direct and indirect GOVERNMENT INTERVENTION (see, e.g., ASSISTED AREA).

regional geography The geographical study of a REGION or regions; the investigation of the spatial differences that occur over the Earth's surface. Ct SYSTEMATIC GEOGRAPHY.

regional imbalance A situation existing within a national territory when there are significant, and problematical, spatial variations (*spatial disequilibria*) in the level of economic wealth, standard of living and QUALITY OF LIFE between its component regions. For example, the imbalance between the prosperous CORE regions and the declining or lagging PERIPHERY regions, as between SE and NE England, between the Eastern Seaboard of the USA and the Appalachians, or between SE Brazil and the Amazon lowlands.

regional inequality See REGIONAL IMBALANCE.

regional policy A term used to described government attempts to redress REGIONAL IMBALANCES. See ASSISTED AREA.

regional specialization The concentration of an economic activity in a particular area or REGION, and the processes or factors contributing to such spatial concentration. Fundamental to the explanation of regional specialization is the concept of COMPARATIVE ADVANTAGE. But there are other considerations to be taken into account. For example, regional specialization was a feature of the distribution of those INDUSTRIES that flourished during the 19th and early 20th centuries, especially those that processed low unit-value materials. Another contributory factor is the great strength of the EXTERNAL ECONOMIES available to firms clustered in a particular area; this was well demonstrated by the textile industry. The greater the degree of regional specialization, the more valuable are these external economies and the greater will become the INDUSTRIAL INERTIA within that region. Nowadays, however, regional specialization is not necessarily sought after, because specialization increases the vulnerability of an area to trade fluctuations.

regolith A layer of decomposed or disintegrated rock debris (that which has been weathered) overlying unweathered BEDROCK. Regolith is well developed in areas of low RELIEF, such as PENEPLAINS, where there has been long-continued WEATHERING and, because of the gentle gradients, slope TRANSPORT is weak. Regolith should not be confused with SOIL, which contains organic matter as well as weathered rock. See DEEP WEATHERING.

regression analysis A technique used to determine whether two VARIABLES are related and how good the relationship is between them. In regression analysis it is assumed that one variable (the independent variable) is responsible for, or causes, changes in the second variable (the DEPENDENT VARIABLE). The data for the two variables are graphed (see GRAPH). Regression then involves fitting a BEST-FIT LINE through the scatter of points in such a way that the sum of the squares of the distance between the points and the line is reduced to a minimum (see LEAST SQUARES). The *regression coefficient* gives the slope of the *regression line*, while the *coefficient of determination* can be used to predict the relationship between the two variables.

[*f* LEAST SQUARES]

regression coefficient See REGRESSION ANALYSIS.

regression line See BEST-FIT LINE, REGRESSION ANALYSIS.

rehabilitation See URBAN RENEWAL.

Reilly's law See LAW OF RETAIL GRAVITATION.

rejuvenation The process whereby a river, as a result of a fall in base-level or a climatic change, regains its powers of downcutting, thus resulting in the formation of valley-in-valley forms, entrenched meanders, KNICKPOINTS and RIVER TERRACES. [*f* RIVER TERRACE]

relative humidity The actual moisture content of a sample of air, expressed as a percentage of that which can be contained in the same volume of saturated air at the same temperature (see ABSOLUTE HUMIDITY). Relative humidity (RH) increases with reduction of temperature, providing the *mixing ratio* (the mass of water vapour present, in grams per kilogram of dry air) remains constant, until the DEW-POINT (at which RH = 100%) is attained. Alternatively, RH increases at the same temperature if the mixing ratio is increased, as a result of evaporation from the ground surface or a water body. High relative humidities are very characteristic of equatorial and tropical climates, where EVAPOTRANSPIRATION is considerable. They have the effect of increasing SENSIBLE TEMPERATURES and having an enervating effect on people.

relative relief The vertical distance between the highest and lowest points in a landscape. In an area dissected by numerous valleys, relative RELIEF is defined by the interval between the INTERFLUVE crests and the valley floors. Low relative relief will be associated with gentle slopes, whereas high relative relief will allow a much steeper gradient to be attained, and the development of FREE FACES and repose slopes will be possible.

relict landform A landform produced by, or bearing the clear imprint of, processes that are no longer operative upon it. The far-reaching changes of climate during and since the PLEISTOCENE have meant that perhaps the majority of landforms in many areas are relict. Examples of relict landforms include RAISED BEACHES, dead CLIFFS, DRY VALLEYS, ancient SAND DUNES and WADI systems.

relict landscape While, in a broad sense, all features in today's LANDSCAPE may be said to be relicts of the past, a relict landscape is one presenting features no longer in active use or undergoing active processes (e.g. strip LYNCHETS, disused canals, abandoned mines and factories, monuments, fortifications, etc.).

relief A general term used to describe the PHYSICAL GEOGRAPHY of an area (including differences in altitude, valley size and shape, and form and steepness of slopes). See also RELATIVE RELIEF.

relocation diffusion A process of SPATIAL DIFFUSION involving not the transmission of an idea (see EXPANSION DIFFUSION) but rather the movement of people and carriers to new locations. The commonest example of relocation diffusion is MIGRATION; the whole geography of early settlement in the USA can be regarded as the relocation diffusion of new immigrants across the face of the country. See GHETTO. [*f*]

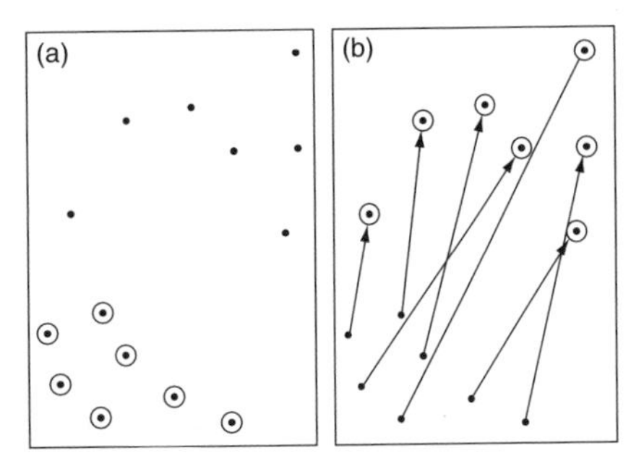

Relocation diffusion: (a) the initial stage and (b) a later stage

remote sensing A group of techniques whereby phenomena at or near the Earth's surface (landforms, SOILS, vegetation, LAND USE, snow cover, meteorological disturbances, etc.) can be 'sensed' by aerial survey, airborne electronic devices and, particularly in the last few decades, orbital satellites. The images obtained can be used to MAP varying surface textures related to differences in land use, natural vegetation, urban development, etc., and also, over a period of time, significant changes.

rendzina An INTRAZONAL SOIL type comprising a thin (20 mm), loamy and friable A-HORIZON, grey in colour and with a high HUMUS content, overlying a C-horizon of broken CHALK or LIMESTONE fragments ('brash'). Rendzinas are widely developed over the chalk uplands of S England, except where these are capped by PLATEAU-drifts or CLAY-WITH-FLINTS.

renewable resources See RESOURCE, NATURAL RESOURCES.

rent See ECONOMIC RENT.

reproduction rate See NET REPRODUCTION RATE.

research and development See R&D.

resequent drainage A type of drainage in which, after a lengthy period of development and change, streams 're-seek' their initial consequent courses (see CONSEQUENT STREAM). For example, in a folded structure the initial synclinal streams will in time be replaced by subsequents occupying anticlinal vales (see BREACHED ANTICLINE). However, if the anticlinal streams, as they cut downwards, encounter very resistant strata in the cores of the folds, they may migrate back to their original synclinal positions. Such resequent streams may be common in fold structures of considerable age.

reserves (i) See NATURE RESERVE. (ii) Part of a RESOURCE considered to be exploitable given current economic and technological conditions. The concept is particularly significant with regard to mineral extraction, where the distinction is made between recoverable reserves and speculative reserves. *Recoverable reserves* are those that it is profitable to work either at present or in the immediate future, given the present state of the market and available technology. *Speculative reserves* are those that are believed to exist, but in areas where as yet no exploration has been undertaken. In short, reserves are inferred on the basis that they have been proven to exist in similar geological structures elsewhere. Within the category of recoverable reserves, somewhat finer distinctions are made between *proven reserves*, *probable reserves* and *possible reserves*. See EXPLOITATION CYCLE.

reservoir An area where water is stored for HEP production, domestic or industrial consumption, or IRRIGATION. Reservoirs are often artificially created by building a DAM at a suitable retaining point across a valley. This is especially necessary when the volume of a river is markedly seasonal, with too much water causing flooding at one time of the year and too little at others; the reservoir thus smoothes out the availability of supply.

residual In STATISTICS it is the difference between an observed and a computed value, as encountered in REGRESSION ANALYSIS (see also LEAST SQUARES).

residual debris A layer of highly decomposed rock, for example GRANITE, resulting from prolonged CHEMICAL WEATHERING in a warm and humid climate.

resource This term is often taken to be synonymous with NATURAL RESOURCE, but it can be extended to embrace *human resources*, such as the manual skills, the innovative ability or the entrepreneurial talents of a population.

resource frontier region See CORE-PERIPHERY MODEL.

resource management A wide range of activities undertaken mainly by public bodies and related broadly to the exploitation of RESOURCES. It includes such matters as: undertaking surveys (e.g. assessing land capability); determining the extent of mineral reserves; PLANNING and evaluating different strategies that might be adopted in the exploitation of a resource; identifying and resolving the potential conflict between different types of resource use (e.g. between mineral working and farming); seeking to minimize the environmental impact of vital activities (e.g. of industrial development and waste disposal). In practice, resource management has to operate between the two frequently conflicting interests of economic expediency (i.e. obtaining resources at the cheapest price and in the shortest time) and of CONSERVATION (i.e. rationing the use of finite resources and protecting the ENVIRONMENT).

response time The time-delay in the reaction of glaciers and ICE SHEETS to short-term climatic changes, causing significant alterations in MASS BALANCE. In small glaciers the response time (before the changes are manifested by the advance or retreat of the ice margins) may be very short, in the order of 30 years or fewer. However, large ice sheets, such as that of Antarctica, may take thousands of years to respond to changes of snowfall in the ACCUMULATION ZONE.

restructuring Making basic alterations to the ECONOMY of a country, principally to counteract adverse changes that threaten stagnation or decline, but also to take advantage of new opportunities. Most often, it involves altering the balance between the ECONOMIC SECTORS. For example, restructuring of the Japanese economy during the last three decades had involved two distinct adjustments: in the SECONDARY SECTOR, the emphasis has shifted from heavy to light industry, particularly consumer industry and HIGH-TECHNOLOGY INDUSTRY (*industrial restructuring*); contraction of the PRIMARY SECTOR has been offset by expansion in the TERTIARY SECTOR.

The first of these was made necessary because of Japan's declining competitiveness on the world markets for heavy industrial goods (e.g. steel, ships, chemicals) and the rising costs of labour (see OFFSHORE ECONOMY). The second was triggered by a massive shake-out of agricultural labour and by the advent of the AFFLUENT SOCIETY in Japan with its greatly increased expenditure on consumer goods and services. See DE-INDUSTRIALIZATION.

resurgence The emergence of an underground river from a cavern, usually at the base of a SCARP face or CLIFF in LIMESTONE (ct SPRING or seepage, in which a smaller quantity of water is involved).

retail gravitation See LAW OF RETAIL GRAVITATION.

retailing The sale of goods in relatively small quantities to the public. The term is sometimes extended to include also the provision of consumer services (e.g. hair-dressing, dry cleaning), PROFESSIONAL SERVICES (e.g. banking, legal advice) and a range of catering and entertainment facilities. Ct WHOLESALING.

retailing ribbon A linear development of shops and consumer services along a major road within the BUILT-UP AREA of a SETTLEMENT. In many CITIES, such ribbons are extremely well developed and deserve to be recognized as a retailing *conformation* that is quite distinct from the retailing centre. Retailing ribbons tend to contain relatively more lower-order shops than are found in the CBD; they seem to exist principally to meet the day-to-day needs of nearby residents. One of the problems posed by these ribbons is that they do not readily fit with CENTRAL PLACE THEORY. This theory assumes that demand focuses on a point and thus generates a centre, whereas a ribbon clearly points to the existence of a flow-like demand. Many of these ribbons appear to come into being as a result of the linear expansion of once-separate, small centres strung out along a major road; through the process of linear expansion, such centres eventually coalesce into a single continuous ribbon. Others, particularly those in the USA, have resulted from simple linear spread along main highways leading out of cities.

retail park A shopping development usually located on the edge of an URBAN area or 'out of town'. A retail park consists of a number of independent retailers in individual buildings served by one vast car park, a petrol station and restaurants. Many such parks have been built in Britain; the Fosse Centre on the outskirts of Leicester and Bluewater near Dartford in Kent are among the most recent examples. Cf HYPERMARKET, OUT-OF-TOWN SHOPPING CENTRE.

reversed drainage A type of drainage pattern in which, as a result of earth movements, rivers are forced to flow in the opposite direction to their original courses. One feature of reversed drainage is the presence of tributaries that join the main stream at an angle pointing upstream. [*f*]

reversed fault See FAULT.

ria A coastal inlet, usually very irregular in outline, resulting from the submergence of a former river valley system (ct FJORD). For example, in S Devon, the Kingsbridge ESTUARY has been produced by the 'drowning' of a dendritic river system by the POSTGLACIAL rise of sea-level. The deepest water is found in the south, at Salcombe; from here, depth decreases 'inland', and the heads of some of the tributary inlets have already been modified by alluviation.

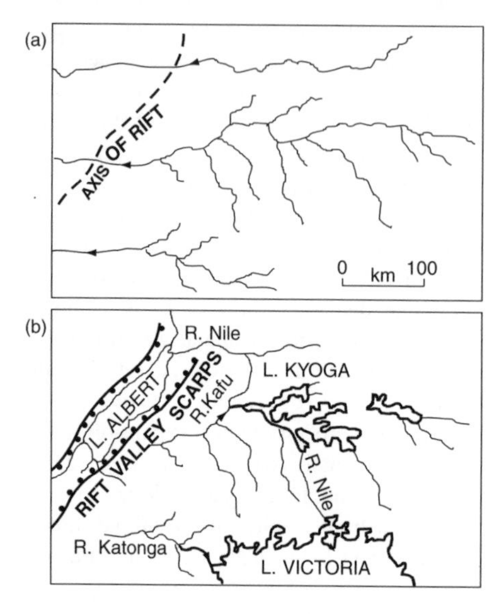

Reversed drainage in Uganda: (a) late-Tertiary river pattern and (b) reversal of drainage following the development of the East African rift valley during the Pleistocene

ribbon development Linear URBAN or suburban growth (mainly residential) occurring along a main road and extending outwards from a TOWN or CITY. Ribbon development was a characteristic feature of urban development in Britain during the inter-war period and before the introduction of stricter development control. Cf RETAILING RIBBON.

ribbon lake See FINGER LAKE.

RIC (recently industrializing country) A second generation of industrializing country emerging during the 1990s and following in the footsteps of the NICS. Representatives include Indonesia, Thailand and Mexico.

Richter scale A scale developed by the American C.F. Richter in 1935, which is used to measure the magnitude of an EARTHQUAKE. There is no limit to the scale – an increase of one point represents a x30 increase in energy released. The strongest earthquake ever recorded is the 1960 Chilean earthquake, which is thought to have registered 9.1 on the Richter scale, although some scientists dispute this, believing it to be closer to 8.9. Those earthquakes causing the greatest destruction usually exceed 7 on the Richter scale; the 1999 Turkish earthquake that killed 14,500 people registered 7.4.

ridge of high pressure A relatively narrow, elongated area of atmospheric high pressure, separating two low-pressure systems. A ridge is less static than an established ANTICYCLONE, and tends to migrate quite rapidly in association with neighbouring depressions. It usually brings a brief spell of dry, bright weather during a longer period of generally unsettled weather; however, an extension of the Azores high-pressure system over S Britain during the summer can last for several days.

riffle See POOL-AND-RIFFLE.

rift valley A major structural landform, resulting from the lowering of a relatively narrow strip of rocks between parallel FAULTS. Rift valleys are *structurally* similar to GRABENS, however, the latter is essentially a geological term describing a downthrown block which – after a long period of DENUDATION – may not coincide with a surface depression. By far the most important and best-known rift valley is that which extends from Jordan in the N, through E Africa, to Mozambique in the S – a distance of 5500 km. At some points the boundary FAULT SCARPS are bold and high, standing up to 600 m or more above the valley floor; at others there are prominent STEP FAULTS. The rift floor is sometimes flat, and sometimes broken by many small subparallel faults. The rift valley as a whole has created many basins of INTERNAL DRAINAGE, and there are several *rift valley lakes.* The RELIEF of the E African rift valley has been made more complicated by the massive outpouring of volcanic LAVAS, both within and on the shoulders of the rift. The E African rift valley is now believed to be the product of tensional forces in the Earth's crust, and is potentially a CONSTRUCTIVE PLATE BOUNDARY. [*f*]

rill erosion The formation of small, subparallel CHANNELS on a slope as sheetwash starts to become concentrated. Rills – often associated with the 'mid-slope' area – may become enlarged on the lower part of the slope to give gullies, owing to the natural downslope increase in surface RUN-OFF. It has been suggested that rills attain widths of a metre or more, and depths of 30–60 cm. Gullies, by contrast, can be up to 15 m in both width and depth. See also GULLY EROSION.

rime A type of frost, strikingly beautiful in effect, resulting from 'freezing fog'. The latter comprises tiny droplets of *supercooled* water (that is, at temperatures below 0°C but remaining unfrozen). In the presence of a gentle breeze, these droplets come into contact with trees, bushes and other objects, and as a result of disturbance immediately freeze, thus forming a covering of white granular ice particles. Ct HOAR FROST.

Rio Earth Summit A landmark conference held in Rio de Janeiro in 1992 attended by the representatives of 178 nations. It was motivated by a growing international concern about the future of planet Earth in the light of ever-increasing environmental degradation and GLOBAL WARMING. Its main achievements were the signing of two treaties – on CLIMATIC CHANGE and BIODIVERSITY – and a declaration setting out principles to guide nations towards SUSTAINABLE DEVELOPMENT.

rip current A sea current, often marked by turbulence and agitation of the water surface, caused by: the meeting of two tidal streams; a tidal stream suddenly entering shallow water; a return flow of water piled up at the shore by high onshore winds and strong waves. The strong offshore current represents a considerable hazard to swimmers, and notices warning of the dangers are often posted on beaches.

rising limb See HYDROGRAPH.

risk assessment Determining or judging the degree of damage that an area may expect from any given ENVIRONMENTAL HAZARD.

risk capital Money put up by investors without any guarantee of return. Long-term funds invested by ENTREPRENEURS, particularly subject to risk, as in a new venture. Sometimes referred to as *equity capital.*

river capture The process by which a river, by a combination of downcutting and HEADWARD EROSION, enlarges its CATCHMENT and is thereby able to 'take over' part of the drainage of a neighbouring river. A simple river capture is shown in the accompanying figure. Note how the tributary *X* to stream *A* extends to capture stream *B*, giving rise to: an *elbow of capture* at the point of diversion; a COL or *abandoned valley* marking the former passage of the captured stream; a MISFIT STREAM resulting from the 'beheading' of stream *B*. [*f*]

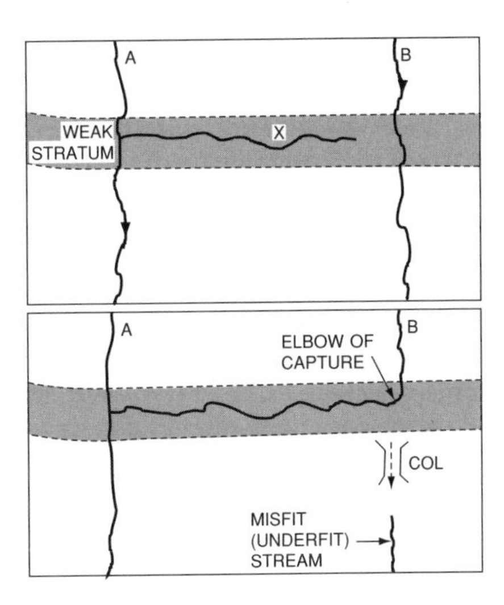

Development of river capture showing the resultant col and misfit (underfit) stream

river terrace A near-level surface in a river valley, usually covered by a thin layer of GRAVEL or ALLUVIUM. At the rear of the terrace there is a steep BLUFF leading up to higher ground (or perhaps to another in a sequence of river terraces). At the front, there is a steep drop either to the river or the FLOOD PLAIN. Many river terraces result from lateral PLANATION at a time when the river was graded to a higher base-level. Subsequently, the river underwent rapid vertical incision, leaving the former valley floor upstanding as the terrace (see REJUVENATION). [*f*]

robber economy The removal or extraction by people of various NATURAL RESOURCES, including MINING and quarrying, in such a

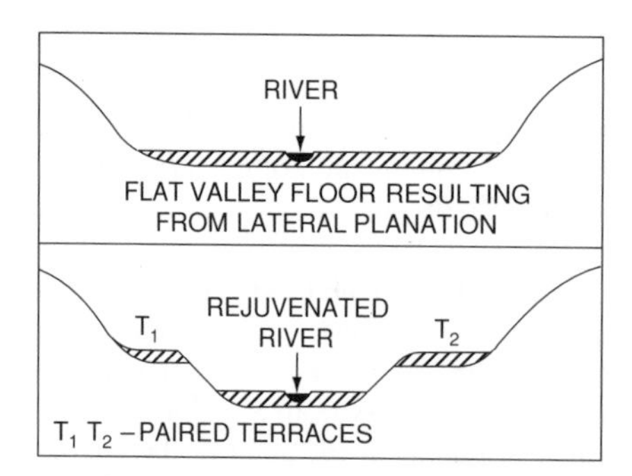

River terraces resulting from rejuvenation

way as to cause the rapid and ruthless destruction of resources for immediate profit, with no thought for the future. This would be exemplified by the overexploitation of renewable resources such as FORESTS, fish and whale stocks, SOIL fertility, etc. Ct SUSTAINABLE DEVELOPMENT.

roche moutonnée A projection of rock from a valley floor or side which has been moulded by the processes of glacial ABRASION and JOINT-BLOCK REMOVAL (otherwise known as plucking). Roche mountonnées commonly reach a height of 3–5 m and have a long axis measuring, say, 10 m. The upglacier face (known as the *stoss slope*) is gently sloping and often convexly rounded, with clear evidence of abrasion in the form of smoothed, polished and striated surfaces. The downglacier face (*lee slope*) is steeper and more rugged, owing to the exploitation of rock jointing by the glacier. [*f*]

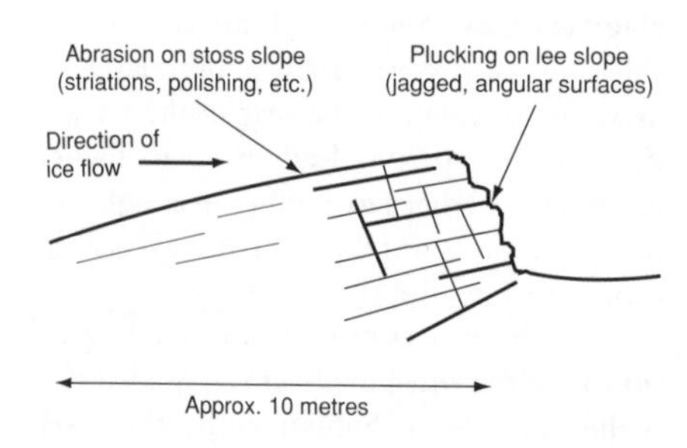

Roche moutonnée

rock fall The free fall of masses of rock detached by WEATHERING or rock failure, to the base of a CLIFF or very steep slope (see FREE FACE). [*f* MASS MOVEMENT]

rock flour The fine products of glacial ABRASION, which when transported by meltwater streams from beneath the ice give to them an opaque, milky appearance (see GLACIER FLOUR).

rock glacier A glacier-like tongue of rock waste, often escaping from CIRQUE-like amphitheatres and undergoing very slow downhill creep. Rock glaciers are sometimes regarded as mixtures of ice and debris, resulting from the successive DEPOSITION of snow (which is transformed into ice crystals) and WEATHERING products from adjacent slopes.

rock pavement See RUWARE.

rock slide The sliding of a detached block of rock along a plane of failure, such as a BEDDING PLANE or well-defined JOINT dipping downslope. The process is usually aided by the presence of water, which reduces friction along the sliding surface. The effects of rock slides can be seen on many coastal CLIFFS, where the strata DIP seawards and stability is reduced by marine undercutting of the cliff base.

rock step A pronounced break in the long-profile of a glaciated mountain valley. Frequently the downvalley face of the step is precipitous, and bears the mark of powerful glacial plucking, while the summit of the step has been smoothed by glacial ABRASION. Rock steps are sometimes related to the outcrops of resistant bands of rock, or relict subtle structural variations. Some rock steps have been ascribed to glacial modification of preglacial KNICKPOINTS resulting from fluvial activity. [*f*]

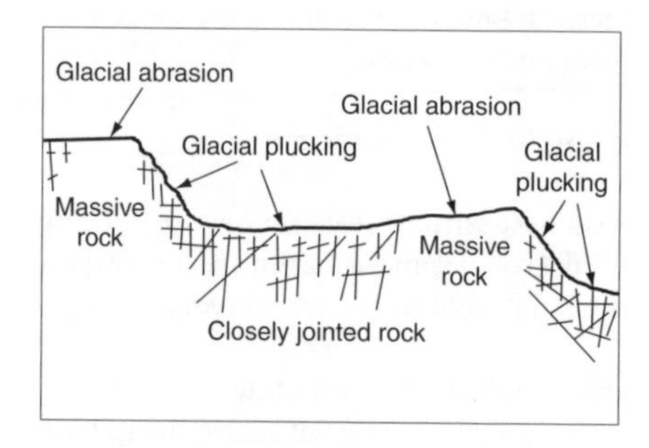

Rock steps in a glacial valley forming part of a glacial stairway

RO-RO (roll-on-roll-off) Specially designed ships that allow road vehicles to be driven on and off them; now widely used on ferry services, as well as for the overseas distribution of new motor vehicles.

Rossby waves Wave-like undulations in the pattern of the westerly winds of the middle TROPOSPHERE in mid and high latitudes. The major 'ridges' and 'troughs' of the Rossby wave system (there are commonly four waves in each hemisphere) reflect the combined influence on upper air pressure and wind systems of major mountain barriers (such as the Rockies), and terrestrial heat sources (warm seas in winter, warm land masses in summer). In the Southern Hemisphere, long waves are comparatively weakly developed. Rossby waves influence both high-level JET STREAMS (which are most intense at the troughs), and the convergence and divergence of air flow near ground level. In the Northern Hemisphere, ANTICYCLONES develop on the south-flowing part of a wave as a result of high-level convergence, while DEPRESSIONS develop on the north-flowing limb due to divergence. [*f*]

Rossby waves in the Northern Hemisphere

Rostow's model See STAGES OF ECONOMIC GROWTH MODEL.

rotational slip (also **rotational slide)** The slow downhill movement of a mass of rock or debris over a curved plane. Rotational slips are particularly associated with a PERMEABLE stratum overlying an unstable IMPERMEABLE stratum, forming part of a near-horizontal structure. The slipped mass, because of the rotational motion, is both lowered and 'back-tilted'. Since rotational slips are often composite, a CLIFF or slope affected by them will become terraced, with each TERRACE 'tread' dipping into the slope. It has also been suggested that CIRQUE glaciers may experience a form of rotational slipping, with the base of the glacier sliding over a BEDROCK surface that is arcuate in long-profile; hence the characteristic basin form of many cirques. In this case, the sliding is more or less continuous, and is promoted by the addition each winter of snow (later to become ice) on the upper slopes of the glacier, and the removal by ABLATION in summer of ice from the lower part of the glacier. [*f*]

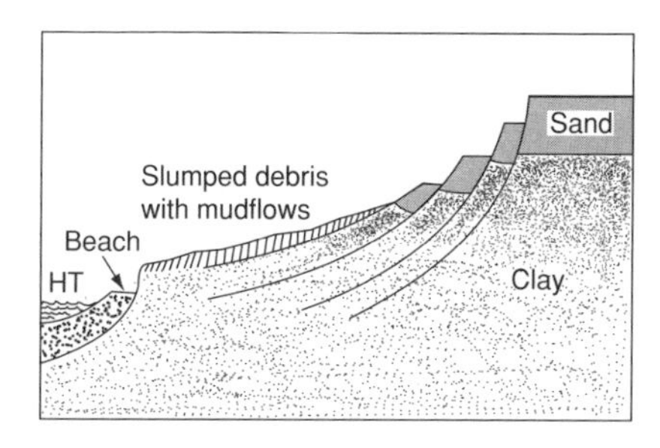

Rotational slips

rough grazing Unimproved grazing land, usually provided by types of semi-natural vegetation such as salt marsh, moorland and mountain pasture.

roughness An expression for the degree to which a stream CHANNEL is marked by irregularities, which by increasing friction slow down stream flow (hence *channel roughness*).

Factors affecting roughness include the size and angularity of BED LOAD particles, rock obstructions on the channel floor, the presence of ripples and SAND bars, channel sinuosity, bank vegetation and man-made obstructions. The degree to which roughness impedes flow is relatively great in small streams, but much less in larger streams owing to the greater depth of the water. In other words there is a tendency for the effects of roughness to be reduced progressively from the source to the mouth of a stream. This helps to explain the increase in mean velocity in a downstream direction, which has been observed in many streams. (see MANNING ROUGHNESS COEFFICIENT)

roundness index An expression of the degree of roundness of a rock fragment. Roundness can be determined by a variety of methods (for example, *Cailleux roundness is* calculated from the formula

$$\frac{2R}{a} \times 1000$$

where R is the minimum radius of curvature in the principal plane, and a is the long axis of the fragment. An alternative, simpler but more subjective measure is the Powers' Scale, which involves matching individual particles to a series of outline shapes. Roundness tends to increase with time, owing to the wear induced by transport processes, unless the fragment becomes broken up by impact or crushing. Thus, study of pebbles in a river channel has shown that these tend to become progressively rounder in a downstream direction, unless the entry of tributaries bringing less worn and more angular material produces a more complex pattern of change.

run-off The amount of water leaving a drainage basin. Expressed simply it is, in effect, the total rainfall falling in a basin minus evaporation from that basin (see WATER BALANCE). It therefore comprises OVERLAND FLOW (water flowing over the land surface), THROUGHFLOW (water flowing through the soil) and GROUNDWATER FLOW (water flowing through rock). The term 'surface run-off' is sometimes used to describe the surface flow of water; however, the term 'overland flow' is now more commonly used.

running mean See MOVING AVERAGE.

rural A word used to describe the character of country areas, and the activities and lifestyles encountered in such areas (ct URBAN). A *rural population* has traditionally been defined as one made up largely of people living in the country and working in AGRICULTURE. However, in many MEDCs, due to the effects of changes such as farm mechanization and COUNTERURBANIZATION, the latter criterion is becoming less valid. Certainly, this distinction between rural and urban is much less clear-cut today. See RURAL–URBAN CONTINUUM, RURBAN.

rural depopulation See DEPOPULATION.

rural deprivation A condition in country areas where the WELL-BEING of local people falls below a level generally regarded as a reasonable minimum. The symptoms include lack of employment, poor healthcare and education, inadequate shops and public transport. It frequently spurs MIGRATION and DEPOPULATION. See CYCLE OF POVERTY.

rural geography The geographical study of selected aspects of the rural environment, including the patterns, functions and morphology of rural SETTLEMENT, the rural ECONOMY (especially AGRICULTURE, RECREATION and TOURISM), the intrusion of URBAN influences into rural areas (by the extension of COMMUTING, SECOND HOMES and the centralization of services in towns), and critical commentary on RURAL PLANNING policies and programmes.

rural planning The definition and implementation of goals related to the rural ENVIRONMENT; the management of change in the countryside towards specified objectives. The principal objectives currently sought by rural planning in Britain include the CONSERVATION of villages and landscapes of outstanding scenic, historic or scientific interest, the provision of recreational opportunities close to URBAN populations (e.g. national parks and country parks), the improvement of services in rural SETTLEMENTS (see KEY SETTLEMENT) and the raising of living standards in remote, RURAL areas of the PERIPHERY.

rural turnaround The revival of more accessible RURAL areas, previously suffering from DEPOPULATION, brought about by the process of COUNTERURBANIZATION. The 'flight from the city' started some time ago with the dispersal of retired people anxious to spend their last years in calmer, cheaper and more attractive locations. They are now followed by growing numbers of commuters, and in turn by various activities such as R&D, offices and health clinics.

rural–urban continuum This term is used to express the fact that in many countries today (particularly in those that are highly urbanized, such as Japan and the UK) there is no longer, either physically or socially, a simple, clear-cut division of town and country. Rather there is a gradation from the one to the other, so that there is no definite point where it can be said that the URBAN way of life ends and the RURAL way of life begins. This blurring of the boundary stems in large measure from the fact that the impact of URBANIZATION reaches well beyond the limits of the BUILT-UP AREA and because that impact is a DISTANCE DECAY phenomenon. Cf RURAL–URBAN FRINGE.

rural–urban fringe A zone of transition between the BUILT-UP AREA of a TOWN or CITY and the surrounding countryside, and in which URBAN activities, uses and structures are mixed with RURAL ones. An essential physical ingredient of the RURAL–URBAN CONTINUUM.

rurban A composite adjective introduced into geographical literature, and applied to the indeterminate and transitional condition between country and TOWN, between the RURAL and URBAN states. See RURAL–URBAN CONTINUUM, RURAL–URBAN FRINGE.

ruware A low, rounded and often elongated exposure of unweathered rock (sometimes referred to as a *rock pavement* or *whaleback*) rising a few metres above the surrounding plain. Ruwares are common in SAVANNA regions. They appear to represent 'conical rises' in the BASAL SURFACE OF WEATHERING, which are in the process of being exposed by the removal of overlying chemically weathered rock. They thus appear to mark an early stage in the formation of BORNHARDTS. See also INSELBERG. [*f*]

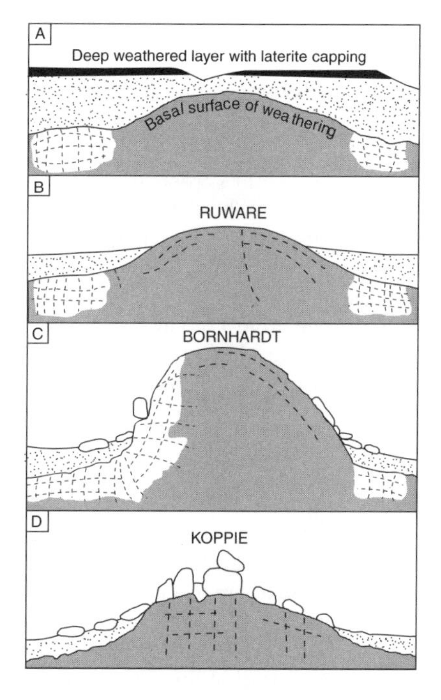

Stages in the formation of ruwares, bornhardts and koppies by the exhumation and modification of the basal surface of weathering

S

Sahel An E–W zone along the southern margins of the Sahara Desert in Africa. Although the mean annual rainfall should be as high as 500 mm in places, the Sahel is subject to severe droughts (like those of the early 1970s and 1980s), bringing crop failures, death of livestock and famine to the inhabitants. There are reasons to believe that these droughts are partly a natural phenomenon, related to long-term climatic changes, but their effects have been emphasized by human activities such as overstocking, destruction of natural vegetation, etc. See DESERTIFICATION.

saline soil A type of SOIL characterized by a high salt content. The salt (mainly sodium chloride and sulphate) is derived either from the evaporation of GROUNDWATER lying just beneath the soil surface, or is drawn towards the surface in SOLUTION by the action of capillarity. In arid regions where rates of evaporation are high, constant IRRIGATION can lead to considerable increases in soil salinity.

salinization The processes whereby sodium, potassium and magnesium salts become concentrated in the SOIL. Natural salinization is characteristic of many DESERT regions, where EVAPOTRANSPIRATION exceeds PRECIPITATION, leading to the capillary rise of salt solutions from GROUNDWATER, and precipitation of salt crystals either within the upper soil layers or as a surface crust. Salinization can also result from excessive IRRIGATION in hot, dry climates; the resulting salt-pan is poisonous to agricultural crops.

salt marsh A coastal marsh, formed within protected ESTUARIES or in the lee of coastal SPITS, bars and SAND DUNES. The salt marsh is gradually built up by the DEPOSITION of fine SILT and mud, which settles out at periods of slack TIDE. These SEDIMENTS are mainly derived from the suspended LOAD of rivers entering the sea, but some may come from the EROSION of CLAY cliffs. When the level of the marsh has been raised sufficiently, it is colonized by plants that can withstand periodic inundation by salt water (see HALOPHYTE). As the vegetation cover extends, the rate of sedimentation accelerates, and the marsh eventually reaches a level at which it is flooded only by high SPRING TIDES. Noteworthy features of fully developed salt marshes are pans (small unvegetated areas) and creek systems (CHANNELS, by way of which salt water enters and drains away from the marsh). People frequently reclaim mature salt marshes by the construction of embankments that exclude the sea, thus providing grazing for animals.

salt weathering A type of WEATHERING, now recognized as particularly important in hot DESERTS but also occurring in tropical humid climates, involving a chemical process but having a physical effect on rocks. Salt weathering results from the crystallization of supersaturated SOLUTIONS of salts occupying pore spaces and cracks. As the crystals form and grow, expansive stresses are exerted on the rock, which will be affected either by GRANULAR DISINTEGRATION or surface flaking. The process may be concentrated at certain points, giving rise to weathering pits or cavernous weathering. Salt weathering is harmful to the foundations and lower walls of buildings in hot deserts and, in the UK, lower walls and bridge supports adjacent to roads have suffered damage where rock salt is used as a de-icer.

saltation A process of transportation in rivers (see TRANSPORT), involving the continual 'jumping' of small particles (usually SAND grains) along the CHANNEL floor. Any particle lying on the river bed will impede water flow, and hydraulic pressure will be built up on the upstream side of the particle. In time, the stress will exceed the resistance of the particle (determined by its weight and shape) and it will then be set in motion, and may be thrust up into the mass of moving water. However, unless the particle is very small and light, gravity will cause it to fall back to the bed, where it may help to dislodge another particle. The term is also used to describe a similar type of transportation by winds.

sample See SAMPLING.

sample estimate See PARAMETER.

sample statistic See PARAMETER.

sampling The essence of sampling lies in the fact that a large number of items, values, individuals or locations (i.e. a statistical POPULATION) may, within specified limits of statistical PROBABILITY, be represented by a small group of items, values, etc. (i.e. by a sample) selected from that original population. A resort to sampling is necessary in many research topics, principally because of: the number of variables involved; the vast dimensions of the phenomena under investigation; the constraints of time, effort and labour. The key to success in sampling lies in adopting a procedure that permits the drawing of satisfactory and valid conclusions about the parent population from a sample of minimum size. There is a variety of sampling methods: see NESTED SAMPLING, RANDOM SAMPLING, SYSTEMATIC SAMPLING. [*f* NESTED SAMPLING]

sampling error The difference between the sample statistic of a POPULATION attribute and the actual PARAMETER. It occurs because SAMPLING is almost inevitably biased either to over- or underestimate the attribute being investigated. In RANDOM SAMPLING limits can be calculated for the sampling error.

sand Small grains, usually composed of quartz, with a diameter of between 0.06 and 2 mm. Sand is divided into fine (0.06–0.2 mm), medium (0.2–0.6 mm), coarse (0.6–1 mm) and very coarse (1–2 mm). Sandy soils comprise a mineral content of sand grains in excess of 90%.

sand dunes Generally smoothly curving features resulting from the action of the wind on deposits of SAND, most commonly found in DESERT environments or at the coast. Sand dunes may take many forms, including crescents (see BARCHAN) and linear ridges (see SEIF). They are very dynamic features whose shape and dimensions reflect the varying directions and strengths of the wind. Over time, vegetation (such as marram grass at the coast) may be able to colonize the dunes and a VEGETATION SUCCESSION takes place. Coastal sand dunes result from SEDIMENT deposited by the sea (see LITTORAL CELL SYSTEM) being reworked by the wind. Extensive sand dune complexes can indicate a relative fall in sea-level (see SEA-LEVEL CHANGE).

sandstone A SEDIMENTARY ROCK comprising mainly grains of quartz (though feldspar may also be present) that have been compacted by pressure and cemented by calcareous, siliceous or ferruginous minerals. Sandstone is thus a more coherent and resistant rock than sands. It is generally regarded as being a porous and PERMEABLE rock and, in some parts of Britain, forms an important AQUIFER.

sandur An Icelandic term for a large glacial OUTWASH PLAIN. In Iceland, these features are very extensive indeed and are criss-crossed by braided rivers. Periodically, they are swept by GLACIAL OUTBURSTS (*jökulhlaups*).

sapping See JOINT-BLOCK REMOVAL, SPRING.

satellite image See REMOTE SENSING.

satellite town See DORMITORY SETTLEMENT, NEW TOWN.

satisficer concept A concept in HUMAN GEOGRAPHY based on the argument that, instead of striving for maximum productivity and profit (to be found at the OPTIMAL LOCATION), the locational decision-maker will be prepared to accept virtually any location that offers a satisfactory level of productivity and profit (see SUBOPTIMAL LOCATION). The concept emerged largely in reaction to industrial location theories based on the assumption that the decision-maker is an all-knowing, optimizing person. Ct OPTIMIZER CONCEPT; see also BOUNDED RATIONALITY.

saturated adiabatic lapse-rate The reduction in temperature of a pocket of saturated air that rises spontaneously (under conditions of atmospheric INSTABILITY), or is forced to rise (up a mountainside or above a frontal surface). As the air pocket encounters progressively decreasing pressure, it expands and cools without exchange of heat with the surrounding air. The saturated adiabatic lapse-rate is variable, depending on the precise temperature and the amount of water vapour in the air. As the latter undergoes CONDENSATION there is release of LATENT HEAT, with the result that the saturated adiabatic lapse-rate is below the DRY ADIABATIC LAPSE-RATE and normally within the range 0.4–0.9°C for every 100 m of ascent. If the air is very moist, the rate will be at the low end of this range, since the adiabatic cooling will be partially offset by the latent heat released by condensation. [*f* DRY ADIABATIC LAPSE-RATE]

saturation-excess overland flow A type of surface RUN-OFF occurring when the SOIL has become saturated and is unable to soak up any additional water. It is most likely to occur in the lower parts of valley slopes where the WATER TABLE intersects the surface or where water from THROUGHFLOW from higher up the slope has accumulated. Ct OVERLAND FLOW.

savanna A type of vegetation that is transitional between the RAINFORESTS of the humid tropics and the short grass and scrub of the hot DESERT margins. It is widely distributed in seasonally arid parts of Africa, S America and Australia. Savanna is often regarded as synonymous with 'tropical grassland'. However, while grasses are certainly important, trees are present in variable numbers, and in the recent past may have dominated many savanna areas. Today the trees are widely scattered or are concentrated into groves and riverine FOREST. The plants of the savanna display clear adaptation to the strongly seasonal climate (for example, the rapid growth of grasses – including the 2–4 m high 'elephant grass' – during the wet season, and the deciduous habit of trees such as the acacia). However, savanna vegetation has also been much modified by fire which, over the thousands of years of occupation and exploitation by pastoralists, has reduced the extent of woodland and encouraged the growth of grasses. Many tree species have been unable to survive, and only fire-resistant trees (such as the baobab, with its thick bark and water-saturated sponge-like wood) are found in many localities. Even today, grassy savanna areas are burned deliberately each year, to clear away dead plant material and pave the way for renewed grass growth when the wet-season rains arrive. Undoubtedly, in some areas, grazing and browsing by vast herds of game (including the highly destructive elephant) have influenced present-day savanna vegetation.

scale (i) The proportion between a length on a MAP and the corresponding length on the ground; it may be expressed in words, shown as a divided line or given as a representative fraction. (ii) Relative spatial extent (i.e. spatial scale), ranging from the macro-scale (e.g. the continent), through the meso-scale (e.g. the country), to the micro-scale (e.g. the local region).

scale economies See ECONOMIES OF SCALE.

scarp A steep edge to an upland such as a CUESTA or PLATEAU. Owing to the removal of rock debris released by WEATHERING, scarps experience relatively rapid recession (hence *scarp retreat*) without significant loss of angle.

scatter diagram (scattergraph) A scatter diagram is a GRAPH used to investigate what sort of relationship, if any, exists between two VARIABLES that occur over a wide area (e.g. the relationship between BIRTH RATE and standard of living in various countries of the world). The two sets of data are plotted on the graph, and any relationship is then deduced from the pattern shown by the scatter of dots. If the dots show a RANDOM scatter with no grouping or obvious trends (A) then no systematic relationship exists between the two variables. On the other hand, the dots may show a marked tendency to form groups or clusters (B), which indicate that there are groups of countries showing similar combinations of characteristics. A rather more usual tendency is for the plotted values to lie along a line or, more commonly, within a narrow linear zone (C). Where this occurs, the calculation of the BEST-FIT LINE allows the relationship between the two variables to be expressed in the form of a mathematical formula (see REGRESSION ANALYSIS). It is possible to introduce a third variable into a scatter diagram by using a proportional symbol when plotting the points based on the first two variables (D). [*f*]

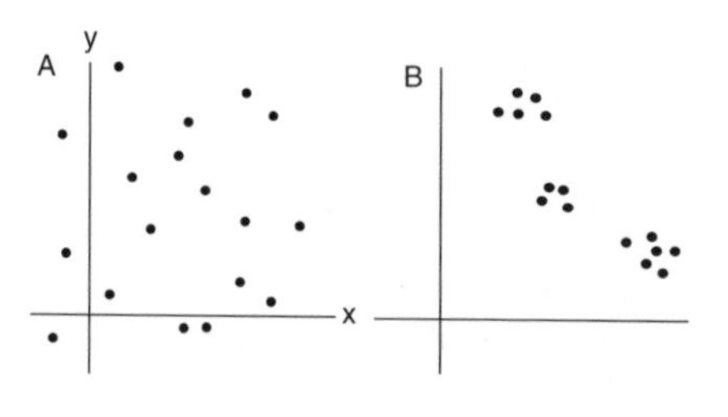

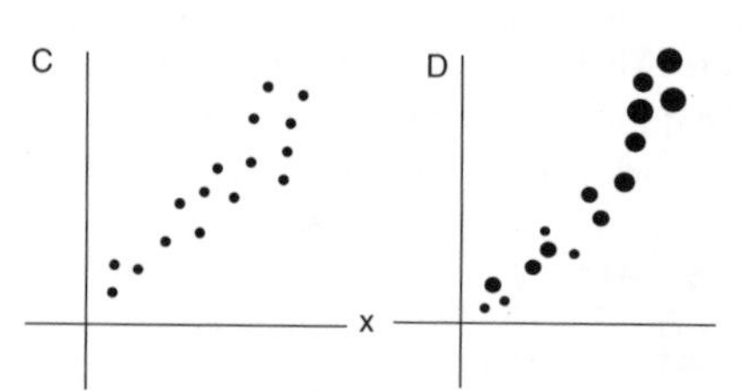

Scatter diagrams

schist A tough and resistant metamorphic rock characterized by foliation – a structure, composed of platy and elongated minerals, resembling tightly packed leaves.

science park A sort of TRADING ESTATE, usually established near a university, for the immediate purpose of encouraging the commercial exploitation of the fruits of academic research, particularly in science and technology. The even broader aim is to foster a profitable collaborative venture between academics and industrialists. Most of the activities found on a science park fall within the categories of R&D and HIGH-TECHNOLOGY INDUSTRY. The science park concept was pioneered in the USA in the 1930s, but it was not until the 1970s that the first parks were established in the UK. Cf BUSINESS PARK. Ct RETAIL PARK.

scree A slope composed of angular rock fragments, resulting from the physical WEATHERING of a FREE FACE above. Screes are usually rectilinear in overall profile, and form at the angle of repose of the constituent fragments. See also TALUS.

sea breeze A gentle cooling wind blowing from the sea to the land, usually in the afternoon or early evening. During the day, as the land is warmed by solar radiation, atmospheric pressure is slightly lowered, by comparison with the relatively high pressure maintained over the cooler sea and this leads to compensation by way of a movement of air landwards. Sea breezes bring welcome relief from the very hot conditions experienced in some parts of the world. Ct LAND BREEZE. [*f*]

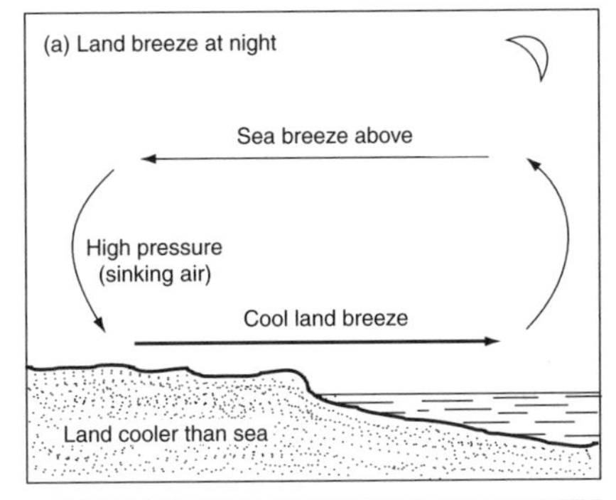

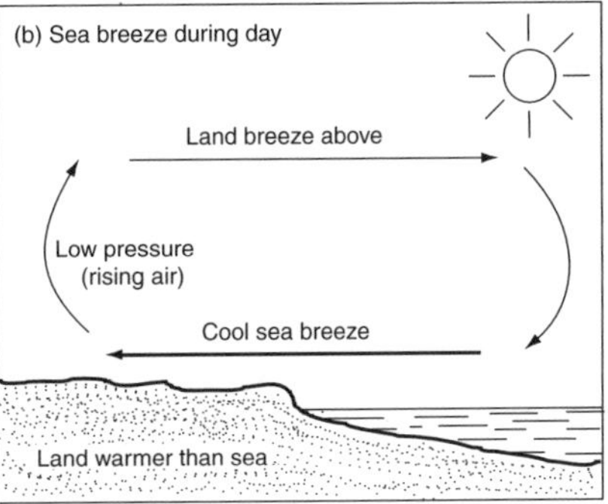

Sea breeze

sea-floor spreading A process occurring where crustal plates move apart (divergent or constructive plate boundaries – see PLATE TECTONICS), as at the Mid-Atlantic ridge and other mid-ocean divergences. Along these boundaries, large fissures open up, to become occupied by BASALT upwelling from the Earth's mantle. Thus volcanic rock is continually added to the margin of each receding plate, and a symmetrical pattern of structures on either side of the central axis is formed. Measurements at the Mid-Atlantic ridge have revealed that the sea-floor is 'spreading' at an approximate rate of 1 cm yr^{-1}. In other words, the Atlantic Ocean is becoming 2 cm wider each year. The processes associated with sea-floor spreading also occur at the splitting of continental areas (as in the Red Sea and the Gulf of California, and the E African RIFT VALLEY). [*f*]

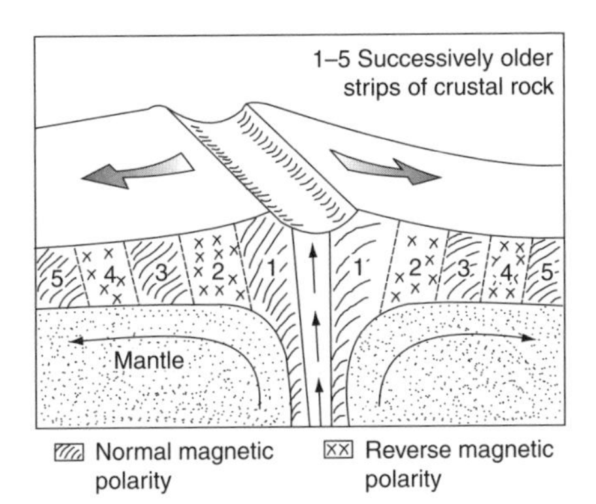

Sea-floor spreading

sea fog See ADVECTION FOG.

sea-level change A relative change in sea-level, resulting either from changes in the amount of water in the oceans (eustatic), or from the rising or falling of the land (isostatic readjustment). An apparent fall in sea-level, is indicated by the occurrence of coastal landforms such as RAISED BEACHES, abandoned cliff-lines and upraised coastal platforms, and by the REJUVENATION of rivers. Negative movements of sea-level may involve an actual lowering of sea-level (owing to the growth of ICE SHEETS and/or the deformation of ocean basins), a rise in the level of the land, or a combination of both. An apparent rise in sea-level is indicated by the occurrence of coastal landforms such as RIAS and FJORDS, coral reefs that 'grow up' from deep water, and SUBMERGED FORESTS. Positive movements of sea-level may involve an actual rise of sea-level (as in the POSTGLACIAL period, when the melting of the PLEISTOCENE ice sheets restored vast quantities of water to the oceans), the subsidence of a land mass, or a combination of the two. There is currently considerable debate about the possible rise in sea-level associated with GLOBAL WARMING.

seamount See GUYOT.

search behaviour The ways in which individuals or groups of people make decisions on particular courses of action (e.g. moving house, changing jobs), including the information on which the decision is based and the manner in which that information is assessed and interpreted.

second home A dwelling that is owned or rented by a householder whose normal residence is elsewhere. Second homes are generally used as holiday or weekend retreats and are located for the most part in RURAL areas that offer recreational opportunities (e.g. coastal and upland areas). In some areas, the incidence of second-home ownership is so high (as in parts of Denmark or by English families in parts of Wales) as to generate local opposition. The local view is that the buying up of properties by wealthy holidaymakers inflates property values and so prices local people out of the housing market. In addition, the existence of quite large numbers of dwellings in SETTLEMENTS that are empty for much of the year tends to work against the maintenance of stable and balanced rural communities. On the other hand, it is argued that properties acquired as second homes would be vacated anyway due to DEPOPULATION and that their occupation as second homes does at least help to keep settlements alive. See COUNTERURBANIZATION, RURAL TURNAROUND.

Second World See THIRD WORLD.

secondary energy See PRIMARY ENERGY.

secondary forest See VEGETATION SUCCESSION.

secondary sector That sector of the ECONOMY concerned with the processing of primary RESOURCES into usable goods (i.e. MANUFACTURING INDUSTRY). Ct PRIMARY SECTOR, TERTIARY SECTOR, QUATERNARY SECTOR; see also DEVELOPMENT-STAGE MODEL.

secondary sources See SOURCES.

secondary succession See SERE.

sector model A model of city structure, proposed by Hoyt, that stresses the importance of axial routeways. Outside the CBD, each radiating routeway is seen as spearheading the outward growth of the BUILT-UP AREA, and encouraging particular LAND USES and social groups to spread out as wedges. Thus each sector is distinctive in terms of function

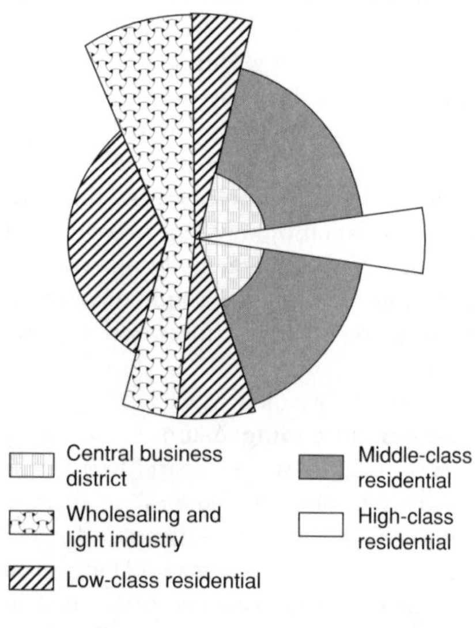

Hoyt's sector model

and, in the case of those sectors devoted to housing, each has a prevailing and distinctive social tone. Hoyt supported his model by the analysis of rent data, which showed that rent levels varied by sectors rather than concentrically. Ct CONCENTRIC ZONE MODEL; see also MANN'S MODEL. [*f*]

sector theory See DEVELOPMENT-STAGE MODEL.

sectoral balance See ECONOMIC GROWTH.

sedentary agriculture AGRICULTURE as practised in one place by a settled farmer. Strictly speaking, virtually all agriculture is now sedentary, but the term was used of early agriculturists in Africa who farmed the same land indefinitely, and therefore in contrast to SHIFTING CULTIVATION.

sediment Particles of rock that have been transported and laid down by agencies such as rivers, glaciers, ICE SHEETS, waves and wind (hence the process of *sedimentation*).

sediment budget The identification and quantification of individual components of a *sediment-transfer system* (a geomorphological process such as beach erosion and deposition, or agent such as a valley glacier). For example, a GLACIO-FLUVIAL sediment-transfer system comprises inputs from SUPRAGLACIAL erosion (*A*) and glacial erosion (*B*); a store of sediment on and within the glacier (*H*), and beyond the glacier snout – the PROGLACIAL store (*K*); and outputs from SUBGLACIAL deposition (*Z*), MORAINE formation (*Y*) and meltwater sediment load (*X*). Thus the sediment budget, or sediment balance equation, may be expressed as follows:

$$A + B = X + Y + Z \pm \text{changes in } H + K$$

For practical purposes it is often difficult to quantify all the components of a sediment budget; thus it is virtually impossible to estimate realistically some sediment stores (such as *H* above), though changes in storage can be determined from successful measurement of INPUTS and OUTPUTS to the system.

sediment yield A calculation of the sediment LOAD (usually, for practical reasons of measurement, the suspended sediment only) transported by a river from its basin. Sediment yield is expressed in terms of volume or weight of sediment per unit area per year ($m^3\ km^{-2}\ yr^{-1}$ or tonnes $km^{-2}\ yr^{-1}$). It is an expression for the rate of denudation over the drainage basin. In a study of sediment yield, erosion rates were found to be particularly high (sometimes exceeding 3000 tonnes $km^{-2}\ yr^{-1}$) in the humid tropics, but much less in the hot DESERT regions where surface RUN-OFF is low (0–10 tonnes $km^{-2}\ yr^{-1}$). In temperate and cold regions, rates were generally in the range 10–60 tonnes $km^{-2}\ yr^{-1}$, except in mountainous areas where substantial increases occurred (60–600 tonnes $km^{-2}\ yr^{-1}$).

sedimentary rock One of the three major groups of rock (see also IGNEOUS ROCK, METAMORPHISM). Sedimentary rocks are usually deposited in distinct layers or strata (hence *stratification*), separated by BEDDING PLANES. They are formed in a variety of ENVIRONMENTS, such as deep oceans and seas, along coasts, and in inland basins, lakes and large river valleys. Most of the material in sedimentary rocks consists of debris from the breakdown of other rocks (igneous, metamorphic and older sedimentaries). The processes by which sediments are compacted, cemented and hardened into coherent rock are referred to as *lithification* (or *lithifaction*).

seed bank A repository of seeds or plant tissue obtained from a wide variety of primitive strains and wild crop varieties, and stored as a future RESOURCE to help maintain BIODIVERSITY. Also known as a *gene bank* or *germplasm bank*.

seedbed growth The spawning of new FIRMS within an AGGLOMERATION of linked industries, and the tendency for such firms to remain there at least during their early stages of existence. The phenomenon is explained partly by the fact that because the new ventures are small and have limited resources, they are encouraged by the availability of relatively cheap rented accommodation and of specialist services that they are able to share with other firms in the agglomeration (see AGGLOMERATION ECONOMIES). While many of these ventures are short-lived and those that do succeed are likely to move out of the agglomeration into better premises, the vacated premises become available for subsequent newcomers. The term *incubator hypothesis* has been used to describe this concentration of births of new firms. See SEEDBED LOCATION.

seedbed location An area where conditions favour the birth of new industrial enterprises. Typically, seedbed locations are to be found in declining INNER CITY areas where there is an abundant supply of cheap premises that can easily be converted into small factories. See SEEDBED GROWTH.

segmented economy An ECONOMY comprising a variety of FIRMS, ranging in size from TNCs to small backyard workshops.

segregation In an ecological sense, the spatial separation of different groups of people and different functions into distinct areas, e.g. the emergence of distinct social areas and different LAND USE regions within the BUILT-UP AREA of a TOWN or CITY. Segregation results partly from the repelling force that operates between different activities (and social groups) and partly from the mutual attraction that exists between similar activities (and people). Cf SEPARATISM, SOCIAL POLARIZATION.

seif A type of longitudinal DUNE found in hot DESERTS and following the direction of the prevailing winds. Like a BARCHAN, a seif is commonly asymmetrical in cross-profile, and it therefore seems that secondary crosswinds play some part in its formation. One theory is that some seifs may actually result from the modification and amalgamation of barchans. Where cross-winds blow from time to time, the windward horn of a barchan will receive greater increments of SAND and will become extended, so much so that it may coalesce with another barchan downwind of it. The steep slip faces of the barchans also become oriented across the path of the secondary winds. When dune amalgamation occurs on a large scale (as is possible in 'fields' of barchans) longitudinal dunes of considerable extent can result. Once in existence, longitudinal dunes are maintained by winds from the prevailing direction 'sweeping clear' sand from the depressions between the dunes. [*f*]

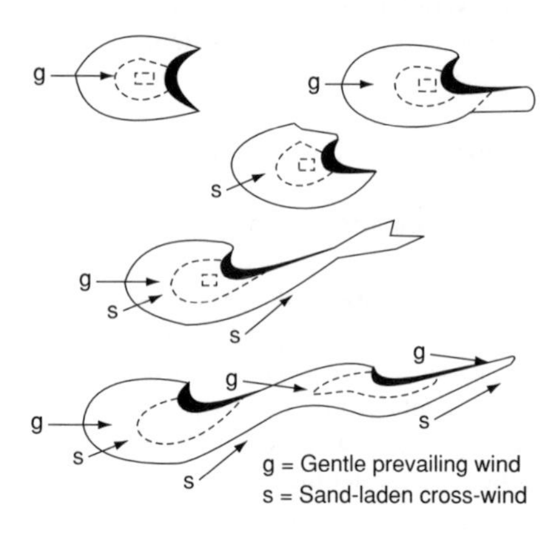

The transformation of barchan into a seif dune by a cross-wind

seismology The scientific study and interpretation of EARTHQUAKES. Hence seismologist (a student of earthquakes), seismic (a descriptive term for earthquake processes) and seismograph (an instrument for recording earthquake shock-waves).

selective migration The MIGRATION of a particular group or type of person, as by young people seeking work or by a persecuted ethnic minority. At an international level, there are governments that pursue policies of selective migration. For example, entry into Australia is carefully controlled, the issue of residential visas being largely restricted to people who command specified job skills and who fall within a prescribed age range. Such policies clearly represent a form of DISCRIMINATION.

self-regulation See NEGATIVE FEEDBACK.

selva An alternative term for tropical RAINFOREST, originally applied to the Amazon Basin but now used more generally.

sensible heat Heat energy within the atmospheric HEAT BALANCE that can be transferred either by direct conduction or by convection involving the rising of warm air.

sensible temperature The temperature, not as recorded by the dry bulb thermometer, but as 'felt' by people. Sensible temperature is considerably influenced by atmospheric humidity and wind speed. Thus, if the humidity is high (in the order of 90%), very warm temperatures of 25°C and above feel extremely uncomfortable and oppressive. Conversely, when cold air is very moist, it gives the impression of being unpleasantly cold and 'raw'. Again, when the wind blows strongly at low temperatures, the so-called 'wind chill' factor comes into play, and the danger of humans suffering from exposure is greatly increased. See COMFORT ZONE.

separatism The wish of a particular ETHNIC GROUP, usually a minority, to form their own sovereign state, e.g. as pursued by the Basques in Spain and the French-speaking people of Quebec (Canada). Cf SEGREGATION.

serac See ICE FALL.

sere The sequential development of PLANT COMMUNITIES in a particular locality (e.g. a SALT MARSH) leading to the attainment of the CLIMATIC CLIMAX VEGETATION. Each stage in the sequence that precedes the climax is known as a seral stage. Primary plant successions (see VEGETATION SUCCESSION) colonizing virgin ground are known as *prisere*. Where the sites are initially very dry, for example, the surface of bare rock *xeroseres* are developed. In very wet sites – for example, along the margins of a lake – *hydroseres* are formed. Where previously vegetated sites are interfered with by burning or by clearance for cultivation, and subsequently recolonized, the vegetation succession will be different (referred to as a *secondary succession*). This is common in tropical RAINFORESTS, leading to the formation of secondary forest. [*f*]

service centre A synonym for a CENTRAL PLACE.

service industry See TERTIARY SECTOR.

Setaside An EXTENSIFICATION scheme currently operating within the EU whereby farmers are paid to take land out of agricultural production. It was introduced as one way of helping to reduce surplus food production within the region. Clearly, there are CONSERVATION benefits (no application of FERTILIZERS and pesticides, undisturbed areas for wildlife) to be gained from this particular retreat from farming. See CAP.

settlement (i) Any form of human habitation, usually implying more than one dwelling, although most would regard a single, isolated building as constituting a settlement. (ii) The opening up, colonizing and settling of a hitherto unpopulated or thinly populated land, especially by immigrants to a 'new' country.

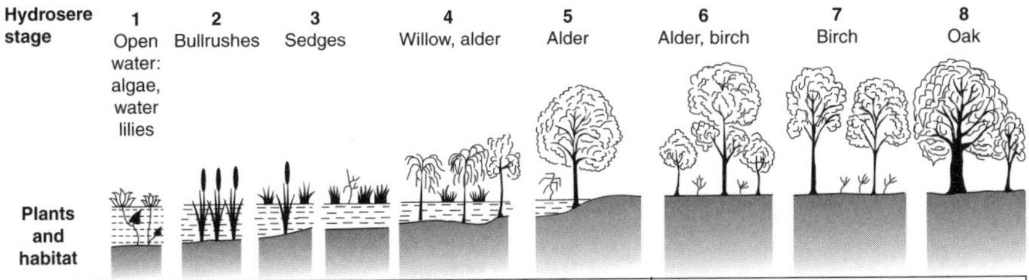

Hydrosere stage	1 Open water: algae, water lilies	2 Bullrushes	3 Sedges	4 Willow, alder	5 Alder	6 Alder, birch	7 Birch	8 Oak
Plants and habitat								

Habitat description	Reed swamp	Marsh or fen	Open wooded fen	Closed wooded fen	Woodland		
Habitat processes	Accelerated deposition of silt and clay. Floating raft of organic matter forms and thickens		Raft now a mat resting on mineral soil	Black mineral soil revealed in patches. Earthworms	Ground level now above water table; oak seedlings	Birch canopy forms; oak saplings	Oak grows through and then over the birch
pH level	–	–	7.3		4.3	3.7	–
Number of species of plant	6	10	14	26	18	14	10

Hydrosere at Sweet Mere, Ellesmere, Shropshire

settlement form The spatial characteristics of an individual settlement, whether it is *nucleated* (as is the case with a VILLAGE or TOWN) or whether it is *dispersed* (where settlement takes the form of isolated dwellings and small HAMLETS). The shape of individual settlements, and their internal arrangement of buildings and activities are other important aspects of settlement form. See DISPERSED SETTLEMENT, NUCLEATED SETTLEMENT; ct SETTLEMENT PATTERN.

settlement pattern This term is strictly applied to the spatial arrangement or DISTRIBUTION of SETTLEMENTS within a given area, as distinct from SETTLEMENT FORM, which relates more to the spatial characteristics of individual settlements. Sometimes, however, the term is taken to embrace both aspects. In the investigation of settlement patterns (using NEAREST-NEIGHBOUR ANALYSIS or QUADRAT ANALYSIS), a threefold classification is commonly adopted: (i) *uniform* or *regular* (where the settlements are evenly spaced and begin to approach the geometric arrangement assumed in CENTRAL PLACE THEORY); (ii) *nucleated* or *clustered* (where settlements are unevenly distributed and tend to cluster in a part or parts of the study area); (iii) *random* (where it would seem that the location of any one settlement is in no way influenced by the location of other settlements in the study area; i.e. the distribution is a chance one).

[*f* NEAREST-NEIGHBOUR ANALYSIS]

settlement-size frequency distribution See CITY-SIZE DISTRIBUTION.

seventh approximation See SOIL CLASSIFICATION.

shale A SEDIMENTARY ROCK formed by the compaction of fine muds and characterized by *laminations* (thin layers that easily split apart). Shale is generally a weak, IMPERMEABLE rock that is easily eroded and tends to form areas of low RELIEF.

shanty town An area of substandard housing, often occupied by SQUATTERS and found mainly in THIRD WORLD cities. Usually constructed either at the city margins or on difficult ground (e.g. steep slopes, areas prone to flooding) within the city, hitherto avoided by the BUILT-UP AREA. While shanty towns are sometimes referred to as *squatter settlements* or *shanties*, they frequently have local names, such as the *barriades* of Peru, the *favelas* of Brazil and the *villas miseries* of Argentina.

Shanty towns result mainly from massive rural–urban MIGRATION and from the inability of city authorities to provide sufficient housing, services and employment for the vast influx of people. Thus the new immigrants are forced to build 'temporary' dwellings for themselves (usually rudely constructed from scrap materials) and thereby become squatters. Such areas of densely packed, shack housing, initially at least, lack basic physical amenities such as piped water, sewerage and power supplies, as well as being unserved by educational and health facilities. Consequently, the shanty town all too often represents a concentration, not just of slum housing, but also of poverty, illiteracy and high mortality. However, there is some evidence from Latin America of a degree of self-improvement in the longer-established shanties, as the residents make attempts to

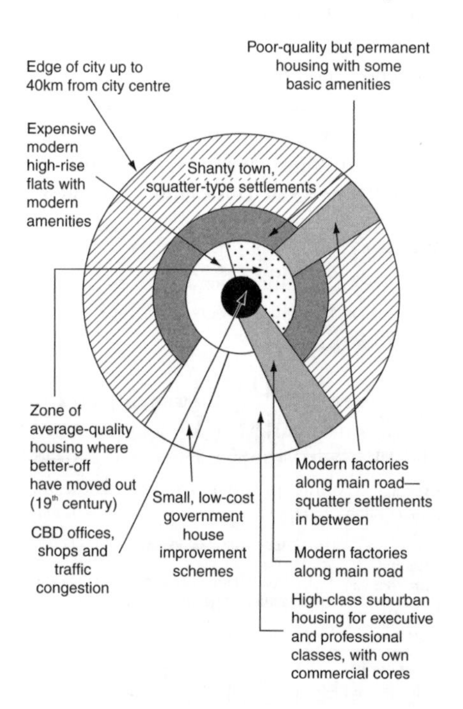

Model of Third World city showing shanty towns and squatter settlements

introduce communal basic services. Another process of change is the deliberate clearance of shanty towns by city authorities and their replacement by government-financed, low-cost housing schemes. [*f*]

shape index A formula for assessing the spatial outline of a geographical phenomenon (e.g. shape of a town's BUILT-UP AREA, of a country or region) and for comparing the shapes of different examples of the same phenomenon. The shape index is

$$\frac{(1.27\ A)}{L}$$

where A is the area of the shape being investigated and L the length of its longest axis. Basically, this index measures the degree to which the individual shape deviates from a circle, with a value of 1 indicating circularity and lesser values increasing elongation.

share cropping A type of agricultural tenancy in which the owner of the land supplies seed and FERTILIZER to the tenant farmer and takes a percentage of the produce. Although associated with THIRD WORLD farming, it is still practised in parts of France where it is known as *métayage.*

shatter belt A clearly defined, narrow zone in which the rocks have been crushed and broken as a result of intense pressures on either side of a FAULT. Where shatter belts are exposed on CLIFF-faces they have been etched by wave attack into narrow inlets or GEOS. Elsewhere, they have been exploited as lines of weakness by rivers.

shear strength The maximum resistance of a material – for example, of a rock or sediment deposit – to the application of stress. When resistance in the form of cohesion and friction are overcome, as for example by an increase in water content or a rise in the WATER TABLE leading to changes in pore water pressure, collapse or 'failure' may occur in the form of a LANDSLIDE or MUDFLOW.

sheet erosion A form of EROSION that affects the whole slope surface, by contrast with the more concentrated forms of slope erosion associated with rills and gullies. Sheet erosion results from the impact of raindrops (RAINDROP EROSION) and the flow of a surface layer of RAINWASH over the ground.

sheet flow (flood) A form of OVERLAND FLOW commonly developed on gently sloping, unchannelled areas in DESERTS (see PEDIMENT) where water is spread widely as a thin layer over the ground surface. On rock PEDIMENTS and BAJADAS at the base of MOUNTAIN FRONTS, sheet flows either form directly from heavy rainstorms of the convectional type, or result indirectly from the transformation of a stream-flood emerging from a valley incised into the mountain front.

sheltered housing Housing for the elderly and/or people with disabilities, where the services of a warden are provided. Each tenant's accommodation is usually self-contained, but there may also be some common or shared facilities. It is a type of housing increasingly being provided in Britain both by private companies and local authorities in response to the needs of a progressively 'greying' (i.e. ageing) population.

shield A large rigid block of ancient rocks, Pre-Cambrian in age. Shields have been affected in the distant geological past by intense folding movements, powerful METAMORPHISM and vulcanicity, but over a long period of time have been reduced to PENEPLAINS. In some cases they have been covered or partially covered by younger SEDIMENTARY ROCKS that have themselves since been stripped away by EROSION.

shield volcano A volcanic cone formed from basic LAVA and characterized by a large basal diameter and very gentle slopes. For example, Mauna Loa in Hawaii has a base 480 km in diameter at the ocean floor, a height from base to summit of 9750 m (of which 4171 m is above sea-level), and slopes that range in angle from 2° near the base to 10° at the summit of the cone. Shield VOLCANOES are generally associated with fiery but non-violent ERUPTIONS.

shifting cultivation Cultivation of a small area of land by a nomadic group for a few years or until the SOIL becomes exhausted. The group then moves on and clears a fresh piece of land, usually by burning the vegetation and digging in the ashes, leaving the abandoned areas to become overgrown. Frequently, the abandoned, unprotected soil may be subject to rapid SOIL EROSION. It is sometimes possible for a site to be re-occupied later, when the soil has had time to recover its fertility by natural means. This type of AGRICULTURE was once very common in E Africa and SE Asia, but is much less so nowadays. Sometimes referred to as *bush fallowing*.

Shimbel index A measure used in NETWORK ANALYSIS. It involves recording on a MATRIX the number of shortest-path links between each NODE and all other nodes in the NETWORK. The total number of shortest-path links in the network is the *gross accessibility index*, while the total for each node is called the Shimbel index. In both cases, the lower the index value, the greater the degree of predicted ACCESSIBILITY.

shingle A mass of stones, rounded to a greater or lesser degree by ATTRITION, forming a coastal BEACH.

shrinkage of distance The reduction in time that it takes to cover a unit of distance brought about by advances in TRANSPORT technology. See TIME–DISTANCE CONVERGENCE.

shopping centre The concentration of RETAILING and other service activities at a nodal and accessible point. The shopping centres found within a large CITY range in scale from the CBD, through the suburban district centre, to the small cluster of shops at the street corner. All such shopping centres may be viewed as intra-urban CENTRAL PLACES and differentiated in a hierarchic manner (see CENTRAL PLACE HIERARCHY) on the basis of the range and quality (i.e. ORDER) of retailing services provided. See also HYPERMARKET, RETAIL PARK. Ct RETAILING RIBBON.

sial The uppermost layer of rocks in the Earth's crust, largely granitic in nature and composed mainly of silica (*si*) and aluminium (*al*) minerals, hence the term sial. The density of these rocks (which are most extensively developed beneath the continental land-masses) is 2.6 to 2.7. See also SIMA.

significance As used in STATISTICS, significance relates to the PROBABILITY that a NULL HYPOTHESIS is true, as determined by a SIGNIFICANCE TEST.

significance level In STATISTICS, a NULL HYPOTHESIS is rejected if the calculated PROBABILITY exceeds a given value of α, which is called the significance level. Depending on the exact context, the result may be deemed to be significant if α is < 0.05 (i.e. at the 95% CONFIDENCE LIMIT) and highly significant if α is < 0.01 (i.e. at the 99% confidence limit).

significance tests These are employed in statistics, usually in conjunction with a NULL HYPOTHESIS, to determine whether a significant relationship exists, as for example between two or more VARIABLES (or data sets) or whether the relationship is one of chance association. They are also used to test hypotheses about POPULATIONS, based on evidence from samples. A variety of tests may be applied, depending on the nature of the data, i.e. whether it is parametric (see PARAMETRIC TESTS) or non-parametric (see NON-PARAMETRIC TESTS). In the case of testing a null hypothesis, if the statistical result shows a value significantly greater at the required level of significance, then it may be assumed, with the appropriate level of PROBABILITY, that the relationship between the two variables, and the sample differences, are not the result of chance.

sill An igneous intrusion, comprising a sheet of MAGMA that has been forced between the BEDDING PLANES of a SEDIMENTARY ROCK. Sills usually consist of relatively hard rock (such as DOLERITE), and give rise to bold and craggy SCARP-faces and (where crossed by rivers) WATERFALLS. A well-known example is the Great Whin Sill in N England which is followed for some distance by Hadrian's Wall. [*f* LACCOLITH]

silt Fine mineral particles with a diameter in the range 0.002 to 0.06 mm (in other words, intermediate between CLAY and SAND). Silt is particularly common in glacial deposits, resulting from the process of ABRASION by glaciers and ICE SHEETS.

sima The layer of rocks in the Earth's crust underlying the SIAL. It is largely basic in composition and is composed mainly of silica (*si*) and magnesium (*ma*) minerals, hence the term sima. The density of the sima rocks (which form the floors of the oceans) is 2.9–3.3.

sink estate A public housing estate in which the local authority concentrates 'difficult' households that have proved socially disruptive elsewhere.

sink hole (shakehole) A generally dry depression formed by SOLUTION processes or subsurface collapse in LIMESTONE. Sink holes should not be confused with SWALLOW HOLES or swallets, where surface water (usually a stream that has flowed off a neighbouring IMPERMEABLE rock) disappears down a shaft into the limestone. See DOLINE.

sinuosity ratio See MEANDER.

site Used in URBAN GEOGRAPHY to denote the ground covered by the BUILT-UP AREA of a SETTLEMENT. Its physical characteristics, such as the alignment of waterfronts, drainage and slopes, can play an important part in moulding the structure of the built-up area. Ct SITUATION.

site of special scientific interest See SSSI.

situation The location of a SETTLEMENT in relation to its wider surroundings. Ct SITE.

sixth-power law See COMPETENCE.

skewed distribution The degree of asymmetry shown by a FREQUENCY DISTRIBUTION curve, i.e. the extent to which the MEAN differs from the MEDIAN. Skew may be either positive (i.e. skewed to the left) or negative (i.e. skewed to the right).

[*f* FREQUENCY DISTRIBUTION]

skid row An American term denoting a SLUM area of a city characterized by, and ostensibly catering for, a high concentration of 'drop-outs' from URBAN society. There is much inadequate single-person accommodation and many cheap eating places, bars and liquor stores, as well as a high incidence of alcoholism, drug abuse, prostitution and crime.

slate A type of fine-grained metamorphic rock, resulting from the intense compression of SHALE or mudstone, and characterized by the development of CLEAVAGE. Slate is a resistant rock commonly associated with areas of mountainous RELIEF, for example, parts of Snowdonia, N Wales.

slash and burn See SHIFTING CULTIVATION.

sleet A form of PRECIPITATION that is transitional between rain and snow, comprising a mixture of snowflakes and raindrops or partially melted snow. It usually occurs when temperatures are slightly above 0°C. For example, showers of sleet in Britain are commonly associated with the advent of polar maritime air during early spring.

slip-off slope A gentle valley slope formed where a river has experienced both downcutting and lateral shifting. Slip-off slopes are also associated with ingrown meanders. As the MEANDER is developed, the river forms a river CLIFF on the outside of the bend, while on the inside of the bend, the slip-off slope is represented by a gently sloping spur, sometimes overlain by alluvial deposits.

slope-over-wall (bevelled) cliff A sea-cliff in which the lower part comprises a steep or vertical FREE FACE, and the upper part a gentler slope, often rectilinear in profile and frequently covered by rock fragments and SOIL. Slope-over-wall cliffs in Britain are most commonly associated with changes in sea-level. During the PLEISTOCENE era, a fall in sea-level led to the cessation of marine EROSION, and the former sea-cliff was reduced by frost WEATHERING and SOLIFLUCTION, only to be revived and undercut by renewed wave attack during a subsequent rise in sea-level. Many examples are found around the western coasts of Britain. A slope-over-wall cliff can also form where the lower cliff is developed in a MASSIVE rock formation that is directly attacked by the waves, and the upper part is formed by a weaker rock that has been weathered subaerially to a gentler angle.

slope replacement A model of slope evolution in which it is envisaged that as a slope develops, some parts of the slope replace other parts. For example, where a steep slope unit occurs below an upper, gentler slope unit, the more rapid recession of the former, resulting from the more effective evacuation of weathered material, will eventually destroy the latter.

slum An overcrowded and squalid neighbourhood of grossly substandard housing and inadequate services. See INNER CITY, INNER-CITY DECLINE, SHANTY TOWN.

slump (i) A type of MASS MOVEMENT, on slopes and sea-CLIFFS, in which material moves over a plane of sliding (or slip plane), but in doing so loses its coherence, usually as a result of high water content. As a result the movement at the bottom of the slump is transformed into a flow. The combination of processes results in an arcuate scar at the head, a linear tongue of mobile material, and a bulging 'toe' produced by flowage. Slumps are frequently observable on newly constructed road cuttings in weak SANDS and CLAYS that are inadequately drained. (ii) See BUSINESS CYCLE.

small business A small business is defined in Britain as one employing between 1 and 100 workers. At present, there is much interest in such FIRMS, principally on three counts: (i) the important part they have to play in the process of change in modern ECONOMIES, since they are regarded as fertile sources of innovation and new technology; (ii) the fact that a healthy population of small firms ensures competition in production and diversification of the demand for different sorts of LABOUR; (iii) most contentiously, the capacity of such firms to create new jobs and thereby counteract unemployment. Although it is clear that the small-business sector does make a contribution to the net increase in jobs, that contribution does tend to be overstated. It has been estimated that they contribute only between 10% and 20% of total new jobs, and that in this respect they do not compare with large firms (employing more than 500 people). The problem is that, while many small businesses are dynamic and flourishing, only a small proportion of them ever graduate into stable, medium-sized enterprises; even so, the progression takes time.

smog A type of RADIATION FOG characterized by a large content of soot particles and atmospheric sulphur dioxide (which may impart a yellowish tinge to the smog). Smogs were once relatively common over large industrial cities in Britain, but have now virtually disappeared as a result of clean air campaigns and the increasing use of smokeless fuels. More recently, smogs have become increasingly associated with atmospheric concentrations of fumes from car exhausts. See PHOTOCHEMICAL SMOG.

snow A type of PRECIPITATION formed when atmospheric water vapour condenses at temperatures below 0°C. The resultant minute spicules of ice join together to form larger crystals (hexagonal plates or prisms), which themselves aggregate into snowflakes. See also SLEET.

snow line The lowest altitude of a more or less continuous cover of SNOW. The snow line is often clearly visible in mountainous regions, though it may be variable in height as a result of differences in PRECIPITATION from place to place. It is also locally influenced by TOPOGRAPHY and aspect – for example, the snow line is usually much lower within deep gullies on north-facing slopes in the Northern Hemisphere. The permanent snow line is developed at the height where ABLATION of snow during the summer thaw is just sufficient to remove winter snow accumulation. Above this level, as reduced temperatures lead to lower ablation rates, some winter snow will persist through the following summer. On glaciers and ICE SHEETS, the permanent snow line (which in the Alps is at an average elevation of about 3000 m) is coincident with the EQUILIBRIUM LINE.

snowpatch erosion See NIVATION.

social area Usually taken to mean a part of the residential area of a TOWN or CITY containing people of the same SOCIAL CLASS sharing the same level of living, the same way of life and the same ethnic background. Thus a person living in one social area may be expected to differ significantly in terms of characteristics, attitudes and behaviour from any person living in another type of social area.

Social Charter Measures to harmonize social legislation within the EU. A total of 12 different areas are covered by the charter: freedom of movement, social protection, vocational training, health and safety, the elderly, the protection of children and adolescents, people with disabilities, sex equality, living and working conditions, employment and remuneration, collective bargaining, and the right to strike.

social class A group of people conscious of certain common traits (e.g. background, education, attitudes, language) and of certain common ways of behaviour that distinguish them from members of other social classes with other traits and ways of behaviour. In Britain, four classes are generally recognized in the social hierarchy: (i) *upper class*: – largely made up of the established landed aristocracy, still rich and powerful, but perhaps not quite so much today as formerly; (ii) *upper middle class* – consisting of the most prosperous and influential sections of the community, frequently professional people and with sources of unearned income; (iii) *lower middle class* – made up of many intelligent and educated people, depending largely on earned income to sustain their standard of living; (iv) *working class* – the most numerous of the four classes, consisting largely of skilled and unskilled manual workers, and depending on wages. The upper and upper middle classes own a large proportion of the nation's capital, and it is this that determines their influence in society, rather than intelligence, education or culture.

social costs The costs of some activity that are borne by society as a whole and that need not be restricted to the costs borne by the individual or FIRM carrying out that activity. For example, the social costs of a car journey are in addition to the costs incurred directly by the motorist, and include such items as contributing to traffic congestion and atmospheric pollution, as well as inflicting wear and tear on the road system.

social geography Sometimes wrongly equated with the whole of HUMAN GEOGRAPHY. Social geography today might be seen as involving the investigation and understanding of the following: the different bases for recognizing social groups (e.g. ethnicity, SOCIO-ECONOMIC STATUS); the patterns produced by the spatial distribution of different social groups; the behaviour of different social groups in the context of space (e.g. residential DECISION-MAKING); the processes that operate in society (e.g. DISCRIMINATION, SEGREGATION); those problems of contemporary society that have a spatial or environmental dimension (e.g. DEPRIVATION, poverty, crime).

social indicators See TERRITORIAL SOCIAL INDICATORS.

social justice See TERRITORIAL JUSTICE.

social mobility See MOBILITY.

social polarization The result of forces of division and SEGREGATION within a society and in which the extremes of the social spectrum are represented by large and very different groups of people.

social satisfaction See QUALITY OF LIFE.

social services These are particularly concerned with helping those who are least able to help themselves, i.e. the young, the elderly, the sick and those with disabilities. They are distinguished from other public and voluntary services, and from commercial enterprises, by the fact that their main aim is promoting the WELL-BEING of individuals or groups. So social services include personal health services, residential and day care for children deprived of normal home life, youth and community services, local authority and SHELTERED HOUSING. They are an essential part of the Welfare State.

social stratification The hierarchic ranking of people within a society on the basis of SOCIAL CLASS and SOCIO-ECONOMIC STATUS.

social welfare See WELFARE.

social well-being See QUALITY OF LIFE, WELL-BEING.

socio-economic status A stratification or classification of society based on a combination of social characteristics (e.g. family background, education, values, prestige of occupation) and on economic standing (i.e. income, STANDARD OF LIVING). On the basis of socio-economic status, the British Census recognizes 17 socio-economic groups: (i) employers and managers in large establishments; (ii) employers and managers in small establishments; (iii) professional workers – self-employed; (iv) professional workers – employees; (v) intermediate non-manual workers; (vi) junior non-manual workers; (vii) personal service workers; (viii) foremen and supervisors – manual; (ix) skilled manual workers; (x) semi-skilled manual workers; (xi) unskilled manual workers; (xii) own-account workers (other than professional); (xiii) farmers – employers and managers; (xiv) farmers – own account; (xv) agricultural workers; (xvi) members of armed forces; (xvii) others.

soil The thin surface layer of the Earth, frequently less than a metre in thickness, comprising closely intermixed mineral and organic substances. Soil is therefore quite different from REGOLITH, which consists of weathered mineral particles only. The principal constituents of soil – present in varying proportions from place to place – are solid mineral particles (for example, SAND, SILT and CLAY derived from WEATHERING), HUMUS (from the decay of plants), dead and living organisms (such as worms and micro-organisms), solutes, water and air. Soil is the product of complex processes, and takes a considerable time to develop to 'maturity'. The factors involved in soil formation can be expressed by the formula

$$s = f(c, p, v, r)t^{o}$$

in which s = soil characteristics, f is a symbol for function of, c is climate, p is PARENT MATERIAL, v is vegetation, r is RELIEF, and t^{o} is time factor.

soil classification The identification and ordering of SOIL types, with the aim of increasing understanding of regional variations in soils and soil-forming processes, and also to aid in the task of soil mapping. The problems of classification are many, since various criteria can be adopted (such as soil depth, SOIL TEXTURE, SOIL STRUCTURE soil chemistry, soil fertility, etc.). Early attempts at classification were based on identification of the main soil-forming environments, and on recognition of ZONAL SOILS related to the principal climatic-vegetation zones of the Earth's surface. In the early 20th century, Russian soil scientists made a fundamental distinction between zonal soils and AZONAL and INTRAZONAL SOILS. The most ambitious and complicated scheme is the US Department of Agriculture Soil Survey *7th Approximation Classification* (so called because it was the seventh and final attempt to produce an 'ideal' classification), in which ten major soil orders, with several subdivisions, were established.

soil creep Probably the most widespread form of MASS MOVEMENT. Soil creep has been defined as the slow downslope movement of SOIL and/or REGOLITH as a result of the net effects of movement of its individual particles. The rate of movement is very slow (in the order of 1 mm yr^{-1} in the uppermost soil layer, and less at depth), but the results are usually evident on steep slopes (the bending of tree trunks, cambering of strata, and the initiation of TERRACETTES). Soil creep is influenced by various factors other than steepness of slope. For example, it is increased by high PORE WATER PRESSURE, and reduced by vegetation, whose roots bind and stabilize the soil. The disturbances that set the soil particles in motion are believed to be (in order of importance in humid temperate climates): wetting and drying (involving the expansion and contraction of CLAY minerals); freeze-thaw activity; the action of worms and burrowing animals; heating and cooling above 0°C; and the growth and decay of plant roots. [*f*]

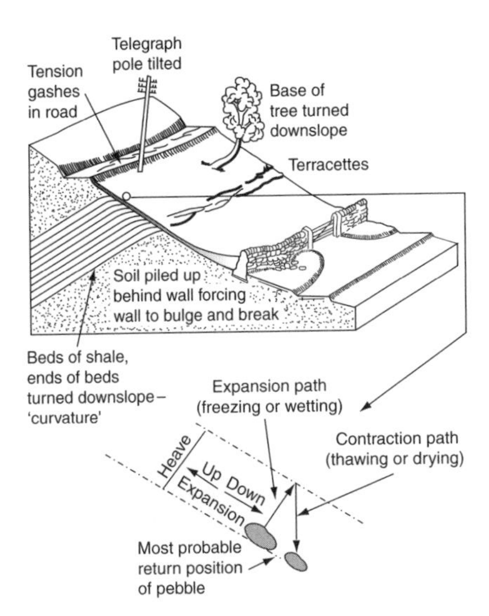

The effects of soil creep

soil erosion The removal of SOIL by processes such as gullying, RAINWASH and DEFLATION. Soil EROSION is thus a form of ACCELERATED EROSION, resulting from such activities as DEFORESTATION, OVERGRAZING and ploughing, which leave the soil surface bare and unprotected from rainfall.

soil horizon A visible 'layer' within a mature SOIL, identifiable in the first instance by its colour but on closer examination seen to reflect differences in mineral composition and HUMUS content. Soil horizons are sometimes clearly demarcated (as in PODSOLS), but on other occasions grade into each other. See also A-HORIZON, B-HORIZON, SOIL PROFILE.

soil moisture The moisture contained within the pore spaces of a SOIL, thus providing one of the vital needs for plant growth. Soil moisture deficit refers to the degree to which soil moisture content falls below FIELD CAPACITY. During late winter in Britain, after considerable rainfall and little EVAPOTRANSPIRATION, zero soil moisture deficit exists. However, through the summer soil moisture deficit increases, as evapotranspiration exceeds cumulative rainfall.

soil profile A vertical section through a soil, revealing its crudely layered structure (see A-HORIZON, B-HORIZON). The individual horizons are identifiable initially by their colour, but on closer examination are seen to be different in mineral composition and HUMUS content. Sometimes the horizons are clearly demarcated (as in a PODSOL), sometimes they grade into each other (as in a BROWN FOREST SOIL), and sometimes the horizons are poorly developed or absent altogether (as in IMMATURE SOILS). The formation of distinctive soil profiles is due to the operation of particular soil-forming processes, particularly LEACHING, ELUVIATION and ILLUVIATION. Since, within larger areas of uniform climate, these processes tend to act more or less uniformly (giving rise to pedogenic regimes), broad regional contrasts in soil profiles are identifiable (see ZONAL SOIL). [*f*]

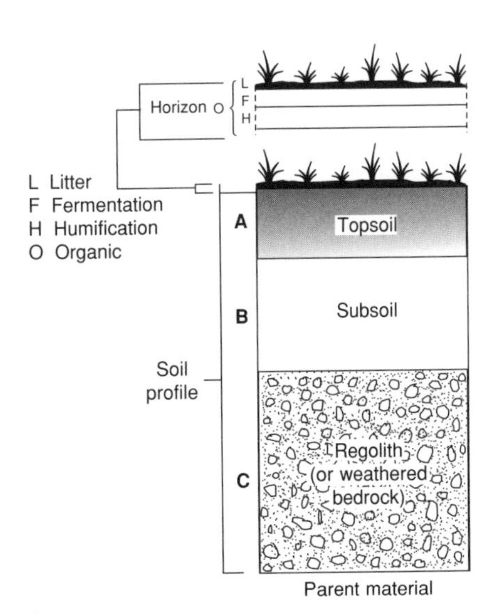

Soil profile

soil structure The manner in which SOIL grains are aggregated together into larger pieces known as *peds*. Blocky (or cubic) structures are commonly found in LOAMS; prismatic (or columnar) structures develop in 'clayey' soils; and 'platy' structures are associated with compacted horizons and impeded soil drainage. A crumb structure, regarded as the most favourable, consists of small, rounded peds, which are very characteristic of HUMUS-rich soils undergoing cultivation.

soil texture The relative coarseness or fineness of the SOIL, as determined by the size of the contained mineral particles. The latter is determined by the mineral composition of the underlying rock or PARENT MATERIAL, and the prevalent WEATHERING processes. For example, CHEMICAL WEATHERING tends to release fine particles in the SILT–CLAY range. In sandy (light-textured) soils, the SAND fraction is dominant, POROSITY is high, LEACHING is rapid, and chemical compounds are not easily retained. On the other hand, these soils are well aerated and warm up rapidly at the start of the growing season. In 'clayey' (heavy-textured) soils, clay particles are dominant, capillary porosity is high, and moisture retention is good – even to the extent that, after heavy rain, waterlogging is likely. Clayey soils are cold, sticky and difficult to cultivate when wet. LOAMS (medium-textured) comprise mixtures of sand, silt and clay, and combine the most favourable features of both sandy and clayey soils. They are quite well aerated but do not drain too freely, warm up fairly rapidly, and can easily be ploughed. [*f*]

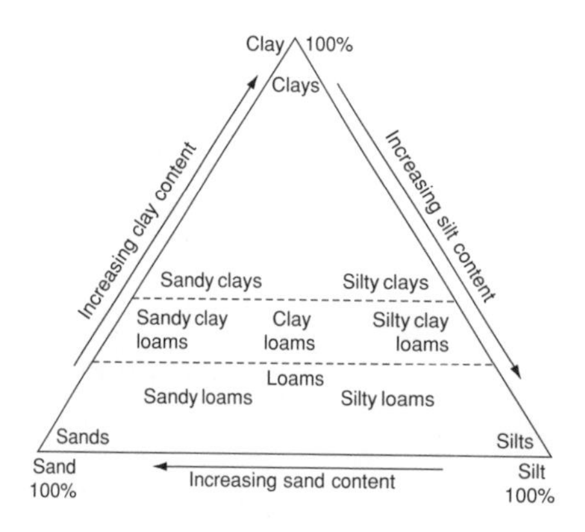

Soil texture related to the variable content of sand, silt and clay

solfatara A volcanic vent emitting sulphurous gases in a non-explosive type of activity. Solfataras are usually associated with the declining phases of volcanic activity. See also FUMAROLE.

solifluction Literally meaning the 'flowage of SOIL' that has become saturated or supersaturated with water. Solifluction is a major transportational process on slopes in PERIGLACIAL regions, owing to the fact that water released by the spring thaw of snow and GROUND ICE cannot percolate downwards in the presence of impermeable PERMAFROST. In addition to true solifluction (now commonly referred to as *gelifluction*) another important MASS MOVEMENT under periglacial climates is *frost creep* (whereby particles on slopes are raised by the formation of ice segregations in the soil at right angles to the slope surface, but fall back more vertically under the influence of gravity when the ice thaws). Periglacial solifluction can be active on slopes as low as 3° in angle; and the annual movement can be measured in cm (by comparison with SOIL CREEP, which operates at about 1 mm y^{-1} in temperate climates). Periglacial slopes are often completely mantled by a sheet of soil and debris, commonly referred to as *head* and often taking the form of *solifluction lobes.*

solstice One of the two dates each year when the overhead Sun is at its furthest declination (angular distance) from the Equator (approximately 23°27'N and S). The Sun reaches the northern solstice, the Tropic of Cancer, about 21 June, and the southern solstice, the Tropic of Capricorn, about 22 December. These dates are known as the *summer* and *winter solstices* in the Northern Hemisphere, where they are associated with maximum and minimum daylight hours respectively. See EQUINOX.

solution The removal by rainwater and percolating GROUNDWATER of dissolved minerals (such as common salt) and the products of other WEATHERING processes. In general, the greater the acidity or alkalinity, the greater its effectiveness. CARBONATION and HYDROLYSIS are specific forms of the process. Solution is a significant process in LIMESTONE regions where it results in the formation of depressions (see DOLINE and SINK HOLE), pipes (see UVALA), and underground caverns and passages. Small solution pipes are very common in CHALK country, where they are exposed in CLIFF sections, quarries and new road sections. The pipes usually penetrate to a depth of a few metres, and are infilled by CLAY, SAND and FLINT gravel. It is believed that pipes in chalk are best developed where there is an overlying layer of sandy SEDIMENT, which may support heathland vegetation and produce humic acid solutions capable of attacking the chalk surface beneath.

sources Where to look for information – as, for example, to help with a PERSONAL ENQUIRY or the preparation of an essay. Basically the sources are of two types: *primary sources* are essentially those that yield unprocessed and often first-hand data, such as fieldwork, travel, questionnaire and other types of survey; *secondary sources* may be described as 'second-hand' in the sense that information has been 'processed' – analysed and presented for general consumption. The potential sources are many and various; they include: textbooks, CD-ROMs, TV programmes, videos, newspapers, journals and the Internet.

South See BRANDT COMMISSION, THIRD WORLD.

space The area or volume occupied by an object or the lateral distances intervening between locations, places and any phenomena distributed over the Earth's surface; the essential dimension and the basic concept of all geography, i.e. geography is, above all else, concerned with SPATIAL DISTRIBUTIONS and SPATIAL RELATIONSHIPS.

space–cost curve A cross-sectional diagram devised by Smith on which the revenue (i.e. the selling price) and cost (i.e. the PRODUCTION COSTS) of a good or service are plotted against distance. In the accompanying figure, both revenue and cost are shown to vary with distance, a major factor being the cost of transport. The space–cost curve may be used to determine the extent of the SPATIAL MARGIN (M to M'), the OPTIMAL LOCATION (O) and the LEAST-COST LOCATION (L). [*f*]

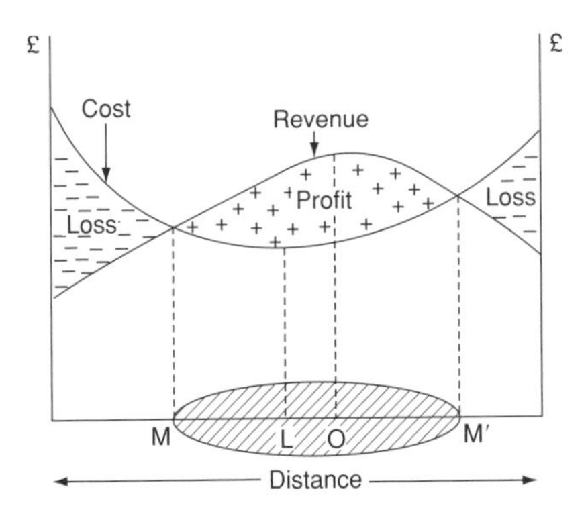

Space–cost curve, showing the spatial margin of profitability

space economy The spatial pattern of an ECONOMY (i.e. the distribution and location of its component activities) and the spatial flows (see SPATIAL INTERACTION) that are an essential part of the workings of that economy (e.g. the movement of goods to consumers, of customers to central places, of farm produce to markets, etc.).

space–revenue curve A variant of the SPACE–COST CURVE, in which the PRODUCTION COSTS of a good are assumed to be the same everywhere, but the selling price (i.e. revenue) is subject to spatial variations. As with the space–cost curve, it is possible to use a cross-sectional diagram to identify the SPATIAL MARGIN and the OPTIMAL LOCATION.

space–time convergence See TIME–DISTANCE CONVERGENCE.

spatial analysis An approach to geography that places emphasis on the investigation of the SPATIAL DISTRIBUTION of phenomena and the factors influencing observed distribution patterns.

spatial autocorrelation This occurs when the observations of a VARIABLE are mapped and the resulting SPATIAL PATTERN shows that neighbouring values in that pattern are either more alike or more dissimilar than would be the case if the pattern were due to RANDOM processes. This clustering of similar or dissimilar values (i.e. the presence of positive or negative autocorrelation) tends to invalidate a basic assumption of many statistical tests, namely that the individual samples of a population are independent (i.e. not autocorrelated). There are various ways of testing for spatial autocorrelation in raw data and in regression residuals.

spatial diffusion The two-dimensional spread of a phenomenon over space and through time; the evolution of its SPATIAL DISTRIBUTION (see DIFFUSION CURVE). Investigations of the processes of diffusion have focused particularly on information flows, the dispersal of INNOVATION and the spread of SETTLEMENT. Two types of spatial diffusion are recognized, namely EXPANSION DIFFUSION [*f*] and RELOCATION DIFFUSION [*f*]; the former can be further subdivided into CONTAGIOUS DIFFUSION and HIERARCHIC DIFFUSION [*f*].

spatial disequilibria See REGIONAL IMBALANCE.

spatial distribution The occurrence of a phenomenon within a given area (see SPACE). Noteworthy aspects include: the spacing and organization of items or objects (e.g. of schools in an URBAN area) relative to others of the same kind (see SPATIAL STRUCTURE); the 'geometry' of that spacing – e.g. of SETTLEMENTS (see, again, SPATIAL STRUCTURE); the density of occurrence (e.g. of POPULATION) per unit area.

spatial interaction The interdependence of areas; the movement of people, capital, goods, information, ideas, etc. between places. Ullman has suggested that the degree of spatial interaction between places is conditioned by three factors: (i) COMPLEMENTARITY; (ii) intervening opportunity (see INTERVENING OPPORTUNITY THEORY); (iii) transferability. GRAVITY MODELS are widely used in the investigation of spatial interaction. See also DISTANCE DECAY.

spatial margin The concept of the spatial margin to profitability was introduced by Rawstron. The limit to profitability is defined where the selling price (i.e. the revenue) of a good or service is equal to the costs of production (*MM′* in the figure referred to below). Although the levels of both revenue and costs, and therefore also of profit, vary within the limits of the spatial margin, any location within its confines offers the FIRM some profit and thereby a degree of locational choice. The concept of the spatial margin thus encourages the idea of SUBOPTIMAL LOCATION.

[*f* SPACE–COST CURVE]

spatial pattern The 'geometry' of the way a particular phenomenon (e.g. SETTLEMENTS, VOLCANOES, etc.) occurs in a given area; the intrinsic character of its SPATIAL DISTRIBUTION (i.e. whether it is uniform, clustered or RANDOM). See SETTLEMENT PATTERN.

spatial preference The DECISION-MAKING process that involves choosing or discriminating between areas. For example, spatial preference figures in the selection of a new area in which to reside or in deciding where to go for a summer holiday. Spatial preference is thus conditioned by the particular values and aspirations of people, and by their perception and evaluation of different areas or places. It creates, as it were, a personal or private geography.

spatial relationship The coincidence and interconnection of two or more VARIABLES in the spatial dimension; e.g. as between climate and vegetation, or unemployment and poverty.

spatial segregation The spatial separation of things that are incompatible or unrelated. The process is well demonstrated within the BUILT-UP AREAS of TOWNS and CITIES by the emergence of distinct LAND USE regions (e.g. the CBD, industrial estates, RETAILING RIBBONS) and clear-cut social areas (e.g. high- and low-class housing districts, GHETTOS).

spatial structure The arrangement and organization of phenomena on the Earth's surface resulting from the operation of physical and/or human processes. Spatial structure may be identified and investigated at a range of spatial scales and in a variety of systematic fields, e.g. from the arrangement of a continent's major physiographic units to the organization of functional and social areas within the BUILT-UP AREA of a TOWN.

Spearman's rank correlation coefficient See RANK CORRELATION.

sphere of influence (i) In a politico-economic sense, an area in which a foreign power has special interests, rights and privileges, e.g. the former USSR in E Europe or the USA in Central America. (ii) An area over which an URBAN centre distributes services (e.g. the delivery areas of shops) and recruits

LABOUR (e.g. the commuter belt) and customers (e.g. the catchment areas of schools), as well as providing that area with a sense of focus through the exercise of various forms of leadership (e.g. publishing a weekly newspaper, possessing a local radio station, functioning as a seat of local government). The term is broadly synonymous with a whole range of terms that occur in studies related to CENTRAL PLACE THEORY, e.g. HINTERLAND, MARKET AREA, tributary area, UMLAND and urban field. One important quality of most spheres of influence is that they display a DISTANCE DECAY characteristic.

spheroidal weathering A process leading to the development of almost spherical boulders from original JOINT-bounded rock masses, frequently composed of GRANITE, GNEISS or BASALT. The rounding is due primarily to CHEMICAL WEATHERING, operating beneath the ground surface within a partially decomposed layer (see REGOLITH). These weathered rocks may eventually be exposed if the finer products are removed by RAINWASH and mass transport.

spit A bank of SAND and/or SHINGLE, projecting from the shoreline into the sea, or partially across the mouth of a river ESTUARY or deep coastal inlet. Spits develop as a result of the LONGSHORE DRIFT of BEACH material, and may extend rapidly, particularly where there is a firm and shallow foundation of SAND or mud. The far point of the spit is commonly fashioned into a *recurved tip*, either by WAVE REFRACTION or the approach of local waves from a direction counter to that of prevalent beach drift. [*f*]

sprawl See URBAN SPRAWL.

spread effect The term coined by Myrdal to denote the transmission and spread of growth throughout the economic system, particularly from the CORE to the PERIPHERY (see also TRICKLE DOWN). It is encouraged in a variety of ways: by increasing demand at the centre for food and RESOURCES produced in the periphery; by DECENTRALIZATION of FIRMS reacting to land costs and congestion in the core; by government encouragement of that decentralization by direct investment in the periphery; by the diffusion of invention and innovation from the core. Ct BACKWASH EFFECT; see also CORE-PERIPHERY MODEL.

spring The emergence of GROUNDWATER at the surface of the ground, usually at a clearly defined point marking a FAULT or major JOINT that has guided the underground flow of the water. Permanent springs occur where the climate is wet throughout the year, or where rainfall in the wet season is sufficient to maintain adequate supplies of groundwater throughout the year. Intermittent springs are

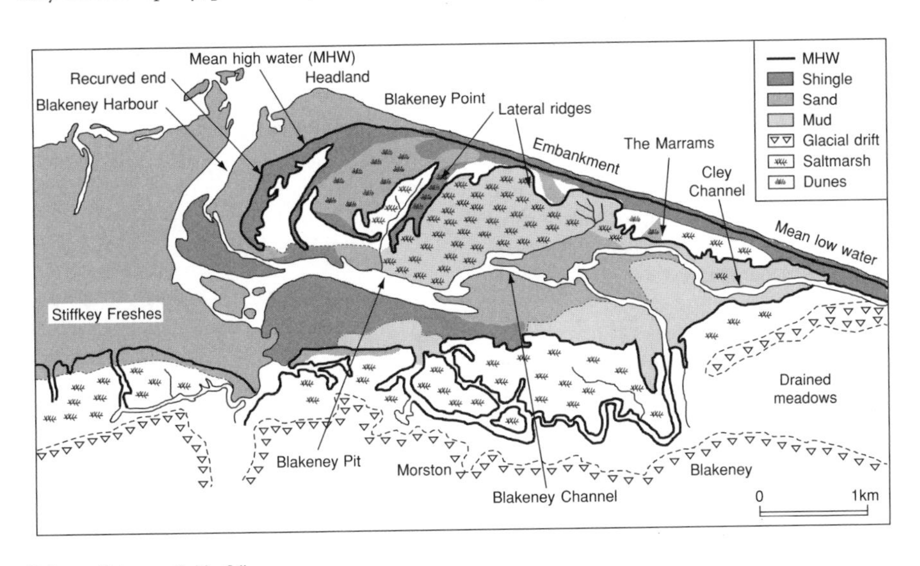

Blakeney Point, north Norfolk

found in climates characterized by long droughts, in which the drain on the water table is such that the springs will cease to flow for a time. Springs occur in a variety of situations – for example, at the base of SCARPS (*scarp-foot springs*) where PERMEABLE rocks overlie IMPERMEABLE rocks. Such scarp-foot springs often form a well-defined SPRING LINE. Springs are also commonly found in the bottoms of valleys incised into the DIP-slopes of permeable rocks, forming CUESTAS. These dip-slope springs are usually more powerful than scarp-foot springs, owing to the greater ease with which underground water moves in a down-dip direction. The points at which springs emerge may be associated with localized EROSION (hence *spring sapping*). It is believed that some steep-headed valleys in CHALK and LIMESTONE scarps may have resulted from this process.

spring line A line of springs or seepages indicating the level at which the WATER TABLE intersects the ground surface. Spring lines are often indicated by lines of SETTLEMENTS originally making use of a local freshwater supply.

spring tide A TIDE with a considerable vertical range, forming every 14.75 days, when the Earth, Sun and Moon are along a straight line – in other words at times of full or new moon. High spring tides are thus especially high, and low spring tides especially low, owing to the fact that the gravitational pull of the Sun and Moon on the Earth's waters is at a maximum. Ct NEAP TIDE. [*f* TIDE]

squall A sudden increase of wind speed, often accompanied by turbulence and showers of rain, but lasting for only a short duration. A series of squalls (LINE-SQUALL) sometimes accompanies the passage of a COLD FRONT.

squatters Those people who occupy property or space, but who do not necessarily have the legal title to do so. Typically, they take up residence in abandoned and vacant dwellings in the inner areas of western CITIES, or erect their own dwellings on vacant areas within THIRD WORLD cities (see SHANTY TOWN).

squatter settlement See SHANTY TOWN.

SSSI (Site of Special Scientific Interest) A site in the UK that has been designated as containing floral, faunal, geological or geomorphological features worthy of conservation (e.g. Fyfield Down near Marlborough, with its remarkable accumulation of several thousand sarsen stones). Approximately 3500 SSSIs have been established, through the work of the former Nature Conservancy Council, since 1949. These are graded 1 to 4 in order of perceived importance – Grade 1 sites being equivalent to National Nature Reserves.

stability The condition of the ATMOSPHERE in which, if a parcel of air is given an upward impulse, it will return to its original position because it remains cooler and heavier than the surrounding air. If an AIR MASS moves over a warm surface, its lower layers become warmer through conduction and it is more likely to rise – it is said to have become more unstable. Conversely, the cooling of an air mass increases its stability and means that it is less likely to rise. See ABSOLUTE STABILITY. [*f*]

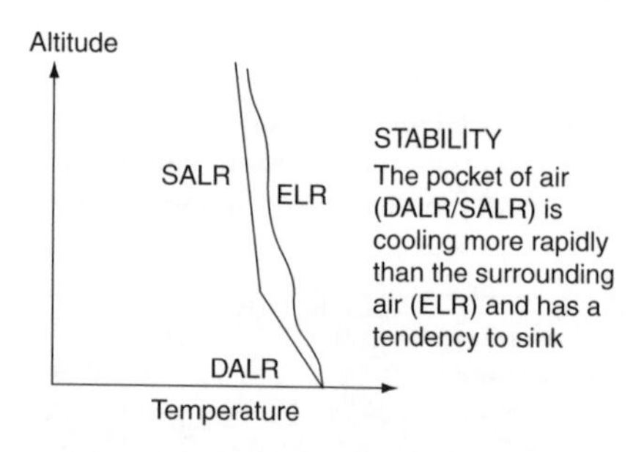

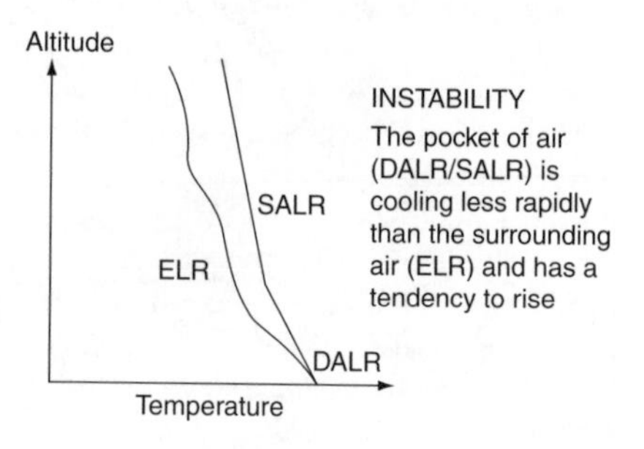

Stability

stable population A POPULATION in which the BIRTH RATE, DEATH RATE and therefore the rate of NATURAL INCREASE, are constant. If the birth and death rates are the same, then the stable population will also be a stationary population.

stack A rocky pinnacle, rising from the sea and isolated from the mainland at high tide. Stacks result from prolonged wave attack along lines of geological weakness (soft strata, FAULTS or SHATTER BELTS). They are often formed following the collapse of the roof of a natural ARCH and they represent one of the final stages in the erosion of a coastal headland. Stacks differ from stumps in that the latter are completely submerged at high tide.

stadial moraine See RECESSIONAL MORAINE.

stages of economic growth model A theory of economic development proposed by Rostow, which stresses the importance of technological innovation as a stimulus to changes in economic wealth over a period of time. Integral to the theory is a five-stage model of economic development. (i) *The traditional society* characterized by the dominance of AGRICULTURE (practised for the most part at a subsistence level) and by the non-realization of potential RESOURCES. (ii) *Preconditions for take-off* – economic growth is speeded up by the introduction of modern methods of agricultural production and by the gradual expansion of TRADE. (iii) *Take-off* – 'the great watershed in the life of modern societies', marked by a rise in investment and national income as one or more vital manufacturing industries begin to dominate the ECONOMY; agriculture becomes even more responsive to innovation and commercialization. (iv) *The drive to maturity* – growth spreads to all sectors of the expanding economy; INDUSTRY becomes more technologically sophisticated, while society acquires a widening range of technical and entrepreneurial skills. (v) *The stage of high mass-consumption* – the continuing rise in the affluence of society causes the leading sectors of the economy to become those concerned with durable consumers' goods and services.

The model has been criticized on a number of grounds. For instance, it is thought to place too much emphasis on CAPITAL formation, while there is no clear mechanism to link, as it were, the five stages. The model is based on the sequence of events observed in ADVANCED COUNTRIES. For this reason, its general applicability to the THIRD WORLD as a basis for forecasting what might subsequently happen there is thought to be highly questionable. But, to be fair to Rostow, he did caution about mistaking the model as a general law, for he recognized that the exact nature of each stage would vary from country to country, depending on the resources available, the pressure of population and the kind of society present. [*f*]

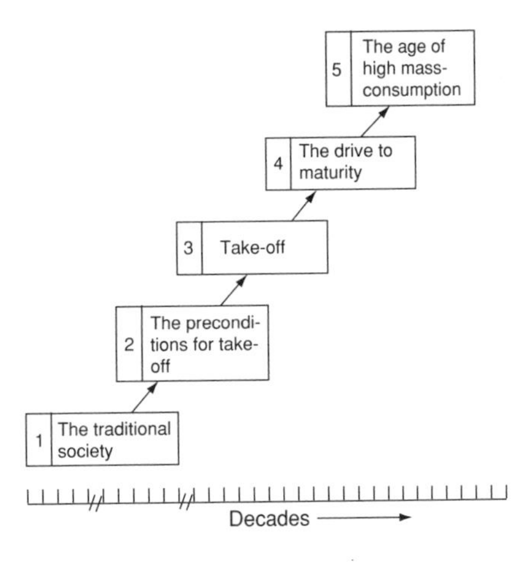

Rostow's stages of economic growth model

stagnant ice See DEAD ICE.

stalactite A mass of calcite, often in columnar form, suspended from the roof of a LIMESTONE cavern. The stalactite is formed by the PRECIPITATION, over a long period of time, of calcium carbonate contained within water that has percolated from above by way of JOINTS and fissures. The process of DEPOSITION involves both evaporation of the water and (more importantly) the escape of carbon dioxide, with the result that dissolved calcium bicarbonate is transformed into soluble calcium carbonate.

stalagmite An accumulation of calcite that has grown upwards from the floor of a LIMESTONE cavern. The processes of formation are similar to those producing STALACTITES. However, the stalagmite is often thicker and less high than the inner, more cylindrical formations of stalactites. Frequently, one escape of water from the cavern roof will form both a stalactite and stalagmite; these will grow towards each other, and eventually join to form a limestone pillar.

standard deviation This DESCRIPTIVE STATISTIC indicates the degree to which the individual values in a data set cluster around the MEAN. It is used as a measure of the variability of a FREQUENCY DISTRIBUTION.

standard distance This SPATIAL DISTRIBUTION index is the equivalent of the STANDARD DEVIATION in a numerical DISTRIBUTION. It measures the degree to which the points of a POINT PATTERN are dispersed about the MEAN CENTRE. It is defined as

$$\text{Standard distance} = \sqrt{\sum \frac{d^2}{n}}$$

where d is the distance to a given point (coordinates x, y) from the mean centre (x, y) and n is the total number of points. Once calculated, the standard distance is of value in that it allows different point patterns to be compared objectively in terms of their degree of dispersion.

standard error The STANDARD DEVIATION of a SAMPLING distribution.

standard hillslope A model hillslope comprising four components: (i) an upper convexity or waxing element; (ii) a FREE FACE; (iii) a debris slope, which is rectilinear in profile; (iv) a basal concavity or waning element. The term was proposed by King to describe the 'fully developed' hillslopes found particularly in areas of massive BEDROCK under semi-arid conditions, where the production of debris by WEATHERING, and its removal by various slope processes, is at an optimum. Where the bedrock is weak and incoherent, or the climate humid, the landscape becomes masked by REGOLITH, and only the waxing and waning elements may occur. [*f*]

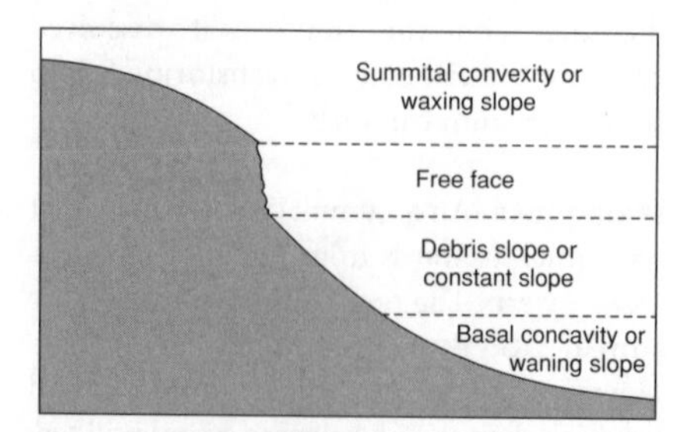

Standard hillslope model

standard of living See LEVEL OF LIVING.

standardized birth rate See BIRTH RATE.

staple (i) A basic item of food; e.g. rice or potato. (ii) A dominant commodity of TRADE (e.g. minerals, manufactured goods); the range of staples produced within a country may be closely related to its level of economic development. (iii) The length of a textile fibre; hence long- and short-staple wool.

starvation An acute condition of hunger that can often prove fatal. Cf FAMINE, MALNUTRITION.

state A group of people occupying a specific territory, organized under one government. It may refer generally to one country, or to a unit of local or regional government within a country. For example, the USA comprises 51 states organized under a system of federal government.

state farm See COLLECTIVE FARMING.

state intervention See GOVERNMENT INTERVENTION.

stationary population See STABLE POPULATION.

statistic Strictly, the sample estimate of a PARAMETER. However, the term is commonly used to denote any numerical fact.

statistics The study concerned with the collection, arrangement and analysis of numerical facts or data, whether relating to human affairs or to natural phenomena. See DESCRIPTIVE STATISTICS, INFERENTIAL STATISTICS.

steady state A condition of EQUILIBRIUM achieved by an open system (see DYNAMIC EQUILIBRIUM, GENERAL SYSTEMS THEORY).

steam fog A type of mist or FOG, usually developed on a small scale, that is associated with the passage of cold air over a warm water body (river, lake or sea). Evaporation from the surface of the water results in 'instant' CONDENSATION in the cold overlying air, to give innumerable tiny water droplets. Thus the water surface gives the appearance of 'steaming'. A similar effect can sometimes be seen after a shower of rain, when dark-coloured road surfaces are heated up rapidly by the sun, causing evaporation of the recently fallen rain and condensation in the relatively cool air above the road surface. (see ARCTIC SMOKE).

step fault One of a series of parallel FAULTS, each associated with a downthrow of the rocks in the same direction. Step faults are characteristic of the margins of RIFT VALLEYS (as in the E African rift valley at Lake Naivasha, Kenya). [*f* RIFT VALLEY, *f* FAULT]

steppe An extensive area of open grassland, in which trees or shrubs are virtually absent except from sheltered moist depressions or along water courses, in the continental interior of E Europe and Asia. The climate of the steppe is characterized by hot summers, cold winters and a relatively low annual rainfall (500–750 mm), occurring mainly in spring and summer. Owing to the high evaporation rates, SOIL MOISTURE is inadequate for tree growth over most of the terrain. The more humid regions in the SW of the former USSR and in the lowlands of Mongolia give rise to meadow steppes, with grasses growing to a metre or more in height; this is the equivalent of the true prairie (see PRAIRIE). However, as aridity increases (for example, to the south of the meadow-steppe in the former USSR) there is a change first to tussock grasses and eventually to steppes characterized by short grasses with many patches of bare SOIL. The soils of the steppes, which derive a good supply of HUMUS from annually decaying grass stems and roots and are affected by the process of calcification, are highly fertile black earths (see CHERNOZEM). They have been cultivated extensively, especially for cereals.

stepped tariffs See FREIGHT RATES.

stepwise migration It is frequently observed that MIGRATION takes place in a step-by-step manner up the settlement HIERARCHY. People move from rural areas into towns, from towns to cities, from cities to the capital. Each step in the progression involves FILL-IN MIGRATION. [*f*]

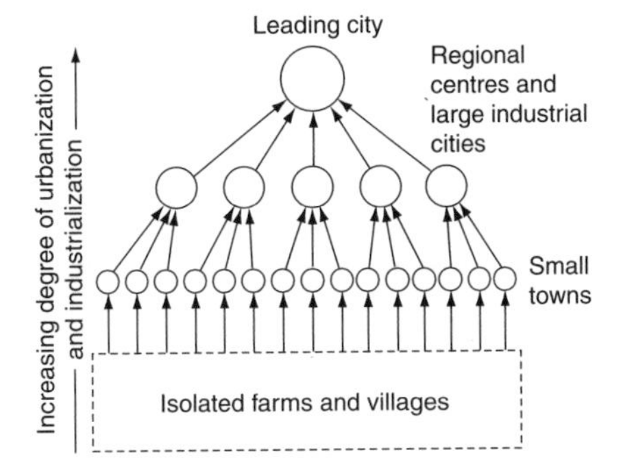

Stepwise migration

stewardship The sympathetic management of the ENVIRONMENT as part of SUSTAINABLE DEVELOPMENT. The idea of successive generations acting as guardians of the environment to ensure its CONSERVATION and survival.

stillstand A period during which the level of the sea in relation to the land remains unchanged. A stillstand is associated with erosion at a particular level, resulting in the formation of landforms such as WAVE-CUT PLATFORMS and PLANATION surfaces.

stochastic process A stochastic process is one that develops in time according to PROBABILITY theory. This means that the future behaviour of the process cannot be predicted with any certainty. In other words, it is a chance or RANDOM element that has a direct influence on the process and direction of change. Awareness of stochastic processes is, therefore, crucial in many aspects of HUMAN GEOGRAPHY, ranging from the evolution of SETTLEMENT PATTERNS to the SPATIAL DIFFUSION of INNOVATION, from industrial location to residential choice.

stone polygon A feature of PATTERNED GROUND in PERIGLACIAL environments. Stone polygons comprise borders of large, uptilted rock fragments and a central 'core' of fine, sticky mud and small stones. Several individual polygons may intersect to give a stone 'net'. It is widely believed that stone polygons result from the processes of FROST HEAVE, which raises coarse fragments in a heterogeneous layer to the ground surface, where they slip sideways into depressions from the centres of low 'cones' and/or frost thrust, the pushing laterally of the larger stones within the debris layer as lenses of GROUND ICE form and expand. See also ICE WEDGE.

[*f* PATTERNED GROUND]

stone stripe A line of coarse debris (stones or even BOULDERS) following the line of maximum gradient on a slope. Stone stripes, which are characteristic of PERIGLACIAL environments, appear to result from a combination of lateral 'sorting' of debris into coarse and fine by frost processes (see STONE POLYGON) and SOLIFLUCTION.

[*f* PATTERNED GROUND]

storm beach A prominent BEACH ridge, composed of coarse particles (SHINGLE, COBBLES, BOULDERS), which is usually developed at the head of a bay subjected to the impact of powerful storm waves (as in western and southern coastal locations in Britain). Storm beaches stand well above the foreshore, and are only overtopped by SWASH under the most severe conditions. It is believed that some storm beaches (as at Chesil Beach, Dorset) comprise beach material that has gradually been 'swept' onshore from the former sea bed as the level of the sea rose during the POST-GLACIAL period. It may well be, therefore, that they represent a 'relict' store of beach material that is gradually being depleted by the process of ATTRITION by waves. See also BERM.

storm surge A rise in relative sea-level resulting from the passage of an intense DEPRESSION, with consequent large and powerful waves. The relatively low atmospheric pressure allows the sea to 'expand' and, if there are strong winds, considerable flooding and erosion of coastal areas may result. In 1953 an intense low-pressure system moved down the North Sea funnelling water southwards. Considerable flooding occurred along the E coast of England and 300 people lost their lives. Some people have suggested that storm surges may become more frequent events as a result of GLOBAL WARMING.

stratified A term applied to deposits that have accumulated in distinct layers (strata). Most SEDIMENTARY ROCKS are stratified.

stratified sampling See RANDOM SAMPLING.

stratosphere The layer of the ATMOSPHERE above the TROPOPAUSE, extending up to the base of the mesosphere at a height of about 50 km. The base of the stratosphere is at 16 km above the Equator, and at 9 km above the poles. At this level, the temperatures range from −80°C to −90°C over the Equator, but from −40°C (in summer) to −80°C above the poles. Within the stratosphere, temperatures rise with altitude to a maximum of about 0°C at the stratopause (marking the junction of the stratosphere and mesosphere).

stratus A layered, frequently unbroken cloud formation, developed mainly in the lower layers of the ATMOSPHERE (at less than 2500 m). Stratus often forms, under stable atmospheric conditions (see STABILITY), from the 'turbulent mixing' by wind of warm and cold air in the surface layer. The result is a low, grey-looking cloud, which gives no rain but occasionally a little drizzle. The cloud may become broken (fracto-stratus) or even disappear as the sun raises the atmospheric temperature. Higher stratus formations are referred to as *alto-stratus*. [*f* CLOUD]

stream order A classification of the component parts (segments) of a stream network. According to the method devised by Strahler, the very smallest headwater streams within a basin are identified and designated as 1st-order streams. Where two such 1st-order streams join, a 2nd-order stream segment results; where two 2nd-order streams join (but not a 1st- and 2nd-order stream), a 3rd-order stream is formed, and so on. Thus the highest-order stream within the drainage basin is the largest in terms of DISCHARGE. Furthermore, the highest order is the value that can then be given to describe the order of the entire basin. When the task of order designation is complete, order analysis can begin. This might involve (for example) calculating the BIFURCATION RATIO or investigating Horton laws, e.g. the relationship between, say, stream order and length of segments. [*f*]

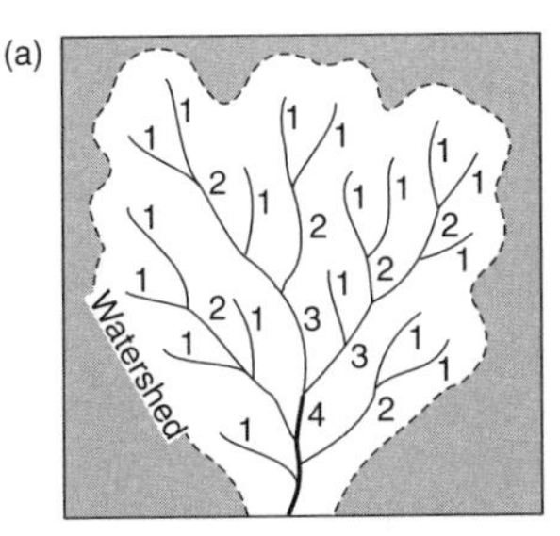

Order	Number of Segments	Bifurcation Ratio*
1	18	
		3
2	6	
		3
3	2	
		2
4	1	

$$\text{*Bifurcation ratio} = \frac{\text{no. of segements of one order}}{\text{no. of segments of next highest order}}$$

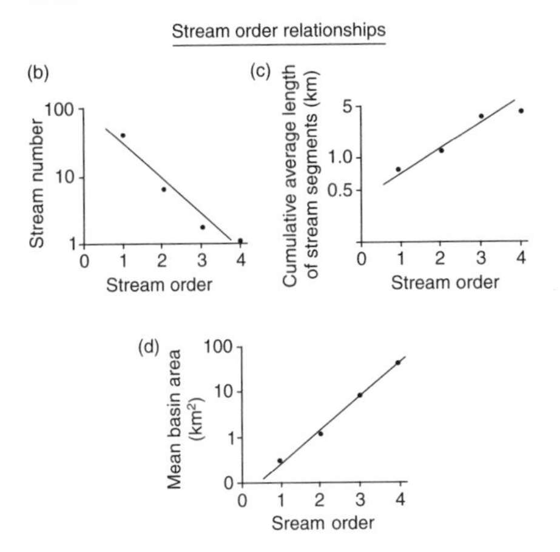

(a) Stream order in a 4th-order drainage basin, (b) stream number and order, (c) cumulative average stream length and stream order and (d) mean basin area and stream order

striation A fine scratch or miniature groove 'engraved' into a hard rock surface by small, hard and angular fragments that are either trapped between sliding ice and the valley floor or actually frozen into the glacier sole (see ABRASION). Striations are often best developed on sloping rock surfaces that the ice is forced to ascend (for example, the *stoss slope* of a ROCHE MOUTONNÉE).

strike The direction along an inclined stratum at right angles to the DIP. Put another way, strike is the direction of a horizontal line along the BEDDING PLANE of the dipping rock layer. [*f* DIP]

strip cultivation (i) A large field cultivated in long strips, each being worked by an individual tenant or owner. This was the basis of the medieval two- and three-field system in England. (ii) See LYNCHET. (iii) The alternation of narrow strips of arable and pasture to check SOIL EROSION, as in parts of the American Midwest.

strip mining See OPENCAST MINING.

structuralism A method of analysis, at present encountered in HUMAN GEOGRAPHY, based on the belief that observed phenomena are not the unique outcome of unique forces or unique events. Rather, observed phenomena are interpreted as the product of much more general and deep-seated mechanisms. For example, the structuralist view of INNER-CITY DECLINE sees it as being neither unique to Britain nor due to particular material circumstances, but instead regards it as the outcome of CAPITALISM as currently operating in many Western countries.

structural unemployment A loss of jobs associated with a basic change in demand or technology, as with the decline in shipbuilding or the closure of coal mines.

Student's *t* test A parametric statistical test used to determine the SIGNIFICANCE of the difference between the MEANS of samples derived from the same POPULATION. This is assessed by comparing the differences between the sample means with the STANDARD ERROR of the difference.

$$t = \frac{\text{difference between sample means}}{\text{standard error of the difference}}$$

The larger the values of *t*, the lower is the PROBABILITY that an assumption of no significant difference (i.e. the NULL HYPOTHESIS) is correct. However, the value of *t* should be checked in Student's *t* tables or on a Student's *t* graph to find the percentage probability that the difference is due to chance. In so doing, account is taken of the DEGREES OF FREEDOM. See also PARAMETRIC TESTS.

subaerial A broad term describing all processes and features occurring on the Earth's surface and, therefore, vulnerable to, among other things, the processes of WEATHERING, EROSION and MASS MOVEMENT. The term is often used to distinguish between 'land-based' processes and marine processes when discussing coastal features.

subcontracting The process whereby a FIRM engages another to provide part of a PRODUCT or SERVICE – for example, when a motor vehicle manufacturer commissions parts to be made by other firms.

subduction zone See PLATE TECTONICS.

subglacial Lying beneath the base of a glacier or ICE SHEET (e.g. subglacial rock surface or subglacial MORAINE, consisting largely of debris produced by active glacial EROSION). Subglacial streams flow in tunnels at the base of a glacier or ice sheet, but emerge at the ice margin from caves. Subglacial streams (which often flow at very high velocities owing to the pressure exerted by the overlying ice) are sometimes capable of spectacular erosion, forming deep POT-HOLE bowls and narrow, winding channels in solid rock. See also ESKER.

submarine canyon A deeply incised, steep-sided trench that crosses the CONTINENTAL SHELF, sometimes continuing the line of major rivers (such as the R Congo). It has been proposed that many canyons are the result of EROSION by turbidity currents (flows of water highly charged with fine SEDIMENT) that are triggered off by disturbances such as EARTHQUAKES.

submarine ridge See MID-OCEAN RIDGE.

submerged coast A coastline that has been affected by SEA-LEVEL CHANGE, resulting in a relative rise of sea-level. For example, during the POSTGLACIAL period, the Pleistocene ICE SHEETS melted, thereby adding water to the oceans and causing a eustatic rise. In detail, submerged coasts will be affected by the form of the land-mass that has been inundated by the sea. A submerged lowland coast will display broad, shallow ESTUARIES, whereas a submerged upland coast will be characterized by features such as FJORDS and RIAS, together with numerous rocky islands. See also DALMATIAN COAST, FJARD.

submerged forest An organic deposit, comprising PEAT, tree trunks and tree roots, overlain and preserved by more recent marine SEDIMENTS. Submerged forests have resulted mainly from the very rapid rise of sea-level that took place between 9000 and 4000BP. Growing trees (represented now by stumps in a position of growth) and other vegetation were suddenly overwhelmed by the sea and smothered by marine SANDS and CLAYS. Today, submerged forests are frequently revealed in excavations of major construction sites.

suboptimal location A location that, for various reasons, is not an OPTIMAL LOCATION, but is none the less one that offers profitability or net benefit, i.e. it occurs within the SPATIAL MARGIN. The notion of suboptimal location is a vital part of the SATISFICER CONCEPT.

subsequent stream A stream that develops as a tributary of a CONSEQUENT STREAM, mainly by the process of HEADWARD EROSION along a line of geological weakness (a stratum of soft rock or a FAULT-line). See also RIVER CAPTURE.

subset Part of a POPULATION or set of objects.

subsidies Grants of money to particular INDUSTRIES or groups of people by the STATE. In an industrial context, the subsidy is often granted in the hope of raising OUTPUT and promoting EXPORTS, while subsidies more directed towards people relate to such things as housing and the cost of food. Subsidies, therefore, represent a form of GOVERNMENT INTERVENTION in the MODE OF PRODUCTION.

subsistence agriculture A type of farming concerned with the production of items to satisfy the food and living requirements of the farmer and his family, and where the emphasis is on self-sufficiency. Much subsistence AGRICULTURE is concerned with the cultivation of basic cereals (e.g. rice in SE Asia, millet in W Africa). In many cases there is a commercial element of selling or bartering, in the sense that part of the agricultural production may be used to trade in other subsistence requirements not produced by the farmer.

suburb The outer or peripheral, mainly residential, parts of a TOWN or CITY (hence *residential suburb, dormitory suburb*) largely dependent on services and employment concentrated in its CBD. Although SUBURBANIZATION is generally regarded as being a 20th-century phenomenon, the word suburb is of much older origin, formerly meaning the territory immediately outside the walls of a town or city (often occupied by craftsmen seeking to escape guild regulations). In the 19th-century city, suburbs were built mainly for occupation by wealthy, middle-class families as they reacted against the undesirable living conditions increasingly characterizing the older parts of the city, and as transport developments (the extension of railway and tram networks) allowed them to live at greater distances from the city centre.

In the 20th century there was a vast spread of suburbs fuelled by: the accelerating and reactive 'flight from the city'; much greater personal MOBILITY (improved public transport services and increased car ownership); higher levels of affluence; the building of large local authority housing estates at the RURAL–URBAN FRINGE. The suburb today is no longer an exclusively residential development, in that suburbanization has encouraged the DECENTRALIZATION of employment to suburban TRADING ESTATES, office and RETAIL PARKS, and the local provision of commercial and welfare services.

Although it is clear that the suburbs offer a distinctive (and sometimes varied) residential ENVIRONMENT, and a particular lifestyle that appeals to a broad spectrum of people, it is extremely difficult to categorize what exactly it is that appeals about this environment, or indeed what exactly it is that constitutes the so-called suburban way of life. The following may figure among the aspects that appeal and pull: the availability of well-equipped housing; the willingness of financial institutions to provide large mortgages on suburban properties; the modern residential layout; the compromise between nearness to the countryside and access to the higher-order services and employment in the CBD; the opportunity to live in neighbourhoods of like-minded people of the same socio-economic group; access to new social and WELFARE services.

suburbanization The DECENTRALIZATION of people, employment and services from the inner and central areas of a TOWN or CITY, and their relocation towards the margins of the BUILT-UP AREA. A process leading to the accretion of SUBURBS.

succession See VEGETATION SUCCESSION.

sunbelt A term coined in the USA to denote those favoured areas of the S and W that, since the early 1970s, have shown rates of population and economic growth far in excess of the national averages. Sunbelt (also called *sunrise*) areas in states like Arizona, California and Texas are seen as the product of the emergence of POSTINDUSTRIAL America (see DE-INDUSTRIALIZATION). The new, rapid-growth, HIGH-TECHNOLOGY INDUSTRIES (many of them defence-oriented) have become located in these sunbelt states, and because of the good job opportunities they offer, are drawing young people from the depressed industrial centres of the N and E (popularly referred to as *sunset areas* or *frost belts*). Other factors encouraging people to remove to such areas are the favourable climate and the whole perception that they offer an attractive lifestyle and a good QUALITY OF LIFE.

The designation of sunbelt has since been applied to certain growth areas in Britain (e.g. the London–Bristol corridor) and W Europe (e.g. along the Rhine valley) which, while they undoubtedly display some of the same characteristics (e.g. concentration of high-tech industry, a pleasant residential environment), also exhibit at least one significant difference. In the USA, the sunbelt areas and cities are located well away from the established national CORES (they are, in this sense, peripheral), while the European examples are really only marginal to, or extensions of, core areas.

sunset areas See SUNBELT.

superimposed drainage A type of drainage in which a river system, initiated on the surface of a younger geological formation, is over a period of time 'let down' on to an underlying older geological formation. Superimposed drainage is a common and widespread phenomenon. For example, it is believed that many of the larger rivers of England and Wales were formed on the surface of a layer of CHALK, uplifted and tilted at the end of the Cretaceous period. In the early part of the Tertiary era, these rivers gradually cut down through the chalk (which has now been removed, except in SE England) into older rocks. However, following superimposition, the discordance of drainage has been reduced as the river patterns have begun to adjust to the newly exposed structures.

supermarket A self-service store selling mainly CONVENIENCE GOODS and having a floor space of at least 185 m^2. Ct HYPERMARKET, RETAIL PARK, SUPERSTORE.

superstore A free-standing, single-storey retail outlet with between 2325 and 4645 m^2 of floor space. The term tends to be used rather loosely and interchangeably with HYPERMARKET, but strictly speaking there is a size difference; a superstore is of smaller dimensions, but is, none the less, larger than a SUPERMARKET. Ct RETAIL PARK.

supply and demand curves A graphical representation of the supply and demand functions. The demand curve indicates how much of a commodity will be bought during a specified period at any given price; the higher the price, the less the demand. The supply curve indicates how much of the commodity will be supplied during a specified period at any given price; the higher the price, the greater the supply. Thus the two curves trend in opposite directions. The intersection of the two curves represents the *equilibrium price*, i.e. the point where demand is sufficient to consume all the supply. If conditions change (perhaps more or less of the commodity might be offered at a given price), the equilibrium price will change, being lowered where more is offered and raised where less is offered. [*f*]

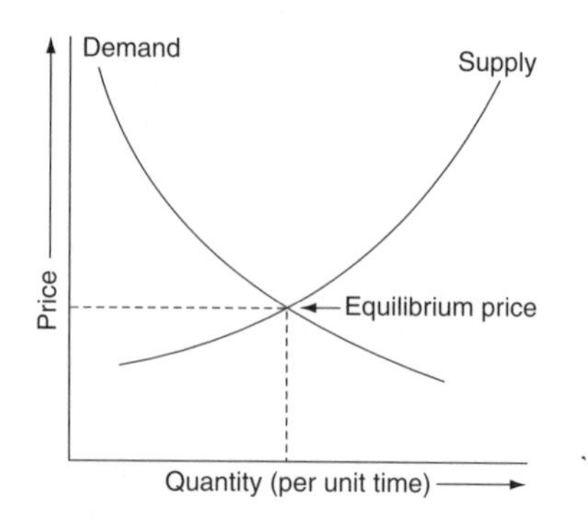

Supply and demand curves

supraglacial Lying on or above the surface of a glacier or ICE SHEET – for example, supraglacial MORAINE, consisting of rock debris that has fallen on to the ice or has melted out from inside the glacier. Supraglacial streams flow over the glacier surface, and derive their water mainly from the melting (ABLATION) of ice. They are most numerous near the ice margins, and show a marked seasonal and/or diurnal variation of flow. For example, in the European Alps the streams cease to flow in winter and, in summer, there is maximum discharge in the late afternoon, and minimum discharge at the end of the night. Many supraglacial streams enter the ice by way of CREVASSES or MOULINS.

surge A relatively fast forward movement of a glacier in which the snout may advance by several km in a very short period of time. A glacial surge is thought to be a sudden, almost catastrophic, response to an accumulation of snow and ice (in effect, a positive MASS BALANCE), or to the build up of SUBGLACIAL meltwater which, on its release, triggers a sudden surge forwards.

suspended sediment load The finer SEDIMENT particles (usually in the CLAY–SILT range) carried along in the body of a stream and supported by the water itself (ct BED LOAD). The suspended sediment load has a tendency to settle out very slowly. The amount of suspended sediment within a stream can be expressed in terms of sediment concentration (milligrams of sediment per litre of water). Particularly high sediment concentrations are found in glacial meltwater streams, which flush silt (the product of ABRASION) from beneath glaciers. The proportion of stream LOAD carried in suspension varies from one stream to another, depending on climate, WEATHERING and the availability of sediment of varying calibre within the stream CATCHMENT.

sustainable development A form of DEVELOPMENT involving a wise use of RESOURCES (particularly *non-renewable* resources) and APPROPRIATE TECHNOLOGY, and that can be sustained without adversely affecting the natural ENVIRONMENT. It is development that meets the needs of the present without compromising the ability of future generations to meet their own needs. POPULATION and the nature of the ECONOMY are other key components. It is a mode of development that is being widely advocated by conservationists concerned about the wholesale destruction of fragile habitats ranging from WETLANDS to tropical RAINFOREST. See RIO EARTH SUMMIT. [*f*]

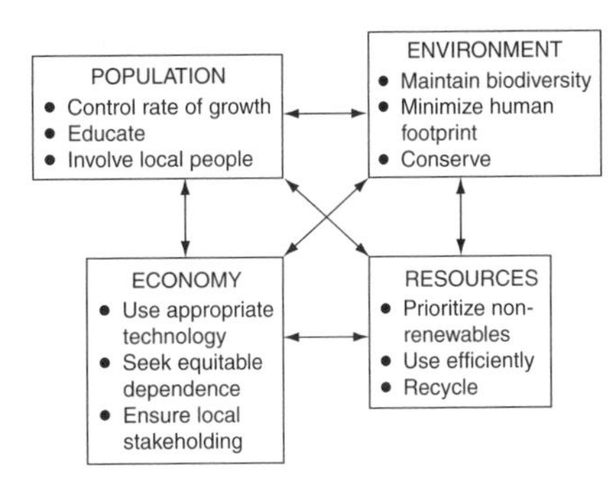

Key components in sustainable development

swallow hole (swallet) A funnel-shaped or vertical shaft down which water flows into LIMESTONE. Most commonly, a swallow hole occurs at the junction of an IMPERMEABLE rock and the limestone, usually at an intersection of JOINTS where the process of SOLUTION has been concentrated. A swallow hole is not to be confused with a SINK HOLE, which tends to be a dry depression or shaft. [*f* LIMESTONE]

swash The turbulent mass of water that flows up a BEACH, following the breaking of a wave. The swash is most powerful when long, surging waves strike the shore. See CONSTRUCTIVE WAVE, BACKWASH.

swash-aligned beach A beach that is developed parallel to the crests of approaching wave. Swash-aligned beaches are characteristic of indented coasts, with many small bays and headlands, on which the LONGSHORE DRIFT of SEDIMENT is impeded. Where the direction of wave approach changes, longshore drift will then rearrange the sediment on the beach, to restore the swash-aligned orientation.

swell Very long, relatively smooth and undisturbed 'waves' formed in the open ocean. Swell waves are generated by local storm conditions, but will then travel immense distances, decaying very slowly as they do so. When swell waves reach gently sloping coasts, the wave form steepens considerably and vast amounts of energy are released as the waves break. Thus swell is a very important geomorphological factor in the evolution of coastal landforms.

syncline A downfold in the rocks resulting from compressive stresses in the Earth's crust. The strata DIP towards the central line, or AXIS, of the syncline. Ct ANTICLINE. [*f* ANTICLINE]

synoptic chart A chart depicting meteorological conditions (isobars, wind speed and direction, cloudiness, temperature, precipitation, fronts) at a moment in time. The construction of a synoptic chart used to be a necessary preliminary to making weather forecasts, but has now been replaced to some extent by satellite images (see REMOTE SENSING).

systematic geography An approach to geography by way of its various contributory aspects; the study of separate aspects of the ENVIRONMENT (e.g. climate, landforms, SOILS, economic activity, SETTLEMENT, etc.) in a predefined area. It is one of two approaches traditionally adopted in geography in order to reduce the study of the Earth's surface to manageable proportions (the other is

REGIONAL GEOGRAPHY). During the second half of the 20th century, it increasingly overshadowed the regional approach.

systematic sampling A method of SAMPLING that employs a regular GRID to determine the pattern of sampling points; thus those points are regularly spaced. Ct NESTED SAMPLING, RANDOM SAMPLING. [*f* NESTED SAMPLING]

systems See GENERAL SYSTEMS THEORY.

systems analysis A search for generalizations based on the whole rather than individual parts; a consideration of a set of objects, and the functional and structural relationships and organizations linking those objects. It is not a replacement for analytical methods, but an alternative, additional line of scientific enquiry. See GENERAL SYSTEMS THEORY.

T

taiga A Russian term for the coniferous FOREST belt extending across the northern part of Eurasia. See BOREAL FOREST.

take-off See ROSTOW'S MODEL.

talik See PINGO.

talus An accumulation of angular fragments on a slope. Talus may comprise a relatively thin veneer of debris, resulting largely from *in situ* WEATHERING of the underlying rock, or a thick deposit of fragments that has collected at the base of a FREE FACE (see SCREE).

tapering See FREIGHT RATES.

tariff A duty or tax charged by a country on its IMPORTS from other countries; a customs duty. The duty may be imposed as a percentage of the value of the goods or as a specific amount per unit of weight or volume. See also TARIFF BARRIER.

tariff barrier The use by a country of a TARIFF in order to protect its own INDUSTRIES from the competition of foreign producers. By imposing a barrier in the form of a high tariff, the price of an imported commodity is increased and thus that commodity becomes less competitive in the market-place. Tariff barriers may also be used by a country to reduce the total volume of imports, especially where there is a BALANCE OF TRADE deficit. Cf QUOTA; see TRADE FRICTION.

taxonomy The scientific CLASSIFICATION of features according to general principles and laws, e.g. the Linnaean classification of plants, Koppen's classification of world climates, a classification of towns on the basis of function or CENTRAL PLACE status (see CENTRALITY).

technological hazard See NATURAL HAZARD.

technology transfer The passing on of innovation and technological advances from one country to another. One of the potentially most important of such transfers is from MEDCS to LEDCS in the context of AID. However, transfers also take place between MEDCs, as for example the transfer of Japanese work practices to European branch plants. In both geographical settings, TNCS are important promoters of such transfers.

technology treadmill This refers to the VICIOUS CIRCLE in which many small farmers particularly in the THIRD WORLD find themselves caught. In order to keep up with larger producers (see AGRIBUSINESS), they are forced to buy either more land or the technology needed to raise productivity. Gradually they are drawn into debt and the only way to relieve that debt burden is to raise productivity still further by buying in still more land or technology. The farmer is therefore stuck on a treadmill from which it is extremely difficult to escape.

tectonic A term describing movements within the Earth's crust, resulting in uplift and depression, lateral sliding, warping, folding and faulting. Landforms produced directly by such movements (e.g. HORSTS and RIFT VALLEYS) are sometimes referred to as *tectonic relief*. See also PLATE TECTONICS.

temperature anomaly See ANOMALY.

tenant capital The equipment, such as seed, livestock, machinery and fertilizers, supplied by a tenant in an agricultural system.

terminal costs The costs of loading and unloading freight at the points of origin and destination, and at BREAK-OF-BULK POINTS; a component of TRANSPORT COSTS.

terminal moraine Also referred to as *end moraine*, a ridge of BOULDERS, GRAVEL, SAND and SILT formed at the terminus of a glacier or ICE SHEET. In many instances the constituent debris consists not only of SUBGLACIAL particles but also of SUPRAGLACIAL and ENGLACIAL sediment, exposed by surface melting of the ice. Terminal moraines are usually asymmetrical in cross-section, with a steep ice-contact face on the side next to the glacier (*proximal slope*) and a gentler slope away from the ice (*distal slope*). In the case of valley glaciers, terminal moraines may merge with LATERAL MORAINES. See also RECESSIONAL MORAINE.

terminal velocity The rate of fall attained by a particle passing through a liquid or gas. The precise velocity is determined by the size and weight of the particle (influenced by the gravitational pull), and the resistance to movement afforded by the liquid or gas. In the ATMOSPHERE the terminal velocity of raindrops and hailstones may be countered by rapidly rising updraughts of air; this is an important factor in their subsequent enlargement. See HAIL.

ternary graph See TRIANGULAR DIAGRAM.

terra rosa A reddish SOIL developed on LIMESTONE in areas of Mediterranean climate; the colour of the soil is due to the presence of iron hydroxides. Terra rosas largely comprise insoluble particles within the limestone that have been exposed as the surface has been lowered by CARBONATION. It is thus essentially a RESIDUAL soil.

terrace A bench-like feature in the landscape, delimited at the 'front' by a relatively steep drop to lower ground and at the 'back' by a pronounced BLUFF. Terraces are either erosional or depositional in origin, and can be formed by a variety of processes (i.e. wave EROSION may produce a platform that, if subsequently upraised, becomes a *marine terrace*). See also ALTIPLANATION, KAME TERRACE, RIVER TERRACE.

terracette A very small TERRACE, up to 30 cm in width and often unvegetated, on a steep grassy slope (usually at an angle of 30° or more). Terracettes occur in parallel series, running approximately along the contours of the hillslope, with a spacing of about a metre or so. It has been suggested that they result from the rupture of the turf mat by active SOIL CREEP. However, there is little doubt that, in many areas, no matter what the process of initiation, terracettes have been greatly exaggerated by trampling animals.

territorial justice A concept encountered in WELFARE GEOGRAPHY that involves relating the level of government spending on such matters as housing and social services in a given area to the needs of that area. Research has demonstrated, particularly in inner-city and peripheral areas, that there is often a mismatch between expenditure and need, and therefore a lack of territorial justice (see INNER-CITY DECLINE). It has also been shown that those areas with the greatest needs not only lack resources and are starved of public funding, but also wield insufficient political power to be able to influence the allocation of public funds. See also DEPRIVATION.

territorial social indicators Social measures that are employed to monitor and assess spatial variations in the QUALITY OF LIFE. A whole range of criteria may be used, relating to such things as income, health, nutrition, education, housing, social order, etc. Some researchers have suggested that the indicators fall into three groups: (i) those that relate to the actual level of territorial WELL-BEING; (ii) those that identify the specific deficiencies or needs of areas; (iii) those that relate to the effectiveness of alternative ways of meeting those needs. See also TERRITORIAL JUSTICE.

territorial waters The coastal waters over which a bordering STATE has jurisdiction. Under international law this was originally defined as a distance of 5 km. In 1958 the

Law of the Sea Convention extended this to 22.2 km (12 nautical miles), as well as clarifying the rights of states to the RESOURCES of the CONTINENTAL SHELF (see LAW OF THE SEA) to a distance of 370 km (200 nautical miles), now known as the *Exclusive Economic Zone.*

tertiary sector One of the four major sectors of the ECONOMY, comprising the *distributive trades*, i.e. RETAILING, WHOLESALING and TRANSPORT. Other *tertiary activities* include PERSONAL SERVICES, the professions and public administration, so that service provision is also a significant aspect of the sector (see OFFICE ACTIVITY). The tertiary sector may be seen as providing a link between many primary and secondary activities and their final customers. Cf PRIMARY SECTOR, SECONDARY SECTOR, QUATERNARY SECTOR; see also DEVELOPMENT-STAGE MODEL.

thaw lake A shallow, rounded depression containing a circular or semi-circular pond, in a lowland PERIGLACIAL region. Thaw lakes are very common features of THERMOKARST, and result from the surface thawing of frozen ground (for example, at points where the vegetation cover has been disrupted). This produces localized collapse, and the formation of an irregular depression that becomes occupied by meltwater. The pond expands laterally rather than vertically through further melting of the PERMAFROST and undercutting of the surrounding vegetation mat. The pond then becomes progressively smoother and more circular in outline.

thermal pollution Pollution of the environment (mainly the atmosphere, rivers and oceans) resulting from the release of heat from human activities, e.g. the escape of industrial and domestic heat into the atmosphere (contributing to the urban HEAT ISLAND effect), or the discharge of hot water into rivers where it causes de-oxygenation, adversely affecting fish and other aquatic fauna.

thermokarst The formation in PERIGLACIAL environments of a highly irregular ground surface, as a result of the thawing of masses of GROUND ICE. Hummocks, pits and larger enclosed depressions may become so numerous that there is a crude resemblance to the karstic features of LIMESTONE (though the processes of formation are totally different). Among the main features of thermokarst are alases (and alas valleys), PINGOS and THAW LAKES. The development of thermokarst landscapes results either from a warming of the climate or from a removal of the vegetation cover that insulates and protects ground ice bodies. Construction work can lead to very rapid thermokarst formation. In a few years the ground surface subsides, and becomes so irregular and marshy as to be virtually impassable to traffic.

Thiessen polygon Frequently used as an alternative to QUADRAT ANALYSIS in the analysis of SPATIAL DISTRIBUTIONS taking the form of POINT PATTERNS. The polygons are created by drawing a straight line between each point and its immediate neighbours, and bisecting those with new lines drawn at right angles. The latter intersect to form polygons. It is assumed that each point dominates the area defined by its polygon. The technique may be used, for example, to calculate the average amount of PRECIPITATION received over the total area of a drainage basin on the basis of data collected at a small number of rain gauges located at different points in the basin. Polygons are constructed around each rain gauge site in the manner described above, and the area of each polygon is calculated and expressed as a fraction of the total drainage basin area. For each rain gauge site the amount of precipitation received is multiplied by its fraction of the total area. The resultant values for all the sites are then summed to give an estimate of the mean precipitation total for the whole basin. [*f*]

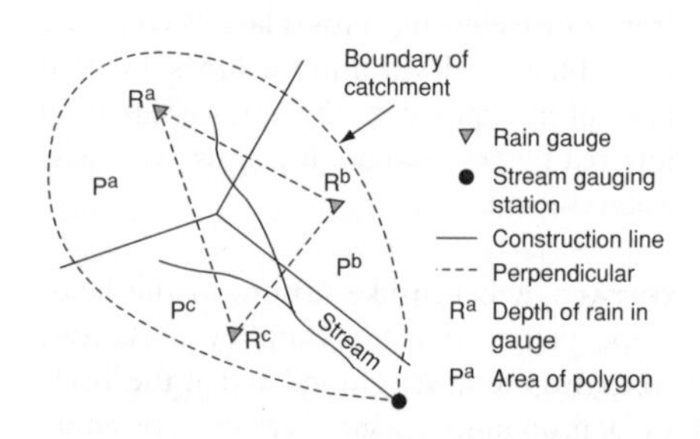

Thiessen polygon

Third World A term used, often rather loosely and along with a range of broadly synonymous terms (e.g. *developing world*, the *South*, LEDCs), to denote relatively poor and under-developed countries located mainly in Africa, Asia and Latin America. The term originated after the Second World War and in the context of the Cold War between the *First World* (capitalist) and the *Second World* (socialist). It refers to a growing group of non-aligned countries, many of which have recently achieved political independence from colonial powers. While the Third World undoubtedly embraces some of the poorest countries of the world (e.g. Ethiopia, Bangladesh), not all of the constituent countries can be regarded in this way. For example, some countries are significant producers of oil (e.g. Nigeria) and minerals (e.g. Brazil); some have quite well-developed industrial sectors (e.g. India). So the Third World might be seen as comprising either what are popularly known as the LEDCs or a mix of *developing countries* (i.e. those in the process of achieving as regards organization, social systems and independence) and *undeveloped countries* (i.e. those that have yet to make significant strides as regards economic development, rising living standards and political independence). Because of this diversity, it is clear that there is no single course of action for dealing with the different economic, political and social challenges facing the Third World. The broad strategy for the future may, in many cases, lie in DEVELOPMENT, but how that development might take place, in what form and at what pace are questions that can only be answered in the light of the circumstances that prevail in the individual countries. See BRANDT COMMISSION, UNDERDEVELOPMENT.

Third World debt See DEBT.

threshold (i) A factor that complicates the self-regulation of systems by NEGATIVE FEEDBACK, and thus the maintenance or restoration of states of EQUILIBRIUM in open systems (see GENERAL SYSTEMS THEORY). When a threshold is crossed, rates of operation of natural processes may be rapidly accelerated, or irreversible changes are set in motion. A simple example is provided by the melting of glaciers early in the ABLATION season. At the end of the winter, glaciers are snow covered (even in their lowermost parts) and, owing to the high ALBEDO of white snow, much solar radiation is reflected and ablation is very slow. However, when the snow is eventually removed, and the darker and often dirt-stained ice surface (with a much lower albedo) is exposed, the rate of melting will be greatly increased.

(ii) In CENTRAL PLACE THEORY the term threshold refers to those conditions that any good or service requires for entry into a CENTRAL PLACE system. Before any good or service is offered for sale at a central place, all the costs involved in its production and provision must be covered by anticipated transactions. This minimum number of sales represents the particular threshold requirement of each good or service, while the *threshold population* is defined as the minimum number of people required to support any central place activity before it can be profitably operated. Although the term threshold is usually interpreted in this way as simply referring to the number of people or customers required, strictly speaking it should relate to that part of the expenditure of each person used to purchase each particular good or service. Cf RANGE.

threshold population See THRESHOLD (ii).

throughflow The movement of water through the soil by PERCOLATION either between interconnecting pore spaces or along discrete pipes. Throughflow is probably a far more important process than OVERLAND FLOW in the disposal of rainwater on hillslopes in humid temperate regions, although it is slower. See HYDROGRAPH.

thrust fault A reversed FAULT at a very low angle. In major geological structures (such as NAPPES), the mass of rocks overlying the almost horizontal fault (also referred to as the *thrust plane*) may have ridden forwards a distance of many kilometres over the rocks beneath.

thufur See INVOLUTION.

thunderstorm A storm characterized by

thunder and lightning and heavy, even violent, PRECIPITATION, resulting from very rapidly rising air currents under conditions of ABSOLUTE INSTABILITY. Thunderstorms are associated with the formation of ANVIL CLOUDS of great vertical extent, within which upcurrents may attain rates of 30 m s^{-1}, and extend as high as 6000–12,000 m. Rapid CONDENSATION leads to the formation of both water droplets, and HAIL and ice crystals. The release of LATENT HEAT is an additional factor promoting rapid uplift of air. Within the thundercloud, positive electrical charges are built up (i.e. by the break-up of large raindrops carried aloft by the updraughts). When these are discharged, either to areas of negative electrical charge within the cloud (associated with hail pellets) or to the negatively charged Earth, lightning results. One important feature of thunderstorms is the formation of downdraughts of cold air (to compensate for the rising currents), which in the later stages of the thunderstorm exceed the updraughts and lead to the degeneration of the 'thunderstorm cell'. These downdraughts carry down, often at a high velocity, the raindrops and hail from the upper part of the cloud, giving rise to violent squalls. [*f*]

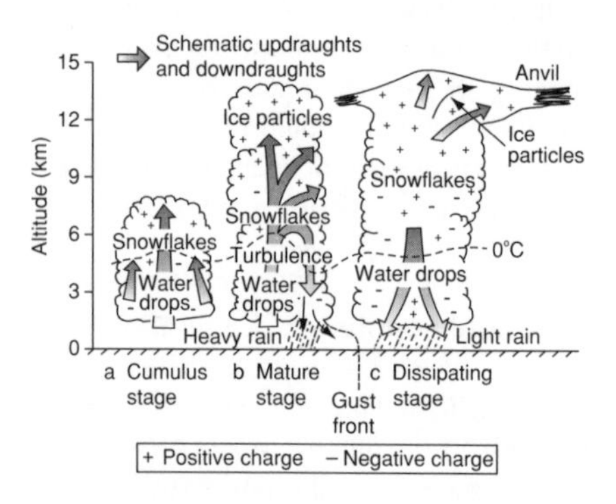

Thunderstorm

tidal barrage See BARRAGE.

tide The (approximately) twice-daily rise and fall in the level of the sea, resulting from the gravitational attraction on the Earth's oceans exerted by the Sun and Moon. The *tidal range* (the vertical interval between high and low tides) varies both from time to time (see NEAP TIDE and SPRING TIDE) and from place to place. *Tidal currents* are streamlike flows of seawater into and out of bays and ESTUARIES with the rising and falling of the tide producing respectively *flood currents* and *ebb currents*. [*f*]

till A deposit laid down by a glacier or ICE SHEET on a land surface. Till is highly variable in character, depending on the precise manner of DEPOSITION, but it is generally highly mixed (with particle sizes ranging from CLAY to BOULDERS) and poorly STRATIFIED. Till may result from the basal melting of debris-rich ice or from the surface melting of the ice. The terms till and MORAINE are sometimes regarded as synonymous; however, strictly speaking, till refers to the deposit itself, and moraine to the surface landforms (such as an elongated ridge) of the till. In being laid down directly by ice, the extent of till can, theoretically at least, be used to suggest the extent of ice flow.

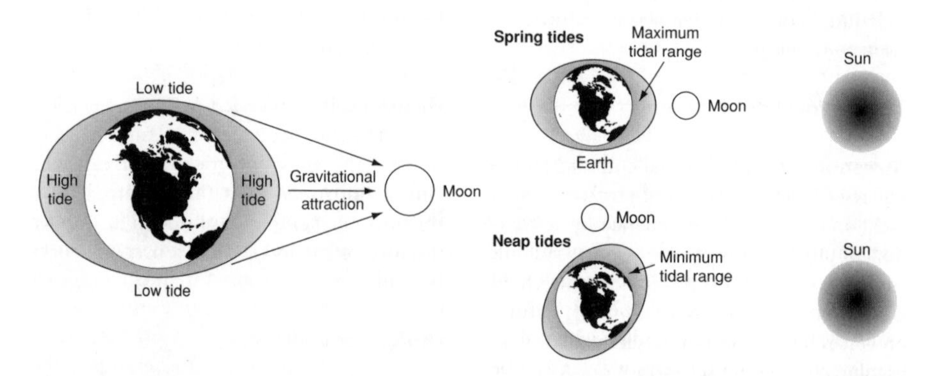

Tides – the gravitational pull of the Moon

till-fabric analysis The measurement and analysis of the constituents of glacial TILLS, with the object of reconstructing the past glacial history of an area. One simple method of till-fabric analysis is to measure the orientation and dip of the long axes ('a axes') of a sample of 50–100 elongated stones contained within a till. These are then plotted on a *rose diagram*, designed to show the preferred orientation (if any) of the stones. The assumption is usually made that, in LODGEMENT TILLS (or GROUND MORAINES) formed by the slow bottom-melting of an ICE SHEET, these stones (which have become aligned, within the ice, parallel to the direction of flow) retain their orientation and dip as deposition proceeds. Thus, till-fabric analysis allows the direction of ice movement to be reconstructed even from very old tills, provided that these have not been disturbed by localized slumping, periglacial freeze-thaw disturbance, or later ice advances.

time–distance convergence The reduction in the travel time between places and the decline in the importance of distance brought about by improvements in transport and COMMUNICATION. By reducing the FRICTION OF DISTANCE, the world is 'shrunk'. It is an important concept in understanding changing SPATIAL DISTRIBUTIONS, SPATIAL INTERACTIONS and SPATIAL RELATIONSHIPS.

timesharing A relatively recent phenomenon in the contexts of RECREATION and TOURISM, which allows people to have a small stake in a holiday property (or SECOND HOME). Developers construct blocks of flats, undertake chalet developments and subdivide large country houses into flatlets, and then sell each unit, fully furnished, on a weekly basis. Each customer buys the unit for a particular week or weeks either for a specified number of years or in perpetuity. The sale price of each week is determined by its timing in the year and therefore the weather expectations at that time. In the case of coastal resorts, high-summer weeks will be at a premium, while for winter ski resorts the guarantee of good snow conditions will be imperative. Similarly, those weeks coinciding with public holidays and school holidays will tend to be more expensive.

time lag On a HYDROGRAPH, the time between peak PRECIPITATION and peak river DISCHARGE. It is a measure of how 'flashy' a river basin is, and how quickly it responds to precipitation.

TNCs (transnational corporations) Enterprises that, because of their size, their arena of operation and their merger of firms scattered in many countries are often known, in short, as the *multinationals*. Not only are they multi-plant enterprises, but they are also multiproduct. The general trend has been for a larger and larger share of economic activity at the regional, national and international scales to be performed by a relatively small number of extremely large business corporations. Direct investment by TNCs today is thought to account for more than one-fifth of total industrial OUTPUT of the non-communist world. They penetrate nearly every country and, in almost all cases, they are increasing their share of GROSS DOMESTIC PRODUCT; in some cases, they are responsible for more than one-third of total manufacturing output. TNCs are more prominent in some branches of economic activity than others. For example, they are dominant in the HIGH-TECHNOLOGY INDUSTRIES (e.g. IBM); they are also very conspicuous in motor vehicles (e.g. Ford), chemicals (e.g. ICI), mechanical and electrical engineering (e.g. GEC) and oil (e.g. Exxon). They are key players in ECONOMIC GLOBALIZATION. See DUAL ECONOMY, INTERNATIONAL DIVISION OF LABOUR.

tombolo A SAND or SHINGLE bar, resulting from the extension of a SPIT by LONGSHORE DRIFT or the migration of an OFFSHORE BAR towards the coast, and linking an island with the mainland.

topography The description of the surface features of a place. Topography, strictly speaking, refers not only to physical features (in which sense it is used frequently though incorrectly), but also human features (SETTLEMENTS, COMMUNICATIONS, etc.). A MAP that shows these features in detail (for example, the 1:50,000 and 1:25,000 Ordnance Survey maps of Great Britain) is known as a *topographical map.*

topological map A MAP based on data that have been subjected to *topological transformation.* On the map, some basic aspects of the real world, like boundaries, the original number of locations and linkages, are faithfully reproduced, but distance and direction are subject to distortion. The transformation or distortion is undertaken to enhance communication of information and to eliminate irrelevant factors. As a result, the topological map assumes a diagrammatic quality. The map of the London Underground is a very famous example of a topological map. Although there is no scale and no accurate portrayal of direction, the map is effective in showing the individual stations and the different lines that make up the underground system. The topological map is widely used in NETWORK ANALYSIS. Sometimes referred to as a *cartogram.* [*f*]

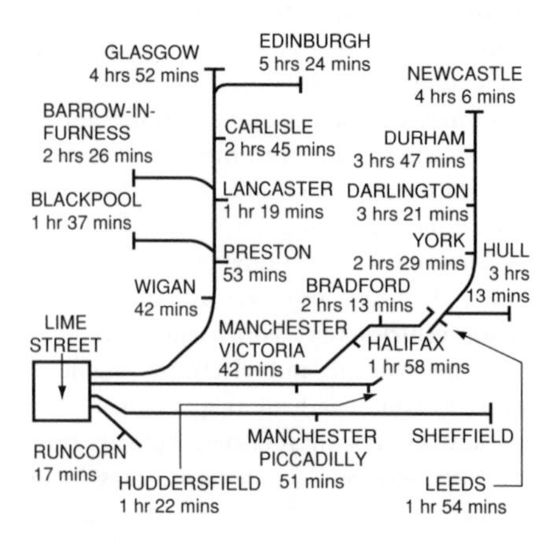

Topological map of rail services from Lime Street Station, Liverpool, England

toponymy The study of place names, particularly of their derivative elements, linguistic origins and meaning. The evidence contained in place names about former aspects of the ENVIRONMENT (physical conditions, patterns of SETTLEMENT and colonization, cultural diffusion, etc.) has been an important source material in HISTORICAL GEOGRAPHY.

tor A rocky outcrop most commonly found on hilltops, varying in scale but usually less than 30 m in height, in which the vertical and horizontal JOINTS are clearly exposed. Tors are characteristic features of GRANITE landscapes, although they are not restricted to these. One theory of tor formation is based on the assumptions of a variably jointed granite and an episode of deep CHEMICAL WEATHERING under warm and humid climatic conditions. Weathering is intense where the joints are closely spaced but much less effective where the joints are more widely spaced (see the accompanying figure). When, in due course, the REGOLITH is removed (by SOLIFLUCTION or fluvial action), the less weathered section of granite is left upstanding to form a tor. [*f*]

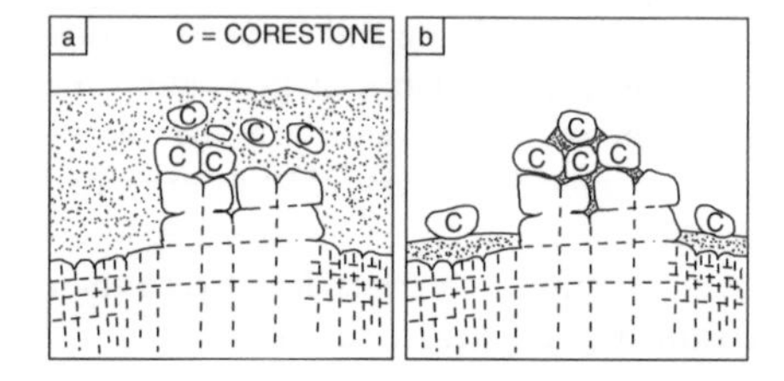

Formation of a tor by (a) deep weathering and (b) stripping of the resultant regolith containing corestones

tornado A counter-clockwise whirling storm, formed around a small cell of very low atmospheric pressure. Wind speeds are exceptionally high (sometimes in excess of 300 km hr^{-1}), causing serious structural damage to buildings along the narrow storm path. Tornadoes develop frequently in the Mississippi Basin, at the line of junction between warm, damp air from the Gulf of

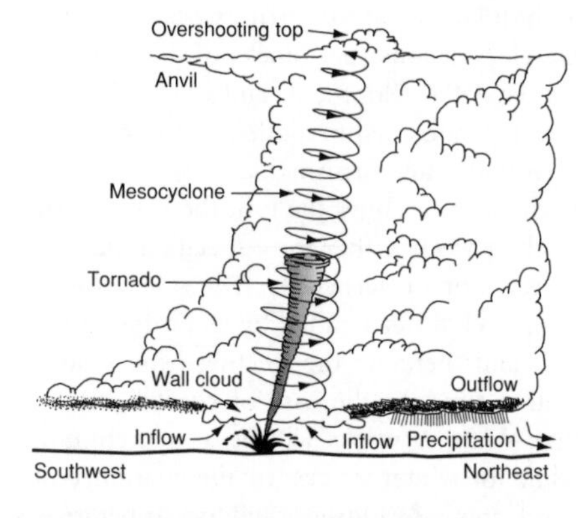

A supercell thunderstorm is an unusually large and intense weather system that features an exceptionally strong updraught. The most violent tornadoes are spawned by supercells

Mexico and cool air from the continental interior, mainly in spring and early summer when there is the additional factor of thermal heating of the ground surface. Tornadoes are usually very short-lived, lasting for only an hour or two, and are limited in size (usually only 100 m or less across). Their destructive effects are also very localized, though they can cut a swathe across a town. [*f*]

tourism LEISURE-time activity generally defined as involving at least an overnight stay away from home, thereby distinguishing it from RECREATION. Tourism frequently involves an important commercial dimension, with investment being made in hotels, motels, caravan sites, etc., in the provision of a diversity of tourist-oriented services, and in the improvement of access to locations favoured by tourists. Thus tourism, through its generation of employment and economic wealth, can make a significant contribution to the local and regional ECONOMY. Indeed, tourism is currently the largest and most rapidly expanding industry in the world. Where tourists are attracted in large numbers from other countries, tourism can create *invisible earnings* (see EXPORTS) and make a notable difference in the BALANCE OF PAYMENTS of a country. See ECOTOURISM. [*f*]

tower karst A type of tropical KARST landscape, more 'advanced' in development than COCKPIT KARST. Tower karst comprises steep-sided, isolated masses of LIMESTONE (MOGOTES), which can be regarded as the equivalent of the INSELBERGS of granitic terrains. The formerly extensive limestone uplands have gradually been undermined by basal WEATHERING, leading to the formation of many individual residual masses, small in extent but of considerable height. These 'towers' rise up to 1000 m or more above a near-level plain, resulting from prolonged SOLUTION at the level of the permanent WATER TABLE. [*f* COCKPIT KARST]

town A compact SETTLEMENT, larger than a VILLAGE, with a community pursuing an URBAN way of life.

townscape In essence, this is the URBAN equivalent of LANDSCAPE and comprises the visible forms of the BUILT-UP AREA, particularly street plan and layout, architectural styles, land and building use. Cf URBAN MORPHOLOGY.

toxic waste Chemical waste that is poisonous to people and wildlife. Until recently, the most common way of to get rid of this was to dump it in LANDFILL SITES. The risks of GROUNDWATER pollution are now such that it is increasingly common for such waste to be disposed of by incineration. However, the POLLUTION risks to the ATMOSPHERE do seem to be considerable. In theory, the movement of hazardous waste across national borders is prohibited.

trace element An element that occurs in the SOIL in minute quantities, but that is of vital importance to growing plants. However, if the trace elements are too abundant, they can actually make the soil poisonous.

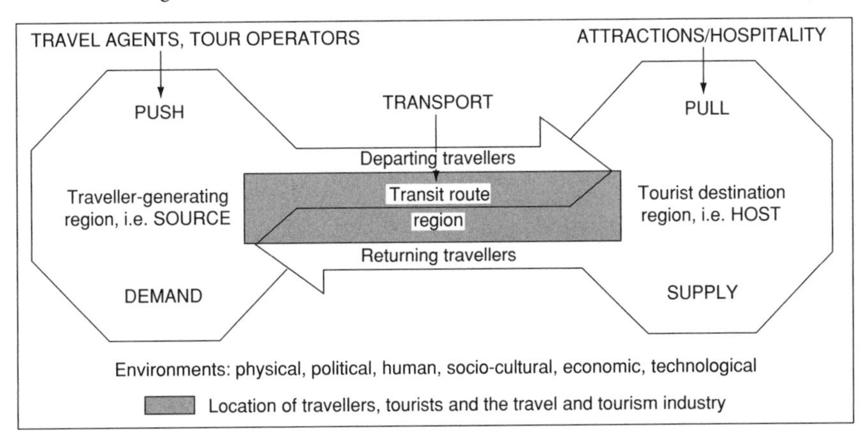

A model of tourism

Valuable trace elements include manganese, boron, molybdenum, copper, cobalt and lead; these are often provided by the WEATHERING of basic IGNEOUS ROCKS.

traction load See BED LOAD.

trade A flow of commodities and services from producers and providers to consumers. The generation of trade may be explained in terms of either COMPARATIVE ADVANTAGE or of exchange relationships within and between different MODES OF PRODUCTION. Trade is a vital aspect of economic DEVELOPMENT as well as a prime example of SPATIAL INTERACTION. See also COMPLEMENTARITY, BALANCE OF TRADE.

trade (trading or **economic) bloc** A group of countries drawn together by trade agreements that promote free trade between them. At the same time, the group as a whole is protected from other countries by TARIFF walls. The EU and NAFTA are two such blocs that account for a large proportion of global trade.

trade cycle See BUSINESS CYCLE.

trade friction A term used to describe the tensions that tend to surface when there is imbalanced TRADE between two trading partners, as has occurred between Japan and the USA. The trading balance between these two countries shows a huge surplus in favour of Japan. As a result, the USA has harboured feelings of resentment; it has also tried to explain its lack of success by accusing Japan of indulging in unfair trading practices. This, in turn, has led to threats of retaliation in the form of PROTECTIONISM. For its part, Japan has felt offended by these accusations and threats.

trade gap The shortfall of EXPORTS compared with IMPORTS.

trade wind See HADLEY CELL.

trading estate A comprehensively planned industrial estate designed by local or national government, primarily to diversify employment opportunities in areas of high unemployment or of unbalanced industrial structure (as where there is undue dependence on a major basic INDUSTRY, e.g. as in NE England). Also constructed around the margins of TOWNS and CITIES to accommodate industry relocated from inner-urban areas (as around the edge of the Greater London CONURBATION during the postwar period). Facilities offered include services, standard designed factories and centralized information, publicity and administrative services. Cf BUSINESS PARK, RETAIL PARK, SCIENCE PARK.

traffic principle One of three principles underlying Christaller's CENTRAL PLACE THEORY and governing the spatial arrangement of CENTRAL PLACES relative to their HINTERLAND. The traffic principle (sometimes also referred to as the *transport principle*) applies where central places are located so that lower-order centres lie along the straight-line paths between higher-order centres. This arrangement, having a K-VALUE of 4, represents the most efficient and rational arrangement from the point of view of the road network required to link the system of central places together. Ct ADMINISTRATIVE PRINCIPLE, MARKET PRINCIPLE. [*f* ADMINISTRATIVE PRINCIPLE]

transfer cost The sum of money that must be offered to attract the supply of a FACTOR OF PRODUCTION (LABOUR, CAPITAL, etc.) away from alternative uses. This sum is the necessary supply price of that factor in its new use. Transferability refers to the TRANSPORT COST characteristics of different commodities and to the handling characteristics of different types of goods. The transport costs of some goods are small relative to their value; some goods are more sensitive to distance than others, while some goods are more easily handled than others. High transferability would indicate that the commodity has a high value relative to its transport costs and that it is readily transportable (e.g. gold bullion). Cf COMPLEMENTARITY, INTERVENING OPPORTUNITY THEORY.

transfluence (glacial transfluence) A type of *glacial watershed breaching* in which ice builds up in a valley whose exit is blocked (for example, by a much larger glacier). The impeded ice will eventually escape, not by way of an ice tongue crossing a COL into a parallel valley (as in DIFFLUENCE), but at the head of the valley across a major WATERSHED.

transform fault See PLATE TECTONICS.

transformation (i) See TOPOLOGICAL MAP. (ii) See LOGARITHMIC TRANSFORMATION.

transhipment The transfer of goods from one mode of transport to another, as at a seaport, airport or railway goods yard. The expenses of loading and unloading between transport modes are referred to as *transhipment costs*. See also BREAK-OF-BULK POINT.

transhumance The seasonal movement of people and animals to fresh pastures. There are three main categories: (i) *alpine* or *mountain* – a movement from valley floors to the high summer pastures or ALPS for the summer, as in Switzerland and Norway; (ii) *Mediterranean* – a movement from the drought and heat of the lowlands in summer into the mountains, as in Spain; (iii) *nomadic pastoralism* – a movement near the borders of DESERTS according to fluctuations in rainfall and therefore pasture, usually following set seasonal tracks.

transition zone See ZONE OF TRANSITION.

transit trade Freight traffic that passes from one country to another across a third. For example, a vast volume of freight enters The Netherlands (via Europoort) and moves up the Rhine to Germany and Switzerland, while Luxembourg derives considerable revenue from transit rail freight across its territory.

translocation The transfer of material within a SOIL by solution or in suspension from one HORIZON to another.

transmigration A government-led programme in Indonesia to encourage people from the overpopulated, fertile islands of Java and Bali to resettle on the outlying, underpopulated islands of Sumatra, Kalimantan and Irian Jaya. It is the world's largest effort to relocate people. The original plan was to move some 68 million people between 1984 and 2008.

transpiration The loss of water vapour from a plant through the minute pores (*stomata*) that cover the leaf surface. The amount of water transpired will be influenced by the structure of the plant (which may be adapted to restrict water losses, especially in arid environments), temperature, humidity and wind speed. A large tree (such as a fully grown oak in S England in summer) may transpire several hundreds of litres of water each day; a maize plant, on the other hand, will transpire only 2–3 litres a day. See also EVAPOTRANSPIRATION.

transport, transportation (i) In the geomorphological sense, the movement of SEDIMENT by an agent such as running water, wind, glaciers, ICE SHEETS, breaking waves and tidal currents. (ii) Conveying people and commodities from one location to another by means of MODES such as road, rail and air.

transport cost The total cost of moving a good, usually proportional to weight or volume and to distance carried (see FREIGHT RATES). Strictly speaking, the cost should also take into account the costs of packaging, insurance and of dealing with the paperwork that normally accompanies the movement of goods (e.g. completing customs forms, invoices, etc.). Transport costs clearly represent a major PRODUCTION COST. For this reason, they figure prominently in location theories, notably in those of Hoover and Weber. While it might be argued that technological advances in transport (pipelines, bulk carriers, etc.) have reduced the general significance of transport costs, the rising costs of fuel might be seen as working in the opposite direction.

transport geography The study of geographical aspects of transport that currently include such diverse themes as spatial aspects of transport (NETWORKS, terminals, flows of commodities and passengers), the part played by transport as an agent of spatial change (especially in the context of DEVELOPMENT) and the impact of transport on ACCESSIBILITY and MOBILITY in specific areas and with reference to different social groups.

transport network The transport routes (roads, railways, canals, etc.) connecting a set of NODES (i.e. junctions and terminals); the links between origins and destinations. See CONNECTIVITY, NETWORK.

transport network development model This model identifies six sequential stages in the evolution of a TRANSPORT NETWORK. It starts with a series of small ports scattered along a coastline, each being isolated and having its own small HINTERLAND. In the next stage, a few of those ports grow more rapidly than the rest and develop longer lines of transport and communication into the interior, while in the following stage intermediate centres grow up along these lines. During the next two stages, there are increasing degrees of integration and interconnection between the two lines, leading eventually in the final stage to the establishment of high-priority routes providing direct connection between the most important centres. [*f*]

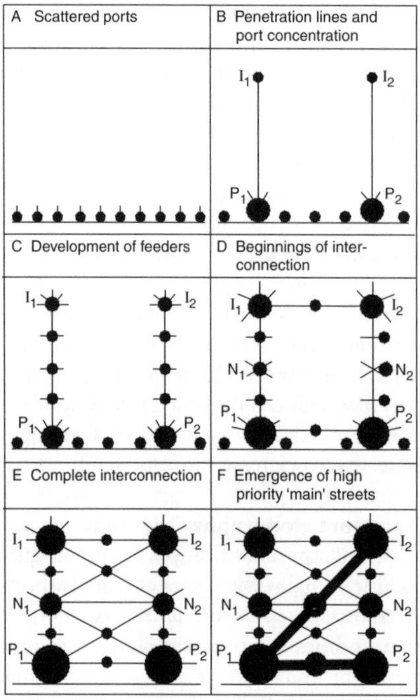

Transport network development model

transverse dune See BARCHAN.

tree line The 'line' marking the limits of tree growth. The tree line refers both to the altitudinal limits (as in mountain areas such as the Alps, where the change from the lower forested slopes to the upland grassy meadows is often quite abrupt) and the latitudinal limits (as on the northern fringes of the BOREAL FOREST, where the tree line is more of a broad transitional zone between the coniferous woodland and the dwarf shrubs of the TUNDRA). In the USA the tree line is known as the *timberline*.

trellised drainage A pattern of drainage often associated with scarp-and-vale landscapes (see CUESTA) and characterized by right-angled stream junctions. The main CONSEQUENT STREAMS flow across bands of resistant and unresistant rocks, following the direction of DIP. The weaker strata are eroded into STRIKE vales by SUBSEQUENT STREAMS, which are often responsible for the capture and disruption of consequents. Trellised drainage is thus well adjusted to geological structure. [*f* DRAINAGE PATTERN]

triangular diagram The plotting of three related or associated aspects of some feature or item on triangular GRAPH paper, ascribing a maximum value for each aspect to each apex of the triangle; e.g. three aspects of climate (pressure, temperature, humidity), population age (young, middle-aged, old), SEDIMENT (SAND, SILT, CLAY), sectoral employment (primary, secondary, tertiary). Sometimes referred to as a *ternary graph*. [*f* SOIL TEXTURE]

tributary area See HINTERLAND, MARKET AREA.

trickle down The term used by Hirshman (1958) to denote the spread of growth from the CORE to the PERIPHERY. Cf SPREAD EFFECT.

trophic level A level of food production or consumption within a FOOD CHAIN or ECOSYSTEM. Most food chains comprise three main trophic levels: *primary producers* (AUTOTROPHS), *primary consumers* (HERBIVORES), and *secondary consumers* (CARNIVORES). [*f*]

tropical cyclone See HURRICANE, TYPHOON.

tropical rainforest See RAINFOREST.

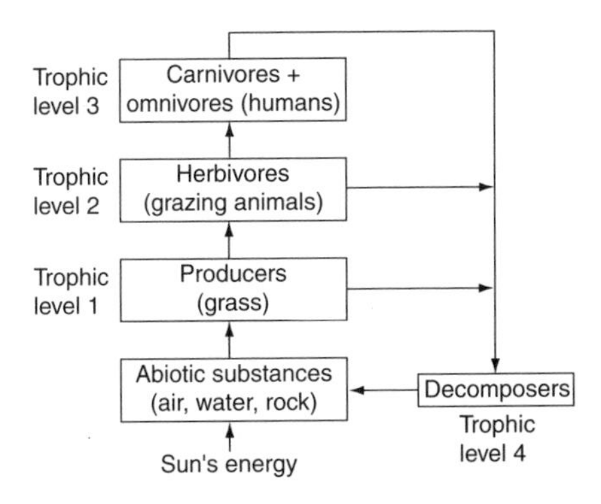

Trophic levels in an ecosystem

tropopause The junction between the TROPOSPHERE beneath and the STRATOSPHERE above.

tropophyte A plant that is adapted to withstanding seasonal periods of cold or drought. Tropophytes thus behave as XEROPHYTES during part of the year (when SOIL MOISTURE is not readily available and/or temperatures are too low for growth) and HYDROPHYTES during the season of growth. Examples of tropophytes are the oak, beech and birch of temperate deciduous woodlands, with their seasonal habit of summer growth and winter 'rest' (achieved by the shedding of leaves).

troposphere The lower part of the ATMOSPHERE, lying beneath the TROPOPAUSE. The troposphere varies in depth, from an average of 18 km above the Equator to only 6 km above the poles. Almost all WEATHER phenomena occur within the troposphere.

trough of low pressure An area of relatively low atmospheric pressure shown on a SYNOPTIC CHART by an extension of the isobars out in one direction from the centre of a DEPRESSION. A trough of low pressure is usually associated with unsettled showery weather.

trough end The abrupt head of a glaciated mountain valley (*glacial trough*), sometimes associated with near-vertical rock faces and spectacular WATERFALLS (as at the head of the Engstligen valley in the Bernese Oberland, Switzerland). Trough ends are believed to result from the immense powers of 'overdeepening' possessed by a large valley glacier that has been formed by the joining together of a number of smaller, much less powerful glaciers flowing out of high-level CIRQUES above the valley head.

truck farming See MARKET GARDENING.

truncated spur A spur, formerly projecting into a preglacial river valley, that has been eroded away in its lower part by a powerful valley glacier, which would have been unable to follow the same sinuous course as the preglacial river. Truncated spurs are common and characteristic landforms of glacial troughs such as the Lauterbrunnen valley, Switzerland.

tsunami A very large seismic sea wave, generated by an EARTHQUAKE shock on the ocean floor. Tsunamis travel for considerable distances across the sea as long, low waves (the wave length may be over 100 km, the wave height as little as 1 m, and the forward velocity as much as 600 km hr^{-1}). As the tsunami approaches the shore, the wave height increases markedly, and sometimes exceeds 15 m; it is thus capable of causing immense destruction to coastal SETTLEMENTS and severe loss of life. For example, the great Krakatoa ERUPTION of 1883, associated with seismic disturbances, caused tsunamis that drowned 36,000 people in coastal villages of Java and Sumatra.

tufa A deposit resulting from the PRECIPITATION of calcium carbonate, or *calcite*, from calcium bicarbonate dissolved in water. The precipitation process, in which carbon dioxide is released, is most effective when there is an increase in temperature or a decrease in pressure. Tufa is therefore commonly formed at points where SPRINGS emerge from underground passages in LIMESTONE or CHALK terrains. Within limestone caverns, features such as STALACTITES and STALAGMITES are formed from tufaceous deposits.

tundra The zone lying between the latitudinal limits of tree growth and the polar ice. (In reality, the tundra is a feature of the Northern Hemisphere only.) Winters are severe, with temperatures falling as low as -30°C at the northern limits. The growing season is brief and cool with the mean temperature of the warmest month rarely exceeding 10°C. Tundra is associated with the extensive development of PERMAFROST, which restricts root penetration and leads to waterlogging of the SOIL in many areas in summer. Soils are mainly skeletal, owing to the predominance of FREEZE-THAW WEATHERING, and the vegetation comprises mosses and lichens, low woody plants and deciduous dwarf shrubs up to 2 m in height (willow, alder, birch) along the southern, less severe margins of the zone.

turbidity current See SUBMARINE CANYON.

turbulent flow A type of flow usually associated with rivers in which there is a variety of secondary movements (eddies) superimposed on to the main forward movement of water. Turbulent flow in rivers is important in maintaining the suspended LOAD, which is constantly being raised by the upward eddies. Turbulent flow also occurs within the lower atmosphere when airflow is disrupted by surface obstructions such as buildings and areas of variable RELIEF. (ct LAMINAR FLOW)

twilight area Normally used with reference to INNER CITY areas in which there is a general deterioration in the condition of the URBAN fabric and in perceived status (see URBAN BLIGHT). The run-down character results from causes such as: the normal obsolescence of buildings with the passage of time; the failure to make proper investment in their maintenance; the increasing intrusion of non-residential uses into formerly residential areas; the progressive accumulation of poorer households in the substandard housing, etc. Twilight areas are frequently characteristic of the ZONE OF TRANSITION.

typhoon A tropical revolving storm, characterized by winds of very high velocity and torrential rainfall, occurring in the China Seas and along the western margins of the Pacific Ocean. Typhoons are a local type of TROPICAL CYCLONE or HURRICANE.

U

ubac A hillslope that, in the Northern Hemisphere, faces northwards or northeastwards, and thus receives the minimum amount of sunshine and warmth. Ct ADRET.

ubiquitous material In WEBER'S THEORY OF INDUSTRIAL LOCATION, this term refers to any material that is available everywhere (e.g. water) and therefore does not exert a pull of industrial location.

umland A German term for the HINTERLAND, MARKET AREA or SPHERE OF INFLUENCE of a CENTRAL PLACE.

unconformity A break in the continuity of a sequence of rocks. The older rocks beneath the unconformity are often folded or tilted, whereas the younger rocks above are usually of simpler structure (even horizontal). The unconformity represents a period of EROSION affecting the surface of the older rocks, leading in some instances to the formation of a PENEPLAIN. If this surface is then, for example, affected by a marine transgression, younger SEDIMENTARY ROCKS will be deposited on it. See also EXHUMATION. [*f*]

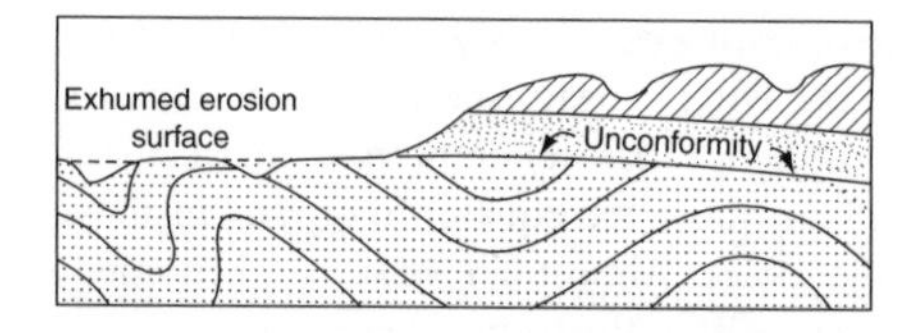

A geological unconformity and associated exhumation of an old erosion surface

undercliff A mass of slumped material at the foot of a sea-CLIFF that is being affected by large-scale MASS MOVEMENTS. The presence of an undercliff (usually in areas of incoherent SANDS and CLAYS, or in PERMEABLE chalk overlying IMPERMEABLE clay) indicates that SUBAERIAL cliff recession is at least as effective as marine EROSION in the strict sense of the term. [*f* ROTATIONAL SLIP]

undercutting LATERAL EXPLOSION at the base of a slope – for example, the undercutting of a stream bank by a meandering stream, the basal EROSION of a valley slope by a laterally shifting river, or the formation of a wave-cut notch at the base of a sea-CLIFF.

underdeveloped countries See LEDC, THIRD WORLD

underdevelopment The economic state of a country or region in which, broadly speaking, there is scope for the fuller exploitation of resources than is currently being achieved. So underdevelopment may be said to prevail in any area that has good potential prospects for using more CAPITAL, more LABOUR and more available RESOURCES to support its present POPULATION at a higher LEVEL OF LIVING or, if its per capita income is already fairly high, to support an even larger population at the same standard of living. Thus underdevelopment in an economic sense is a relative state that exists at a range of spatial scales. It undoubtedly exists in the THIRD WORLD, but equally it may be said to prevail in areas and regions of what would be described as developed countries (e.g. in the PERIPHERY areas of such countries). Some would argue that underdevelopment also has important political and social dimensions.

underground stream A stream flowing through an underground passage, usually within LIMESTONE. Underground streams usually commence as surface flows on IMPERMEABLE rocks adjacent to the limestone outcrop. They enter the limestone by way of SWALLOW HOLES and eventually re-emerge, often at the head of steep-sided gorges, in the form of RESURGENCES.

underpopulation This exists where RESOURCES and DEVELOPMENT could support a larger population without any lowering of the LEVEL OF LIVING or where a population is too small to develop its resources effectively. The former situation occurs in parts of Australia and New Zealand, where EXTENSIVE AGRICULTURE is capable of supporting quite a high level of living; areas of DEPOPULATION in W Europe would also fall into this category. Examples of the latter situation occur where low technical levels prevail, as among pastoral nomads (see PASTORALISM, NOMADISM) in semi-arid regions, or in areas of PIONEER SETTLEMENT, as in the remoter parts of Australia and Siberia. Ct OVERPOPULATION, OPTIMUM POPULATION.

unemployment rate The number of people of working age willing to work but unable to find jobs expressed as a percentage of all those of working age available for employment at a given time.

uniclinal structure A geological structure comprising a sequence of sedimentary strata dipping more or less uniformly in one direction. Uniclinal structures comprising alternating resistant and unresistant beds provide ideal conditions for the development of *scarp-and-vale landscapes*. See CUESTA.

uniformitarianism A fundamental geological principle, which states that Earth processes observed today are essentially the same as those that operated in the past. In short, rocks are being formed today in the same ways as they were hundreds of millions of years ago. However, uniformitarianism must not be interpreted in too simple a fashion. It does not imply that there have never been changes in the magnitude and frequency of Earth processes. Indeed, nowadays, many such processes are being speeded up by human activities (see, for example, DESERTIFICATION).

UNO (United Nations Organization) This formally came into being at the end of the Second World War in 1945, with the aim of maintaining international peace and security, and of establishing the right sort of political, economic and social conditions for the realization of that objective. There are now about 160 signatories to the United Nations Charter. Within the UNO a range of specialist agencies have been established, the most significant from a geographical viewpoint include: the FAO (Food and Agriculture Organization), GATT (the General Agreement on Tariffs and Trade, now the WTO), the IMF (International Monetary Fund), UNESCO (the *United Nations Education, Scientific and Cultural Organization*), the WHO (World Health Organization), the WMO (*World Meteorological Organization*) and the WORLD BANK.

unstratified A term applied to deposits that are not clearly layered – for example, TILL.

upward spiral See VIRTUOUS CIRCLE.

upward transition region See CORE-PERIPHERY MODEL.

urban Although widely used in the description of places, POPULATIONS and SETTLEMENTS, it is difficult to provide a simple, clear-cut definition beyond 'relating to, characteristic of, a TOWN or CITY (ct RURAL). This is partly explained by the fact that the urban condition relates to a number of different, but interrelated aspects, of which four might be regarded as being particularly significant. (i) Urban settlements generally have larger populations than rural villages, although the size threshold will vary from country to country. (ii) Urban places are characterized by higher POPULATION DENSITIES created by the spatial concentration of activities and buildings. (iii) In economic terms, urban settlements and places are concerned almost wholly with non-agricultural activities and, in many instances, with the provision of goods and services for HINTERLANDS (see CENTRAL PLACE THEORY). (iv) The people who live in urban places are thought to participate in a distinctive way of life (see URBANISM).

The degree and precise mix of these characteristics vary from place to place (e.g. THIRD WORLD cities exhibit features not encountered in MEDC cities) and from time to time (i.e. urban Britain today shows quite different characteristics than those that marked the urban growth of 100 years ago).

urban blight A deterioration in the physical condition of an urban area or a reduction in its general standing (see TWILIGHT AREA). It most frequently occurs when there is some doubt or uncertainty about the future. Rumours about possible redevelopment schemes or compulsory purchase, for example, might fuel such uncertainty, which in its turn dissuades property owners from making proper investment in the maintenance of buildings. In the case of residential areas, urban blight might have more of a social dimension and involve some reduction in social status or desirability. This might, for example, be caused by the intrusion of non-residential activities, by the concentration of a MINORITY group or by the conversion of single-family dwellings into multi-family use. Established families are increasingly persuaded to move elsewhere, only to be replaced by residents or landlords who, generally, are either less able or less willing to invest in property maintenance. Thus obsolescence and deterioration of fabric set in and so the blight spreads.

urban climate The distinctive climate associated with a large URBAN area. One major feature of urban climates is the occurrence of significantly higher temperatures than in the RURAL surroundings (see HEAT ISLAND). Other effects of BUILT-UP AREAS are: lower atmospheric humidity (owing to rapid RUN-OFF of rainfall over IMPERMEABLE surfaces and reduced EVAPOTRANSPIRATION); an increase in the ENVIRONMENTAL LAPSE-RATE (resulting from the presence of rapidly heated road surfaces and roofs of buildings), making convectional rainfall and even THUNDERSTORMS more likely; local changes in the pattern and speed of winds, as the air is funnelled between tall buildings and along narrow streets. In cities where there are marked concentrations of impurities in the air, RADIATION FOG, or even SMOG, is likely to form under conditions of atmospheric stability and INVERSION OF TEMPERATURE.

urban conservation This generally aims at retaining or protecting the traditional appearance of a TOWN's fabric, i.e. maintaining, renovating and enhancing those parts that have character and represent good examples of past achievements in urban design. At the same time, however, urban conservation will often involve finding new functions for old buildings, adapting old structures to new uses, as well as allowing a certain amount of DEVELOPMENT, provided it is in harmony with what already exists. In this respect, it is a form of URBAN RENEWAL, and it is important to stress that good CONSERVATION does not seek to preserve in a 'fossilizing' sense.

urban continuum See CONTINUUM, URBAN HIERARCHY.

urban density gradient See DENSITY GRADIENT.

urban ecology See CONCENTRIC ZONE MODEL.

urban fallow URBAN land that is currently unused, possibly because it is in the process of REDEVELOPMENT or because it is the victim of INNER-CITY DECLINE. Cf BROWNFIELD SITE.

urban field A term formerly used by British geographers when referring to the area located around, and functionally linked to, a TOWN or CITY. Other terms are now more widely used, e.g. HINTERLAND, MARKET AREA, SPHERE OF INFLUENCE, *tributary area.*

urban geography That branch of geography which concentrates on the location and spatial arrangement of TOWNS and CITIES, seeking to describe and explain both the DISTRIBUTION of URBAN places, and the similarities and contrasts (in economic and social terms) that exist between them. It is also concerned with the internal arrangements of towns and cities, identifying both the patterns (especially of LAND USE and social areas) and the processes that have moulded those patterns over time. A final area of interest lies in those contemporary issues that have a spatial dimension, as for example INNER-CITY DECLINE, housing and traffic congestion (see URBAN PLANNING).

urban hierarchy The vertical CLASSIFICATION of TOWNS and CITIES according to a single VARIABLE, such as POPULATION size, extent of BUILT-UP AREA or CENTRAL PLACE status. If each urban SETTLEMENT within a given country or REGION is plotted on a DISPERSION DIAGRAM according to the value it records for the chosen variable, the array of plotted values may reveal a degree of either even spacing or clustering. The former situation would be described as indicating the existence of an *urban continuum* (with each urban centre occupying a unique position along the variable). In the latter situation, each cluster might be regarded as a distinct class or order within the urban hierarchy; the lower the mean value of each cluster, the lower its status or standing in the hierarchy. The character of regional and national URBAN SYSTEMS may be such that the urban hierarchy is often perceived as a pyramidal structure (see CENTRAL PLACE HIERARCHY), so that the lower the status, the greater the number of representative settlements (i.e. the larger the cluster on the dispersion diagram). Crudely put, in most urban hierarchies, it will be found that there will be more towns than cities, and more cities than regional or provincial capitals. Ct CONTINUUM. [*f*]

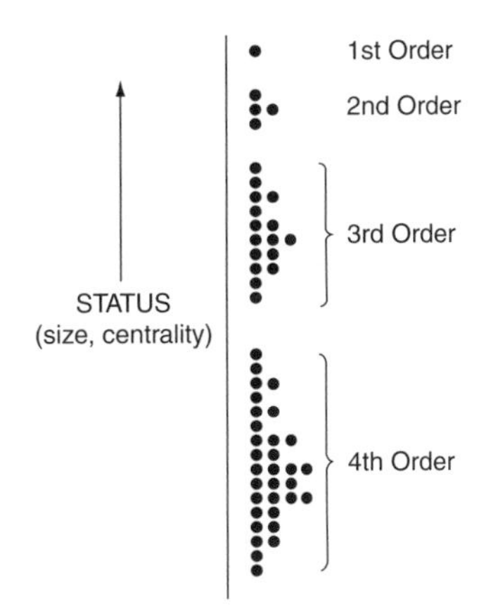

The urban hierarchy

urban land-value surface Spatial variations in rents or land values within the BUILT-UP AREA, created by the bidding process in what is normally assumed to be a free-market situation (see BID-RENT THEORY). It is prompted by spatial variations in the level of demand and the locational qualities of individual sites. The urban land-value surface is thought to exert a very powerful influence over the spatial structure of TOWNS and CITIES, especially the arrangement of LAND USES and different social groups. The accompanying figure depicts some of its salient characteristics: land values peak at the centre, where the demand for sites is highest and where CENTRALITY is greatest (i.e. the CBD); land values decline towards the city margins, but the land-value gradients vary in different directions from the peak at the centre (thus the surface is somewhat asymmetrical overall); in any one direction, the land-value gradient is unlikely to be uniform (e.g. a significant break or fall may be expected at the margins of the CBD); relatively high land values will be maintained along the major axial routes leading to the centre and also along ring routes (such routes being perceived as offering enhanced accessibility); at the intersections between axial and ring routes, the appearance of minor land value peaks reflects the prized NODALITY of such points. [*f*]

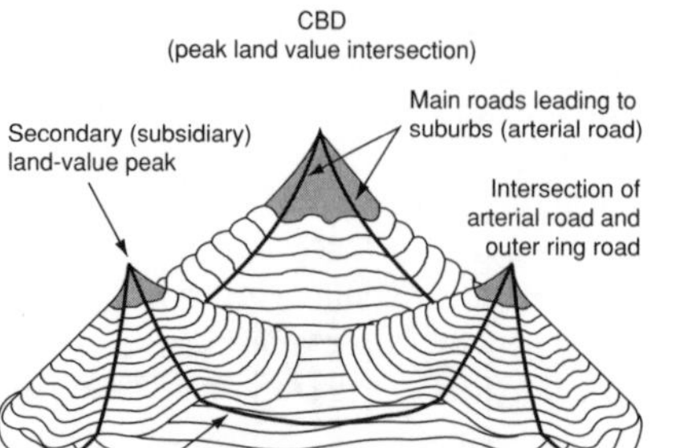

Urban land-value surface

urban managers Bureaucrats (i.e. planners, housing and welfare officers, councillors) who, along with GATEKEEPERS, exert considerable influence over the allocation of scarce resources in URBAN areas, as between different groups and between different localities. Their control is most keenly felt in relation to PUBLIC GOODS AND SERVICES, particularly with regard to such matters as access to housing (notably council housing), and the provision of social and welfare services (schools, clinics, community halls, etc.) in different parts of the city. See also TERRITORIAL JUSTICE. [*f*]

urban mesh The network of TOWNS and CITIES found in a given area, viewed especially in terms of spacing, spatial pattern, functional linkages and connectivity. See also URBAN SYSTEM.

urban morphology Analysis of the built fabric of TOWNS and CITIES (its layout, form, functions, etc.) and of the ways in which this fabric has evolved over time. Cf TOWNSCAPE.

urban planning The process of managing changes in order to achieve particular objectives as regards the URBAN SYSTEM. PLANNING activity may assume a number of different complexions, as well as operate at different spatial scales. For example, much urban planning is undoubtedly concerned with the amelioration of urban problems, such as trying to correct imbalances within the urban system (e.g. devising programmes of DECENTRALIZATION to transfer growth from PRIMATE CITIES to lower orders of urban centre in less favoured areas) or to improve conditions within individual TOWNS and CITIES (by introducing housing programmes and traffic schemes, and providing better social and welfare services). In contrast, urban planning can be much more forward-looking in the sense of projecting current trends, forecasting problems likely to arise and then devising appropriate planning strategies to minimize those problems and maximize the benefits. Such a planning exercise might involve projecting the trends of INNER-CITY DECLINE and then formulating a programme of action. The programme might not necessarily seek to resist the decline, but might instead aim to carefully phase the removal of people and jobs, and the rundown of services. By so

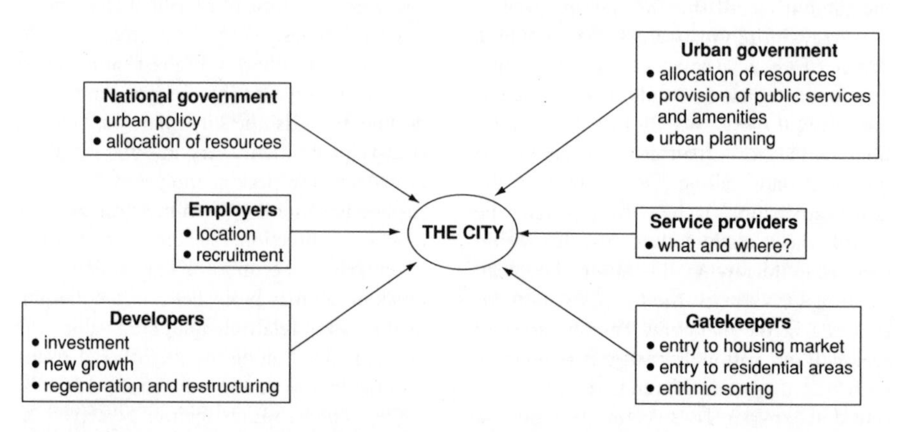

Urban managers and some of their responsibilities

doing, and by ensuring their efficient relocation and accommodation elsewhere, the impact of the decline might be minimized. Finally, it is significant to note that the traditional segregation of town and country planning has given way today to a much more comprehensive approach to planning issues, for it has been recognized that the solution of urban problems will often involve RURAL areas and vice versa.

urban primacy See PRIMATE CITY.

urban renewal The renovation and rehabilitation of obsolescent URBAN areas by means of either *improvement* (e.g. by installing modern facilities in old dwellings, by road widening, etc.) or *redevelopment* (i.e. demolishing all existing structures and starting afresh). In the early postwar period, urban renewal programmes undertaken in British cities tended to concentrate on REDEVELOPMENT (no doubt encouraged by the ravages of bomb damage suffered during the Second World War), but subsequent realization of its SOCIAL COSTS (especially where residential areas are concerned) prompted a shift of emphasis towards improvement both of individual structures and of the urban ENVIRONMENT.

urban rent theory See BID-RENT THEORY.

urban–rural continuum See RURAL–URBAN CONTINUUM.

urban sprawl A largely unplanned, straggling and low-density form of URBAN or SUBURBAN growth occurring around the margins of a TOWN or CITY, particularly along radial routeways (see RIBBON DEVELOPMENT) and often leading to the coalescence of once-separate settlements (see CONURBATION). Sprawl was a characteristic of much suburban growth in Britain during the inter-war period when, without the strict planning controls that now apply, large amounts of farmland were sold off to speculative builders. In the USA areas of such development are often referred to as *slurbia*. See also SUBURB, SUBURBANIZATION.

urban system The NETWORK of URBAN settlements found in a given area, but with each urban SETTLEMENT and its dependent HINTERLAND seen as constituting a distinct, urban-centred REGION. An essential feature of the urban system is the interdependence of its constituent settlements. No one TOWN or CITY is wholly self-sufficient – rather it relies, to varying degrees, on goods or services produced and provided elsewhere, i.e. by other urban centres. The complex functioning of the urban system relies critically upon the development of a connecting TRANSPORT NETWORK, for it is through this that the vital movement of people, goods, CAPITAL, etc. takes place, thereby facilitating interaction between urban settlements, and between urban settlements and their hinterlands. Also crucial to this integration is the development of efficient COMMUNICATIONS systems.

urban village A VILLAGE that has become engulfed by the BUILT-UP AREA of an expanding CITY, but which still retains part of its original character and identity (e.g. Fulham, Hampstead and Islington in London). A residential district within a TOWN or city and in which the inhabitants share a strong sense of community and local attachment. In many instances, this sense of identity stems from the grouping of households with similar ethnic characteristics or sharing a common national extraction. It is further reinforced by the provision of specialist services (shops, places of worship, clubs) to meet the particular needs of the group. Urban villages are well exemplified in the inner areas of N American cities, due to the concentration of minority immigrant groups in specific locations (see GHETTO). The existence and social cohesion of such urban villages tend to contradict the general claim that the urban way of life (URBANISM) is characterized by anonymity and impersonality. [*f* GHETTO]

urbanism The lifestyles, values and attitudes that characterize people who inhabit TOWNS and CITIES. The question is increasingly being raised as to the degree to which URBAN and RURAL populations really differ with respect to these three criteria. The drift of opinion is that, both in ADVANCED COUNTRIES and THIRD WORLD, nations the differences are blurred, if not insignificant, particularly since urban-based mass media serve to promote urbanism in country districts, well beyond the limits of the BUILT-UP AREA.

urbanization The process of becoming URBAN; a complex process of change affecting both people and places. Its main dimensions are: a progressive concentration of people and activities in TOWNS and CITIES, thereby increasing the general scale of urban SETTLEMENT; a change in the ECONOMY of a country or region, whereby non-agricultural activities become dominant; a change in the 'structural' characteristics of populations (e.g. lower BIRTH RATES, higher DEATH RATES, positive MIGRATION balances); a spread of URBANISM beyond the BUILT-UP AREAS of towns and cities, thereby inducing *rural dilution*; the transmission or diffusion of change (economic, social, technological) down the URBAN HIERARCHY and into RURAL areas.

Urbanization does not always take the same form, nor does it progress at the same rate everywhere (see URBANIZATION CURVE). In MEDCs, urbanization has now reached the stage of being rather more dispersed, through the proliferation of towns and cities, through DECENTRALIZATION and through the spread of URBANISM beyond the built-up area (see COUNTERURBANIZATION). In LEDCs, urbanization tends to be more concentrated, with large rural-to-urban migration flows converging on a limited number of large cities (see OVERURBANIZATION).

The rate of urbanization varies from place to place. Although the rate is broadly related to the speed and scale of economic DEVELOPMENT, suffice it to say that, in some of the most advanced MEDCs, the rate is beginning to slacken off (see, again, URBANIZATION CURVE). The degree of urbanization also shows marked spatial variations, with the MEDCs having more than three-quarters of their populations living in urban areas, while for some LEDCs the figure is less than one-quarter.

Urbanization brings both costs and benefits. On the cost or debit side, there are undesirable by-products such as poor housing, congestion, ENVIRONMENTAL POLLUTION and encroachment on agricultural land. On the other hand, for many people, urbanization brings material, social and economic progress, higher living standards, and the provision of diverse services (commercial, social, cultural) that contribute to the overall QUALITY OF LIFE.

urbanization curve A model that may be used to chart the degree and progress of URBANIZATION in a given country or region. The curve takes the form of an attenuated S and may be subdivided into three segments or stages. (i) The *initial stage:* during this phase the rate of urbanization is extremely slow and only a small proportion (less than about 25%) of the POPULATION live in URBAN settlements. During this phase, the traditional society persists, with the emphasis placed on the agrarian sector of the economy and with a largely dispersed RURAL population. (ii) The *acceleration stage*: during this phase there is a profound redistribution of population brought about by massive rural-to-urban migration. Population thus becomes progressively concentrated in a proliferation of TOWNS and CITIES. This demographic change is largely prompted by a fundamental restructuring of the ECONOMY, in that the SECONDARY SECTOR and the TERTIARY SECTOR become dominant. During this time, the urban component in the total population increases to between 60% and 70%. (iii) The *mature stage*: the curve begins to level off as the rate of increase in the urban population begins to fall back to match the overall increase in population. Further sectoral shifts occur in the economy, with the tertiary and QUATERNARY SECTORS becoming more and more important.

The recent onset in some highly urbanized countries of DECENTRALIZATION and COUNTERURBANIZATION might suggest that a possible fourth stage is about to be initiated, in which there is the prospect that the urbanization curve might eventually begin to fall away. For the moment, however, the losses appear to be hitting only the largest cities and to be matched by compensating gains in those urban settlements within the small to medium size range.

The urbanization curve is a highly generalized depiction of the sequence of change. At any point in time, different countries will occupy different positions along that curve, while, over time, countries will progress at different rates. Thus, for those countries experiencing particularly rapid urbanization, the curve will be very compressed along the timescale, while for those making slow progress, the curve will be even more attenuated. [*f*]

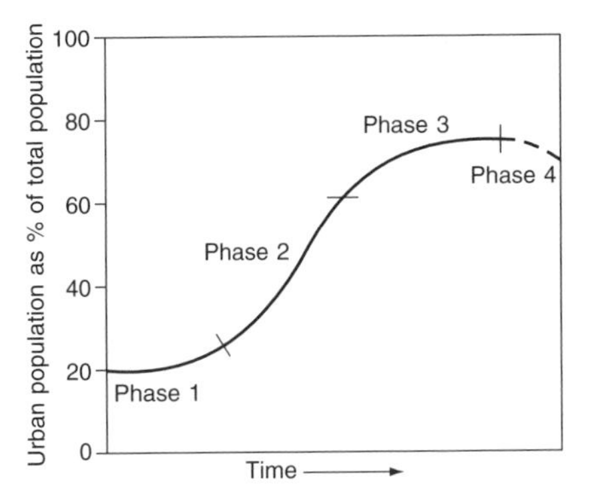

Urbanization curve

U-shaped valley (glacial trough) A 'typical' glaciated upland valley, with steep or near-vertical walls and a relatively flat floor (which is sometimes further emphasized by POSTGLACIAL alluvial deposits). The U-shaped valley is, in effect, the former 'channel' of a valley glacier. It has been suggested that the cross-profiles of glaciated valleys often approximate to mathematical curves (for example, catenary curves); however, in reality profiles are greatly variable, particularly in the UK where 10,000 years' worth of postglacial processes have left their mark. [*f*]

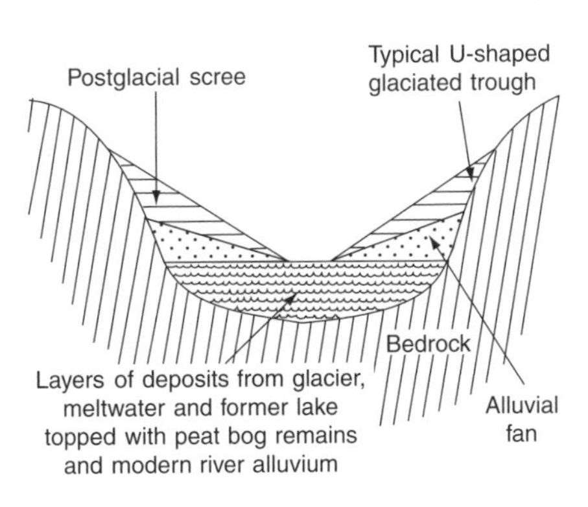

Cross-section through a glaciated valley

uvala An enclosed depression in LIMESTONE country, usually irregular in outline and resulting from the amalgamation of a number of DOLINES.

V

vadose water A term for water percolating downwards through the rock to the zone of saturation (see WATER TABLE). Hence the *vadose zone*, which is that lying above the *phreatic* zone (see PHREATIC WATER). [*f* WATER TABLE]

Valentin's coastal classification A threefold classification of coasts. (i) Advancing coasts, where either deposition or uplift of the land is dominant. Typical coastal features include SPITS, BARS, RAISED BEACHES, DUNES and SALT MARSHES. (ii) Retreating coasts, where erosion or submergence of the land is dominant. Typical features include STACKS, FJORDS, RIAS and WAVE-CUT PLATFORMS. (iii) Intermediate coasts – a stationary situation in which either DEPOSITION equals submergence or EROSION equals uplift.

valley glacier See ALPINE GLACIER.

value added The difference between the revenue gained from the sale of a commodity and the cost incurred in producing it. It is the value the production process adds to the INPUTS (FACTORS OF PRODUCTION). Value added is used as a basis for taxation in the EU. However, rather than being levied on the difference between the sale price and the costs of production, the tax tends to be a fixed percentage of the price charged for a particular good or service.

value judgement A decision made by a person in the light of their particular perceptions, prejudices, beliefs, etc.

Van't Hoff's rule The 'law' stating the rate of increase of chemical reactions with rise in temperature. In broad terms the increase is 2-3 times for every 10°C rise. Van't Hoff's rule helps to explain the considerable importance of CHEMICAL WEATHERING processes in tropical landscapes.

variable Any item or phenomenon that can

assume a range of individual values. A *continuous variable* may be defined as one in which there are no clear-cut or sharp breaks between the values. Variables such as length, weight, temperature and time are examples of this type in that any value within a prescribed range may be assumed. A *discrete* or *discontinuous variable* is one that can only be measured in terms of whole numbers or *integers*, as for example the number of children per family or the number of goals scored at a football match. See DEPENDENT VARIABLE, INDEPENDENT VARIABLE.

variable cost analysis An approach to the study of industrial location that concentrates on those costs (e.g. PROCUREMENT COSTS, PRODUCTION COSTS) that are subject to spatial variation (see VARIABLE COSTS). On the basis of variable cost input, a composite COST SURFACE is produced, and on this surface may be identified LEAST-COST LOCATIONS.

variable costs (i) Costs that vary with the scale of production; however, ECONOMIES OF SCALE may disrupt the simple arithmetical relationship between production costs and volume of production. Ct FIXED COSTS. (ii) In geographical studies, the term is usually employed in order to indicate costs that are subject to spatial variation; such variations are likely to have a strong influence on locational choice. See also VARIABLE COST ANALYSIS.

variable-k hierarchy See K-VALUE.

variable revenue analysis An approach to the investigation of economic location that concentrates on the demand rather than the cost side of the industrial location equation. See MARKET AREA ANALYSIS; ct VARIABLE COST ANALYSIS.

variance See ANALYSIS OF VARIANCE.

variate Any one value of a VARIABLE; an individual observation.

Varignon frame A mechanical model that may be used in the application of WEBER'S THEORY OF INDUSTRIAL LOCATION to determine the point of minimum TRANSPORT COSTS. The model simulates the LOCATIONAL POLYGON by appropriately scaled weights and pulleys connected by wires. The respective weights represent the strength of the attraction force of each corner of the polygon, while wire lengths are proportional to distance. The point of balance where the connected wires come to rest is then assumed to be the LEAST-COST LOCATION (i.e. the OPTIMAL LOCATION). [*f*]

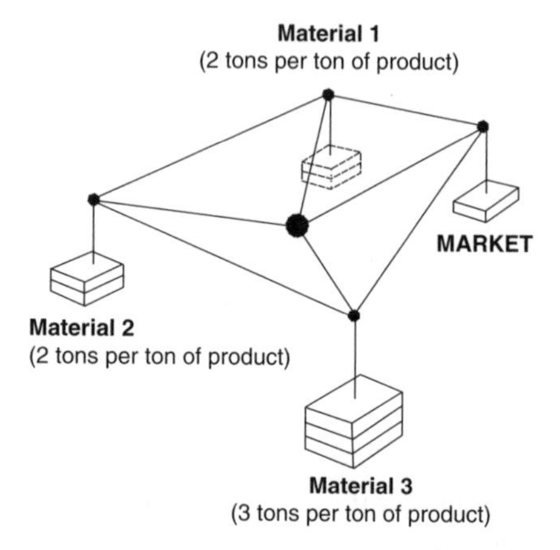

Mechanical solution to the multi-point location problem, using a Varignon frame

varve A distinct band of SEDIMENT deposited on the floor of a lake close to an ice margin. Most verve deposits comprise an alternating sequence of coarser (SAND, SILT) and finer (CLAY) bands, each a few mm in thickness. The coarser sediment is deposited in summer, when meltwater is abundant and stream transport is active; the finer sediment settles out slowly during the winter, when little or no coarse sediment is being washed into the lake. Thus each pair of varves represents a year's accumulation. The age (or duration of existence) of the lake can therefore be discovered by counting the varves on the lake floor. The varve deposits for individual years will vary somewhat in thickness, according to the year's weather and its effects on the rate of TRANSPORT and sedimentation. A particularly active year, represented by an abnormally thick pair of varves, might be identified in the varves of a number of separate lakes. Such information can tell us much about CLIMATIC CHANGE.

vauclusian spring See RESURGENCE.

vector analysis Vector analysis provides a method for investigating a moving phenomenon that experiences changes both in direction and velocity (e.g. the tide or wind). A *wind vector* may be produced by summing the frequency of wind-force records from eight different directions cumulatively. The bold arrow in the accompanying figure indicates the resultant *vector*, which is from WSW. [*f*]

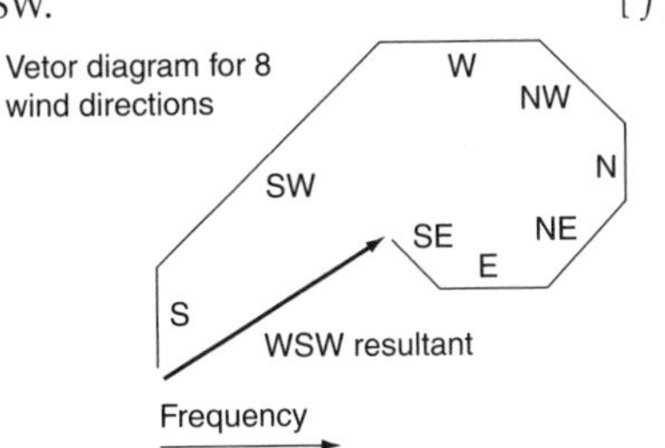

Vector diagram

vegetation (plant) succession The sequential development of vegetation types (PLANT COMMUNITIES), as part of the process whereby all vegetation, from its initial establishment, undergoes a series of modifications before the 'ultimate' vegetation cover (CLIMATIC CLIMAX VEGETATION) is attained. When a bare surface is created (for example, by glaciation, volcanic activity or coastal sedimentation) it is at first hostile to plant colonization, as it lacks SOIL cover and plant nutrients, and may be excessively wet or dry. Thus the *pioneer plant cover* will struggle to establish itself, and will comprise small plants (mosses, lichens and herbs). Subsequently, as soil begins to form, a series of plant communities (each known as a *seral stage* – see SERE) will develop. With the passage of time, the seral communities will tend to become more complex and to comprise larger plants (there is a natural succession: grasses–shrubs–trees), with each community actually helping to destroy that which precedes it. With each seral stage, soil and moisture conditions progressively improve, as HUMUS from the expanding plant cover is provided in increasing quantities, and WEATHERING adds to soil depth and releases plant nutrients. The development of vegetation on a 'natural' site (that is, one not created by people) is referred to as a *primary plant succession*. Where people create a new site by the destruction of the existing vegetation (as in SHIFTING CULTIVATION), recolonization by plants will be more rapid and (since soil and moisture conditions will from the start be more favourable), in some respects, quite different. This process is referred to as *secondary plant succession* (hence 'secondary forest'). See PIONEER COMMUNITY, PLAGIOCLIMAX.

Venn diagram A simple way of visually representing sets and subsets using circles drawn within rectangles. For example, if the rectangle is taken as representing a set or NETWORK of CENTRAL PLACES a circle might be drawn within it to represent the subset of central places that possess at least one bank. A further subset might be recognized, made up of those central places having a building society office. Since some central places will possess both establishments, the two subset circles will be shown on the Venn diagram as intersecting or overlapping. Thus the diagram represents a fourfold classification of central places: (i) subset *A*, comprising those that have both a bank and a building society office; (ii) subset *B*, made up of those with only a bank; (iii) subset *C*, those with only a building society office; (iv) subset *D*, those possessing neither facility. [*f*]

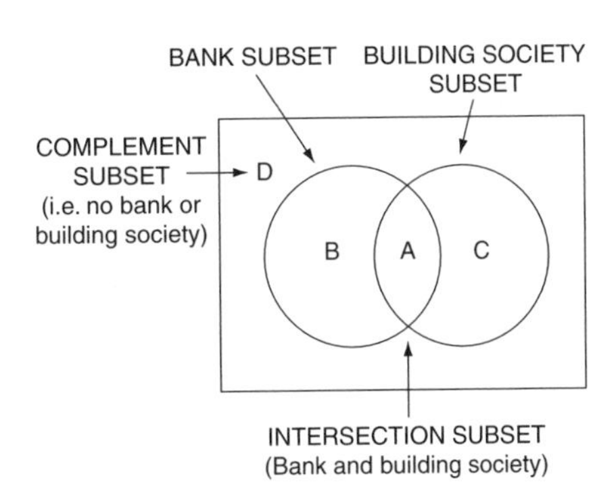

Venn diagram

ventifact A stone that has been shaped and 'polished' by wind ABRASION under DESERT conditions. Initially the ventifact will be worn away on the upwind side only, resulting in a facet or bevel. If the wind direction changes seasonally, or the stone is turned over, other bevels may be formed. This is believed to be the explanation of *dreikanter* (three-faceted stones).

vertical erosion The *in situ* downcutting by a river, in response to a rapid land uplift, fall of sea-level or an excess of energy due to increased DISCHARGE without an equivalent addition of SEDIMENT LOAD. Vertical erosion results in the formation of deep V-SHAPED VALLEYS and INCISED MEANDERS. Ct LATERAL EROSION.

vertical expansion One of three ways in which an enterprise may expand (ct DIVERSIFIED EXPANSION, HORIZONTAL EXPANSION), in this instance by involving itself further in the same production sequence. For example, a brewing company might enlarge its field of operation either by taking on the malting of barley (*backward vertical expansion*), or by controlling the distribution and retailing of its beer (*forward vertical expansion*).

vertical integration Vertical integration is achieved when the different stages of a production process are located on the same site, as in an integrated steelworks (where the refining and smelting of iron ore might take place alongside the rolling of sheet steel) or in a pulp and paper mill. The major benefit of having a succession of production stages in one location is the potential saving in the time and costs of transport. Vertical integration also offers ECONOMIES OF SCALE.

vicious circle In the context of REGIONAL DEVELOPMENT theory, this term is used to describe the sequence of consequences for the PERIPHERY of the increasing spatial concentration of RESOURCES and growth in the CORE (see also BACKWASH EFFECT). The term *downward spiral* is also applicable (ct VIRTUOUS CIRCLE; see also CYCLE OF POVERTY). The accompanying figure shows the nature of two vicious circles: one operating with respect to labour and the other to investment. [*f*]

village A grouping of buildings (houses, farms, shops, places of worship, etc.) in RURAL surroundings, smaller than a TOWN, larger than a HAMLET and without a municipal government. Villages were usually founded as agricultural SETTLEMENTS (although there are examples of planned *industrial villages*; e.g. New Lanark, Scotland, and Saltaire, Yorkshire), but they may not be so today, particularly those located within the COMMUTING orbits of towns and CITIES (i.e. *dormitory villages*).

virtuous circle The term adopted by Myrdal (1957) with reference to the circular process associated with the spatial concentration of resources in CORE areas, whereby they maintain their initial advantage. The term *upward spiral* is sometimes used. Ct VICIOUS CIRCLE.

visible trade The import and export of goods. Ct INVISIBLE TRADE.

vital statistics Numerical data dealing with births, marriages, deaths and other recorded information about local, national and international POPULATIONS. Cf POPULATION STRUCTURE.

viticulture Cultivation of the vine, usually with the aim of producing wine.

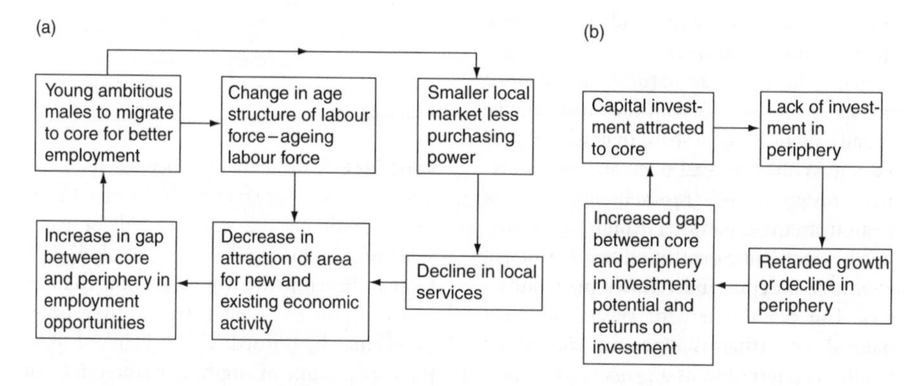

Vicious circles or downward spirals typical of the periphery (a) labour migration, (b) investment

volcanic rock See EXTRUSIVE ROCK.

volcano A conical hill or mountain (volcanic peak) built up by the ejection of igneous materials from a vent. The form of a volcano is influenced by many factors, such as the nature of the materials extruded (LAVA, cinders or ash); the chemical composition of the lava (ACID LAVA, rich in silica and relatively viscous, forms steep-sided, dome-like volcanoes; BASIC LAVA, containing little quartz, is more mobile, giving rise to more gently sloping volcanic cones); the extent to which either powerful explosive activity or central collapse has formed a large CRATER (see CALDERA); and the subsequent erosional history of the volcano, involving deep dissection by radial streams of its flanks or, in some instances, modification by glacial erosion. Some volcanoes form along a crack or fissure in the Earth's crust. These are termed *fissure volcanoes* and tend to lack the distinctive individual 'volcano' shape. Volcanoes are also classified as *active, dormant* or *extinct.*

volumetric symbols 'Three-dimensional' symbols whose volumes are proportional to the quantity being portrayed graphically (or cartographically, when located on a MAP). The construction of proportional spheres, cubes and columns is based on the cube root of the value being represented.

von Thunen's model A model, published in 1826, by a German economist-landowner, of the pattern of agricultural production and related LAND USES around a market town. The model makes a number of important simplifying assumptions, namely that the market town is situated in a physically uniform region and that TRANSPORT COSTS are directly proportional to distance. It is also assumed that each farmer in the region will sell his surplus produce only in that town, that he bears the total costs of transport himself and that he always aims to practise the type of farming that will yield maximum profit.

The model is based on the principle of ECONOMIC RENT, whereby different types of agricultural land use produce different net returns per unit area, by reason of different yields and different transport costs. As a result, the model is made up of a series of concentric zones, with each zone characterized by a particular type of agricultural production. Market gardening and dairying are located closest to the market town, since they require most labour, involve the highest transport costs and produce a perishable commodity. They yield the highest return per unit area, but as distance from the market increases, so the return falls to a point when it becomes more profitable to pursue another type of AGRICULTURE. Profitability and intensity of agriculture continue to decline outward from the market, with the result that one type of farming succeeds another until, eventually, cattle-grazing is reached. Since this requires least labour and involves least transport costs (the livestock are moved to market on the hoof), stock grazing becomes the dominant activity at the periphery of the region. Although technology and market conditions have changed greatly since the early 19th century, and although some of von Thunen's simplifying assumptions must be questioned, the model does remain a useful one, if only that it links economic concepts with spatial locations. [*f*]

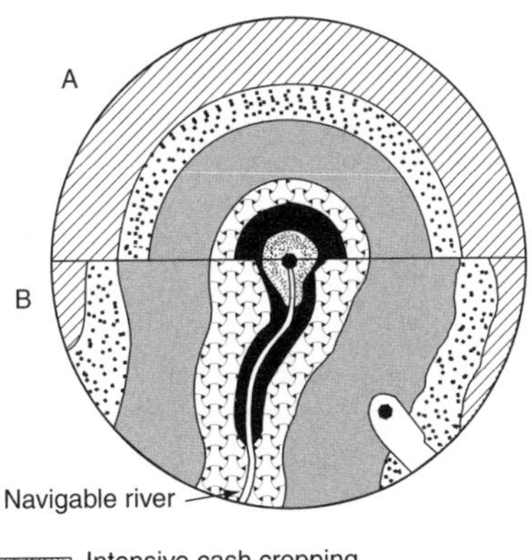

Land use in von Thunen's 'Isolated State': (a) the simple case and (b) the more complex case

V-shaped valley A 'typical' river valley, by contrast with a glaciated U-SHAPED VALLEY. The precise angle of the V will depend on the relative rates of VERTICAL EROSION by the river, and valley slope recession by WEATHERING, RAINWASH and mass transport. If river downcutting is relatively rapid, the V will be narrow, whereas if slope recession is relatively rapid (as in an area of unresistant rocks or a humid climate), the V will be more open. In reality, by no means all river valleys are V-shaped in cross-section. Where lateral stream EROSION is effective, the valley will be flat-floored. A similar effect will result from extensive DEPOSITION of ALLUVIUM on the valley floor – for example, as a result of a rise in the BASE-LEVEL OF EROSION.

vulcanicity The processes by which solid, liquid and gaseous materials are either forced into the Earth's crust or escape on to the surface via vents and fissures. Thus the term embraces igneous activity generally, not merely the formation of VOLCANOES.

W

wadi A steep-sided, flat-floored and usually dry valley in a hot DESERT. Wadis display many of the features of 'normal' river valleys (dendritic patterns and smoothly concave long-profiles), and are occasionally occupied by running water under present-day conditions. However, there is little doubt that the wadis of the Sahara and Arabian deserts were mainly eroded in the recent past, when deserts experienced more humid climates. See also PLUVIAL.

waning slope See CONCAVE SLOPE.

warm front The well-defined boundary between a warm and cold air mass, where the former is advancing and overriding the cold air, as at the leading edge of the WARM SECTOR of a depression. The gradient of the warm front is much less steep than that of the COLD FRONT and the relatively slow ascent of warm moist air at the front thus results in cooling and CONDENSATION over a wide area, and the development of NIMBUS clouds. The passage of a warm front is usually marked by a long period of steady rainfall lasting for several hours. [*f* FRONTAL DEPRESSION]

warm (warm-based) glacier A glacier characterized throughout its depth by temperatures very close to 0°C (PRESSURE MELTING POINT). Warm glaciers are sometimes referred to as *temperate glaciers*, from their occurrence in mid-latitude mountain regions such as the Alps and Rockies. In summer, warm glaciers generate large quantities of meltwater, which enters the glaciers by way of MOULINS and CREVASSES, forming an ENGLACIAL and SUBGLACIAL drainage system (which is absent from COLD GLACIERS). The presence of meltwater at the base of a warm glacier allows the ice to slide over BEDROCK, thus favouring effective ABRASION. In winter, the upper few metres of a warm glacier may be chilled below 0°C, but the major part of the ice remains at, or close to, pressure melting point.

warm occlusion See OCCLUDED FRONT.

warm sector A 'wedge' of warm, moist air, tapering northwards in the Northern Hemisphere and southwards in the Southern Hemisphere and contained within a mid-latitude FRONTAL DEPRESSION. In advance of the warm sector lies the WARM FRONT (with its continuous cloud cover and lengthy period of rainfall) and to the rear is the COLD FRONT (with its CUMULUS CLOUDS and heavy showers). The warm sector itself frequently gives rise to an interlude of pleasant, mild weather, with 'fair weather' cumulus clouds and sunny periods. With the passage of time, the extent of the warm sector is reduced, owing to the relatively more rapid forward movement of the cold front, which gradually overtakes the warm front (see OCCLUDED FRONT). [*f* FRONTAL DEPRESSION]

waste disposal Getting rid of the rubbish or by-products (liquid and solid) produced by people and the human use of RESOURCES. Today, the dumping of waste involves the following locations: landfill sites, particularly for domestic rubbish and some chemical waste; rivers for treated water from sewage works and for low-level, non-toxic chemical waste; the sea for all types of waste – a matter of growing international concern; some LEDCS that, contrary to international law, import waste from MEDCS and are paid large sums of money for doing so. The safe disposal of nuclear waste is currently a controversial issue. See ENVIRONMENTAL POLLUTION, TOXIC WASTE.

water balance The manner in which the PRECIPITATION received at a place is accounted for by EVAPOTRANSPIRATION, RUN-OFF and changes in the amounts of water stored within the soil and in rocks (GROUNDWATER). The water balance (sometimes referred to as the *hydrological balance budget*) is calculated from the formula

$$P = E + R \pm S$$

where P is precipitation, E is evapotranspiration, R is run-off and S represents changes in storage over the period of measurement (usually one year). If a study of water balance is made over a period of years, S may be, to all intents and purposes, constant and is therefore sometimes omitted from the water balance equation. It is important to realize that significant changes in water balance may occur within the space of a year. For example, in Britain a much greater proportion of the precipitation is lost to run-off in winter than in summer, whereas evapotranspiration is at a maximum during summer (when it may actually exceed precipitation), but negligible in winter. Moreover, water tends to pass into storage during winter (when underground water is replenished by PERCOLATION), but out of storage during summer, to provide the BASE FLOW of rivers.

waterfall A vertical or near-vertical fall of water or a series of step-like falls, developed where a river course is interrupted by a marked break of gradient – as at the edge of a PLATEAU, along a FAULT SCARP, at the junction of a HANGING VALLEY with a major glacial trough, or – occasionally – on a sea-CLIFF. Some waterfalls are the product of differential EROSION – for example, along a FAULT-line, soft rocks may be brought against hard rocks, and will be rapidly eroded by fluvial activity, leading to the formation of a waterfall at the fault. However, even where initiated by structures such as faults, waterfalls may become dissociated from them by the process of HEADWARD EROSION. See also RAPIDS, PLUNGE-POOL. [*f*]

water gap A valley through a CUESTA or ridge eroded by a river that continues to occupy the gap. See GAP TOWN.

water table The upper surface of the zone of saturation in a PERMEABLE rock (see also PHREATIC WATER). Rainwater percolates to the water table (which commonly lies at a depth of tens or even hundreds of metres beneath the surface) whenever PRECIPITATION exceeds EVAPOTRANSPIRATION. In Britain this occurs mainly in winter, with the result that the water table rises to a maximum elevation in

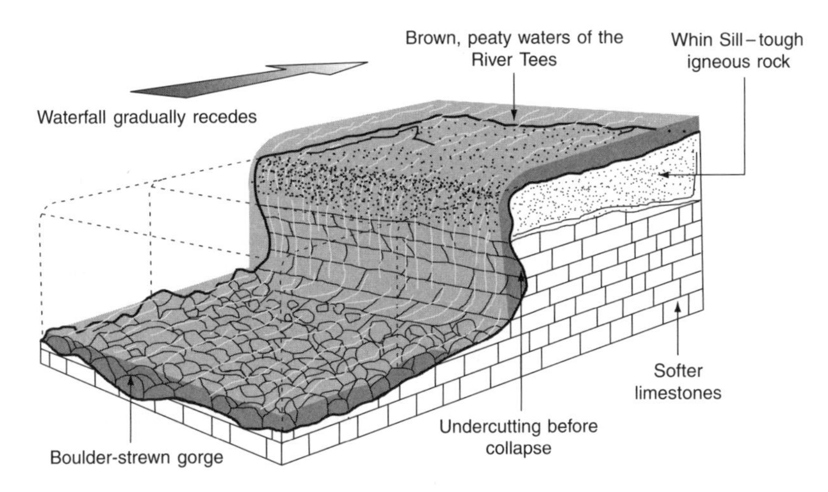

High Force, River Tees

early spring (March–April). However, during summer, PERCOLATION effectively ceases, and the zone of saturation is depleted by way of SPRINGS and seepages (which develop where the water table intersects the land surface, at the base of SCARPS and in deep valley bottoms). The slope of the water table varies from place to place (it is steeper where the rock is less permeable, and gentler where the rock is highly permeable), and in general reflects in a subdued fashion the shape of the surface relief, though there are important exceptions to this rule (see ABSTRACTION). Where the percolation of rainwater to the main water table is locally impeded (e.g. by a CLAY layer of limited extent) subsidiary areas of saturation, or PERCHED WATER TABLES, are formed. [*f*]

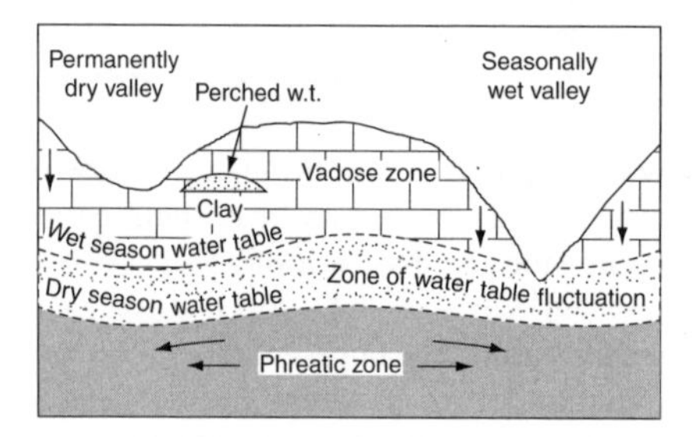

Water table and associated features

watershed The line separating the head streams that flow into different drainage basins; in effect, it is the line that marks the edge of a DRAINAGE BASIN. Watersheds may be sharply defined (by the crest of a ridge) or indeterminate (in areas of low RELIEF where the 'divide' between river basins is broad and gentle). See also CATCHMENT.

wave A surface feature of oceans, seas and inland water bodies, comprising a linear crest separating parallel troughs. Most waves are generated by winds blowing over the water. The moving air exerts a frictional drag on the surface water particles, setting up a series of orbital water movements. At the wave crest these are 'forward', and in the trough 'backward', giving rise to *oscillations.* The wave form as a whole moves in the direction of air movement. Waves can be accurately defined in terms of: *height* – the vertical distance between crest and trough; *length* – the distance between two successive crests; *period* – the time taken for a wave to move forwards by one wave length; *frequency* – the number of waves passing a given point in a minute; *velocity* – the forward speed of movement.

A fundamental distinction can be made between long, low waves (see SWELL) and short, steep waves. The main controls over wave form and size are wind speed, wind duration, and distance from a lee shore (see FETCH). When waves approach the shore, they break when the orbital velocity of the water particles exceeds the forward velocity of the wave, which is reduced by friction with the sea bed in shallow water. See CONSTRUCTIVE WAVE, DESTRUCTIVE WAVE, BACKWASH, SWASH.

wave-cut platform A near-level surface eroded in solid rock by wave action at the base of a retreating sea-CLIFF. Wave-cut platforms vary in width from a few metres to hundreds of metres, depending on rock type and resistance, and the duration of marine EROSION at its present level. In detail, wave-cut platforms vary considerably. Some are highly irregular, with grooves and depressions eroded along FAULTS, JOINT-lines and weak strata. Around the coastline of Britain wave-cut platforms often appear to be 'composite', with two or three distinct levels (a few metres apart vertically), reflecting slight changes of sea-level during the late PLEISTOCENE and the POSTGLACIAL periods.

wave refraction The process by which waves undergo a change of direction as they approach headlands and BEACHES, and pass the distal ends of SPITS and bars. Refraction results from the shallowing of the sea-floor in these situations, and the effect of this in reducing wave velocity. Thus, waves that approach the shore obliquely are 'turned', so that their crests are nearly parallel to the shore when wave-break occurs. The effect of refraction around headlands is to concentrate wave energy (shown by lines called *orthogonals*), and thus erosive potential, on the promontory, which becomes cliffed, with features such as ARCHES, STACKS and stumps, and to reduce wave energy in the intervening bays, where DEPOSITION dominates to form beaches. At the far ends of spits, refraction may be

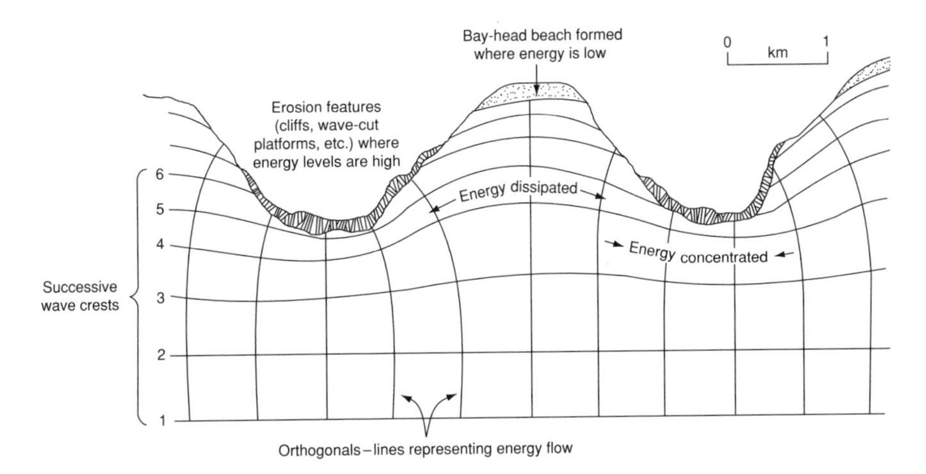

Wave refraction on a headland and bay coastline

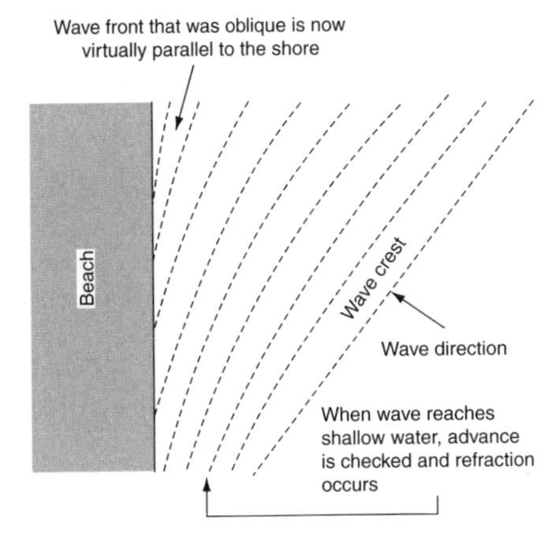

Wave refraction on a straight section of coastline

largely responsible for the development of recurved tips. [*f*]

waxing slope See CONVEX SLOPE.

weathering The breakdown and decay of rocks *in situ*, giving rise to a mantle of waste (see REGOLITH) or loose debris that may be removed by the processes of TRANSPORT. Weathering is divided into two main types: CHEMICAL WEATHERING; and MECHANICAL (PHYSICAL) WEATHERING. However, a third type, BIOLOGICAL or ORGANIC WEATHERING, is also sometimes recognized.

Weber's theory of industrial location This theory, published in 1909, holds that industries become sited at LEAST-COST LOCATIONS and that, more specifically, such sites are frequently the points of minimum TRANSPORT COSTS. As with most theories, Weber made a number of simplifying assumptions, such as a uniformity of terrain, that some RAW MATERIALS occur in fixed locations while others are found everywhere, standardized wage rates, the existence of PERFECT COMPETITION and economic man, and that transport costs are determined by weight of load and distance. The accompanying figure shows the simple case of two raw-material sources (R_1 and R_2) and a single market (M). If transport costs are assumed to be the same for both raw materials and finished products, then the theory states that the least-cost location will be at the centre of the *locational triangle* (P_i), equidistant from R_1, R_2 and M. At this juncture, Weber introduced a complication in the form of his MATERIAL INDEX, used to distinguish between *weight-losing* and *weight-gaining* industries. Clearly, it may be argued that in the case of a weight-losing industry, the least-cost location will be nearer the raw material sources (at P_2 rather than P_1) because those materials effectively contain waste. Conversely, where an industry is weight-gaining, the least-cost location will lie nearer the market (at P_3 rather than P_1). See also ISODAPANE.

Weber's theory has been criticized on a number of different counts, most of which relate to his original assumptions. Admittedly, Weber himself confessed later to the need to make allowances for such things as spatial variations in LABOUR availability and

LABOUR COSTS, as well as the magnetic effect on industrial location of existing AGGLOMERATIONS. Even so, the principal criticism remains, namely that the theory puts undue emphasis on transport costs, and incorrectly assumes that such costs are directly related to distance and weight. [*f*]

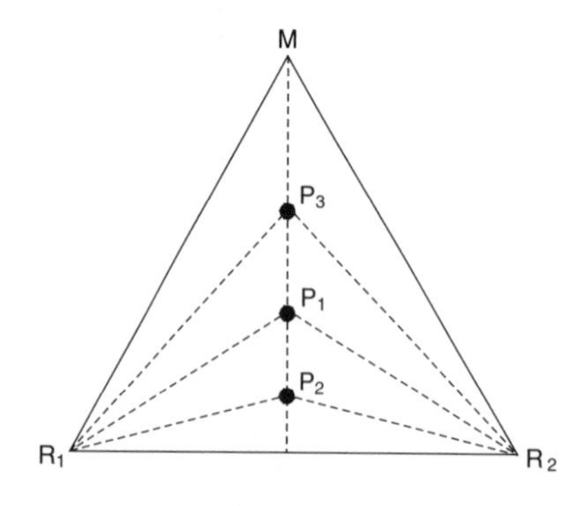

Weber's locational triangle

weight-loss ratio The relationship between the weight of raw materials used during a production process and the weight of the finished product. The weight-loss ratio is held to be significant in industrial location in that the higher the ratio, the more likely production will be located close to RAW MATERIAL sources. See WEBER'S THEORY OF INDUSTRIAL LOCATION.

welfare (i) Welfare may be defined as the state or condition of society at large, as in *welfare state*. It is a relatively new focal point in HUMAN GEOGRAPHY (see WELFARE GEOGRAPHY) and embraces such things as DIET, housing, medical care, education, employment and so on, from which human satisfaction is derived. A distinction may be drawn between *economic welfare* and *social welfare*. The former usually refers to what people get from the consumption of goods and services, while the latter includes those things contributing to the quality of human existence. In this respect, there appears to be some overlap or confusion with QUALITY OF LIFE, but one possible clarification would be to regard welfare as simply one important dimension of it. (ii) In the USA the term welfare refers to supplementary benefit given by the government to needy households.

welfare geography An aspect of HUMAN GEOGRAPHY that first developed during the 1970s as interest in that general field shifted from model-building and quantification to a growing concern about such broad issues as QUALITY OF LIFE, TERRITORIAL JUSTICE and WELFARE. More specifically, it focuses on contemporary problems like poverty, hunger, crime, differential access to housing and social services. One of the leading proponents of welfare geography has defined it as being about 'who gets what, where, and how'. There is no doubt that much welfare geography has been influenced by MARXISM and that it has a strongly radical character.

well-being A generic term for a group of overlapping concepts, which includes LEVEL OF LIVING, QUALITY OF LIFE, SOCIAL SATISFACTION, *standard of living* and WELFARE. See NEED-SATISFACTION CURVE.

wetland Land that is periodically or permanently waterlogged, e.g. the tidal mudflats of an ESTUARY, a marsh or bog. Wetlands represent important feeding and breeding areas for birds, and many are threatened by RECLAMATION, as in Hong Kong, Japan and Singapore.

whaleback See RUWARE.

white-collar worker See BLUE-COLLAR WORKER.

wholesaling An intermediary activity between the producer and the retailer (see RETAILING). The main functions of wholesaling are: the breaking of bulk (see BREAK-OF-BULK POINT); *warehousing* (holding stocks to meet fluctuations in demand); helping to finance distribution by allowing credit to retailers; in some instances, preparing a commodity for sale by grading, packing and branding.

wilderness Used in CONSERVATION to indicate an area left untouched in a natural state, with little or no human control and interference (as, for example, mountains and large areas of DESERT, of TAIGA and TUNDRA). The recreational potential of such areas is being increasingly realized, providing as they do a range of opportunities, from adventure to

Windchill equivalent temperature

Wind Speed (m/sec)	Air temperature (°C)															
	6	3	0	−3	−6	−9	−12	−15	−18	−21	−24	−27	−30	−33	−36	−39
3	3	−1	−4	−7	−11	−14	−18	−21	−24	−28	−31	−34	−38	−41	−45	−48
6	−2	−6	−10	−14	−18	−22	−26	−30	−34	−38	−42	−46	−50	−54	−58	−62
9	−6	−10	−14	−18	−23	−27	−31	−35	−40	−44	−48	−53	−57	−61	−65	−70
12	−8	−12	−17	−21	−26	−30	−35	−39	−44	−48	−53	−57	−62	−66	−71	−75
15	−9	−14	−18	−23	−27	−32	−37	−41	−46	−51	−55	−60	−65	−69	−74	−79
18	−10	−14	−19	−24	−29	−33	−38	−43	−48	−52	−57	−62	−67	−71	−76	−81
21	−10	−15	−20	−25	−29	−34	−39	−44	−49	−53	−58	−63	−68	−73	−77	−82
24	−10	−15	−20	−25	−30	−35	−39	−44	−49	−54	−59	−63	−68	−73	−78	−83

Windchill

birdwatching, from camping to exploring, or simply the chance to 'get away from it all'. Cf NATURE RESERVE.

wind chill The effect of the wind on the 'feel' of temperature (see SENSIBLE TEMPERATURE), causing it to feel bitterly cold even if the absolute temperature is above freezing. Wind chill influences people's vulnerability to exposure because it increases the rate of loss of body heat, thereby speeding up the onset of hypothermia. [*f*]

wind deposition See DEPOSITION.

wind erosion See EROSION.

wind gap See COL.

windbreak An obstacle, usually a hedge or belt of trees, planted across the direction of the prevailing wind to provide shelter for dwellings, soil, crops or animals. Windbreaks are important in exposed areas – for example, The Netherlands, the Rhône valley in France (see MISTRAL) and parts of Australia – to reduce SOIL EROSION and wind damage to crops.

wind power One of the few RENEWABLE sources of ENERGY that are being developed commercially on a large scale. This is achieved by wind turbines grouped in *wind farms* on suitably exposed and windy sites.

workers' cooperative See COOPERATIVE.

World Bank Formerly known as the International Bank for Reconstruction and Development, the World Bank is an agency of the United Nations. Set up in 1945, it was originally concerned with the encouragement of postwar reconstruction in Europe. However, it subsequently turned towards assisting the developing world. In the latter context, it has made loans to THIRD WORLD countries amounting to US$203 billion. Although well intentioned, these loans have undoubtedly played a part in the generation of THIRD WORLD DEBT. This weakness has been remedied to a limited extent by the move towards the granting of interest-free credits.

WTO (World Trade Organization) See GATT.

WWF (World Wide Fund for Nature) An NGO founded in 1961 with the aim of protecting ENDANGERED SPECIES and dealing with any serious threats to wildlife.

xenophobia Fear, dislike or distrust of foreigners and foreign things.

xerophyte A plant that is adapted to withstand seasonal or perennial drought (ct HYDROPHYTE). Xerophytes (such as DESERT cacti and the thorn bushes of desert margins) are characterized by exceptionally long roots (to tap GROUNDWATER), thick bark, small glossy leaves (to reduce transpiration), and a capacity to store water when it becomes available (as in *succulents*, which retain moisture in a spongy substance in their stems).

xerosere A VEGETATION SUCCESSION developed in a dry ENVIRONMENT – for example, a bare rock surface or an area of loose SAND.

Y

yardang A desert landform produced by wind ABRASION. Yardangs are elongated ridges, formed parallel to the prevailing wind direction, and displaying clear signs of basal undercutting by the impact of wind-borne SAND grains, giving a 'blasting effect' on their upwind sides only.

yield (i) The rate of return from an investment of capital over a specified period, usually expressed in percentage terms. (ii) OUTPUT or production expressed in relation to one of the INPUTS, e.g. cereal production per ha, industrial output per manhour.

young fold mountains Fold mountains created by earth movements of the Alpine OROGENY of mid-Tertiary times (ct OLD FOLD MOUNTAINS). Young fold mountains are characterized by their great elevations (as in the Himalayas, Andes and Alps), resulting partly from their limited age and the lack of time for peneplanation to be achieved, and partly from continued isostatic uplift since the folding movements. Their RELIEF is highly irregular, as a result of both deep vertical incision by rivers and intense glacial EROSION, mainly during the PLEISTOCENE – though, in many young fold mountains, glaciation is still active. See PLATE TECTONICS.

yuppy Not quite an acronym, but standing for a 'young, upwardly mobile person', i.e. someone who is successfully embarked on a dynamic career and who enjoys a high level of financial remuneration; most likely someone at stage one in the LIFE CYCLE.

Z

Zelinsky See MOBILITY TRANSITION.

zero population growth Where the balance of births, deaths and net MIGRATION is such as to produce a stationary demographic situation. It is a state that many MEDCs are beginning to approach as a result of a marked decline in fertility. No doubt, it will require adding a further stage to the DEMOGRAPHIC TRANSITION.

zeugen Tabular masses of hard SEDIMENTARY ROCK (often SANDSTONE) resulting from selective wind EROSION in DESERTS. Zeugen stand up to 30 m in height, and are separated from each other by depressions that have been 'scoured out' where the wind has been able to attack and remove weak underlying SHALES.

Zipf See LEAST EFFORT, RANK-SIZE RULE.

zonal model See CONCENTRIC ZONE MODEL.

zonal soil A type of SOIL that has undergone advanced development. Over a long period of time it has been affected by soil-forming processes such as humifaction (see HUMUS), LEACHING and ELUVIATION, and acidification (see PH VALUE), which have resulted in a well-developed SOIL PROFILE, and a greatly reduced influence of the parent material. Thus zonal soils reflect broad climatic and vegetational controls (as in the case of PODSOLS developed in cool temperate climates where PRECIPITATION is adequate for coniferous forest growth). On a world scale, SOIL CLASSIFICATION is based on the identification of zonal soils. See also AZONAL and INTRAZONAL SOILS.

zone of assimilation A transitional zone created by the advancing front of a moving CBD, usually involving the invasion of residential areas by various types of central business (notably small offices and RETAILING firms). Within the transition from the actual front of the zone to the 'core' of the CBD, it is possible to recognize two subzones: (i) an initial phase, where dwellings are simply converted into business premises; and (ii), nearer the core, a phase of REDEVELOPMENT and consolidation, where the converted dwellings are gradually replaced by structures purpose-built to accommodate central businesses. This second phase tends to occur when real estate investors and property developers are convinced that the risk perceived to be associated with the movement of the CBD is reduced to an acceptable level. Ct ZONE OF DISCARD. [*f* CBD]

zone of discard A transitional zone created in the wake of a moving CBD, where there is progressive abandonment of premises as central business FIRMS endeavour to maintain a location near to the CBD's shifting centre of gravity. The degree of withdrawal clearly increases in the direction opposite to that in which the CBD is moving. Where the abandonment by central business firms is complete, opportunities arise for the conversion of existing properties to some new use or for the wholesale REDEVELOPMENT of the area to make way for some new activity. Ct ZONE OF ASSIMILATION. [*f* CBD]

zone of transition The second ring in the CONCENTRIC ZONE MODEL of CITY structure. Although originally a residential area, its close proximity to the city centre makes it attractive to commercial and industrial development, particularly as the CBD expands. This invasion by non-residential activities, together with the general ageing of housing and the urban INFRASTRUCTURE, eventually leads to a decline in residential desirability (see URBAN BLIGHT). As households leave in search of better housing and more attractive residential environments elsewhere in the growing city, so poorer households and ethnic MINORITIES take their place to become increasingly concentrated there (see GHETTO). Dwellings are subdivided; they become overcrowded and even more dilapidated. With the general deterioration both in the fabric and the ENVIRONMENT, the zone becomes increasingly marked by high levels of vice and crime. Cf TWILIGHT AREA. [*f* CONCENTRIC ZONE MODEL]